Footprints in Cambridge and Aviation Industries of China

Yanzhong Zhang

Footprints in Cambridge and Aviation Industries of China

Scientific Papers of Yanzhong Zhang

ZHEJIANG UNIVERSITY PRESS
浙江大学出版社

Springer

Yanzhong Zhang
Beihang University
Beijing, China

ISBN 978-981-16-3175-7 ISBN 978-981-16-3176-4 (eBook)
https://doi.org/10.1007/978-981-16-3176-4

Jointly published with Zhejiang University Press, Hangzhou, China
The print edition is not for sale in China Mainland. Customers from China Mainland please order the print book from: Zhejiang University Press.

This Springer imprint is published by the registered company Springer Nature Singapore Pte Ltd.
The registered company address is: 152 Beach Road, #21-01/04 Gateway East, Singapore 189721, Singapore

Preface

The book has selected some of my academic papers in English, which fall into four parts: information technique, aeronautical engineering, strategy issue of development and PhD thesis. The Chinese Section has been published separately by Zhejiang University Press.

Over the past six decades, I studied theoretical physics (undergraduate), mechanical vibration, information technique, aeronautical system engineering and strategy issue of development successively. My lifelong academic career separates into the following four phases.

In the 1960s and 1970s, I dedicated myself to vibration, impact and mechanical failure research upon graduation and delivered a few papers. What could be found at the moment were only some summaries and work reports with ordinary quality. A couple of papers are taken in the aeronautical engineering part of the book as a record of the work completed in that period.

In the 1980s, I studied at Cambridge University mainly focusing on the information technique, such as signal processing, digital circuit and advanced algorithm, and published a number of papers at home and abroad. Though part of them was lost, dozens of papers are incorporated in the information technique part of the book.

In the 1990s, I was engaged in civil aircraft system engineering, airborne system engineering and avionics system. A portion of papers written during this period of time are added in the aeronautical engineering part of the book.

In the past decade, I was involved in strategy issue of development and presided over the implementation of national projects: "Large Aircraft" and "Aeroengine & Gas Turbine". Apart from that, I also headed the research on construction of a conservation-minded society, building an emerging industry system and promotion of R&D globalization. Considering the length of these project reports, only a few excerpts are included in the strategy issue of development part of the book.

Owing to my limited abilities, mistakes would inevitably exist in the papers and your corrections and suggestions are mostly welcome and appreciated.

In the end, I'd like to extend my gratitude to editors in Zhejiang University Press for their hard work. They've devoted considerable time and energy to resetting or retyping those papers that lack the electronic versions, especially the figures and equations that call for recharting. And hereby I also express my deep appreciation to all those who participate in the publication of the book for their sweat and effort.

ZHANG Yanzhong
Dec. 2018
West Lake, Hangzhou

Introduction

Mr. ZHANG Yanzhong was born in the County of Sanyuan, Shaanxi Province of China in 1940. It is a cultural ancient county with more than 1,500 years' history. A small river running from west to east divides the county into northern and southern parts, which are connected by an ancient stone bridge named Loong Bridge (the Bridge of Dragon). Born in a doctors' family at the northern town near the bridge, ZHANG regards **three bridges** as the landmarks of his life. He was born near the **Loong Bridge** in the Sanyuan County, studied in **Cambridge** of England, and has been dedicated to Chinas' commercial aircraft assembly line near the **Zhu Bridge** (the Bridge of Congratulation) in Pudong, Shanghai. Both his grandfather and father were famous doctors. ZHANG received very good family education since childhood. He won the first place among thousands of primary school graduates in the examination of the whole county and then went to secondary school. Later with distinction, he was admitted to the Department of Physics, Northwest University.

Mr. ZHANG graduated from university and entered China's aviation industry in 1962, engaging in vibration, shock and fault diagnosis research. He was in charge of developing the first set of large acceleration calibration system of China, which solved the calibration problem of accelerometers. During the 1970s, he took the lead in the research of aircraft fault diagnosis in China, and removed the vibration fault from dozens of imported aircraft.

Mr. ZHANG was admitted to University of Cambridge in 1980, then, entered Trinity College of Cambridge as a postgraduate the following year. At that time, the Master of Trinity College was the Nobel Prize winner, Sir A. L. Hodgkin, and ZHANG's tutor was Dr. M. D. Cowley. Mr. ZHANG studied in the Department of Engineering, Information Engineering Division, under the guidance of his supervisor, Prof. P. J. W. Rayner. He devoted himself to the research on digital signal processing, digital circuits and digital systems. Mr. ZHANG has been a diligent scholar. In the laboratory, library of Cambridge and Cambridge Riverside left his footprints almost everywhere. His papers on **information technique** were published on Cambridge journals and other renowned international periodicals. In 1984, he completed his PhD thesis titled "Digital Signal Processing System with Finite State Machine Realization". After passing the oral examination, he received his PhD degree and returned to China the same year. Cambridge, where he lived longer than anywhere else except China, left him an unforgettable memory. In 1989, he met with Sir M. F. Atiyah, famous mathematician, the new Master of Trinity College, while attending an international conference in Britain. In 2009, on the 800th anniversary of Cambridge University, Prof. ZHANG and his wife were invited back to their alma mater to attend the university celebration.

In 1985, Dr. ZHANG entered the headquarters of China Aviation Industries, and took charge of **aeronautical engineering**, including development of commercial aircraft, helicopters and airborne equipment systems, as well as the fundamental research of aviation. He established the Aviation Science Foundation, the Aviation Scholarship, and proposed to build the Aviation Key Laboratory to strengthen the crucial technology research. Dr. ZHANG led the development of various aircraft, helicopters and airborne systems, and most importantly, directed the implementation of two national projects of "Large Aircraft" and "Aeroengine & Gasturbine". Also he hosted crucial work of C919 and other large aircraft, including the argumentation of the research on project approval, technical reviews in different stages, and technical reviews for the first flight, to ensure their first flight successfully. His footprints have been scattered over a hundred of institutions in the aviation industry of China, and meanwhile, hundreds of papers on aviation have been published home and abroad.

Prof. ZHANG was elected as an academician of the Chinese Academy of Engineering (CAE) in 2001. Later the focus of his research has moved to **strategy issues of development**, such as the construction of a conservation-minded society, the promotion of R&D globalization, the establishment of aviation power, etc.

Having been devoted to China's aviation industry for nearly six decades, Academician ZHANG Yanzhong with his academic career and works, to a large extent, is a reflection of the main course of China's aviation industry since its establishment for the past seven decades (1951—2021).

Contents

INFORMATION TECHNIQUE

AERONAUTICAL ENGINEERING

STRATEGY ISSUE OF DEVELOPMENT

PHD THESIS

INFORMATION TECHNIQUE

Fast DFT Algorithm with [N−1]/2 Multiplications*

Zhang Yanzhong
Ministry of Aero-Space Industry

A fast algorithm is proposed for recursively computing the DFTs of prime length. Only $(N-1)/2$ real multiplications are required to compute all N frequency components in terms of permuting the input data. The multiplication in recursive computation is replaced by shifting. Complexity of the algorithm is studied. A factor η is introduced and presented. When the ratio of multiplier's period T_m to adder's period T_a is greater than the factor η (i. e. $T_m/T_a>\eta$), the new algorithm is faster than FFT. The necessary condition and error of the algorithm are studied. The signal-to-noise ratio for different length N is presented. A high accuracy scheme is proposed for improving the SNR about 20～30 dB.

1. Introduction

Discrete Fourier Transformation (DFT) is a central operation in digital signal processing. It can transform a signal from time domain to frequency domain and vice versa. DFT is widely used to analyse and estimate spectrum of signals, to design and implement finite impulse response digital filters, and to compute convolution and correction functions, etc.

The direct evaluation of an N-point DFT needs about $4N^2$ real multiplications and additions. Thus, for reasonably large value of N, direct evaluation of DFTs requires an inordinate amount of computations. In 1965, Cooley and Tukey[1] proposed Fast Fourier Transformation (FFT) algorithm which requires about $2N\log_2 N$ real multiplications for computing the DFT of length $N=2^M$. Recently, the Prime Factor Algorithm (PFA), Winograd Fourier Transform Algorithm (WFTA), etc[2-4] were proposed. These algorithms need about $0(N)$ multiplications. On the other hand, FFT and the other fast DFT algorithm have to use a large number of Fourier coefficients in the systems. This increases the size of memory and the complexity of hardware.

The new algorithm with only single coefficient and $(N-1)/2$ real multiplications is introduced in the next Section. Its complexity and accuracy are discussed and a high accuracy scheme is presented in subsequent Sections.

---

* Published in Chinese Journal of Aeronautics, Vol. 3, No. 2, May 1990: 131-139.

Y. Zhang, *Footprints in Cambridge and Aviation Industries of China*,
https://doi.org/10.1007/978-981-16-3176-4_1

2. Principle of Fast Recursive DFTs

Consider a DFT of length N:

$$X(k) = \sum_{a=0}^{N-1} x(n) W_N^{nk} \tag{1}$$
$$k=0,1,\cdots,(N-1)$$
$$W_N = e^{-j2\pi/N}$$

If the transform length $N = P$ prime, then the DFT may be computed by the following recursive form[5-6]

$$y_k(n) = W_P^D y_k(n-1) + u_k(n) \tag{2}$$
$$y_k(0) = u_k(0)$$
$$k=1,2,\cdots,(P-1) \quad n=0,1,\cdots,(P-1)$$

where

$$u_k(P-1-n) = x(\langle nk^{-1}D\rangle_P) \tag{3}$$

is a permutation of input sequences;

D is any one of the integers $(1 \leqslant D < P)$;

k^{-1} is an inverse element of k in GF(P);

$k^{-1}k = 1 \bmod P$.

The DFT output is as follows:

$$X(0) = \sum_{n=0}^{P-1} x(n)$$
$$X(k) = y_k(P-1)$$
$$k = 1,2,\cdots,(P-1) \tag{4}$$

The recursive Equation (2) may be viewed as a first order IIR digital filter with a complex coefficient. Its z-transform function is as follows:

$$H(z) = \frac{1}{1 - W_P^D z^{-1}} \tag{5}$$

The above equation can be written as the following real form:

$$H(z) = \frac{1 - \cos(2\pi D/P) z^{-1} - j\sin(2\pi D/P) z^{-1}}{1 - 2\cos(2\pi D/P) z^{-1} + z^{-2}} \tag{6}$$

The above equation may be considered as a second order IIR digital filter with real coefficients as shown in Fig. 1.

In order to compute a frequency output, the section within the dotted line in Fig. 1 has to do $(P-1)$ real multiplications with the constant $2\cos(2\pi D/P)$, and the section beyond the dotted line only needs to do two real multiplications for computing the real frequency component $X_R(k)$ and imaginary frequency component $X_I(k)$. Since the recursive coefficient $2\cos(2\pi D/P)$ is independent of the frequency index k, only one coefficient can be used to compute all P frequency components in terms of permuting the input sequences.

Because the figure D in the recursive coefficient $2\cos(2\pi D/P)$ can be any one of the integer set $[1, 2,\cdots,(P-1)]$, the proper choice of D may make the coefficient $\cos(2\pi D/P)$ have the approximate form of $\pm 2^{-m}$; then the multiplication with the constant $2\cos(2\pi D/P)$ in the recursive loop can be replaced by shifting $(m-1)$ steps. Shift is much faster than conventional multiplication. This algorithm may possess a very high speed. For example, for a 31-point DFT, if $D=9$ then $\cos(2\pi D/31) = -0.250653 \approx -2^{-2}$; one shift can be used to replace the multiplication with the constant $2\cos(2\pi D/31)$ in the recursive loop. Some coefficients with the form $\pm 2^{-m}$ are shown in Table 1.

Table 1 Coefficients and errors of FRDFTs

P	D	$\cos(2\pi D/P)$	$\pm 2^{-m}$	error $\pm\delta$
3	1	−0.500000	-2^{-1}	0.0000×2^{-1}
5	1	0.309017	2^{2}	0.9443×2^{4}
7	2	−0.222521	-2^{-2}	0.8793×2^{-5}
11	3	−0.142315	-2^{-1}	-0.5541×2^{-5}
13	3	0.120537	2^{-3}	-0.5713×2^{-7}
17	5	−0.237663	-2^{-2}	-0.7572×2^{-5}
19	4	0.245485	2^{-2}	-0.5779×2^{-7}
23	6	−0.068242	-2^{-4}	-0.7350×2^{-7}
29	7	0.054139	2^{-4}	-0.5351×2^{-6}
31	9	−0.250653	-2^{-2}	-0.6682×2^{-10}
37	10	−0.127018	-2^{-3}	-0.5166×2^{-8}
41	10	0.038303	2^{-5}	0.9027×2^{-7}
43	9	0.252933	2^{-2}	0.7509×2^{-8}
47	12	−0.033415	-2^{-5}	-0.5542×2^{-8}
53	13	0.029633	2^{-5}	-0.8277×2^{-9}
59	15	−0.026621	-2^{-5}	0.5926×2^{-7}
61	14	0.128398	2^{-1}	0.8700×2^{-8}
67	14	0.255043	2^{-2}	0.6455×2^{-7}
71	17	0.066323	2^{-4}	0.9787×2^{-8}
73	19	−0.064508	-2^{-4}	-0.5142×2^{-8}
79	19	0.059615	2^{-4}	-0.7385×2^{-8}
83	21	−0.018924	-2^{-6}	-0.8446×2^{-8}
89	24	−0.123232	-2^{-3}	0.9054×2^{-9}
97	24	0.016193	2^{-6}	0.5817×2^{-10}
101	25	0.015552	2^{-6}	-0.5996×2^{-13}

3. Complexity of Algorithm

As shown in Fig. 1, to compute a pair of frequency components $X_R(k)$ and $X_I(k)$, $(P-1)$

recursive operations within the dotted line are needed, i. e. , $(2P-3)$ additions and $(P-1)(m-1)$ shifts are required. Because the multiplication with $\cos(2\pi D/P)$ beyond the dotted line for computing $X_R(k)$ can also be replaced by shifting m steps, only one multiplication with $\sin(2\pi D/P)$ is required to compute an $X_I(k)$. An extra real addition is needed to compute the $X_R(k)$. To sum up, $(2P-2)$ real additions and one real multiplication are required for computing a pair of frequency components $X_R(k)$ and $X_I(k)$. For a real input, the DFT output has the following conjugate relationship[7]:

$$X(P-k)=X^*(k), k=1,2,\cdots,\frac{(P-1)}{2} \tag{7}$$

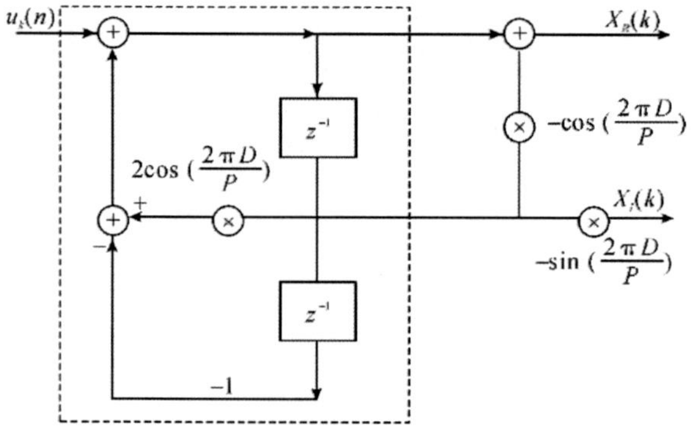

Fig. 1 Second recursive form of DFTs

So only half of the frequency components are needed to calculate. The total real addition number A and multiplication number M for computing all P frequency components (including $(P-1)$ real additions for computing $X(0)$ in Eq. (4)) are

$$A=(2P-2)(P-1)/2+(P+1)=P(P-1)$$
$$M=(P-1)/2$$

For complex input, as shown in Fig. 2, two parallel filters are architectured. Only $2(P-1)$ real additions are increased for combining the two filter operations. The total amounts of real calculations for the new algorithm are as follows:

$$A=P(P-1)+2(P-1)=P^2+P-2 \tag{8}$$

$$M=\frac{P-1}{2} \tag{9}$$

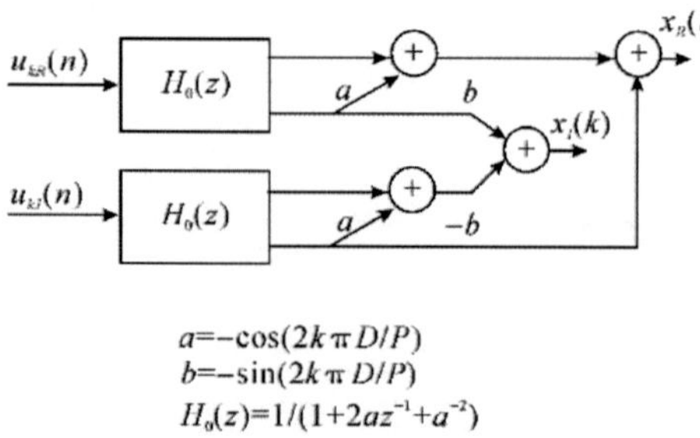

Fig. 2 Scheme of complex FRDFT

The real addition number A_F and real multiplication number M_F for radix 2 FFT algorithm[8] are

$$A_F=3N\log_2 N-2N+2 \tag{10}$$

$$M_F=2N\log_2 N-4N+4 \tag{11}$$

It is obvious that the multiplication number of FRDFT is less than that of FFT, but the addition number of FRDFT is more than that of FFT for long transform length. The actual speed of the new algorithm depends on the multiplier's and adder's period. In general, if the word length is B bits, then the multiplier's period T_m is B times of the adder's period T_a. Shifting is the fastest one among three operations. Ignore the shifting time, compare the computing time between FFT and FRDFT algorithms, and a criterion factor η is introduced by

$$\eta=\frac{P^2+3P-4-3\log_2 P}{2P\log_2 P-9(P-1)/2} \tag{12}$$

If the period ratio T_m/T_a is greater than the criterion factor η, i. e.

$$\frac{T_m}{T_a}>\eta \tag{13}$$

then the new algorithm FRDFT is faster than FFT. Some criterion factor η are shown in Table 2.

Table 2 Criterion factors η

P	3	5	7	11	13	17	19	23	29	31	37	43
η	0.00	0.22	0.57	1.15	1.41	1.90	2.13	2.58	3.22	3.42	4.02	4.59
P	47	53	59	61	67	71	73	79	83	89	97	101
η	4.96	5.50	6.03	6.21	6.72	7.06	7.22	7.72	8.04	8.53	9.16	9.49

It is shown in Table 2 that the FRDFT algorithm is efficient for short transform length.

4. Accuracy and S/M Ratio of Algorithm

Because $\pm 2^{-m}$ is used to replace the coefficient $\cos(2\pi D/P)$ in the recursive operations, some phase and amplitude errors are introduced in the DFT output. If the amplitude error of the coefficient is

$$\delta=\cos(2\pi D/P)-2^{-m} \tag{14}$$

Let φ_0 and φ be ideal phase and actual phase respectively; then

$$\varphi_0=\frac{2\pi D}{P}=D\theta_0,\theta_0\ \frac{2\pi}{P} \tag{15}$$

$$\varphi=\cos^{-1}(2^{-m})=\varphi_0+\Delta\varphi \tag{16}$$

where $\Delta\varphi$ is phase difference.

From the above two equations, it follows that

$$\delta=\cos(\varphi_0)-\cos(\varphi_0+\Delta\varphi)\approx\sin\varphi_0\Delta\varphi \tag{17}$$

The phase difference is

$$\Delta\varphi=\frac{\delta}{\sin\varphi_0} \tag{18}$$

The phase difference between two adjacent frequency components is

$$\Delta\theta_0 = \frac{\delta}{\sin\varphi_0} \tag{19}$$

The maximum phase difference for the Pth frequency component is

$$P\Delta\theta_0 = \frac{P\delta}{D\sin\varphi_0} \tag{20}$$

In order to satisfy the prime condition of the transform length, the maximum phase difference must be

$$P\Delta\theta_0 \ll \theta_0 \tag{21}$$

From Eqs. (15), (20) and (21), the necessary condition for the FRDFT algorithm of prime length is as follows:

$$\delta \ll \frac{2\pi D\sin\varphi_0}{P^2}, \varphi_0 = \frac{2\pi D}{P} \tag{22}$$

To evaluate the amplitude accuracy of FRDFT algorithm, the recursive section within the dotted line in Fig. 1 can be expressed as a second order all-pole IIR filter as follows:

$$w(n) = kw(n-1) - w(n-2) + u_k(n) \tag{23}$$

$$k = 2^{-(m-1)} = k_0 + \Delta k$$

$$k_0 = 2\cos\left(\frac{2\pi D}{P}\right), \Delta k = 2\delta$$

Here, only the amplitude error caused by the coefficient error is considered. Let

$$w(n) = w_0(n) + \Delta(n)$$

$$w(n-1) = w_0(n-1) + \Delta(n-1)$$

$$w(n-2) = w_0(n-2) + \Delta(n-2) \tag{24}$$

where $\Delta(n)$ is amplitude error signal; $w_0(n)$ is determined by the following free-error equation:

$$w_0(n) = k_0 w_0(n-1) - w_0(n-2) + u_k(n) \tag{25}$$

Replace Eqs. (24) and (25) into Eq. (23), and neglect the second order obtained by miniterm $\Delta k * \Delta(n-1)$, and a linear difference equation for determining the amplitude error is obtained by

$$\Delta(n) = k_0\Delta(n-1) - \Delta(n-2) + \Delta k w_0(n-1) \tag{26}$$

The z-transform of the above equation is

$$\Delta(z)(1 - k_0 z^{-1} + z^{-2}) = \Delta k w_0(z) z^{-1} \tag{27}$$

$$\frac{\Delta(z)}{w_0(z)} = \frac{\Delta k z^{-1}}{1 - k_0 z^{-1} + z^{-2}}$$

$$k_0 = 2\cos\varphi_0 \tag{28}$$

This is a 2nd order IIR filter with 2 poles and single zero. The RMS output of this filter under a unity input is the RMS error of FRDFT amplitudes. In order to estimate the noise value, the impulse response of the above error equation is presented as follows:

$$h(n) = \Delta k \frac{\sin(n\varphi_0)}{\sin\varphi_0} \tag{29}$$

Assume the noise signal is uncorrected. The output power of the error equation for unity input is

$$\frac{e^2}{w^2} = \sum_{n=1}^{P} |h(n)|^2 \tag{30}$$

From the above two equations, it follows that

$$\frac{e^2}{w^2} = \frac{\Delta k^2}{\sin^2\varphi_0} \sum_{n=1}^{P} |\sin\varphi_0|^2 = \frac{P\Delta k^2}{2\sin^2\varphi_0} \tag{31}$$

Considering $\Delta k = 2\delta$, the S/N ratio is obtained by

$$S/N = 10\log_{10}\left(\frac{\sin^2 \varphi_0}{2P\delta^2}\right)(\text{dB})$$

$$\varphi_0 = \frac{2\pi D}{P} \tag{32}$$

For example, for 31-point FRDFT, when $D=9$, the coefficient error $\delta = 0.6682 \times 2^{-10}$. The noise caused by the coefficient error is 45 dB. The S/N ratio of some coefficients are shown in Table 3.

Table 3 S/N ratio of some coefficients

P	D	$\pm 2^{-m}$	S/N(dB)
3	1	-2^{-1}	∞
5	1	2^{-2}	14
7	2	2^{-2}	19
11	3	-2^{-2}	22
13	3	-2^{-3}	32
17	5	2^{-3}	17
19	4	-2^{-2}	30
23	6	2^{-2}	28
29	7	-2^{-4}	23
31	9	2^{-4}	45
37	10	-2^{-2}	35
41	10	2^{-5}	24
43	9	2^{-2}	31
47	12	-2^{-5}	32
53	13	-2^{-5}	29
59	15	-2^{-5}	26
61	14	2^{-3}	28
67	14	2^{-2}	24
71	17	2^{-4}	27
73	19	-2^{-4}	32
79	19	2^{-4}	29
83	21	-2^{-6}	27
89	24	-2^{-3}	32
97	24	2^{-6}	42
101	25	2^{-6}	59

5. High Accuracy Implementation

Table 3 shows that the accuracy and S/N ratio of many coefficients are not very high. For requirement of high precision application, the accuracy of coefficients should be improved. One of the improved schemes is to use a sum of two binary integers to approximate the coefficient cos $(2\pi D/P)$, i. e. let

$$\cos(2\pi D/P) = \pm 2^{-m} \pm 2^{-n} \tag{33}$$

Because there are a lot of m and n combinations, the accuracy of approximate coefficients can be greatly improved. For example, for a 43-point DFT, let $D=10$; then

$$\cos\left(\frac{2\pi D}{43}\right) = 2^{-3} - 2^{-6} - 0.9940 \times 2^{-18}$$

Its S/N ratio is 89 dB.

The multiplication of the coefficient $(\pm 2^{-m} \pm 2^{-n})$ with data can be implemented in terms of two shift registers and an adder as shown in Fig. 3. One of registers shifts $(n-1)$ steps; another shifts $(m-1)$ steps; the results are added. The recursive multiplication in this scheme is replaced by two shifts and an addition. Its speed is a little lower than that of the simple scheme. However, its accuracy is very high; the S/N ratio increases about 20~30 dB. These high accuracy coefficients and their S/N ratios are shown in Table 4.

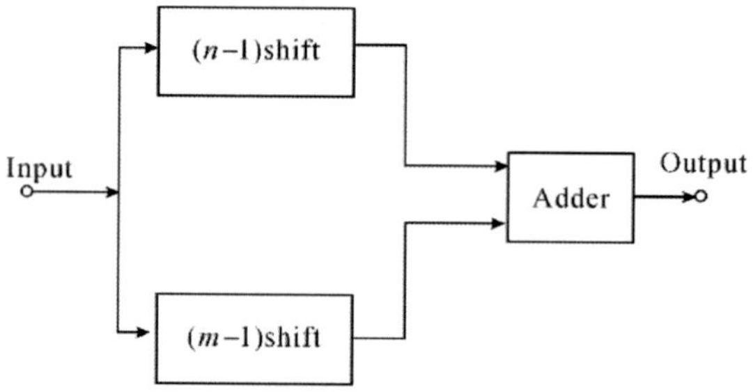

Fig. 3 High accuracy scheme

Table 4 High accuracy coefficients and SNR

P	D	$\pm 2^{-n} \pm 2^{-m}$	errors $\pm\delta$	SNR(dB)
3	1	-2^{-1}	0	∞
5	1	$2^{-2}+2^{-4}$	-0.8916×2^{-8}	39
7	2	$2^{-1}+2^{-4}$	-0.7732×2^{-9}	44
11	3	$-2^{-3}-2^{-4}$	-0.8652×2^{-9}	42
13	3	$-2^{-3}-2^{-8}$	-0.5704×2^{-10}	51
17	5	$2^{-3}-2^{-5}$	-0.7586×2^{-9}	41
19	4	$-2^{-2}-2^{-8}$	-0.6229×2^{-10}	48

(To be continued)

Table 4

P	D	$\pm 2^{-n} \pm 2^{-m}$	errors$\pm\delta$	SNR(dB)
23	6	$2^{-2}-2^{-8}$	-0.9401×2^{-9}	38
29	7	$-2^{-1}-2^{-5}$	-0.6995×2^{-11}	51
31	9	$2^{-2}-2^{-11}$	-0.6728×2^{-12}	58
37	10	$-2^{-3}-2^{-9}$	-0.5300×2^{-13}	65
41	10	$-2^{-2}-2^{-6}$	0.6589×2^{-10}	45
43	9	$2^{-3}-2^{-6}$	-0.9940×2^{-18}	89
47	12	$2^{-3}-2^{-9}$	-0.8677×2^{-12}	54
53	13	$-2^{-3}-2^{-9}$	0.6891×2^{-11}	49
59	15	$2^{-3}-2^{-7}$	0.7125×2^{-13}	61
61	14	$-2^{-3}-2^{-8}$	-0.5201×2^{-10}	45
67	14	$2^{-6}-2^{-7}$	-0.6650×2^{-17}	84
71	17	$2^{-1}+2^{-4}$	-0.8309×2^{-14}	64
73	19	$2^{-4}-2^{-9}$	-0.9604×2^{-14}	64
79	19	$-2^{-6}-2^{-8}$	-0.7187×2^{-11}	47
83	21	$2^{-6}-2^{-8}$	0.6217×2^{-10}	42
89	24	$-2^{-6}+2^{-9}$	0.5764×2^{-13}	61
97	24	$2^{-6}+2^{-11}$	0.6536×2^{-13}	59
101	25	$2^{-6}-2^{-14}$	-0.7965×2^{-16}	75

6. Conclusion

A Fast Recursive DFT algorithm has been proposed in this paper. The advantages of FRDFT are that it has a very simple structure and needs only $(N-1)/2$ real multiplications for computing all N frequency components. The study of complexity shows that if the ratio of the multiplier's period T_m to the adder's period T_a is greater than the criterion factor $\eta(T_m/T_a>\eta)$, FRDFT is faster than FFT. The new algorithm is efficient for a short transform length. A long DFT can be factored into two or more short transforms with relative prime length. These short transforms can be calculated by using FRDFT algorithm. The combination operations of those short transforms need no extra twiddle factor multiplications. The necessary condition and S/N ratio of the algorithm are presented. A high accuracy scheme is proposed. It can increase 20~30 dB in SNR.

References

[1] Cooley, J. W., and Tukey, J. W., An Algorithm for Machine Calculation of Complex Fourier Series, Math. Comput., Vol. 19, No. 90, April 1965, pp. 297-301.

[2] Winograd, S. , On Computing the Discrete Fourier Transform, Proc. Acad. Sci. , U. S. A. , vol. 73(Mathematics), No. 4 April 1976, pp. 1005-1006.

[3] Burrus, C. S. , Index Mapping for Multidimensional Formulation of DFT and Convolutions, IEEE Trans. , vol. ASSP-25, No. 3, June 1977, pp. 237-242.

[4] Winograd, S. , Some Bilinear Forms Whose Multiplicative Complexity Depends on the Fields of Constants, Mathematical Systems Theory, vol. 10, 1977, pp. 169-180.

[5] Zhang, Y. Z. , Optimum Recursive Computation of Prime Length DFTs, Proc. of the 20th International Electronics Convention of IREE, Melbourne, Australia, October 1985, pp. 583-586.

[6] Zhang, Y. Z. , Fast Implementation of Recursive DFTs, Proc. of IEEE International Conference on ASSP, Glasgow, U. K. , May 1989, pp. 1071-1-74.

[7] Gold. B. and Rader, C. , Digital Processing of Signals, McGraw-hill, New York, 1969.

[8] Brigham, E. O. , The Fast Fourier Transform, Prentice-Hall, Englewood Cliffs, N. J. , 1974.

Fast DFT Algorithm Using Subgroup Convolutions*

Zhang Yan-zhong and Rayner, P. J. W.
Department of Engineering, University of Cambridge, U. K.

A method is presented for converting a DFT of length (p^M-1) into several subgroup circular convolutions of length M or divisor M. This decomposition leads to efficient methods for computing the DFT.

1. Introduction

The computation of the Discrete Fourier Transform (DFT) of N points

$$x(K) = \sum_{n=0}^{N-1} x(n)\ w^{nk}\ K = 0,1,\cdots,N-1 \qquad (1.1)$$
$$w = e^{-j2\pi/N}$$

is an important operation in digital signal processing. In 1965 Cooley and Tukey[1] proposed the Fast Fourier Transform (FFT) when $N(=2^M)$ is highly composite. In 1968 Rader[2] showed that the DFT, when its length $N=p$ is a prime, can be represented as a circular convolution of length $(p-1)$ by rearranging the data. Later Rader[3] and Winograd[4] showed that DFT's of length $N-p^M$, where p is prime, can be converted into one convolution of length $p^{k-1}(p-1)$, two convolutions of length $p^{M-1}(p-1)$, four convolutions of length $p^{M-3}(p-1),\cdots$, terminating with p^{M-1} convolutions of length $(p-1)$.

The computation of small N convolutions with minimum number of multiplications has been reported by Winograd[5]. Agarwal and Cooley[6] showed that a one-dimensional circular convolution, whose length is the product of relatively prime integers, can be converted in a multi-dimensional circular convolution. The long sequence DFT may be computed by the Winograd nested algorithm[7] or prime factor algorithm[8] that requires less multiplications than FFT algorithm.

these algorithms require the computation of different length circular convolutions. The different circular convolution lengths lead to a rather complex program structure.

In this paper it is shown that DFT's of length (p^M-1) may be converted in several subgroup circular convolutions of length M or divisor of M. Most of these convolutions for the DFT have the same length which can give considerable program simplification.

---

* Published in Signal Processing Ⅱ: Theories & Applications edited by H. M. Schüssler, Elsevier Science Publishers B. V., 1983: 717-724.

Y. Zhang, *Footprints in Cambridge and Aviation Industries of China*,
https://doi.org/10.1007/978-981-16-3176-4_2

The cyclic subgroup properties of Ring $Z(p^M-1)$ are introduced in section 2. In section 3, it is shown that any DFT of length (p^M-1) may be computed in terms of subgroup circular convolutions.

2. Cyclic Subgroup Properties of Ring $Z(p^M-1)$

It is well known that all the non zero elements of $GF(p^M)$ may be separated into a number of subsets of length M or divisor of M. For example. the non zero elements of $GF(2^4)$ may be separated into the following subsets.

$$
\begin{array}{cccc}
\alpha & \alpha^2 & \alpha^4 & \alpha^8 \\
\alpha^{11} & \alpha^7 & \alpha^{14} & \alpha^{13} \\
\alpha^6 & \alpha^{12} & \alpha^9 & \alpha^3 \\
\alpha^5 & \alpha^{10} & &
\end{array}
$$

where a is a primitive element of $GF(2^4)$, $\alpha^{15}=1$.

The above subsets may be obtained by computing the power of a modulo 15. The powers of a are the elements of Ring $Z(15)$.

2.1 When $N=p^M-1$ is a prime

The non zero elements of $Z(N)$ may be represented as m subsets of length M as follows:

$$
\begin{array}{ccccc}
p^0 & p^1 & p^2 & \cdots & p^{M-1} \\
gp^0 & gp^1 & gp^2 & \cdots & gp^{M-1} \\
g^{m-1}p^0 & g^{m-1}p^1 & g^{m-1}p^2 & \cdots & g^{m-1}p^{M-1}
\end{array}
\quad \text{mod } N \tag{2.1}
$$

where

$$m=\emptyset(p^M-1)/M=\frac{N-1}{M} \tag{2.2}$$

$\emptyset$-Euler function g is a unity root of order m, where m is the least integer such that $g^M=1 \bmod N$. Since $N=p^M-1$ is a prime, $Z(N)$ is a prime, $Z(N)$ is a field, and there exists a primitive element β such that

$$\beta^{\phi(N)}=\beta^{N-1}=1 \bmod N \tag{2.3}$$

Let

$$g=\beta^{\phi(N)}/m=\beta^M \tag{2.4}$$

then g is a unity root of order m on $Z(N)$.

For example, Ring Z_{31},

$$N=2^5-1=31,\ m=\emptyset(31)/5=6$$

The unity root of order 6 is $g=26$. From eqn. (2.1), the non-zero elements of Z_{31} may be separated into 6 subsets of length 5 as follows

1	2	4	8	16	
26	21	11	22	13	
25	19	7	14	28	mod 31
30	29	27	23	15	
5	10	20	9	18	
6	12	24	17	3	

2.2 $N=p^M-1$ is not a prime

There are $\varnothing(N)$ elements that are prime to N in $Z(N)$. These $\varnothing(N)$ elements may be represented as m_1 Subsets of length M as follows:

$$\begin{matrix} p^0 & p^1 & p^2 & \cdots & p^{M-1} \\ gp^0 & gp^1 & gp^2 & \cdots & gp^{M-1} \\ g^{m_1-1}p^0 & g^{m_1-1}p^1 & g^{m_1-1}p^2 & \cdots & g^{m_1-1}p^{M-1} \end{matrix} \mod N \tag{2.5}$$

where $m_1=\varnothing(N)/M$, g is a unity root of order m_1.

The remaining $N-Q(N)$ elements that are not prime to N may be represented as m_2 subsets.

$$\begin{matrix} g_1p^0 & g_1p^1 & g_1p^2 & \cdots & g_1p^{d_1-2} \\ g_2p^0 & g_2p^1 & g_2p^2 & \cdots & g_2p^{d_1-2} \\ g_{m_2}p^0 & g_{m_2}p^1 & g_{m_2}p^2 & \cdots & g_{m_2}p^{d_{m_2}-2} \end{matrix} \mod N \tag{2.6}$$

where

d_i is equal to M or divisor of M;

g_i is the factor of N or the product of the factors;

$m=m_1+m_2$ is the total number of subsets and equal to the total number of irreducible polynomials in GF(p^M). For example,

$$N=2^4-1=15$$

$$\varnothing(15)=\varnothing(3)\ \varnothing(5)=8\ m_1=\varnothing(15)/4=2$$

$11^2=1 \mod 15$ is a unity root of order 2, from eqn. (2.5), the 8 elements that are prime to 15 may be separated into 2 subsets of length 4 as follows.

$$\begin{matrix} 1 & 2 & 4 & 8 \\ 11 & 7 & 14 & 13 \end{matrix} \mod 15$$

From eqn. (2.6) the remaining 6 elements may be divided into 2 subsets of length 4 and 2 respectively.

$$\begin{matrix} 3 & 6 & 12 & 9 \\ 5 & 10 & & \end{matrix} \mod 15$$

3. DFT of Length $N=(p^M-1)$, with N Prime, Using Subgroup Convolution

From eqn. (1.1) the DFT of length $N=p^M-1$ may be written as follows

$$x(0)=\sum_{n=0}^{N-1}x(n)$$

$$x(k)=x(10)+\bar{x}(k) \quad k=1,2,\cdots,N-1 \tag{3.1}$$

where

$$x(k)=\sum_{n=1}^{N-1}x(n)w^{nk} \quad w=e^{-j2\pi/N} \tag{3.2}$$

When $N=p^M-1$ is prime eqn. (3.2) may be converted into circular convolutions.

3.1 $N=p^M-1$ is prime, i.e., N is a prime Mersenne number

For example, $N=3,7,31,127,8191$, etc.

From eqn. (2.1), n and k in eqn. (3.2) may be written as:

$$\begin{aligned} n=(n_1,n_2)=g^{n1}p^{n2} \quad & n_1=0,1,\cdots,(m-1) \\ & n_2=0,1,\cdots,(M-1) \\ k=(k_1,k_2)=g^{k1}p^{k2} \quad & k_1=0,1,\cdots,(m-1) \\ & k_2=0,1,\cdots,(M-1) \end{aligned} \tag{3.3}$$

where $m=\phi(n)/M$ g is a unity root of order m, Using eqn. (3.3), eqn. (3.2) may be expressed as follows

$$\bar{x}(k_1,k_2)=\sum_{n_1=0}^{m-1}\sum_{n_2=0}^{M-1}x(n_1,n_2)w^{(g^{k_1+n_1}p^{k_2+n_2})} \tag{3.4}$$

where $k_1=0,1,\cdots,(m-1)$; $k_2=0,1,\cdots,(M-1)$. eqn. (3.4) may be considered as a 2-D circular convolutions of length M. For example,

$$N=2^3-1=7 \text{ is prime}$$

$$M=\phi(7)/3=2 \quad M=3$$

Since $6^2=1 \bmod 7$, 6 is a unity root of order 2. From eqn. (2.1) the two subsets are as follows

1 2 4 6 5 3

Rearranging the order in accordance with the above subsets eqn. (3.2), may be written matrix from, as:

$$\begin{vmatrix} \bar{x}(1) \\ \bar{x}(2) \\ \bar{x}(4) \\ \bar{x}(6) \\ \bar{x}(5) \\ \bar{x}(3) \end{vmatrix} = \begin{vmatrix} w^1 & w^2 & w^4 & w^6 & w^5 & w^3 \\ w^2 & w^4 & w^1 & w^5 & w^3 & w^6 \\ w^4 & w^1 & w^2 & w^3 & w^6 & w^5 \\ w^6 & w^5 & w^3 & w^1 & w^2 & w^4 \\ w^5 & w^3 & w^6 & w^2 & w^4 & w^1 \\ w^3 & w^6 & w^5 & w^4 & w^1 & w^2 \end{vmatrix} = \begin{vmatrix} x(1) \\ x(2) \\ x(4) \\ x(6) \\ x(5) \\ x(3) \end{vmatrix}$$

where $w=e^{-j2\pi 7}$.

There above may be computed in terms of the circular convolutions.

The regular structure of the decomposition can lead to appreciable advantages in terms of programming complexity and computation speed for large N.

3.2 Reordering the rows and columns

The order may be rearranged from the following equation

$$g^ip^0, g^ip^1, g^ip^2, \cdots, g^ip^{M-1} \tag{3.5}$$

where $\{i\}$ is a permutation of $0,1,2,\cdots,(m-1)$.

If $m=\phi(p^M-1)/M$ can be factored into several prime factors, say,

$$m=m_1 \cdot m_2, \cdots, m_5 \tag{3.6}$$

then the permutation $\{i\}$ may be determined by the Chinese Remainder Theorem.

Step 1

Determine the permutation $\{i\}$.

Let $j=0,1,2,\cdots,(m-1)$ and

$$(j_1,j_2,\cdots,j_s)=(j \bmod m_1,\ j \bmod m_2,\cdots,\ j \bmod m_s) \tag{3.7}$$

then, the permutation $\{i\}$ may be obtained from lexicographic ordering of $(j_1,j_2,\cdots,j_s)$ by the Chinese Remainder Theorem(3)

$$i=\sum_{k=1}^{s} j_k N_k M_k \tag{3.8}$$

where $M_k=\frac{M}{M_k}(N_k,M_k)=1 \bmod m_k \quad k=1,2,\cdots,s$

Step 2

And a unity root g of order m, as in eqn. (2. 4) the generating element of the subsets is as follows

$$g^i \quad i \text{ is a permutation of } (0,1,\cdots,m-1) \tag{3.9}$$

Step 3

Determine the order of rows and columns in terms of eqn. (3. 5). For example,

$$N=2^5-1=31 \text{ is prime}$$
$$m=\phi(31)/5=6=3\times 2, m_1=3, m_2=2$$

Step 1

Determine permutation $\{i\}$.

Let $j=0,1,2,3,4,5$. From eqn. (3. 7),

$$(j_1,j_2)=(0,0),(1,1),(2,0),(0,1),(1,0),(2,1)$$

The permutation $\{i\}$ may be obtained from the lexicographic order of (j_1,j_2) by the Chinese Remainder Theorem as follows:

$$\{i\}=\{0,3,4,1,2,5\}$$

Step 2

Find a unity root of order 6.

Since 3 is a primitive root of GF(31), say $3^{30}=1 \bmod 31$, so $26=3^5 \bmod 31$ is a unity root of order 6.

$$26^6=1 \quad \bmod 31$$

From eqn. (3. 9), the generating elements of subsets are as follows:

$$\{g\}=\{2,30,5,26,25,6\}$$

Step 3

Compute the orders of rows and columns. From eqn. (3. 5), the orders of the rows and columns of the matrix are as follows:

$$k,n=\begin{matrix}1,2,4,8,16,30,29,27,23,15,5,10,20,9,18,\\ 26,21,11,22,13,25,19,7,14,28,6,12,24,17,3\end{matrix}$$

Then, the matrix may be written as follows:

$$w=\begin{vmatrix} A & B & C & D & E & F \\ B & A & D & C & F & E \\ C & D & E & F & A & B \\ D & C & F & E & B & A \\ E & F & A & B & C & D \\ F & E & B & A & D & C \end{vmatrix}$$

where

$$A=\begin{vmatrix} w^{1} & w^{2} & w^{4} & w^{8} & w^{16} \\ w^{2} & w^{4} & w^{8} & w^{16} & w^{1} \\ w^{4} & w^{8} & w^{16} & w^{1} & w^{2} \\ w^{8} & w^{16} & w^{1} & w^{2} & w^{4} \\ w^{16} & w^{1} & w^{2} & w^{4} & w^{8} \end{vmatrix} \qquad B=\begin{vmatrix} w^{30} & w^{29} & w^{27} & w^{23} & w^{15} \\ w^{29} & w^{27} & w^{23} & w^{15} & w^{30} \\ w^{27} & w^{23} & w^{15} & w^{30} & w^{29} \\ w^{23} & w^{15} & w^{30} & w^{29} & w^{27} \\ w^{15} & w^{30} & w^{29} & w^{27} & w^{23} \end{vmatrix}$$

$$C=\begin{vmatrix} w^{5} & w^{10} & w^{20} & w^{9} & w^{18} \\ w^{10} & w^{20} & w^{9} & w^{18} & w^{5} \\ w^{20} & w^{9} & w^{18} & w^{8} & w^{10} \\ w^{9} & w^{18} & w^{5} & w^{10} & w^{20} \\ w^{18} & w^{5} & w^{10} & w^{20} & w^{9} \end{vmatrix} \qquad D=\begin{vmatrix} w^{26} & w^{21} & w^{11} & w^{22} & w^{13} \\ w^{21} & w^{11} & w^{22} & w^{13} & w^{26} \\ w^{11} & w^{22} & w^{13} & w^{26} & w^{21} \\ w^{22} & w^{13} & w^{26} & w^{21} & w^{11} \\ w^{13} & w^{26} & w^{21} & w^{11} & w^{22} \end{vmatrix}$$

$$E=\begin{vmatrix} w^{25} & w^{19} & w^{7} & w^{14} & w^{28} \\ w^{19} & w^{7} & w^{14} & w^{28} & w^{25} \\ w^{7} & w^{14} & w^{28} & w^{25} & w^{19} \\ w^{14} & w^{28} & w^{25} & w^{19} & w^{7} \\ w^{28} & w^{25} & w^{19} & w^{7} & w^{14} \end{vmatrix} \qquad F=\begin{vmatrix} w^{6} & w^{12} & w^{24} & w^{17} & w^{3} \\ w^{12} & w^{24} & w^{17} & w^{3} & w^{6} \\ w^{24} & w^{17} & w^{3} & w^{6} & w^{12} \\ w^{17} & w^{3} & w^{6} & w^{12} & w^{24} \\ w^{3} & w^{6} & w^{12} & w^{24} & w^{17} \end{vmatrix}$$

The transform can be computed in terms of the circular convolutions shown above.

4. $N=P^M-1$ is Not a Prime

There are $\phi(N)$ elements that are prime to N on Ring Z_N. Eqn. (3.2) may be expressed in the two cases: $(k,N)=1$ and $(k,N)>1$.

$$\overline{x}(k)=\sum_{n=1}^{N-1}x(n)w^{nk} \quad (k,N)>1 \tag{4.1}$$

$$\overline{x}(k)=\sum_{(n,N)>1}x(n)w^{nk}+\sum_{(n,N)=1}x(n)w^{nk} \quad (k,N)=1 \tag{4.2}$$

Since $(k,N)=1$ and $(n,N)=1$ in eqn. (4.3) from eqn. (2.5). n and k may be expressed as

$$n=(n_1,n_2)=g^{n1}p^{n2} \tag{4.3}$$

$$k=(k_1,k_2)=g^{k1}p^{k2} \tag{4.4}$$

where $n_1,k_1=0,1,2,\cdots,(m-1)$

$n_2,k_2=0,1,2,\cdots,(M-1)$

$m_1=\phi(N)/M$

g is a unity root of order m_1, say

$g^{w_1}=1 \bmod N$

Using eqn. (4.4), eqn. (4.3) may be written as follows:

$$\bar{x}_3(k_1,k_2) = \sum_{\substack{(n,N)=1 \\ (k,N)=1}} x(n)w^{nk} = \sum_{n_1=0}^{w_1-1}\sum_{n_2=0}^{M-1} x(n_1,n_2)w^{(g^{k_1+n_1}p^{k_2+n_2})} \tag{4.5}$$

$$k_1=0,1,\cdots,(m_1-1), \quad k_2=0,1,\cdots,(M-1)$$

Eqn. (4.5) may be considered as a 2-D circular convolution.

Since $(k,N)=1$, and $(n,N)>1$ in eqn. (4.2), from eqn. (2.6) and (2.5), n and k may be expressed as

$$\begin{aligned} n=(n_1,n_2)=g^{n1}p^{n2} \quad & n_1=0,1,\cdots,(m_2-1) \\ & n_2=0,1,\cdots,(d-1) \\ k=(k_1,k_2)=g^{k1}p^{k2} \quad & k_1=0,1,\cdots,(m_1-1) \\ & k_2=0,1,\cdots,(M-1) \end{aligned} \tag{4.6}$$

Using eqn. (4.6), eqn. (4.2) may be written as follows:

$$\bar{x}_2(k_1,k_2) = \sum_{\substack{(n,N)>1 \\ (k,N)=1}} x(n)w^{nk} = \sum_{n_1=0}^{m_2-1}\sum_{n_2=0}^{d-1} x(n_1,n_2)\times(w^{g^{n1}g^{k1}p^{k_2+n_2}}) \tag{4.7}$$

$$k_1=0,1,\cdots,(m_1-1), k_2=0,1,\cdots,(M-1)$$

Eqn. (4.7) may be considered as $(m_1\times m_2)$ circular convolutions.

Since $(k,N)>1$ in eqn. (4.1), from eqn. (2.6) k may be expressed as follows:

$$k=(k_1,k_2)=g^{k1}p^{k2} \quad \begin{aligned} & k_1=0,1,\cdots,(m_2-1) \\ & k_2=0,1,\cdots,(d-1) \end{aligned} \tag{4.8}$$

n in eqn. (4.1) may be separated into 2 subsets as follows:

$$n=\begin{cases} n' & (n',N)=1 \\ n'' & (n'',N)>1 \end{cases} \tag{4.9}$$

From eqn. (2.5) and (2.6), n' and n'' may be written as follows

$$n'=(n'_1,n'_2)=g^{n'_1}p^{n'_2} \quad \begin{aligned} & n'_1=0,1,\cdots,(m_1-1) \\ & n'_2=0,1,\cdots,(M-1) \end{aligned} \tag{4.10}$$

$$n''=(n''_1,n''_2)=g^{n''_1}p^{n''_2} \quad \begin{aligned} & n''_1=0,1,\cdots,(m_2-1) \\ & n''_2=0,1,\cdots,(d-1) \end{aligned} \tag{4.11}$$

Using eqn. (4.8), (4.10) and (4.11), eqn. (4.1) may be written as follows:

$$\begin{aligned} \bar{x}_1(k_1,k_2) = \sum_{\substack{n=0 \\ (k,N)>1}}^{N-1} x(n)w^{nk} & \sum_{n'_1=0}^{m_1-1}\sum_{n''_2=0}^{M-1} x(n'_1,n'_2)\times(w^{g^{n'_1}g^{k_1}p^{k_2+n'_2}}) \\ & + \sum_{n''_1=0}^{m_2-1}\sum_{n''_2=0}^{d-1} x(n''_1,n''_2)\times(w^{g^{n''_1}g^{k_1}})^{p^{k_2+n''_2}} \end{aligned} \tag{4.12}$$

$$k_1=0,1,\cdots,(m_2-1), k_2=0,1,\cdots,(d-1)$$

Eqn. (4.11) may be considered as $(m_1\times m_2)$ circular convolutions of length M. Eqn. (4.12) may be considered as m_2^2 circular convolutions.

For example,

$$N=2^4-1=15=3\times5 \text{ is not a prime.}$$

From eqn. (3.2) it follows

$$\bar{x}(k) = \sum_{n=1}^{14} c(n)w^{nk} \qquad \begin{aligned} & k=0,1,\cdots,14 \\ & w=e^{-j2\pi/15} \end{aligned}$$

Rearrange the order as follows

$$1,2,4,8,11,7,14,13,6,12,9,3,5,10$$

The above equation may be written as a matrix form

$$\begin{vmatrix} x(1) \\ x(2) \\ x(4) \\ x(8) \\ x(11) \\ x(7) \\ x(14) \\ x(13) \\ x(6) \\ x(12) \\ x(9) \\ x(3) \\ x(5) \\ x(10) \end{vmatrix} \begin{vmatrix} w^1 & w^2 & w^4 & w^8 & w^{11} & w^7 & w^{14} & w^{13} & w^6 & w^{12} & w^9 & w^3 & w^5 & w^{10} \\ w^2 & w^4 & w^8 & w^1 & w^7 & w^{14} & w^{13} & w^{11} & w^{12} & w^9 & w^3 & w^6 & w^{10} & w^5 \\ w^4 & w^8 & w^1 & w^2 & w^{14} & w^{13} & w^{11} & w^7 & w^9 & w^3 & w^6 & w^{12} & w^5 & w^{10} \\ w^8 & w^1 & w^2 & w^4 & w^{13} & w^{11} & w^7 & w^{14} & w^3 & w^6 & w^{12} & w^9 & w^{10} & w^5 \\ w^{11} & w^7 & w^{14} & w^{13} & w^1 & w^2 & w^4 & w^8 & w^6 & w^{12} & w^9 & w^3 & w^5 & w^{10} \\ w^7 & w^{14} & w^{13} & w^{11} & w^2 & w^2 & w^8 & w^1 & w^{12} & w^9 & w^3 & w^{10} & w^5 & w^5 \\ w^{14} & w^{13} & w^{11} & w^7 & w^4 & w^8 & w^1 & w^2 & w^9 & w^3 & w^6 & w^{12} & w^5 & w^{10} \\ w^{13} & w^{11} & w^7 & w^4 & w^8 & w^1 & w^2 & w^4 & w^3 & w^6 & w^{12} & w & w^{10} & w^5 \\ w^6 & w^{12} & w^9 & w^3 & w^6 & w^{12} & w^9 & w^3 & w^6 & w^{12} & w^9 & w^3 & 1 & 1 \\ w^{12} & w^9 & w^3 & w^6 & w^{12} & w^9 & w^3 & w^6 & w^{12} & w^9 & w^3 & w^6 & 1 & 1 \\ w^9 & w^3 & w^6 & w^{12} & w^1 & w^3 & w^6 & w^{12} & w^9 & w^3 & w^6 & w^{12} & 1 & 1 \\ w^3 & w^6 & w^{12} & w^9 & w^3 & w^6 & w^{12} & w^1 & w^3 & w^6 & w^{12} & w^9 & 1 & 1 \\ w^5 & w^{10} & w^5 & w^{10} & w^5 & w^{10} & w^5 & w^{10} & 1 & 1 & 1 & 1 & w^{10} & w^5 \\ w^{10} & w^5 & w^{10} & w^5 & w^{10} & w^5 & w^{10} & w^5 & 1 & 1 & 1 & 1 & w^5 & w^{10} \end{vmatrix} \begin{vmatrix} x(1) \\ x(2) \\ x(4) \\ x(8) \\ x(11) \\ x(7) \\ x(14) \\ x(13) \\ x(6) \\ x(12) \\ x(9) \\ x(3) \\ x(5) \\ x(10) \end{vmatrix}$$

5. Conclusions

It has been proved that any DFT of length $N=(p^M-1)$, where p is prime, may be converted into several subgroup circular convolutions of length M or divisor of M. The length M circular may be computed by Winograd algorithm[5]. Most circular convolutions for this DFT have the same length, which allows a more simple realisation than the algorithms of Rader[2] and Winograd[4].

6. Acknowledgement

One of the authors, Y. Z. Zhang, wishes to thank the Government of the People's Republic of China for financial support by a scholarship, by ORS awards of the British Government and by a grant from Trinity College of Cambridge University.

References

[1] Cooley, J. W and Turkey,J. W. , An Algorithm for the Machine Calculation of Complex Fourier Series, Math. Comput. 19, (1965) 297-301.

[2] Rader, C. M. ,Discrete Fourier Transforms when the Number of Data Samples is Prime, Proc. IEEE, 56, (1968) 1107-1108.

[3] McClellan, J. H. and Rader, C. M. , Number Theory in Digital Signal Processing (Prentice-Hall, 1979).

[4] Winograd, S. , On Computing the Discrete Fourier Transform, Proc. Nat. Acad. Sci. U. S.

A. , 73, (1976) 1005-1006.
[5] Winograd, S. , On Computing the Discrete Fourier Transform, Math. Comput, 32, (1978) 175-199.
[6] Agawal, R. C. and Cooley, J. W. , New Algorithm for Digital Convolution, IEEE. Trans. ASSP-25 (1977) 392-410.
[7] Silverman, H. , An Introduction to Programming the Winograd Fourier Transform Algorithm (WFTA), IEEE Trans. ASSP-25 (1977) 152-165.
[8] Kolba, D. P. and Parks, T. W. , A Prime Factor FFT Algorithm Using High-Speed Convolution. IEEE Trans. ASSP-25 (1977) 281-294.

Minimisation of Reed-Muller Polynomials with Fixed Polarity*

Y. Z. Zhang and P. J. W. Rayner, M. A., Ph. D.
Department of Engineering, University of Cambridge, U. K.

Abstract: An efficient algorithm for minimisation of Reed-Muller polynomials with fixed polarities is presented. The common terms of multiple-output polynomials are considered by applying a number of logical operations on their coefficients. The minimisation of the polynomials over extension Galois fields $GF(2^M)$ is considered. The average number of field multiplications for mapping a set of coefficients is reduced to less than $M \cdot 2^{M-2}$.

Keywords: polynomials; logic; algorithms; Reed-Muller polynomials

1. Introduction

The advantage of using a Reed-Muller polynomial for realisation of a switching function is that it may be more economical than the conventional Boolean function realisation, either in the number of gates or in the number of gate interconnections[1]. Furthermore, the exclusive-OR realisation of switching functions can be easily tested[2]. The main problem of realising modulo-2 expansions is how to minimise the number of product terms in the representation. This problem has long been, and still is, an open one. With the advent of LSI and VLSI techniques, the development of RM polynomial minimisation algorithms is becoming important in synthesising a function using cellular logic array.

Many authors[3-9] have been concerned with the minimisation of RM polynomials. Saluja and Ong[3] proposed an exhaustive algorithm to compute all the fixed-polarity modulo-2 expressions by matrix multiplication. This algorithm needs 2^n matrix multiplications and $(2^n - 1)$ permutations of function output vectors.

Mukhopadhyay and Schmitz[4] gave the polarity functions of RM polynomial coefficients. The best polarity for the minimum RM polynomial can be obtained by finding the maximum clique of the polarity-compatibility graph. This graph possesses very large numbers of vertices in most applications. Robinson and Yeh[5] proposed a local minimisation procedure for mixed-polarity RM polynomials using the minimisation polynomial coefficients with fixed polarities, but it does not guarantee global minimisation. Recently, based on Wu, Chen and Hurst's work[6], an efficient computer

* Published in IEE Proceedings, Pt. E Vol. 131, No. 5, Sept. 1984: 176-186.

Y. Zhang, *Footprints in Cambridge and Aviation Industries of China*,
https://doi.org/10.1007/978-981-16-3176-4_3

method for single-output exclusive-OR logic design has been developed by Besslich[7]. This work is not concerned with multiple-output functions and does not deal with the minimisation of extension field RM polynomials.

In Section 2, an efficient algorithm is presented for the minimisation of single-output RM polynomials with fixed polarities. In Section 3, the minimisation of multi-output RM polynomials, by considering the effect of common terms, is developed. The minimisation of RM polynomials over extension Galois fields is proposed in Section 4.

2. Efficient Algorithm for Minimisation of Single-Output Functions

Any switching function of n variables may be represented as a Reed-Muller polynomial with fixed polarity as follows:

$$f(x_0, x_1, \cdots, x_{n-1}) = a_0 \oplus a_1 \overset{*}{x}_0 \oplus a_2 \overset{*}{x}_1 \oplus a_3 \overset{*}{x}_0 \overset{*}{x}_1 \oplus \cdots \oplus a_r \overset{*}{x}_0 \overset{*}{x}_1 \cdots \overset{*}{x}_{n-1} \qquad (1)$$

where

$\oplus$ denotes modulo-2 addition

$a_i \in (0,1)\ i=0,1,2,\cdots,2^n-1$

x_i^* represents either complemented $\bar{x}_i$ for polarity 1 or uncomplemented x_i, for polarity 0; but not both

$$r = 2^n - 1$$

If one knows all the possible function outputs $f_0, f_1, \cdots, f_r$, then the coefficients of the above polynomial may be obtained[10] by

$$\mathbf{A} = \mathbf{S}_n \mathbf{F} \qquad (2)$$

where

$\mathbf{A} = (a_0, a_1, \cdots, a_r)^t$

$\mathbf{F} = (f_0, f_1, \cdots, f_r)^t$, for $f_0 = f(0,0,\cdots,0), f_1 = f(0,0,\cdots,1)$, etc.

$\mathbf{S}$ is a Reed-Muller transform matrix, recursively defined by

$$\mathbf{S}_n = \begin{vmatrix} \mathbf{S}_{n-1} & \mathbf{0} \\ \mathbf{S}_{n-1} & \mathbf{S}_{n-1} \end{vmatrix} \quad \mathbf{S}_1 = \begin{vmatrix} 1 & 0 \\ 1 & 1 \end{vmatrix} \qquad (3)$$

or

$$\mathbf{S}_n = \mathbf{S}_1 \otimes \mathbf{S}_1 \otimes \cdots \otimes \mathbf{S}_1 \qquad (4)$$

where $\otimes$ denotes the Kronecker product.

For example, the RM polynomial of three variables with positive polarity (x_2, x_1, x_0) may be written as

$$f(x_2, x_1, x_0) = a_0 \oplus a_1 x_0 \oplus a_2 x_1 \oplus a_3 x_0 x_1 \oplus a_4 x_2 \oplus a_5 x_0 x_2 \oplus a_6 x_1 x_2 \oplus a_7 x_0 x_1 x_2$$

Its coefficients may be obtained by

$$
\begin{bmatrix} a_0 \\ a_1 \\ a_2 \\ a_3 \\ a_4 \\ a_5 \\ a_6 \\ a_7 \end{bmatrix}
=
\left[\begin{array}{cccc:cccc}
1 & & & & & & & \\
1 & 1 & & & & & & \\
1 & 0 & 1 & & & & & \\
1 & 1 & 1 & 1 & & & & \\ \hdashline
1 & 0 & 0 & 0 & 1 & & & \\
1 & 1 & 0 & 0 & 1 & 1 & & \\
1 & 0 & 1 & 0 & 1 & 0 & 1 & \\
1 & 1 & 1 & 1 & 1 & 1 & 1 & 1
\end{array}\right]
\begin{bmatrix} f(0,0,0) \\ f(0,0,1) \\ f(0,1,0) \\ f(0,1,1) \\ f(1,0,0) \\ f(1,0,1) \\ f(1,1,0) \\ f(1,1,1) \end{bmatrix}
$$

It is obvious that the above matrix possesses 3^3 'ones' and it needs $3^3 - 2^3 = 19$ modulo-2 additions to compute a set of coefficients. However, the matrix may be factored into the Kronecker product form as follows:

$$
S_3 =
\left[\begin{array}{cccc:cccc}
1 & & & & & & & \\
1 & 1 & & & & & & \\
 & & 1 & & & & & \\
 & & 1 & 1 & & & & \\ \hdashline
 & & & & 1 & & & \\
 & & & & 1 & 1 & & \\
 & & & & & & 1 & \\
 & & & & & & 1 & 1
\end{array}\right]
\left[\begin{array}{cccc:cccc}
1 & & & & & & & \\
 & 1 & & & & & & \\
1 & & 1 & & & & & \\
 & 1 & & 1 & & & & \\ \hdashline
 & & & & 1 & & & \\
 & & & & & 1 & & \\
 & & & & 1 & & 1 & \\
 & & & & & 1 & & 1
\end{array}\right]
\left[\begin{array}{cccc:cccc}
1 & & & & & & & \\
 & 1 & & & & & & \\
 & & 1 & & & & & \\
 & & & 1 & & & & \\ \hdashline
1 & & & & 1 & & & \\
 & 1 & & & & 1 & & \\
 & & 1 & & & & 1 & \\
 & & & 1 & & & & 1
\end{array}\right]
$$

Only $3 * 2^2$ modulo-2 additions are required to compute a set of coefficients for this algorithm, which is called the Fast Reed-Muller Transform (FRMT) algorithm.

For the case of n variables, $n * 2^{n-1}$ modulo-2 additions are required to obtain a set of coefficients from its output function vector. There exist 2^n different polarities for n variables. The minimisation of RM polynomials is the process of finding the polarity from 2^n possible polarities, such that the polynomial possesses the minimum number of modulo-2 additions.

Saluja and Ong[3] proposed that the function output vector for a new polarity is a permutation of another polarity function output vector. All 2^n sets of polynomial coefficients may be obtained in terms of multiplying the successive modified output vectors by a RM transform matrix. One or more among them are minimum. If the FRMT algorithm is used, then the total number of modulo-2 additions for this exhaustive algorithm is

$$2^n * n * 2^{n-1} \tag{5}$$

It will now be shown that a set of polynomial coefficients with a new polarity may be obtained directly from another set of polynomial coefficients, without permuting the output function vectors and computing eqn. (2). We begin with the adjacent polarity polynomials. The polarity of $(\overset{*}{x}_{n-1}, \cdots, \overset{*}{x}_k, \cdots, \overset{*}{x}_0)$ is said to be adjacent to the polarity of $(\overset{*}{x}_{n-1}, \cdots, \overline{\overset{*}{x}_k}, \cdots, \overset{*}{x}_0)$, as only one variable x_k has different polarity between the two sets of variables.

Theorem 1

The coefficients of a RM polynomial with n variables may be obtained directly from the coefficients of its adjacent polarity polynomial, in terms of the map with 2^{n-1} modulo-2 additions. There exist n adjacent polarity maps for a n variable polynomial.

Proof

From eqn. (1), the RM polynomial with n variables may be rewritten as

$$f(\overset{*}{x}_0, \overset{*}{x}_1, \cdots, \overset{*}{x}_{n-1}) = a_{0,\cdots,00} \oplus a_{0,\cdots,01}\overset{*}{x}_0 \oplus a_{0,\cdots,10}\overset{*}{x}_1 \oplus a_{0,\cdots,11}\overset{*}{x}_0\overset{*}{x}_1 \oplus \cdots \oplus a_{1,\cdots,11}\overset{*}{x}_0\overset{*}{x}_1, \cdots, \overset{*}{x}_{n-1} \tag{6}$$

This may be separated into two parts, according to whether it contains x_0 or not, as follows:

$$f(\overset{*}{x}_0,\overset{*}{x}_1,\cdots,\overset{*}{x}_{n-1})=f_0(\overset{*}{x}_1,\overset{*}{x}_2,\cdots,\overset{*}{x}_{n-1})\oplus\overset{*}{x}_0 f_1(\overset{*}{x}_1,\overset{*}{x}_2,\cdots,\overset{*}{x}_{n-1}) \tag{7}$$

where

$$\begin{cases} f_0(\overset{*}{x}_1,\overset{*}{x}_2,\cdots,\overset{*}{x}_{n-1})=a_{0,\cdots,00}\oplus a_{0,\cdots,10}\overset{*}{x}_1\oplus\cdots\oplus a_{1,\cdots,10}\overset{*}{x}_1\overset{*}{x}_2,\cdots,\overset{*}{x}_{n-1} \\ f_1(\overset{*}{x}_1,\overset{*}{x}_2,\cdots,\overset{*}{x}_{n-1})=a_{0,\cdots,01}\oplus a_{0,\cdots,11}\overset{*}{x}_1\oplus\cdots\oplus a_{1,\cdots,11}\overset{*}{x}_1\overset{*}{x}_2,\cdots,\overset{*}{x}_{n-1} \end{cases} \tag{8}$$

If one changed the polarity of $\overset{*}{x}_0$ into $\overline{\overset{*}{x}}_0$, and used the relation

$$\overset{*}{x}_0=1\oplus\overline{\overset{*}{x}}_0 \tag{9}$$

then eqn. (7) may be changed into

$$f'(\overline{\overset{*}{x}}_0,\overset{*}{x}_1,\cdots,\overset{*}{x}_{n-1})=f'_0(\overset{*}{x}_1,\overset{*}{x}_2,\cdots,\overset{*}{x}_{n-1})\oplus\overline{\overset{*}{x}}_0 f_1(\overset{*}{x}_1,\overset{*}{x}_2,\cdots,\overset{*}{x}_{n-1}) \tag{10}$$

where

$$\begin{aligned} f'_0(\overset{*}{x}_1,\overset{*}{x}_2,\cdots,\overset{*}{x}_{n-1})&=(a_{0,\cdots,00}\oplus a_{0,\cdots,01})\oplus(a_{0,\cdots,10}\oplus a_{0,\cdots,11})\overset{*}{x}_1 \\ &\oplus\cdots\oplus(a_{1,\cdots,10}\oplus a_{1,\cdots,11})\overset{*}{x}_1\overset{*}{x}_2,\cdots,\overset{*}{x}_{n-1} \end{aligned} \tag{11}$$

Note that $f_0(\overset{*}{x}_1,\overset{*}{x}_2,\cdots,\overset{*}{x}_{n-1})$ is not changed.

Here, $f'(\overline{\overset{*}{x}}_0,\overset{*}{x}_1,\cdots,\overset{*}{x}_{n-1})$ is the adjacent polarity polynomial of $f(\overset{*}{x}_0,\overset{*}{x}_1,\cdots,\overset{*}{x}_{n-1})$, and can be represented by

$$\begin{aligned} f'(\overline{\overset{*}{x}}_0,\overset{*}{x}_1,\cdots,\overset{*}{x}_{n-1})&=a'_{0,\cdots,01}\oplus a'_{0,\cdots,01}\overline{\overset{*}{x}}_0\oplus a'_{0,\cdots,10}x_1\oplus a'_{0,\cdots,11}\overline{\overset{*}{x}}_0\overset{*}{x}_1 \\ &\oplus\cdots\oplus a'_{1,\cdots,11}\overline{\overset{*}{x}}_0\overset{*}{x}_1,\cdots,\overset{*}{x}_{n-1} \end{aligned} \tag{12}$$

Comparing eqns. (10), (11) and (12), the coefficient of $f'(\overset{*}{x}_1,\overset{*}{x}_2,\cdots,\overset{*}{x}_{n-1})$ is

$$\begin{cases} a'_{xx,\cdots,x1}=a_{xx,\cdots,x1} \\ a'_{xx,\cdots,x0}=a_{xx,\cdots,x0}\oplus a_{xx,\cdots,x1} \end{cases} \tag{13}$$

where $(XX,\cdots,X)$ denotes$(00,\cdots,00),(00,\cdots,1),\cdots,(11,\cdots,1)$.

This is the map of RM polynomial coefficients from the polarity $(\overset{*}{x}_0,\overset{*}{x}_1,\cdots,\overset{*}{x}_{n-1})$ into the polarity $(\overline{\overset{*}{x}}_0,\overset{*}{x}_1,\cdots,\overset{*}{x}_{n-1})$. The map needs 2^{n-1} modulo-2 additions.

Similarly, the polarity of x_1 may be changed so there are n adjacent polarity maps for n variables. For example, the coefficients of three-variable RM polynomials can be mapped from the polarity $(\overset{*}{x}_2\overset{*}{x}_1\overset{*}{x}_0)$ into the polarity $(\overset{*}{x}_2\overset{*}{x}_1\overline{\overset{*}{x}}_0)$ by the following adjacent polarity map:

$$\begin{vmatrix} a_0(1) \\ a_1(1) \\ a_2(1) \\ a_3(1) \\ a_4(1) \\ a_5(1) \\ a_6(1) \\ a_7(1) \end{vmatrix} = \left|\begin{array}{cccc:cccc} 1 & 1 & & & & & & \\ & 1 & & & & & & \\ & & 1 & 1 & & & & \\ & & & 1 & & & & \\ \hdashline & & & & 1 & 1 & & \\ & & & & & 1 & & \\ & & & & & & 1 & 1 \\ & & & & & & & 1 \end{array}\right| \begin{vmatrix} a_0(0) \\ a_1(0) \\ a_2(0) \\ a_3(0) \\ a_4(0) \\ a_5(0) \\ a_6(0) \\ a_7(0) \end{vmatrix}$$

where

$a_i(1)$ denotes the coefficient with the polarity $(\overset{*}{x}_2\overset{*}{x}_1\overline{\overset{*}{x}}_0)$

$a_i(0)$ denotes the coefficient with polarity $(\overset{*}{x}_2\overset{*}{x}_1\overset{*}{x}_0)i=0,1,\cdots,7$

Similarly, the coefficients with polarities $(\overset{*}{x}_2\overline{\overset{*}{x}}_1\overset{*}{x}_0)$ and $(\overline{\overset{*}{x}}_2\overset{*}{x}_1\overset{*}{x}_0)$ can be mapped from the coefficient with the polarity ($\overset{*}{x}_2\overset{*}{x}_1\overset{*}{x}_0$) by the following adjacent maps, respectively:

$$\begin{vmatrix} a_0(2) \\ a_1(2) \\ a_2(2) \\ a_3(2) \\ a_4(2) \\ a_5(2) \\ a_6(2) \\ a_7(2) \end{vmatrix} = \left|\begin{array}{cccc:cccc} 1 & & 1 & & & & & \\ & 1 & & 1 & & & & \\ & & 1 & & & & & \\ & & & 1 & & & & \\ \hdashline & & & & 1 & & 1 & \\ & & & & & 1 & & 1 \\ & & & & & & 1 & \\ & & & & & & & 1 \end{array}\right| \begin{vmatrix} a_0(0) \\ a_1(0) \\ a_2(0) \\ a_3(0) \\ a_4(0) \\ a_5(0) \\ a_6(0) \\ a_7(0) \end{vmatrix}$$

and

$$\begin{vmatrix} a_0(4) \\ a_1(4) \\ a_2(4) \\ a_3(4) \\ a_4(4) \\ a_5(4) \\ a_6(4) \\ a_7(4) \end{vmatrix} = \left|\begin{array}{cccc:cccc} 1 & & & & 1 & & & \\ & 1 & & & & 1 & & \\ & & 1 & & & & 1 & \\ & & & 1 & & & & 1 \\ \hdashline & & & & 1 & & & \\ & & & & & 1 & & \\ & & & & & & 1 & 1 \\ & & & & & & & 1 \end{array}\right| \begin{vmatrix} a_0(0) \\ a_1(0) \\ a_2(0) \\ a_3(0) \\ a_4(0) \\ a_5(0) \\ a_6(0) \\ a_7(0) \end{vmatrix}$$

where $a_i(2)$ and $a_i(4)$ denote, respectively, the coefficients of polarities ($\overset{*}{x}_2\,\overline{\overset{*}{x}}_1\overset{*}{x}_0$) and ($\overline{\overset{*}{x}}_2\overset{*}{x}_1\overset{*}{x}_0$), for $i=0,1,\cdots,7$.

There exist 3 adjacent polarity maps for 3-variable RM polynomials. Each map needs 4 modulo-2 additions.

If all the polarities of n variables are rearranged according to Gray codes, then each polarity is adjacent to the next one, as the Gray code is a reflective code, i. e. in changing from one value to the next increment only one bit is changed at a time. Therefore, all 2^n sets of the RM polynomial coefficients may be mapped by adjacent polarity maps based on Gray-code ordering. An efficient algorithm for exhaustive search of the minimum polynomial coefficients can be obtained by (2^n-1) adjacent polarity maps.

For example, all the possible polarity coefficients of a 3-variable RM polynomial may be obtained in terms of 7 adjacent polarity maps, according to the following Gray code ordering:

0 0 0	0 0 1	0 1 1	0 1 0	1 1 0	1 1 1	1 0 1	1 0 0
$x_2x_1x_0$	$x_2x_1\overline{x}_0$	$x_2\overline{x}_1\overline{x}_0$	$x_2\overline{x}_1x_0$	$\overline{x}_2\overline{x}_1x_0$	$\overline{x}_2\overline{x}_1\overline{x}_0$	$\overline{x}_2x_1\overline{x}_0$	$\overline{x}_2x_1x_0$

This efficient algorithm may be represented by the flow graph in Fig. 1.

In Fig. 1, it is shown that $7 * 2^2=28$ modulo-2 additions are required to compute 7 sets of the polarity coefficients, and that the first set of the coefficient may be obtained from its output function vector by the FRMT algorithm, which needs 12 modulo-2 additions. The total number of modulo-2 additions for computing all the polynomial coefficients is 40. For the case of n variables, $(2^n-1)2^{n-1}$ modulo-2 additions are required to compute (2^n-1) sets of the polynomial polarity coefficients. $n * 2^{n-1}$ modulo-2 additions are required to compute the first set of coefficients by the FRMT algorithm. The total number of modulo-2 additions for computing all the polynomial coefficients is

$$(2^n+n-1) * 2^{n-1} \tag{14}$$

One or more among the polynomials are minimum. If only the minimal set of coefficients are required, then one can use the 'in place' algorithm, and only 2^n memory locations are required.

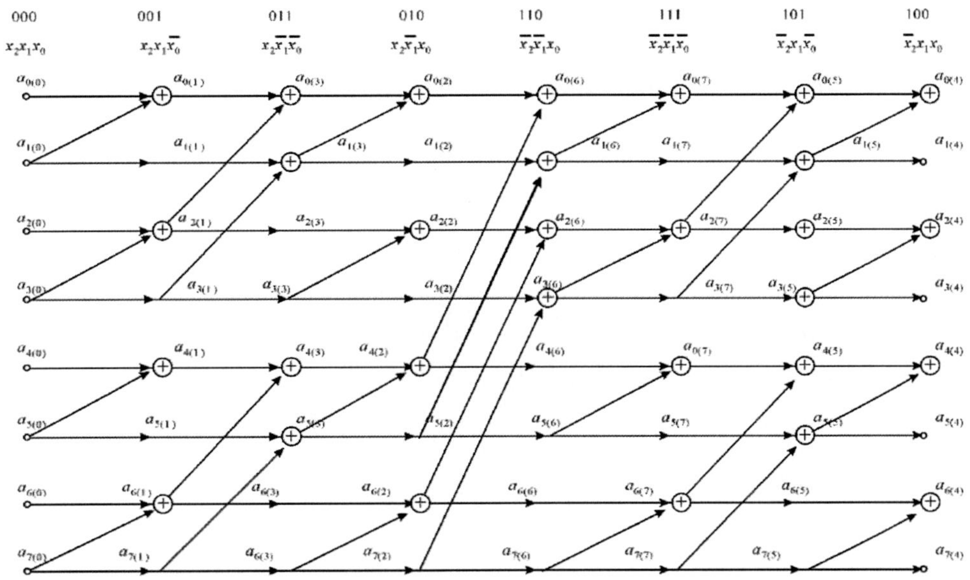

Fig. 1 Flow graph of the algorithm for a single function

From eqns. (5) and (14), the amount of modulo-2 additions for this efficient algorithm is $n/(1+(n-1)2^{-n})$ times less than that of the FRMT algorithm. The comparison of the two algorithms is shown in Table 1.

Table 1 Comparison of the two algorithms

Variable number n	Number of $\oplus(*2^{n-1})$		Times
	New	FRMT	
2	5	8	1.600
3	10	24	2.400
4	19	64	3.368
5	36	160	4.444
6	69	384	5.565
7	134	896	6.686
8	263	2048	7.787
9	520	4608	8.861
10	1033	10240	9.912
12	4107	49152	11.967
16	65551	1048576	15.996

3. Minimisation of Multiple-Output Functions

The minimum polarity of a single-output function can be obtained by the efficient algorithm proposed in Section 2. The minimum number $w(j)$ of exclusive-OR gates which are required to realise the function is determined by the nonzero coefficient number of the polynomial at the minimum polarity (j).

For the case of m-output functions, it is very obvious that the minimum polarity for one of the output functions is not likely to be optimum for the whole m-output function, as the polarity j that makes the exclusive-OR gate number $w_i(j)$ of the ith function minimum cannot guarantee to make the total number of exclusive-OR gates for all the m functions minimum. Furthermore, the polarity j which makes the sum $w = \sum_{i=0}^{m-1} w_i(j)$ minimum is also unlikely to be optimum for all m-output functions, because. some product terms will be common to a number of the output functions. One common term among s-output polynomials may be realised in terms of only one exclusive-OR gate, and $(s-1)$ exclusive-OR gates may be saved. Hence the optimum polarity for m-output functions should be searched with regard to all possible common terms.

For example, the truth table of a 3-output function is shown in Table 2. The coefficients of the polynomials with polarity (000) may be computed from eqn. (2), and are also shown in Table 2.

Therefore, the 3-output polynomials with polarity (000) are as follows:

$$f_0 = x_1 \oplus x_0 x_1 \oplus x_2 \oplus x_0 x_1 x_2$$
$$f_1 = x_1 \oplus x_0 x_1 \oplus x_0 x_2 \oplus x_0 x_1 x_2$$
$$f_2 = 1 \oplus x_0 \oplus x_0 x_2 \oplus x_0 x_1 x_2$$

Table 2 Example of 3-output functions

Input			Output			Polynomial coeff.		
x_2	x_1	x_0	f_2	f_1	f_0	c_i	b_i	a_i
0	0	0	1	0	0	1	0	0
0	0	1	0	0	0	1	0	0
0	1	0	1	1	1	0	1	1
0	1	1	0	0	0	0	1	1
1	0	0	1	0	1	0	0	1
1	0	1	1	1	1	1	1	0
1	1	0	1	1	0	0	0	0
1	1	1	0	0	0	1	1	1

Using the algorithm in Section 2, the coefficients of 3 polynomials for all possible polarities may be directly mapped from the coefficients **A**(0), **B**(0) and **C**(0). The flow graph is shown in Fig. 2.

In Fig. 2, w'_i denotes the number of nonzero coefficients of the ith polynomial, f_i, referred to

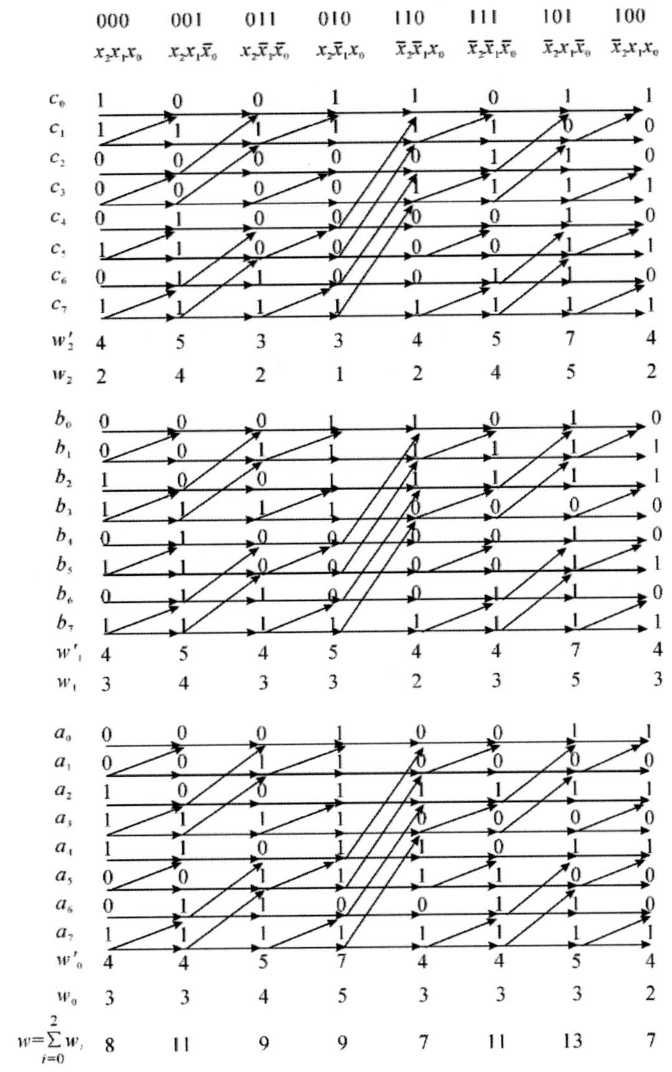

Fig. 2 **Flow graph for multiple-output functions**

as the weight of the ith polynomial, and w_i denotes the number of exclusive-OR gates for realisation of the ith polynomial. Since

$$1 \oplus f = \overline{f} \tag{15}$$

the constant term of polynomials may be realised by means of a NOT gate at the output instead of an exclusive-OR gate in circuits. Hence, the number of exclusive-OR gates for realisation of the ith polynomial is as follows:

$$w_i = w'_i - a_{i0} - 1 \tag{16}$$

where

$a_{i0} \in (0, 1)$ is the constant term of the ith polynomial

$w = \sum_{i=0}^{m-1} w_i$ represents the total number of exclusive-OR gates for realisation of m-output functions.

Here, w, w_i, and w_i' are all functions of polarities $j(j=0, 1, \cdots, 2^n-1)$. One or more polarities may make w minimum. Fig. 2 shows that the sum w has the minimum value 7 when the polarity is (100) or (110). The output polynomials with input polarity (100) are

$$f_0 = x_1 \oplus \bar{x}_2 \oplus x_0 x_1 \bar{x}_2$$
$$f_1 = x_0 \oplus x_1 \oplus x_0 \bar{x}_2 \oplus x_0 x_1 \bar{x}_2$$
$$f_2 = x_0 x_1 \oplus x_0 \bar{x}_2 \oplus x_0 x_1 \bar{x}_2$$

Seven exclusive-OR gates are required to realise the 3 output functions. But this is not minimum, because $(x_0 \bar{x}_2 \oplus x_0 x_1 \bar{x}_2)$ is the common term between the functions f_1 and f_2. One exclusive-OR gate may be commonly used by the functions f_1 and f_2, and another exclusive-OR gate can be saved. Only 6 exclusive-OR gates are needed to realise these 3-output functions.

Is this an optimum realisation for the example? It is unlikely to be optimum, as the number of common terms may be greater in the other polarity polynomials. Although $w = \sum_{i=0}^{2} w_i$ is not minimum for that polarity, the final required number of exclusive-OR gates with regard to the common terms is possibly less than that of the polarity (100). For example, the polarity (011) has $w=9>7$, but the polynomials are

$$f_0 = \bar{x}_0 \oplus \bar{x}_0 \bar{x}_1 \oplus \bar{x}_0 x_2 \oplus \bar{x}_1 x_2 \oplus \bar{x}_0 \bar{x}_1 x_2$$
$$f_1 = \bar{x}_0 \oplus \bar{x}_0 \bar{x}_1 \oplus \bar{x}_0 x_2 \oplus \bar{x}_0 \bar{x}_1 x_2$$
$$f_2 = \bar{x}_0 \oplus \bar{x}_1 x_2 \oplus \bar{x}_0 \bar{x}_1 x_2$$

where $(\bar{x}_0 \oplus \bar{x}_1 x_2)$ is the common term for 3-output polynomials, 4 exclusive-OR gates may be saved, and as $(\oplus \bar{x}_0 \bar{x}_1)$ is the common term of the functions f_0 and f_1, one exclusive-OR gate may be saved. Only $9-(4+1)=4$ exclusive-OR gates are required to realise the 3 polynomials. Although w for the polarity (011) is larger than that for the polarity (100), the polarity (011) needs less exclusive-OR gates as a result of the common terms.

From the above example, the optimum polarity of m-output functions should be searched by considering the common terms. Theorem 2 will give a method for determining the common terms among the multiple-output polynomials.

Before giving the theorem some definitions are introduced.

Definition 1

The function consisting of the common terms among s functions is called a 's-common function'. For example, the 3-common function of the above example is

$$f_{012} = \bar{x}_0 \oplus \bar{x}_1 x_2 \oplus \bar{x}_0 \bar{x}_1 x_2$$

The coefficients of common functions may be directly obtained by logical AND operations among the s function coefficients:

$$\mathbf{A}_{j_1 j_2, \cdots, j_s} = \mathbf{A}_{j_1} \wedge \mathbf{A}_{j_2} \wedge \cdots \wedge \mathbf{A}_{j_s} \tag{17}$$

where

$\mathbf{A}_{j_1 j_2, \cdots, j_s}$ denotes the coefficient vector of the s-common function

$\mathbf{A}_{j_1}, \mathbf{A}_{j_2}, \cdots, \mathbf{A}_{j_s}$ are the coefficient vectors of the output functions $j_1, j_2, \cdots, j_s$, respectively

$\wedge$ denotes logical AND operations

For the above example of polarity (011), the coefficient of the 3-common polynomial is

$$\mathbf{A}_{012} = \mathbf{A}_0 \wedge \mathbf{A}_1 \wedge \mathbf{A}_2 = (01000011)^t$$

Let $w'_{j_1 j_2, \cdots, j_s}$ denote the weight of the s-common function; then the number $w_{j_1 j_2, \cdots, j_s}$ of exclusive-OR gates may be obtained from eqn. (16). For the above example of the polarity (011), we have

$$w'_{012} = 3$$

$$w_{012} = 2$$

Because $\mathbf{A}_{j_1}, \mathbf{A}_{j_2}, \cdots, \mathbf{A}_{j_s}$ are functions of the polarity, so the number $w_{j_1 j_2, \cdots, j_s}$ of exclusive-OR gates of the common function is also a function of the polarity. The number of s-common functions among m output functions is

$$C_m^s \text{ for } s = 2, 3, \cdots, m \tag{18}$$

The total number of common functions among m output functions is

$$C_m^2 + C_m^3 + \cdots + C_m^{m-1} + C_m^m = 2^m - (m+1) \tag{19}$$

For the above example, the common function numbers for 3 and 2 polynomials are $C_3^3 = 1$ and $C_3^2 = 3$, respectively. The total number of common functions among 3-output functions is $2^3 - (3+1) = 4$.

To avoid the repeated counting of common exclusive-OR gates in different common functions, the residue function is introduced.

Definition 2

The residue function of a function f_i is the subfunction which consists of the terms of without $(w_{j_1 j_2, \cdots, j_s} - 1)$ common exclusive-OR terms of the s-common function, where $i = j_1, j_2, \cdots, j_s$.

For the above example, the 3-common function is

$$f_{012} = \bar{x}_0 \oplus \bar{x}_1 x_2 \oplus \bar{x}_0 \bar{x}_1 x_2$$

The 3 residue functions may be

$$f_0^{012} = \bar{x}_0 \oplus \bar{x}_0 \bar{x}_1 \oplus \bar{x}_1 x_2$$

$$f_1^{012} = \bar{x}_0 \oplus \bar{x}_0 \bar{x}_1$$

$$f_2^{012} = \bar{x}_0$$

The coefficients of residue functions may be obtained by modulo-2 additions of the coefficients of the function f_i, the s-common function $f_{j_1 j_2, \cdots, j_s}$ and the remainder term $\mathbf{C}_s$, as follows:

$$\mathbf{A}_i^{(s)} = \mathbf{A}_i \oplus \mathbf{A}_{j_1 j_2, \cdots, j_s} \oplus \mathbf{C}_s \tag{20}$$

where

$\oplus$ denotes modulo-2 addition

$\mathbf{A}_i$ is the coefficient vector of f_j

$\mathbf{A}_{j_1 j_2, \cdots, j_s}$ is the coefficient vector of the s-common function $f_{j_1 j_2, \cdots, j_s}$

$\mathbf{C}_s$ is a column vector, with a single 'one' remaining from $\mathbf{A}_{j_1 j_2, \cdots, j_s}$ and all other elements in $\mathbf{C}_s$ at zero.

In the above example, the remainder term is $\mathbf{C}_3 = (01000000)^t$. The existence of $\mathbf{C}_s$ means that one of the common function terms is remained in the residue functions. In the above example, the common term $\bar{x}_0$ is remained in the residue functions f_0^{012}, f_1^{012} and f_2^{012}. It is named the 'remainder term'.

As the common function possesses $w_{j_1 j_2, \cdots, j_s}$ terms, so $\mathbf{C}_s$ has $w_{j_1 j_2, \cdots, j_s}$ forms. For the example, if we let $\mathbf{C}_3 = (00000001)^t$, then the remainder term will be $x_0 x_1 x_2$, and the residue functions will be

$$f_0^{012} = \bar{x}_0 \oplus \bar{x}_0 x_2 \oplus \bar{x}_0 \bar{x}_1 x_2$$

$$f_1^{012} = \bar{x}_0 \bar{x}_1 \oplus \bar{x}_0 \bar{x}_1 x_2$$

$$f_2^{012} = \bar{x}_0 \bar{x}_1 x_2$$

It is obvious that the choice of the remainder term and the form of $\mathbf{C}_s$ does not change the weight and number of exclusive-OR gates of the residue functions. Therefore, any term of the s-common function may be the remainder term. However, the remainder term of the s-common function must not be the remainder term of the $(s-1)$-common function, i. e.

$$\mathbf{C}_s \neq \mathbf{C}_{s-1} \tag{21}$$

If $\mathbf{C}_s = \mathbf{C}_{s-1}$, then $\mathbf{C}_s \oplus \mathbf{C}_{s-1} = 0$, and the incorrect common gates may be introduced.

For the above example, if is the 3-remainder term, then $\overline{x}_0$ must not be the 2-remainder term: the 2-remainder term should be $\overline{x}_0\overline{x}_1$.

After the common function is produced, the output function should be substituted by its residue function for further operations. The coefficient $\mathbf{A}_i$ is substituted by $\mathbf{A}_i^{(s)}$ in memory after $\mathbf{A}_{j_1 j_2,\cdots,j_s}$ is produced.

For an example of a 4-output function, if the output functions are f_0, f_1, f_2 and f_3, then the 4-common function f_{0123} is

$$\mathbf{A}_{0123} = \mathbf{A}_0 \wedge \mathbf{A}_1 \wedge \mathbf{A}_2 \wedge \mathbf{A}_3$$

The 4-residue function is

$$\mathbf{A}_i^{(4)} = \mathbf{A}_i \oplus \mathbf{A}_{0123} \oplus \mathbf{C}_4 \quad (i=0,1,2,3)$$

The 3-common function f_{012} is

$$\mathbf{A}_{012} = \mathbf{A}_0^{(4)} \wedge \mathbf{A}_1^{(4)} \wedge \mathbf{A}_2^{(4)}$$

The residue function is

$$\mathbf{A}_i^{012} = \mathbf{A}_i^{(4)} \oplus \mathbf{A}_{012} \oplus \mathbf{C}_3 \quad (i=0,1,2)$$

But the 3-common function f_{123} is

$$\mathbf{A}_{123} = \mathbf{A}_1^{012} \wedge \mathbf{A}_2^{012} \wedge \mathbf{A}_3^{(4)}$$

Here, $\mathbf{A}_1^{012}$ and $\mathbf{A}_2^{012}$ are substituted for $\mathbf{A}_1^{(4)}$ and $\mathbf{A}_2^{(4)}$.

The residue functions are therefore

$$\mathbf{A}_1^{123} = \mathbf{A}_1^{012} \oplus \mathbf{A}_{123} \oplus \mathbf{C}'_3$$

$$\mathbf{A}_2^{123} = \mathbf{A}_2^{012} \oplus \mathbf{A}_{123} \oplus \mathbf{C}'_3$$

$$\mathbf{A}_3^{123} = \mathbf{A}_3^{(4)} \oplus \mathbf{A}_{123} \oplus \mathbf{C}'_3$$

Here $\mathbf{A}_1^{(4)}$ and $\mathbf{A}_2^{(4)}$ were substituted by $\mathbf{A}_1^{012}$ and $\mathbf{A}_2^{012}$ in the first two equations, and $\mathbf{C}'_3 \neq \mathbf{C}_3$. This substitution may be easily realised by an 'in place' algorithm.

After giving the definition and the algorithm for finding common and residue functions, the number of gates which may be saved is given by the following theorem.

Theorem 2

The number of exclusive-OR gates that may be saved for m-output functions is as follows:

$$N_s = (m-1)w^{(m)} + (m-2)w^{(m-1)} + \cdots + 2w^{(3)} + w^{(2)} \tag{22}$$

where

N_s denotes the number of gates that may be saved

$w^{(m)}$ is the number of common gates of m output functions

$w^{(m-1)} = w_{01,\cdots,(m-2)} + w_{01,\cdots,(m-3)(m-1)} + \cdots + w_{12,\cdots,(m-1)}$, and denotes the number of common gates of C_m^1, and the $(m-1)$-common functions

$w^{(3)} = \sum_{i \neq j \neq k}^{m-1} w_{ijk}$ denotes the gate number of all the 3-common functions

$w^{(2)} = \sum_{i \neq j} w_{ij}$ denotes the gate number of all the 2-common functions

Corollary 1

The total number of exclusive-OR gates for the realisation of m-output functions with regard to the common terms is

$$N_{eor} = \sum_{i=0}^{m-1} w_i - N_s \tag{23}$$

where

N_{eor} denotes the required total number of exclusive-OR gates

w_i is the number of ExOR gates for individual realisation of each single function f_i

N_s is the number of ExOR gates that may be saved by common terms

As w_i and N_s are functions of the polarities, so N_{exor} is also a function of polarities. The optimum polarity may be obtained by finding the minimum N_{exor}.

The optimum polarity and the number of ExOR gates for m-output functions may be obtained by the algorithm shown in Fig. 3.

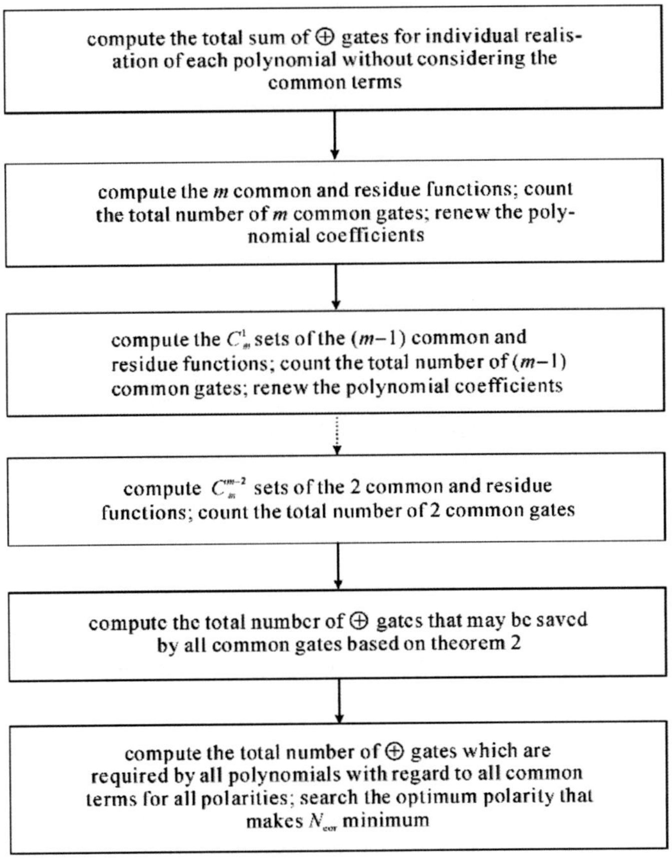

Fig. 3 flow chart of the minimisation algorithm

Step 1

Compute the common gate number $w^{(m)}$ of the m functions:

(i) compute the m-common function coefficient $\mathbf{A}_{01\cdots m-1}$ based on eqn. (17) in terms of logical AND operations;

(ii) count the number of common gates $w^{(m)}$;

(iii) choose $\mathbf{C}_m$, compute m residue function coefficients $\mathbf{A}_i^{(m)}$ ($i=0, 1, \cdots, m-1$) based on eqn. (20) in terms of modulo-2 additions;

(iv) substitute $\mathbf{A}_i^{(m)}$ for $\mathbf{A}_i$ in the computer memory ($i=0, 1, \ldots, m-1$).

Step 2

Compute the number of common gates $w^{(m-1)}$ for all sets of $(m-1)$ functions:

(i) choose the first set of $(m-1)$ functions, as step 1, compute the $(m-1)$-common function coefficients $\mathbf{A}_{01,\cdots,(m-2)}$ and the number of common gates $w_{01\cdots,(m-2)}$, choose $\mathbf{C}_{m-1} \neq \mathbf{C}_m$, compute $(m-1)$ residue function coefficients $\mathbf{A}_i^{(m-1)}$ ($i=0, 1, \cdots (m-2)$), and substitute $\mathbf{A}_i^{(m)}$ by $\mathbf{A}_i^{(m-1)}$ in the computer memory;

(ii) choose the second set of $(m-1)$ functions, as in step 2(i), compute $\mathbf{A}_{01,\cdots,(m-3)(m-1)}$, $w_{01,\cdots,(m-3)(m-1)}$ and residue functions $\mathbf{A}'^{m-1}_i$ ($i=0,1,\cdots,(m-3),(m-1)$), and substitute $\mathbf{A}'_i{}^{(}m-1)$ for $\mathbf{A}_i^{(m-1)}$ and $\mathbf{A}_i^{(m)}$, but the second set of $(m-1)$ function coefficients have been modified by the above step, and they will be modified by $\mathbf{A}'^{(m-1)}_i$ ($i=0, 1, \cdots, m-3, m-1$) again in this step;

⋮

(m) choose the mth set of $(m-1)$ functions and compute $\mathbf{A}_{12,\cdots,m-1}$ and $w_{12,\cdots,m-1}$ etc., as in the above step, and modify the function coefficients;

(n) compute the total number of all $(m-1)$-common gates using

$$w^{(m-1)} = w_{01,\cdots,(m-2)} + w_{01,\cdots,(m-3)(m-1)} + \cdots + w_{12,\cdots,(m-1)}$$

Step 3

Compute the number of common gates $w^{(m-2)}$, $w^{(m-3)}$, ..., $w^{(2)}$ as above steps.

Step 4

Compute the total number N_s of the ExOR gates that may be saved based on eqn. (22), and compute the total gate number N_{exor} for one polarity by eqn. (23).

Step 5

Compute the total number N_{exor} of ExOR gates for all possible polarities. The polarity that makes N_{exor} minimum is optimum.

For the above example, the total number of ExOR gates for different polarities is shown in Table 3.

From Table 3, it may be seen that the polarity (011) is the optimum one for realisation of the 3-output functions with regard to the common terms. Only 4 ExOR gates are required to realise the circuit. The logical circuit of the example is shown in Fig. 4.

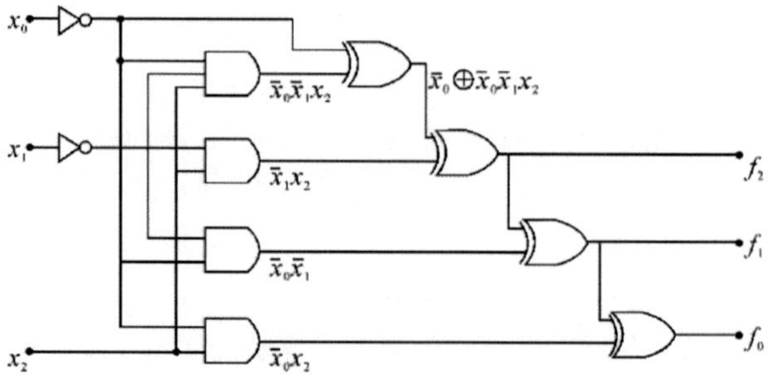

Fig. 4 Optimum logical circuits for the example

The number N_{in} of input NOT gates depends on the polarities. The numbers of output NOT gates and AND gates are given by the following corollaries.

Corollary 2

The number N_{out} of output NOT gates is

$$N_{out} = \sum_{i=0}^{m-1} a_{io} \tag{24}$$

where

N_{out} denotes the number of NOT gates at output end

a_{io} is the constant term of the ith output function

It is obvious that N_{out} is a function of the polarities.

Corollary 3

The number N_{and} of AND gates for realisation of m-output functions is

$$N_{and} = w_d - \sum_{i=0}^{n-1} d_{2^i} - d_0 \tag{25}$$

where

w_d is the weight of the OR polynomial **D** for the m-output functions, where

$$\mathbf{D} = \mathbf{A}_0 \vee \mathbf{A}_1 \vee \cdots \vee \mathbf{A}_{m-1} \tag{26}$$

($\vee$ denotes the logical OR operation)

d_{2^i} denotes the coefficients of single variable terms of the **D** polynomial

N_{and} denotes the number of AND gates.

For the example in Table 2, the numbers of input NOT gates, output NOT gates and ExOR gates are shown in Table 4.

Table 3 The number of ExOR gates for the example

Gray Code	000	001	011	010	110	111	101	100
polarity	$x_2 x_1 x_0$	$x_2 x_1 \bar{x}_0$	$x_2 \bar{x}_1 \bar{x}_0$	$x_2 \bar{x}_1 x_0$	$\bar{x}_2 \bar{x}_1 x_0$	$\bar{x}_2 \bar{x}_1 \bar{x}_0$	$\bar{x}_2 x_1 \bar{x}_0$	$\bar{x}_2 x_1 x_0$
w_0	2	4	2	1	2	4	5	2
w_1	3	4	3	3	2	3	5	3
w_2	3	3	4	5	3	3	3	2
$w = \sum_{i=0}^{2} w_i$	8	11	9	9	7	11	13	7
$w^{(3)}$	0	2	2	1	0	2	3	0
$w^{(2)}$	2	1	1	2	1	1	1	1
N_s	2	5	5	4	1	5	7	1
N_{exor}	6	6	4	5	6	6	6	6

Table 4 Number of NOT, AND&ExOR gates for the example

Gray Code	000	001	011	010	110	111	101	100
polarity	$x_2 x_1 x_0$	$x_2 x_1 \bar{x}_0$	$x_2 \bar{x}_1 \bar{x}_0$	$x_2 \bar{x}_1 x_0$	$\bar{x}_2 \bar{x}_1 x_0$	$\bar{x}_2 \bar{x}_1 \bar{x}_0$	$\bar{x}_2 x_1 \bar{x}_0$	$\bar{x}_2 x_1 x_0$
N_{in}	0	1	2	1	2	3	2	1
N_{out}	1	0	0	3	2	0	3	2
N_{not}	1	1	2	4	4	3	5	3
N_{and}	3	4	4	3	3	4	4	3
N_{exor}	6	6	4	5	6	6	6	6

If the price of NOT, AND and ExOR gates is P_1, P_2 and P_3, respectively, then the total price of the circuits is

$$P = P_1 N_{\text{not}} + P_2 N_{\text{and}} + P_3 N_{\text{exor}} \tag{27}$$

The optimum polarities should be obtained by minimum P.

4. Minimisation of Polynomials Over Extension Field $GF(2^M)$

Any multiple-valued logic function may be represented as a polynomial over the extension Galois field $GF(2^M)$, as follows[11]:

$$y = f(x) = \sum_{i=0}^{r} a_i x^i \tag{28}$$

where

$$a_i, x, y \in GF(2^M) \quad i = 0, 1, \cdots, (2^M - 1) \quad r = 2^M - 1$$

The polynomial coefficients may be obtained from the function output vector by

$$\begin{cases} a_0 = f(0) \\ a_i = \sum_{x \in GF(2^M)} f(x) x^{r-i} \end{cases} \tag{29}$$

where

$i = 1, 2, \cdots, (2^M - 1)$

$\sum_{x \in GF(2^M)}$ denotes the sum for all the possible elements of $GF(2^M)$

If α is a primitive root of a primitive polynomial $g_M(\alpha)$ over $GF(2)$, then

$$g_M(\alpha) = d_M \alpha^M + d_{M-1} \alpha^{M-1} + \cdots + d_1 \alpha + d_0 \tag{30}$$

where

$$d_i \in GF(2) \quad i = 0, 1, \cdots, M$$

Then any nonzero element x of $GF(2^M)$ may be uniquely represented as a power of α or a polynomial mod. $g_M(\alpha)$ as follows:

$$x = \alpha^i = x_{M-1} \alpha^{M-1} + x_{M-2} \alpha^{M-2} + \cdots + x_1 \alpha + x_0 \bmod g_M(\alpha) \tag{31}$$

where

$i = 0, 1, .., (2^M - 2)$

$x_j \in GF(2) \quad j = 0, 1, \cdots, (M-1)$

$x \in GF(2^M)$

Therefore, the coefficent a_i of eqn. (29) may be represented as the powers of α.

For the example of Table 2, the primitive polynomial of $GF(2^3)$ is

$$g_3(\alpha)=\alpha^3+\alpha+1$$

The nonzero elements of $GF(2^3)$ are

$$\alpha^i \bmod (\alpha^3+\alpha+1) \text{ for } i=0,1,2,3,4,5,6$$

Table 2 may be rewritten in terms of powers of α, as is shown in Table 5.

Table 5 Truth table over $GF(2^3)$

Input				Output			
x	x_2	x_1	x_0	y_2	y_1	y_0	y
0	0	0	0	1	0	0	α^2
1	0	0	1	0	0	0	0
α	0	1	0	1	1	1	α^5
α^2	1	0	0	1	0	1	α^6
α^3	0	1	1	0	0	0	0
α^4	1	1	0	1	1	0	α^4
α^5	1	1	1	0	0	0	0
α^6	1	0	1	1	1	1	α^5

The coefficients of the polynomial are computed from eqn. (29). The truth table may be represented by the polynomial over $GF(2^3)$ as follows:

$$y=f(x)=\alpha^2+\alpha^2x+\alpha^6x^2+\alpha x^3+\alpha^6x^4+\alpha^4x^5+\alpha^3x^6+\alpha^5x^7$$

where $x, y \in GF(2^3)$, and where

$$x=x_2\alpha^2+x_1\alpha+x_0$$

$$y=y_2\alpha^2+y_1\alpha+y_0(\text{mod}(\alpha^3+\alpha+1))$$

$$x_i, y_i \in (0,1) \text{ for } i=0,1,2$$

Seven multipliers and adders are required to realise the polynomial.

The multipliers over $GF(2^M)$ may be implemented by an M-stage feedback shift register[12], and $GF(2^M)$-adders are implemented by M modulo-2 adders.

It is desirable to reduce the number of nonzero coefficients in the polynomials in eqn. (28), thus reducing the number of multiplications. Obviously, the number of nonzero coefficients is affected by polarities of input and output variables. For the example of Table 5, let the input polarity be changed to $(\bar{x}_2, x_1, \bar{x}_0)$. Then the truth table is changed to Table 6.

Let $x'=\bar{x}_2\alpha^2+x_1\alpha+\bar{x}_0(\text{mod}(\alpha^3+\alpha+1))$. From eqn. (29), the polynomial is

$$y=\alpha^5+\alpha^6x'+\alpha^6x'^5+\alpha^6x'^6+\alpha^5x'^7$$

The polynomial possesses 5 nonzero coefficients which are less than that of the polarity (000). The minimum polarity may be obtained by finding of all the possible polarity coefficients.

Table 6 Truth table with new input polarity

Input				Output			
x'	$\bar{x}_2$	x_1	$\bar{x}_0$	y_2	y_1	y_0	y
α^6	1	0	1	1	0	0	α^2
α^2	1	0	0	0	0	0	0
α^5	1	1	1	1	1	1	α^5
1	0	0	1	1	0	1	α^6
α^4	1	1	0	0	0	0	0
α^3	0	1	1	1	1	0	α^4
α	0	1	0	0	0	0	0
0	0	0	0	1	1	1	α^5

The minimisation of polynomials over extension Galois fields $GF(2^M)$ is a special problem of the minimisation of M-output functions over $GF(2)$ with the constraint of a minimum number of nonzero coefficients for the extension Galois field polynomial. Certain properties of Galois fields lead to an efficient algorithm for minimisation of the extension field polynomials. Eqn. (29) shows that the polynomial's coefficients may be computed from its function output vectors. The direct computation of all the polarity coefficients needs $2^M(2^M-1)$ field multiplications and additions, and (2^M-1) permutations of the truth table. It will be shown that the polynomial coefficient with a new polarity may be mapped from another polarity coefficient without permuting the truth table and recomputing eqn. (29).

Actually, any change of input polarities corresponds to an input variable transformation as follows:

$$\begin{aligned} x' &= \overset{*}{x}_{M-1}\alpha^{M-1}+\cdots+\overset{*}{x}_1\alpha+\overset{*}{x}_0 \\ &= x+b_{M-1}\alpha^{M-1}+\cdots+b_1\alpha+b_0 = x+B \end{aligned} \tag{32}$$

where $b_i=\begin{cases}1 \text{ for } \overset{*}{x}_i=\bar{x}_i \\ \text{for } \overset{*}{x}=x_i, i=0,1,\cdots,(M-1)\end{cases}$

Substituting eqn. (32) into eqn. (28), it follows that

$$y = f(x'+B) = \sum_{i=0}^{r} a_i(x'+B)^i = \sum_{i=0}^{r} a'_i x'^i$$

Comparing the coefficients of variable x', the new polarity coefficient can be mapped from another polarity coefficient by

$$\mathbf{A}'=\mathbf{T}_M(B)\mathbf{A} \tag{33}$$

where $\mathbf{A}'=(a'_0 a'_1,\cdots,a'_{M-1})^t$, $\mathbf{A}=(a_0 a_1,\cdots,a_{M-1})^t$ and $\mathbf{T}_M(B)$ is recursively computed by

$$\mathbf{T}_M(B)=\begin{vmatrix}\mathbf{T}_{M-1}(B) & B^{2M-1}\mathbf{T}_{M-1}(B) \\ 0 & \mathbf{T}_{M-1}(B)\end{vmatrix} \quad \mathbf{T}_1(B)=\begin{vmatrix}1 & B \\ 0 & 1\end{vmatrix} \tag{34}$$

For the polynomial over $GF(2^3)$, eqn. 33 may be written in matrix form as

$$\begin{vmatrix} a'_0 \\ a'_1 \\ a'_2 \\ a'_3 \\ a'_4 \\ a'_5 \\ a'_6 \\ a'_7 \end{vmatrix} = \left|\begin{array}{cccc:cccc} 1 & B & B^2 & B^3 & B^4 & B^5 & B^6 & B^7 \\ & 1 & 0 & B^2 & & B^4 & 0 & B^6 \\ & & 1 & B & & & B^4 & B^5 \\ & & & 1 & & & & B^4 \\ \hdashline & & & & 1 & B & B^2 & B^3 \\ & & & & & 1 & 0 & B^2 \\ & & & & & & 1 & B \\ & & & & & & & 1 \end{array}\right| \begin{vmatrix} a_0 \\ a_1 \\ a_2 \\ a_3 \\ a_4 \\ a_5 \\ a_6 \\ a_7 \end{vmatrix}$$

In general, any input polarity map corresponds to a polynomial in the transformed variable $(x+B)$, where B may be any element of $GF(2^M)$. The number of field multiplications and additions of $GF(2^M)$ for one polarity map is

$$(3^M-2^M) \tag{35}$$

The above mapping matrix may be factored as Kronecker products as follows:

$$\left|\begin{array}{cccc:cccc} 1 & B & & & & & & \\ & 1 & & & & & & \\ & & 1 & B & & & & \\ & & & 1 & & & & \\ \hdashline & & & & 1 & B & & \\ & & & & & 1 & & \\ & & & & & & 1 & B \\ & & & & & & & 1 \end{array}\right| \left|\begin{array}{cccc:cccc} 1 & & B^2 & & & & & \\ & 1 & & B^2 & & & & \\ & & 1 & & & & & \\ & & & 1 & & & & \\ \hdashline & & & & 1 & & B^2 & \\ & & & & & 1 & & B^2 \\ & & & & & & 1 & \\ & & & & & & & 1 \end{array}\right| \left|\begin{array}{cccc:cccc} 1 & & & & B^4 & & & \\ & 1 & & & & B^4 & & \\ & & 1 & & & & B^4 & \\ & & & 1 & & & & B^4 \\ \hdashline & & & & 1 & & & \\ & & & & & 1 & & \\ & & & & & & 1 & \\ & & & & & & & 1 \end{array}\right|$$

Using the above factorisation, the number of field multiplications and additions for each mapping is reduced to

$$M * 2^{M-1} \tag{36}$$

An efficient algorithm for further reducing the number of multiplications is developed by introducing adjacent polarity mappings and rearranging the polarity orderings. The polarity $(\overset{*}{x}_{M-1},\cdots,\overline{\overset{*}{x}}_k,\cdots,\overset{*}{x}_0)$ is the adjacent polarity of the polarity $(\overset{*}{x}_{M-1},\cdots,\overset{*}{x}_k,\cdots,\overset{*}{x}_0)$, as only one variable $\overset{*}{x}_k$ between them is in different polarity. From eqn. (32), any adjacent polarity change corresponds to a low-order transformed variable of $(x+\alpha^m)$, where $m<M$. The most important is the adjacent polarity change of variable x_0, which corresponds to the transformed variable $(x+1)$, i. e. $B=1$. No field multiplications are required for computing this polarity mapping. If the 2^M sets of polarities can be arranged in an ordering such that the transformed variable $(x+1)$ appears as often as possible, then the exhaustive search of polarity coefficients needs a minimum of multiplications. If all 2^M sets of polarities are rearranged according to the Gray code ordering, then 2^{M-1} sets of polarity changes out of a total of (2^M-1) sets of polarity changes correspond to the transformed variable of $(x+1)$. No field multiplications are needed for these 2^{M-1} mappings. The other $(2^{M-1}-1)$ sets of polarity changes correspond to low-order transformed variables. The total number of field multiplications for all (2^M-1) adjacent polarity mappings is $M * 2^{M-1}(2^{M-1}-1)$. The average number of field multiplications for one mapping is less than

$$M * 2^{M-2} \tag{37}$$

For the example shown in Table 5, the flow chart of the algorithm is shown in Fig. 5.

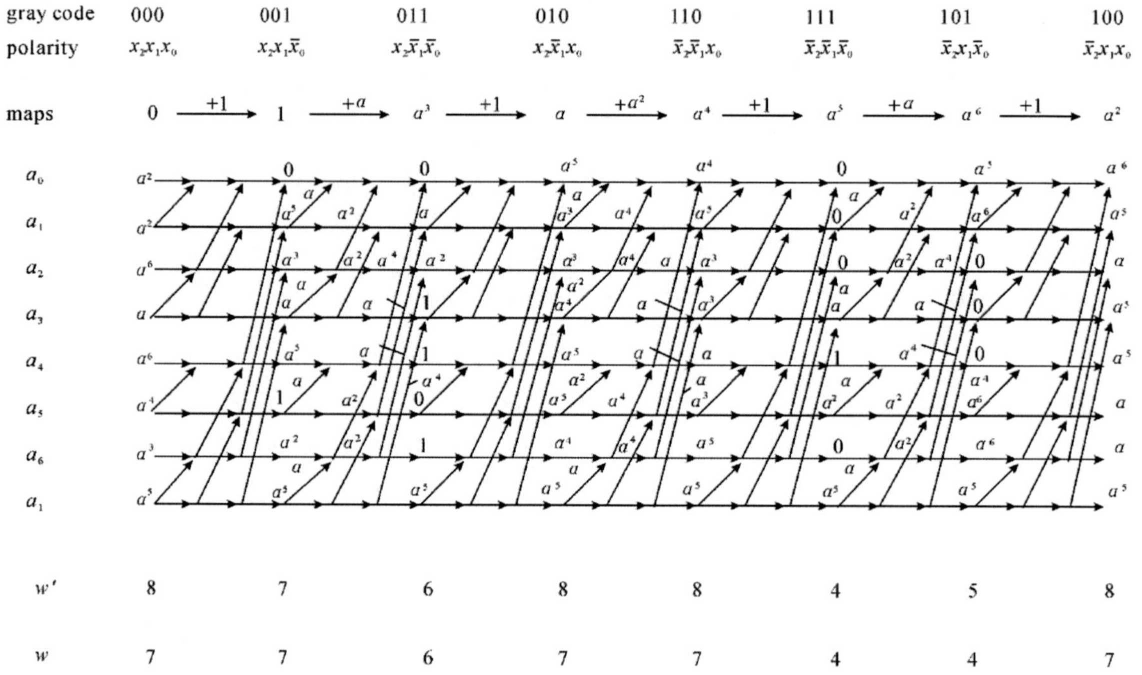

Fig. 5 Algorithm for extension field polynomials over $GF(2^3)$

In Fig. 5, w' is the weight of the polynomial coefficients and w is the number of nonconstant terms.

Furthermore, the constant term of the output polynomial may be saved by changing the output variable polarity. For the above example, if the output polarity is changed from (000) to (111), the truth table in Table 7 results.

Table 7 Truth table with new output polarity

Input				Output			
x'	$\bar{x}_2$	x_1	$\bar{x}_0$	$\bar{y}_2$	$\bar{y}_1$	$\bar{y}_0$	y'
α^6	1	0	1	0	1	1	α^3
α^2	1	0	0	1	1	1	α^5
α^5	1	1	1	0	0	0	0
1	0	0	1	0	1	0	α
α^4	1	1	0	1	1	1	α^5
α^3	0	1	1	0	0	1	1
α	0	1	0	1	1	1	α^5
0	0	0	0	0	0	0	0

Let $y' = \bar{y}_2\alpha^2 + \bar{y}_1\alpha + \bar{y}_0 (\bmod(\alpha^3+\alpha+1))$. Then $y' = y + \alpha^2 + \alpha + 1 = y + \alpha^5$. From eqn. (32), the polynomial becomes

$$y' = \alpha^6 x' + \alpha^6 x'^5 + \alpha^6 x'^6 + \alpha^5 x'^7$$

The constant term α^5 is removed by the output map (y_2, y_1, y_0) to $(\bar{y}_2, \bar{y}_1, \bar{y}_0)$. The input polarity $(\bar{x}_2, x_1, \bar{x}_0)$ and output polarity $(\bar{y}_2, \bar{y}_1, \bar{y}_0)$ are optimum. Only 4 nonzero terms are in the optimum polarity polynomial.

From the above example, it may be seen that any nonzero constant term $(a_0 \neq 0)$ in the polynomial may be removed be the output map

$$y' = y + a_0 \tag{39}$$

The optimum input and output polarity may be obtained by the above algorithms.

5. Conclusions

An efficient algorithm for minimisation of Reed-Muller polynomials with fixed polarities has been presented. The number of modulo-2 additions for this algorithm is $n/(1+(n-1)2^{-n})$ times less than that of fast RM transforms.

With regard to the common terms, the minimisation of multiple-output functions is developed. The common functions, residue functions and remainder terms are introduced. An algorithm for computing the common terms of multiple-output functions is presented. The common ExOR gates may be obtained by modulo-2 additions and logical AND operations among m sets of polynomial coefficients. The number of AND gates may be computed by logical OR operations.

The minimisation of extension Galois field polynomials is studied, and an efficient algorithm that directly maps the polynomial coefficient from one polarity to the adjacent one, without permuting the truth table and computing the RM transformations, is presented. The average number of field multiplications for each mapping is reduced to $(M * 2^{M-2})$. This algorithm can be easily extended to use in the case of multiple-output polynomials over extension Galois fields.

This paper is concerned only with the fixed-polarity Reed-Muller polynomials. The mixed-polarity polynomials may give a smaller number of nonzero coefficients than fixed-polarity polynomials. However, so far as the authors know, no general minimisation algorithm for mixed-polarity polynomials has been proposed for the case of variable number $n>6$[8].

The algorithms proposed in this paper are exhaustive, effective and can be computed 'in place'. They can be efficiently used in computer-aided logic design. The algorithms require only about 2^M memory locations and are efficient in terms of the multiplication and addition numbers.

6. Acknowledgment

One of the authors, Y. Z. Zhang, wishes to thank Trinity College, Cambridge, for financial support during this period.

References

[1] HELLERMAN, L.: 'A measure of computational work', *IEEE Trans.*, 1972, C-21, pp. 439-446.

[2] REDDY, S. M.: 'Easily testable realization for logical functions', *ibid.*, 1972, C-21, pp. 1183-1188.

[3] SALUJA, K. K., and ONG, E. H.: 'Minimization of Reed-Muller canonic expansion', *ibid.*, 1979, C-28, pp. 535-537.

[4] MUKHOPADHYAY, A., and SCHMITZ, G.: 'Minimization of exclusive-OR and logical equivalence switching circuits', *ibid.*, 1970, C-19, pp. 132-140.

[5] ROBINSON, J. P., and YEH, C-H.: 'Methods for modulo-2 minimization', *ibid.*, 1982, C-31, pp. 800-801.

[6] WU, X., CHEN, X., and HURST, S. L.: 'Mapping of Reed-Muller coefficients and the minimisation of exclusive OR-switching functions', *IEE Proc. E, Comput. & Digital Tech.*, 1982, 129, (1), pp. 15-20.

[7] BESSLICH, Ph. W.: 'Efficient computer method for ExOR logic design', *ibid.*, 1983, 130, (6), pp. 203-206

[8] BIOUL, G., DAVIO, M., and DESCHAMPS, J. P.: 'Minimization of ring-sum expansions of Boolean functions', *Philips Res. Rep.*, 1973, 28, pp. 17-36.

[9] PAPAKONSTANTINOU, G.: 'Minimization of modulo-2 sum of product', *IEEE Trans.*, 1979, C-28, pp. 163-167.

[10] GREEN, D. H., and TAYLOR, I. S.: 'Modular representation of multiple-valued logic systems', *Proc. IEE*, 1974, 121, (6), pp. 409-418.

[11] RAYNER, P. J. W.: 'The application of finite arithmetic structures to design of digital processing systems'. Proceedings of international conference on DSP, Florence, Italy, Sept. 1978.

[12] TANAKA, H., KESABARA, M., TEZUKA, Y., and KASAHAVA, Y.: 'Computation over Galois Fields using shift registers', *Inf. & Control*, 1968, 13, pp. 75-84.

A Direct Algorithm for Synthesis of Stable Feedback Shift Registers*

ZHANG Yan-zhong

Department of Engineering, Cambridge University, U. K.

The sufficient and necessary conditions for generating a stable sequence by means of a stable feedback shift register (FSR) is proved in this paper. A direct algorithm is presented for synthesis of stable FSRs with a minimum number of stages n.

1. Introduction

Stable feedback shift registers (FSRs) are useful for code theory and sequential machines. Golomb (1967) systematically studied the properties of nonlinear FSRs, especially cyclic properties. Mowle (1966) proposed the relation between cyclic and stable FSRs, but did not give feedback function forms. Lempel (1969) dealt with K-stable FSRs and presented a direct realization procedure. Cohn and Even (1969) developed a design procedure for finite sequences. The theory and an algorithm for synthesis of stable and infinite sequences using a stable FSR with a minimum number of stages n is presented.

2. Stable FSRs

The general form of a FSR is shown in Fig. 1. In Fig. 1, the n squares denote binary storage elements, feedback logic $f(x_0, x_1, \cdots, x_{n-1})$ is a combinational circuit, in general this is non-linear.

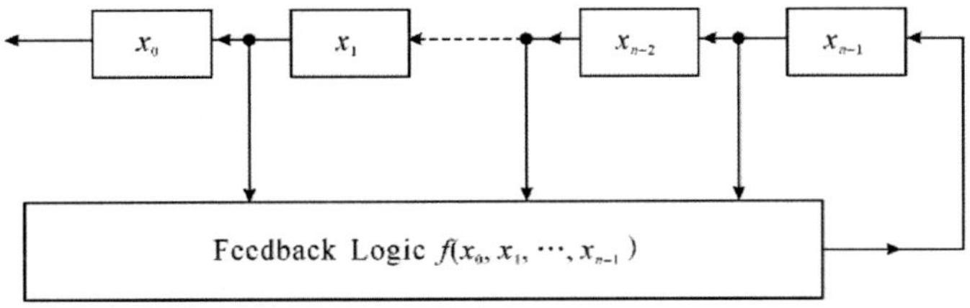

Fig. 1 General feedback shift registers

* Published in International Journal of Electronics, Vol 57, No. 1, 1984: 79-84.

Y. Zhang, *Footprints in Cambridge and Aviation Industries of China*,
https://doi.org/10.1007/978-981-16-3176-4_4

Any FSR may be considered as a finite state machine (FSM). Let the present state of a finite state machine be

$$\mathbf{S}=(x_0,x_1,\cdots,x_{n-1}) \tag{1}$$

then the next state of the machine is

$$\mathbf{S}'=(x'_0,x'_1,\cdots,x'_{n-1}) \tag{2}$$

where

$$\left.\begin{aligned} x'_0&=x_1\\ x'_1&=x_2\\ &\vdots\\ x'_{n-2}&=x_{n-1}\\ x'_{n-1}&=f(x_0,x_1,\cdots,x_{n-1}) \end{aligned}\right\} \tag{3}$$

Let $\mathbf{T}$ be a next state operator, the next state $\mathbf{S}'$ may be expressed by

$$\mathbf{S}'=\mathbf{T}(\mathbf{S})=(x_1,x_2,\cdots x_{n-1},f(x_0,x_1,\cdots,x_{n-1})) \tag{4}$$

The state after i shifts will be denoted by

$$\mathbf{T}^i(\mathbf{S}) \tag{5}$$

Definition

An FSR is stable if there exists an integer $q=q(\mathbf{S})$ for any state $\mathbf{S}$ of the machine, such that

$$\mathbf{T}^q(\mathbf{S})=\mathbf{0} \tag{6}$$

where $\mathbf{0}=(0, 0,\cdots,0)$ is the zero state. The state $\mathbf{0}$ is the only cyclic state and there are no other loops in the stable FSM state graph,

The output sequence of a stable FSR is as follows

$$c_0,c_1,\cdots,c_{q-1},0,0,0,\cdots \tag{7}$$

where $c_i\in(0,1)$ and $i=0,1,\cdots,q-1$. This is called a stable sequence. After a finite sequence of length q, the stable sequences result in all zeros.

3. Synthesis Theory of Stable FSRs

Any stable sequence may be generated by an FSR of stage $n\geqslant q$, but in the synthesis of stable FSRs it is desirable to design an FSR with a minimum number of stages n which generates the stable sequence (7). If the feedback function of the required stable FSR is

$$f(x_0,x_1,\cdots x_{n-1}) \tag{8}$$

then from Fig. 1 and eqn. (3) it follows

$$\left\{\begin{aligned} &f(c_k,c_{k+1},\cdots,c_{k+n-1})=c_{k+n}\\ &f(0,0,\cdots,0)=0 \end{aligned}\right\},\text{for } k=0,1,\cdots,q \tag{9}$$

Actually, eqns. (8) and (9) is a combinational function. The combination function (8) with minimum variable number n may be obtained from the stable sequence (7) by the following theorem.

Theorem

A stable sequence $(c_0,c_1,\cdots,c_{q-1},0,0,0\cdots)$ may be generated by a stable FSR of stage n, if and only if all $(q+1)$ sets $(c_k,c_{k+1},\cdots,c_{k+n-1})$ of length n for $k=0,1,\cdots$, q are distinct.

Proof

For the necessary condition, suppose a stable FSR. of stage n can generate the stable sequence (7), if two sets of length n are identical, say, there exists k_1 and k_2 such that

$$(c_{k_1}, c_{k_1+1}, \cdots, c_{k_1+n-1}) = (c_{k_2}, c_{k_2+1}, \cdots, c_{k_2+n-1}) \tag{10}$$

where $k_1 \neq k_2$, and $0 \leqslant k_1, k_2 \leqslant q$. Then from eqn. (9) it must hold that

$$\begin{aligned} c_{k_1+n} &= f(c_{k_1}, c_{k_1+1}, \cdots, c_{k_1+n-1}) \\ &= f(c_{k_2}, c_{k_2+1}, \cdots, c_{k_2+n-1}) = c_{k_2+n} \end{aligned} \tag{11}$$

therefore

$$(c_{k_1+1}, c_{k_1+2}, \cdots, c_{k_1+n}) = (c_{k_2+1}, c_{k_2+2}, \cdots, c_{k_2+n}) \tag{12}$$

Similarly

$$\begin{aligned} c_{k_1+n+1} &= c_{k_2+n+1} \\ c_{k_1+j} &= c_{k_2+j} \text{ for } j = n, (n+1), \cdots \end{aligned} \tag{13}$$

Hence the sequence is cyclic and the FSR is unstable, this is in contradiction with the definition.

The sufficient condition is obvious. If all the $(q+1)$ sets of length n from the sequence (7) are distinct, then a truth table of $(q+1)$ rows and $(n+1)$ columns may be set up from the sequence as follows.

Table 1 Truth table of distinct sets of sequences

x_0	x_1	$\cdots$	x_{n-1}	f
c_0	c_1	$\cdots$	c_{n-1}	c_n
c_1	c_2	$\cdots$	c_n	c_{n+1}
$\vdots$	$\vdots$	$\cdots$	$\vdots$	$\vdots$
c_{q-n-1}	c_{q-n}	$\cdots$	c_{q-2}	c_{q-1}
c_{q-n}	c_{q-n+1}	$\cdots$	c_{q-1}	0
$\vdots$	$\vdots$	$\cdots$	$\vdots$	$\vdots$
c_{q-1}	0	$\cdots$	0	0
0	0	$\cdots$	0	0

Since the $(q+1)$ sets of $(x_0, x_1, \cdots, x_{n-1})$ in Table 1 are distinct, so the combinational function f can be uniquely determined by the above truth table. From Fig. 1, this function is the feedback function of stable FSRs which generate the output sequence (7).

4. Algorithm and Example

The theorem presented not only the proof of necessary and sufficient conditions, but also a procedure for synthesis of stable FSRs. A direct synthesis algorithm of stable FSRs from the stable sequence (7) is shown below. Step 1 determines the minimum number of stages n of the FSR. Step 2 is used in finding the feedback functions

Step 1

Let $n=2,3,\cdots$, determine the minimum n such that any sets of length n $(c_k, c_{k+1}, c_{k+n-1})$ for $k=0,1,2,\cdots,q$, appear once and only once in the stable sequence $(c_0, c_1, \cdots, c_{q-1}, 0, 0, 0, \cdots)$. The flow chart is shown in Fig. 2.

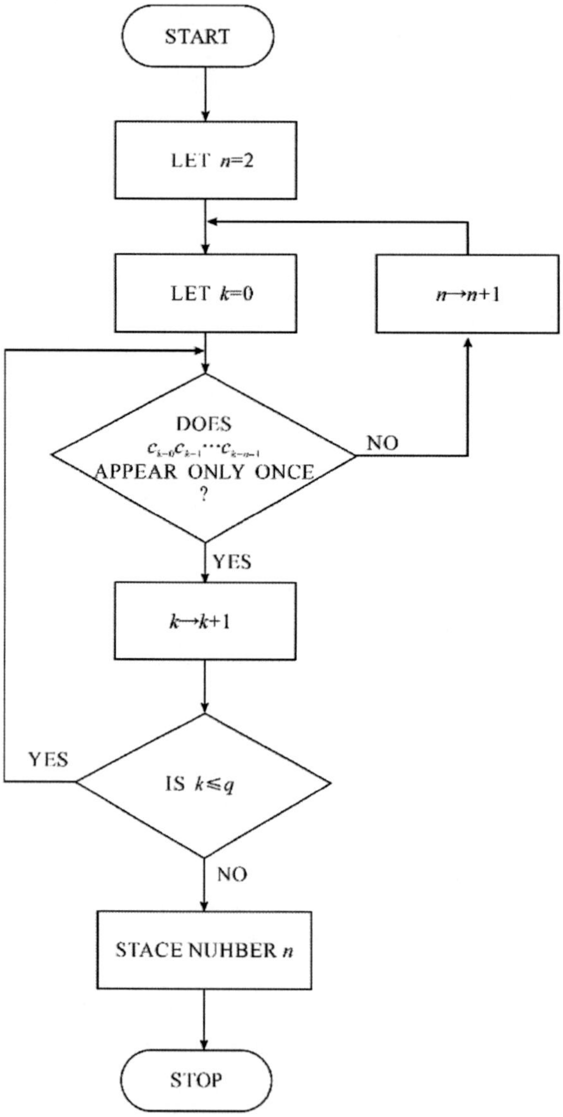

Fig. 2 Flow chart for determining the minimum number of stages *n*

Step 2

After finding the minimum stage n, the truth table of the feedback function may be obtained as shown in Table 1. Since

$$c_q = c_{q+1} = \cdots = 0 \tag{14}$$

the feedback function may be computed from the first $(q-n)$ rows of the truth table as follows

$$f(x_0, x_1, \cdots, x_{n-1}) = \bigvee_{k=0}^{q-n-1} c_{k+n} x_0^{c_k} x_1^{c_{k+1}} \cdots x_n^{c_{k+n-1}} \tag{15}$$

where

$$x_i^{c_{k+1}} = \begin{cases} x_i, & \text{for } x_{k+i}=1 \\ \overline{x}_i, & \text{for } c_{k+i}=0, i=0,1,\cdots,(n-1) \end{cases}$$

$c_{k+n}=0$ or 1, $k=0,1,\cdots,(q-n-1)$

$\vee$ denotes logical OR operation

For example, a stable sequence of length 12 is shown below

1	1	1	0	1	1	0	0	0	1	0	1	0	0	0	…
c_0	c_1	c_2	c_3	c_4	c_5	c_6	c_7	c_8	c_9	c_{10}	c_{11}	c_{12}	c_{13}	c_{14}	…

Step 1

Find the minimum number of stages n.

$n=2$ (1 1)=(c_0 c_1)=(c_1 c_2)=(c_4 c_5) appears three times

$n=3$ (1 1 1) appears once

(1 1 0)=(c_1 c_2 c_3)=(c_4 c_5 c_6) appears twice

$n=4$ (1 1 1 0), (1 1 0 1), (1 0 1 1), (0 1 1 0), (1 1 0 0) appears once

(1 0 0 0)=(c_5 c_6 c_7 c_8)=(c_{11} c_{12} c_{13} c_{14}) appears twice

$n=5$ (1 1 1 0 1), (1 1 0 1 1), (1 0 1 1 0), (0 1 1 0 0), (1 1 0 0 0)

(1 0 0 0 1), (0 0 0 1 0), (0 0 1 0 1), (0 1 0 1 0), (1 0 1 0 0)

(0 1 0 0 0), (1 0 0 0 0), (0 0 0 0 0)

Each of the 13 sets of length 5 appears only once in the sequence, so the minimum number of stages of the FSR which can generate the above sequence is 5.

Step 2

Determine the feedback function. From Table 1, the first (12−5)=7 rows of the truth table are as follows.

Table 2 Truth table of the example

x_0	x_1	x_2	x_3	x_4	f
1	1	1	0	1	1
1	1	0	1	1	0
1	0	1	1	0	0
0	1	1	0	0	0
1	1	0	0	0	1
1	0	0	0	1	0
0	0	0	1	0	1

The feedback function is as follows:

$$f(x_0, x_1, x_2, x_3, x_4) = x_0 x_1 x_2 \overline{x}_3 x_4 \vee x_0 x_1 \overline{x}_2 \overline{x}_3 \overline{x}_4 \vee \overline{x}_0 \overline{x}_1 \overline{x}_2 x_3 \overline{x}_4$$

5. Discussion

In the above example, 3 AND gates of 5 inputs, 2 OR gates, 5 NOT gates and 5 binary storage elements are required to implement the stable FSR. It is easy to prove that if all the $(q+1)$ sets of length n are distinct for the stable sequence (7), then all the $(q+1)$ sets of length larger than n are also distinct. The stable sequence (7) can be also implemented by a stable FSR with greater than n stages. For the above example, all the 13 sets of length 6 are distinct, the truth table consisting of the first $(12-6)=6$ rows from the sequence is as follows.

Table 3 Truth table of 6 variables

x_0	x_1	x_2	x_3	x_4	x_5	f
1	1	1	0	1	1	0
1	1	0	1	1	0	0
1	0	1	1	0	0	0
0	1	1	0	0	0	1
1	1	0	0	0	1	0
1	0	0	0	1	0	1

The feedback function with 6 variables for the example is

$$f(x_0, x_1, x_2, x_3, x_4, x_5) = \overline{x}_0 x_1 x_2 \overline{x}_3 \overline{x}_4 \overline{x}_5 \vee x_0 \overline{x}_1 \overline{x}_2 \overline{x}_3 x_4 \overline{x}_5$$

Two AND gates of 6 inputs, 1 OR gate, 6 NOT gates and 6 binary storage elements are required to realize the stable FSR. The optimum stage number may be obtained bv the number and price of AND, OR, NOT gates and binary storage elements.

6. Conclusion

Any stable FSR may be synthesized from the required stable sequences by the direct algorithm proposed in this paper. The optimum realization may be obtained by certain constrained conditions.

References

[1] COHN, M. , and EVEN, S. , 1969, The design of shift register generators for finite sequences. *I. E. E. E. Trans. Comput.* , 18,660-662.

[2] GOLOMB, S. W. , 1967, *Shift Register Sequences*, Part 111 (Holden. Day Inc).

[3] KOHAVI, Z. , 1978, Switching and Finite Automata Theory (McGraw-Hill Co.). LEMPEL, A. , 1969, On *k*-stable feedback shift registers. *I. E. E. E. Trans. Comput.* , 18, 652-660.

[4] MOWLE, F. J. , 1966, Relation between Pn cycles and stable feedback shift registers. *I. E. E. E. Trans. Comput.* , 15, 375-377.

Indirect Synthesis of Stable Binary Sequences*

ZHANG Yan-zhong

Department of Engineering, Cambridge University, U. K.

An indirect method is presented for the synthesis of stable binary sequences by means of maximum transient feedback shift registers (FSRs). An algorithm for the synthesis of maximum transient FSRs and output mapping is developed. The feedback functions of maximum transient FSRs are presented.

1. Introduction

Zhang (1984) proposed a direct algorithm for the synthesis of stable feedback shift registers from the required output sequences. The minimum number of stages n and the feedback function of stable FSRs which can generate the required output sequences may be obtained by means of that algorithm. However, for some stable binary sequences the number of stages obtained from the direct algorithm is rather large. For example, for a stable sequence of length 15

$$1000,0000,0000,0010,0000\cdots \tag{1}$$

the direct synthesis method leads to a 15-stage stable FSR.

In general, for any stable binary sequences of length q, let n be the minimum integer which satisfies

$$2^{n-1} < q < 2^n \tag{2}$$

Then the possible number of stages m of the required stable FSRs is within

$$n \leqslant m \leqslant q \tag{3}$$

If a stable FSR of n stages generates a stable sequence of length $N = 2^n$, then it is called a *maximum transient feedback shift register* which outputs the longest sequence among all the stable FSRs of n stages. If any stable binary sequence of length less than 2^n may be mapped from a maximum transient FSR of n stages, then an indirect synthesis method which uses the minimum stage FSR may be developed.

The synthesis of maximum transient FSRs is first presented in § 2. Section 3 introduces the theory of indirect synthesis, and the algorithm and an example are presented in § 4 and § 5.

* Published in International Journal of Electronics, Vol 57, No. 4, 1984: 569-576.

Y. Zhang, *Footprints in Cambridge and Aviation Industries of China*,
https://doi.org/10.1007/978-981-16-3176-4_5

2. Synthesis of Maximum Transient Feedback Shift Registers

A non-linear FSR of n stages is shown in Fig. 1.

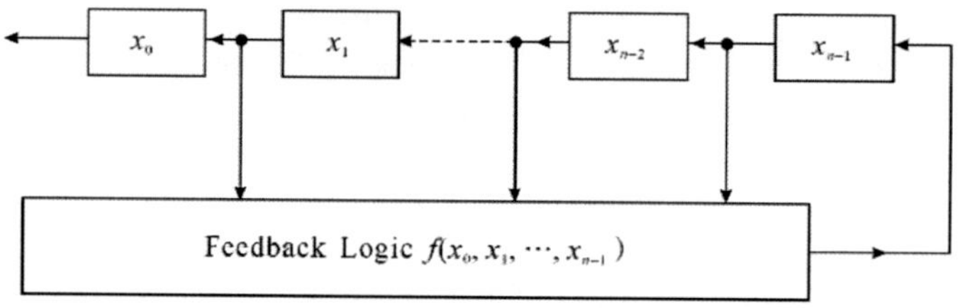

Fig. 1 General forms of FSRs

Any FSR may be viewed as a finite state machine (FSM). Let the present state of an FSM be $\mathbf{S}=(x_0, x_1, \cdots, x_{n-1})$; then from Fig. 1 its next state may be represented $\mathbf{S}'=(x'_0, x'_1, \cdots, x'_{n-1})$ may be represented as follows

$$\mathbf{S}'=T(\mathbf{S})=(x_1, x_2, \cdots, x_{n-1}, f(x_0, x_1, \cdots, x_{n-1})) \tag{4}$$

where T denotes a next-state operator, f is the feedback function, and $x_i, f \in (0, 1)$ with $i=0$, 1, $n-1$.

The state graph of maximum transient. FSRs is a maximum single tree of length 2^n which contains all the 2^n distinct states of FSRs of n stages. Before we give the synthesis of maximum transient FSRs, the following two lemmas must be introduced.

Lemma 1

The possible successors of state $(x_0, x_1 \ldots, x_{n-1})$ are the states

$$(x_1, x_2, \cdots, x_{n-1}, 0) \text{ or } (x_1, x_2, \cdots, x_{n-1}).$$

This is easily proved from (4) and Fig. 1 by using $f=0$ or 1. For example

$$T(0,0,\cdots\cdots,0,0)=(0,0,\cdots\cdots,0,0) \text{ or } (0,0,\cdots\cdots,0,1) \tag{5}$$

$$T(1,0,\cdots\cdots,0,0)=(0,0,\cdots\cdots,0,0) \text{ or } (0,0,\cdots\cdots,0,1) \tag{6}$$

Lemma 2

If $p(x)=1+a_1x+a_2x^2+\cdots+a_nx^n$ with $a_n=1$ is a primitive polynomial of degree n over GF(2), then the linear FSR with feedback function

$$p(x_0, x_1, \cdots, x_{n-1})=a_nx_0 \oplus a_{n-1}x_1 \oplus \cdots \oplus a_1x_{n-1} \tag{7}$$

is a primitive linear FSR of order n ($\oplus$ denotes modulo-2 addition).

The proof of this lemma was given by Stone (1974). A primitive LFSR has two cycles in its state graph: all the non-zero states on a pure cycle of length (2^n-1), and the state $\mathbf{0}$ forms a pure cycle of length 1. No branch is in the state graph (see Fig. 2). A large class of feedback functions for maximum transient FSRs may be derived from primitive linear FSRs by the following theorem.

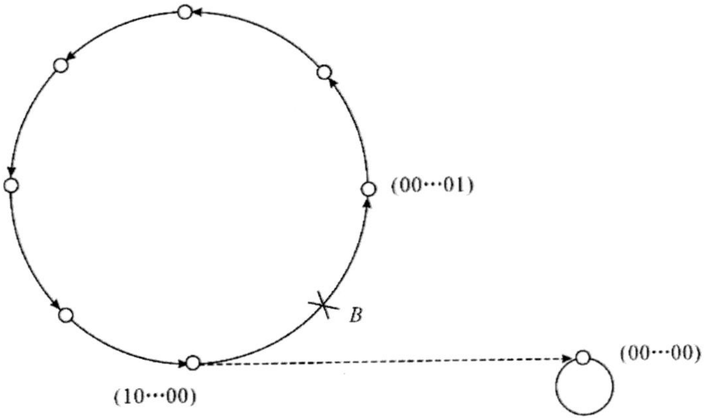

Fig. 2 Primitive linear FSRs

Theorem 1

If $p(x)=1+a_1x+a_2x^2+\cdots+a_nx^n$ is a primitive polynomial over GF(2), where $a_n=1$ and $a_i \in (0,1)$ with $i=1,2,\cdots,n$; then a large class of maximum transient FSRs may be obtained as follows

$$f(x_0,x_1,\cdots,x_{n-1})=a_0x_0 \oplus a_{n-1}x_1 \oplus \cdots \oplus a_1x_{n-1} \oplus x_0\overline{x}_1\cdots\overline{x}_{n-1} \tag{8}$$

Proof

Since $p(x)$ is a primitive polynomial of degree n over GF(2), from Lemma 2 the linear FSR with feedback function $p(x_0, x_1,\cdots, x_{n-1})$ is a primitive linear FSR. The state 0 forms a pure cycle of length 1. Its predecessor is only itself, From (6), the successor of the state (1,0,. . ,0) must be (0, 0,. . ,1), say, $x'_{n-1}=1$. From (4) we have

$$p(1,0,\cdots,0,0)=1 \tag{9}$$

Since

$$x_0\overline{x}_1\cdots\overline{x}_{n-1}=\begin{cases}1 & \text{only for } x_0=1\text{, and } x_1=x_2=\cdots=x_{n-1}=0\\ 0 & \text{otherwise}\end{cases} \tag{10}$$

Eqn. (8) becomes as follows

$$f(x_0,x_1,\cdots,x_{n-1})=\begin{cases}0 & \text{for } x_0=1\text{, and } x_1=x_2=\cdots=x_{n-1}=0\\ p(x_0,x_1,\cdots,x_{n-1}) & \text{otherwise}\end{cases} \tag{11}$$

The large cycle in the state graph of the primitive linear FSR is broken at point B (Fig. 2), and the state $(1,0,\cdots,0,0)$ is connected to the state $(0, 0,\cdots,0, 0)$; all the other states form a maximum single tree.

Table 1 The state table of the example in § 2

S	x_0	x_1	x_2	x_3	$f(x_0, x_1, x_2, x_3)$	x'_0	x'_1	x'_2	x'_3	S'
S_0	0	0	0	0	0	0	0	0	0	S_0
S_1	0	0	0	1	1	0	0	1	1	S_3
S_2	0	0	1	0	0	0	1	0	0	S_4
S_3	0	0	1	1	1	0	1	1	1	S_7
S_4	0	1	0	0	0	1	0	0	0	S_8
S_5	0	1	0	1	1	1	0	1	1	S_{11}
S_6	0	1	1	0	0	1	1	0	0	S_{12}
S_7	0	1	1	1	1	1	1	1	1	S_{15}
S_8	1	0	0	0	0	0	0	0	0	S_0
S_9	1	0	0	1	0	0	0	1	0	S_2
S_{10}	1	0	1	0	1	0	1	0	1	S_5
S_{11}	1	0	1	1	0	0	1	1	0	S_6
S_{12}	1	1	0	0	1	1	0	0	1	S_9
S_{13}	1	1	0	1	0	1	0	1	0	S_{10}
S_{14}	1	1	1	0	1	1	1	0	1	S_{13}
S_{15}	1	1	1	1	0	1	1	1	0	S_{14}

For example, the primitive polynomial of degree 4 over GF(2) is given by

$$p(x) = 1 + x + x^4$$

Then the following feedback function

$$f(x_0, x_1, x_2, x_3) = x_0 \oplus x_3 \oplus x_0 \bar{x}_1 \bar{x}_2 \bar{x}_3$$

generates a maximum transient FSR of 4 stages, its state table and state graph being as shown in Table 1 and Fig. 3 respectively.

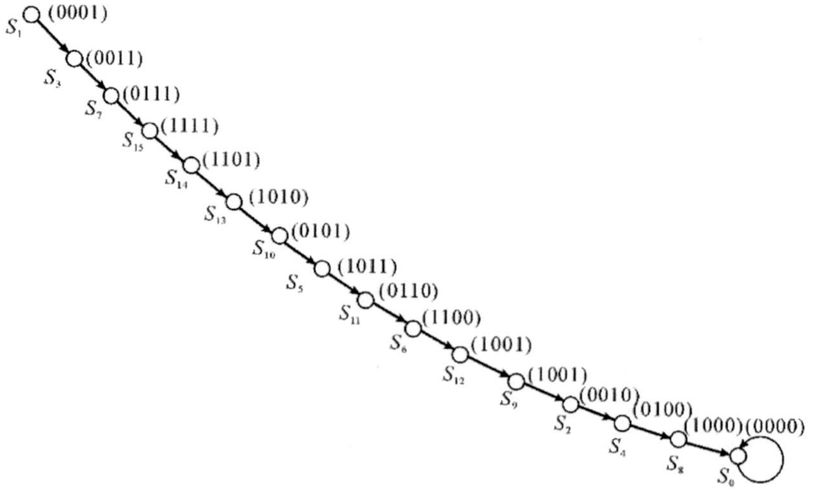

Fig. 3 Example of stable maximum length FSRs

A large class of feedback functions of maximum transient FSRs is shown in the Appendix.

3. Indirect Synthesis of Stable Binary Sequences

The above section has shown a method for the synthesis of maximum transient FSRs. The maximum transient FSR is the minimum stage FSR for generating stable sequences of length 2^n. All the 2^n states of maximum transient FSRs form a maximum single tree. An important property of a maximum transient FSR is that each of its 2^n states appears once and only once in the state graph. Therefore a maximum transient FSR may be employed as a field clement generator; any stable sequences of length less than 2^n may be generated by means of maximum transient FSRs and an output logic as shown in Fig. 4.

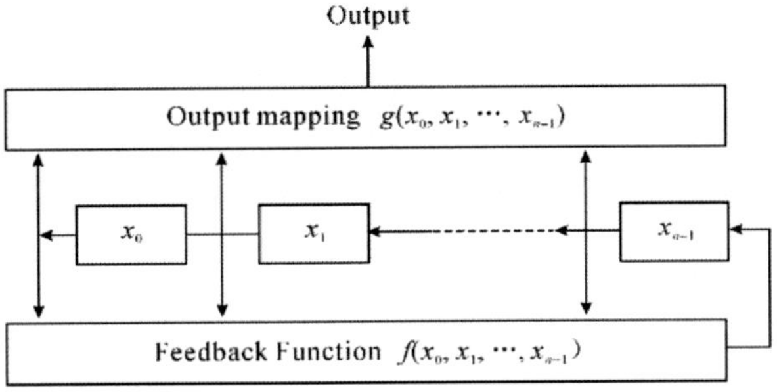

Fig. 4 FSRs with output mapping

Theorem 2

If a stable binary sequence of length q is as follows

$$d_0, d_1, \cdots, d_{q-1}, 0, 0, \cdots \tag{12}$$

where $d_i \in (0,1)$ with $i=0,1,\cdots,(q-1)$; and a maximum transient sequence of length 2^n ($2^{n-1} < q < 2^n$) is as follows

$$c_0, c_1, \cdots, c_{2^n-1}, 0, 0, 0, \cdots \tag{13}$$

where $c_i \in (0,1)$ with $i=0,1,\cdots,(2^n-1)$; then the stable sequence (12) may be generated by using the following output mapping logic

$$g(x_0, x_1, \cdots, x_{n-1}) = \sum_{k=0}^{q-1} d_{k}{}_{0}^{c_k} x_1^{c_{k+1}} \cdots x_{n-1}^{c_{k+n-1}} \tag{14}$$

where

$$x_j^{c_{k+j-1}} = \begin{cases} x_j & \text{for } c_{k+j-1}=1 \\ \overline{x}_j & \text{for } c_{k+j-1}=0 \end{cases}$$
$$j=1,2,\cdots,n \tag{15}$$

Proof

From (4), the state s_k of maximum transient FSRs may be expressed by

$$s_k = (c_k, c_{k+1}, \cdots, c_{k+n-1}) \tag{16}$$

where $k=0,1,\cdots,(2^n-1)$. Because a maximum transient FSR is stable, its last state ($k=2^n-1$) is the 0 state and forms a cycle of length 1

$$s_{2^n-1}=(0,0,\cdots,0) \text{ and } c_{2^n-1}=0 \tag{17}$$

and

$$T(0,0,\cdots,0)=(0,0,\cdots,0) \tag{18}$$

Since the 2^n states of maximum transient FSRs are distinct, a truth table of $(n+1)$ columns and 2^n rows (Table 2) can be set up from the sequences (12) and (13).

Table 2 The truth table of output mapping

x_0	x_1	$\cdots$	x_{n-1}	
c_0	c_1	$\cdots$	c_{n-1}	d_0
c_1	c_2	$\cdots$	c_n	d_1
$\vdots$	$\vdots$	$\cdots$	$\vdots$	$\vdots$
c_{q-1}	c_q	$\cdots$	c_{q+n-2}	d_{q-1}
c_q	c_{q+1}	$\cdots$	c_{q+n-1}	0
$\vdots$	$\vdots$	$\cdots$	$\vdots$	$\vdots$
c_{2^n-2}	0	$\cdots$	0	0
0	0	$\cdots$	0	0

As 2^n sets of variables $(x_0,x_1,\cdots,x_{n-1})$ and distinct in Table 2, so the combination function $g(x_0,x_1,\cdots,x_{n-1})$ can be uniquely determined from the table in tho form of (14). From the last row of Table 2, it follows that $g(0,\ 0,\ldots,0)=0$. So, from (18), the output of g is stable,

4. Algorithm

The proof of Theorem 2 also presented a procedure for generating a stable binary sequence from a maximum transient FSR. If a stable binary sequence of length q is given by (12), then the algorithm for indirect synthesis of the required sequence from a maximum transient FSR is as follows.

Step 1. Choose n such that

$$2^{n-1}<q<2^n \tag{19}$$

Step 2. Choose a primitive polynomial $p(x)$ of degree n over GF(2)

$$p(x)=1+a_1x+a_2x^2+\cdots+a_nx^n \tag{20}$$

Step 3. Design a maximum transient FSR with feedback function f as follows

$$f(x_0,x_1,\cdots,x_{n-1})=a_nx_0\oplus a_{n-1}x_1\oplus\cdots\oplus a_1x_{n-1}\oplus x_0\overline{x}_1\cdots\overline{x}_{-1} \tag{21}$$

Step 4. Generate the output sequence of the maximum transient FSR

$$c_0,c_1,\cdots,c_{2^n-2},0,0,0,\cdots \tag{22}$$

Step 5. Construct a truth table of an output mapping function from the sequences (12) and (22); design the output mapping function g as follows

$$g(x_0,x_1,\cdots,x_{n-1}) = \sum_{k=0}^{q-1} d_k x_0^{c_k} x_1^{c_{k+1}} \cdots x_{n-1}^{c_{k+n-1}} \tag{23}$$

The flow chart of this indirect algorithm is shown in Fig. 5.

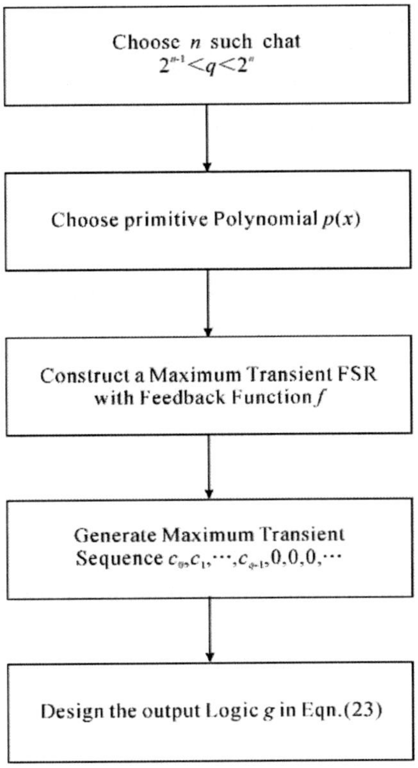

Fig. 5 Flow chart of indirect algorithm

5. Example

Consider a stable binary sequence of length 15 as follows

$$1\,0\,0\,0, 0\,0\,0\,0, 0\,0\,0\,0, 0\,0\,1\,0, 0\,0\,0\cdots$$

where $d_0 = d_{14} = 1$ *and* $d_1 = d_2 = \cdots = d_{13} = 0$.

Step 1. Choose $n=4$, as $2^3 < 15 < 2^4$.

Step 2. Choose a primitive polynomial of degree 4 over GF(2) as follows

$$p(x) = 1 + x + x^4$$

Step 3. From Theorem 1, design a maximum transient FSR with the feedback function

$$f(x_0, x_1, x_2, x_3) = x_0 \oplus x_3 \oplus x_0 \overline{x_1}\,\overline{x_2}\,\overline{x_3}$$

Step 4. The output sequence of the above maximum transient FSR is

$$0\,0\,0\,1,\ 1\,1\,1\,0,\ 1\,0\,1\,1,\ 0\,0\,1\,0,\ 0\,0\,0\ \cdots$$

Step 5. As $d_0 = d_{14} = 1$, the corresponding sets $s_0 = (0\ 0\ 0\ 1)$ and $s_{14} = (1\ 0\ 0\ 0)$.

From Theorem 2, the required sequence may be obtained from the following output mapping

$$g(x_0, x_1, x_2, x_3) = x_0 \overline{x_1}\,\overline{x_2} \oplus \overline{x_1}\,\overline{x_2}\, x_3$$

6. Discussion and Conclusions

The indirect synthesis of stable binary sequences uses a maximum transient FSR which possesses a minimum of n stages, so it requires the least number of binary memory elements. However, it needs more logic to realize the output mapping, which increases the complexity of the circuits. The direct synthesis of stable sequences need not have any output mapping, but it usually requires more stages than the indirect method. The optimum synthesis may be a compromise.

We have seen that any stable binary sequences may be indirectly generated from a maximum transient FSR by means of an output mapping. Maximum transient FSRs which possess the minimum number of stages can be obtained from primitive linear FSRs.

Appendix

Feedback functions of maximum transient FSRs

n	Feedback functions $f(x_0, x_1, \cdots, x_{n-1})$	Length
1		
2	$x_0 \oplus x_1 \oplus x_0 \bar{x}_1$	4
3	$x_0 \oplus x_2 \oplus x_0 \bar{x}_1 \bar{x}_2$	8
4	$x_0 \oplus x_3 \oplus x_0 \bar{x}_1 \bar{x}_2 \bar{x}_3$	16
5	$x_0 \oplus x_3 \oplus x_0 \bar{x}_1 \bar{x}_2 \bar{x}_3 \bar{x}_4$	32
6	$x_0 \oplus x_5 \oplus x_0 \bar{x}_1 \bar{x}_2 \bar{x}_3 \bar{x}_4 \bar{x}_5$	64
7	$x_0 \oplus x_6 \oplus x_0 \bar{x}_1 \bar{x}_2 \bar{x}_3 \bar{x}_4 \bar{x}^2 \bar{x}_6$	128
8	$x_0 \oplus x_4 \oplus x_5 \oplus x_6 \oplus x_0 \bar{x}_1 \bar{x}_2 \bar{x}_3 \bar{x}_4 \bar{x}^2 \bar{x}_6 \bar{x}_7$	256
9	$x_0 \oplus x_5 \oplus x_0 \bar{x}_1 \bar{x}_2 \bar{x}_3 \bar{x}_4 \bar{x}_5 \bar{x}_6 \bar{x}_7 \bar{x}_8$	512
10	$x_0 \oplus x_7 \oplus x_0 \bar{x}_1 \bar{x}_2 \bar{x}_3 \bar{x}_4 \bar{x}_5 \bar{x}_6 \bar{x}_7 \bar{x}_8 \bar{x}_9$	1024
11	$x_0 \oplus x_9 \oplus x_0 \bar{x}_1 \bar{x}_2 \bar{x}_3 \bar{x}_4 \bar{x}_5 \bar{x}_6 \bar{x}_7 \bar{x}_8 \bar{x}_9 \bar{x}_{10}$	2048
12	$x_0 \oplus x_6 \oplus x_8 \oplus x_{11} \oplus x_0 \bar{x}_1 \bar{x}_2 \bar{x}_3 \bar{x}_4 \bar{x}^2 \bar{x}_6 \bar{x}_7 \bar{x}_8 \bar{x}_1 \bar{x}_{10} \bar{x}_{11}$	4096
13	$x_0 \oplus x_9 \oplus x_{10} \oplus x_{11} \oplus x_0 \bar{x}_1 \bar{x}_2 \bar{x}_3 \bar{x}_4 \bar{x}_5 \bar{x}_6 \bar{x}_7 \bar{x}_8 \bar{x}_1 \bar{x}_{10} \bar{x}_{11} \bar{x}_{12}$	8192
14	$x_0 \oplus x_9 \oplus x_{11} \oplus x_{13} \oplus x_0 \bar{x}_1 \bar{x}_2 \bar{x}_3 \bar{x}_4 \bar{x}_5 \bar{x}_6 \bar{x}_7 \bar{x}_8 \bar{x}_1 \bar{x}_{10} \bar{x}_{11} \bar{x}_{12} \bar{x}_{13}$	16384
15	$x_0 \oplus x_{14} \oplus x_0 \bar{x}_1 \bar{x}_2 \bar{x}_3 \bar{x}_4 \bar{x}_5 \bar{x}_6 \bar{x}_7 \bar{x}_8 \bar{x}_9 \bar{x}_{10} \bar{x}_{11} \bar{x}_{12} \bar{x}_{13} \bar{x}_{14}$	32768
16	$x_0 \oplus x_{11} \oplus x_{13} \oplus x_{14} \oplus x_0 \bar{x}_1 \bar{x}_2 \bar{x}_3 \bar{x}_4 \bar{x}_5 \bar{x}_6 \bar{x}_7 \bar{x}_8 \bar{x}_1 \bar{x}_{10} \bar{x}_{11} \bar{x}_{12} \bar{x}_{13} \bar{x}_{14} \bar{x}_{15}$	65536

References

[1] KOUAVI, Z. , 1978, *Switching and Finite Automata Theory* (New York: McGraw-Hill). LEMFKL, A. , 1969, On k-stable feedback shift registers. *I. E. E. E. Trans. Comput.* , 18, 652-660.

[2] STONE, H. S. , 1974, *Discrete Mathematical Structures and their Applications* (Science Research Associates).

[3] ZHANG, Y. Z. , 1984 A direct algorithm for synthesis of stable feedback shift registers. *Int. J. Electron.* , 57, 79-84.

Galois Field Representation of Finite State Machines*

ZHANG Yan-zhong
Engineering Department, Cambridge University, U. K.

A systematic study for representing a finite state machine as a polynomial over Galois Field is presented. A Discrete Fourier Transform method is developed for computing the coefficients of polynomials. The representation of the machines which inputs and outputs possess different word lengths is presented. The cyclic subgroup property of the polynomial coefficients is revealed. This property considerably reduces the amount of computation. The conversion between one and multiple-variable machines is discussed. The Ring-Sum representation and Fast Reed-Muller Transformation are also discussed.

1. Introduction

1.1 Finite State Machine

A finite state machine is a mathematical model of sequential systems. It comprises a finite input set $U \in (U_0, U_1, \cdots, U_{N_U-1})$, a finite output set $Y \in (Y_0, Y_1, \cdots, Y_{N_Y-1})$, a finite state $S \in (S_0, S_1, \cdots, S_{N_S-1})$ set and two mappings $\underline{f}$ and $\underline{g}$ defined by[21]

$$\underline{s}' = \underline{f}(\underline{s}, \underline{u}) \tag{1a}$$

$$\underline{y} = \underline{g}(\underline{s}, \underline{u}) \tag{1b}$$

where $\underline{u} \in U$, $\underline{y} \in Y$ and $\underline{s} \in S$ are respectively the input, output and present state of systems at the nth clock instant. $\underline{s}' \in S$ denotes the state at the $(n+1)$th clock instant and is named the next state, The mapping $\underline{f}$ maps the present state $\underline{s}$ and the input $\underline{u}$ into the next state $\underline{s}'$, it is called the next state mapping, $\underline{g}$ is referred to as the output mapping which produces the present output $\underline{y}$ from only the present state $\underline{S}$. This model is called the Moore model. Another finite state machine model is the Mealy model which produces the output $\underline{y}$ from both the present state $\underline{s}$ and the input $\underline{u}$ as follows.

$$\underline{y} = \underline{g}(\underline{s}, \underline{u})$$

It has been proved that any Mealy model machine is equivalent to a Moore machine. The Moore machine might need more states than the Mealy machine for performing the same computation[21],

* Published in Cambridge University Engineering Department CUED/B-Elect/TRTO, 1984:1-44.

Y. Zhang, *Footprints in Cambridge and Aviation Industries of China*,
https://doi.org/10.1007/978-981-16-3176-4_6

Because we are particularly interested in the next-state behaviour of machines, it is convenient to use the model with the simplest possible output functions. Therefore only Moore machines are used in this paper. A Moore finite state machine is shown in Fig. 1, In Fig. 1,

$\underline{s}=(s_0, s_1, \cdots, s_{m-1})^t$ is a present state vector of m dimensions

$\underline{s}'=(s'_0, s'_1, \cdots, s'_{m-1})^t$ is a next state vector of m dimensions

$\underline{u}=(u_0, u_1, \cdots, u_{n-1})^t$ is a n-dimensional input vector

$\underline{y}=(y_0, y_1, \cdots, y_{k-1})^t$ is a k-dimensional output vector

$\underline{f}$ denotes a m-dimensional next state mapping

$\underline{g}$ denotes a k-dimensional output mapping

If there are no inputs to a finite state machine, e. g, $\underline{u}=0$, then the machine is called autonomous and the next state function becomes

$$\underline{s}'=\underline{f}(\underline{s}) \tag{2}$$

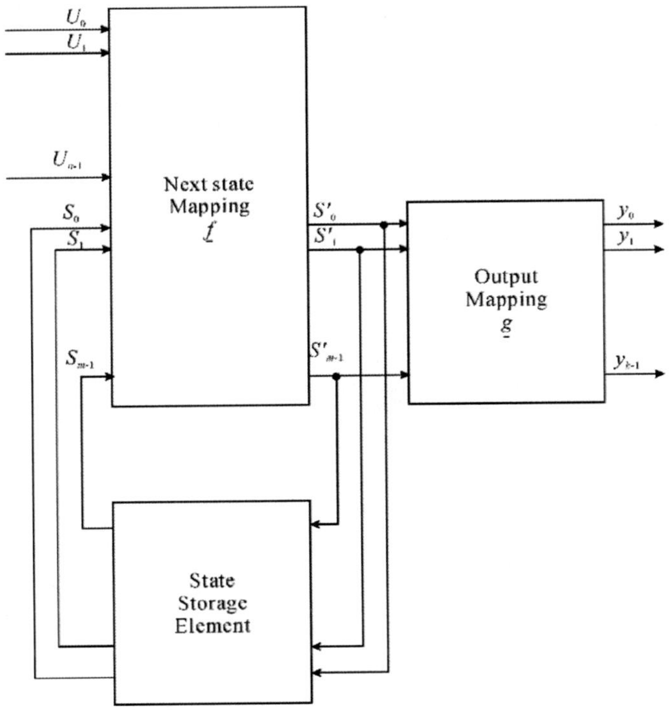

Fig. 1 Moore model of finite state machines

1.2 Description Methods for Finite State Machine: The State Table and the State Transition Graph

A finite state machine may be described by a state table or by a state transition graph. The state table has N_S rows and (N_U+1) columns, each row corresponds to each member of the finite state set S, and the first N_U columns correspond to the N_U members of the finite input set U. The entries in the $N_S \times N_U$ section of the table denote the next states s'. The last column specifies the machine's output y.

The state transition graph is a directed graph which has N_S vertices and $N_S \times N_U$ arcs. Each vertex represents a member of the finite state set S and the arcs with arrays represent the state transitions from the present states to the next states with the corresponding inputs U_i and outputs Y_i indicated by the label U_i/Y_i. An example of a Moore model is shown in Fig. 2.

	Input U_0	Input U_1	Output
S_0	S_0	S_3	Y_0
S_1	S_0	S_2	Y_1
S_2	S_1	S_0	Y_1
S_3	S_2	S_2	Y_0

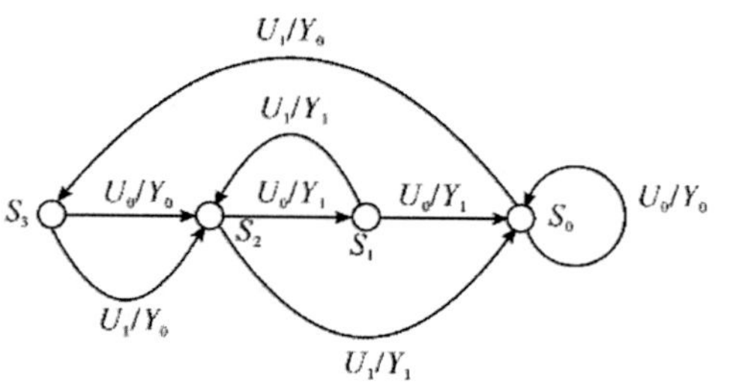

Fig. 2 The state table and the state transition graph of the example

The state table may be obtained from some physical or mathematical models and it can be stored on computers. The disadvantage of state table representation is that it cannot show clearly the dynamic behaviour and stability of systems. The state transition graph, can clearly show the dynamic behaviour and the stability of systems, but it is difficult to draw a state transition graph when the state set S and the input set U is large. An analytical representation of finite state machines is necessary for realising the systems and studying the stability of machines.

1.3 Introduction to Galois Field Representation

The analytical form of a system may be obtained from its state table; conversely, the state table should be easily found from its analytical representation. Boolean OR/AND/NOT functions may be used to represent the next state and the output function of a finite state machine from its state table. However, the FSM representation of digital filters requires a large number of Boolean variables, the Boolean function description of systems is very complex, and it is not easily tested and minimised for the number of variables $n>6$[22]. Moreover, Boolean function are not well-suited to investigating the dynamic performance of systems. One algebraic structure that is well-suited to the analytical representation of FSMs is the Galois Field.

Galois Field GF(p^M) is a finite algebraic structure which has $q=p^M$ elements for any prime $p>1$ and for any integer $M>0$, and two operations, which are called addition and multiplication, defined on all p^M elements. Both operations are closed, commutative and associative, and multiplication is distributive over addition. Each field contains two special elements: 0 and 1, which act as null and identity elements for addition and multiplication operations respectively. For every element of GF(p^M) there exists an inverse element in the same fields.

All finite fields of order p^M are isomorphic[21]. If $p_M(x)$ is a primitive polynomial of order M over GF(p), then its root α is a primitive element of GF(p^M). All the (p^M-1) non-zero elements of GF(p^M) may be represented as powers of α, or polynomials of order ($M-1$) as follows[21]:

$$\alpha^i = x_{M-1}\alpha^{M-1} + x_{M-2}\alpha^{M-2} + \cdots + x_1\alpha + x_0 \mod p_M(\alpha)$$
$$i=0,1,2,\cdots,(q-2) \quad \alpha^{q-1}=1 \tag{3}$$

From the above equation, an important property of Galois Field can be obtained as follows:

$$x^{n(q-1)+1} = x \text{ for } \forall x \in GF(q), n\text{-integer} \tag{4}$$

This property is very useful for the representation of finite state machines. The arithmetic operations of Galois Fields may be implemented in terms of Exclusive-OR and AND logic circuits[22].

This paper is concerned with the Galois Field representation of finite state machines. The advantages of Galois Field representations are as follows:

a) Galois Fields possess properties which may be efficiently employed in computing the polynomial coefficients, minimising machines and investigating the dynamic behaviour of systems.

b) The functions over Galois Fields may be implemented in terms of Exclusive-OR and AND logic circuits which are easier tested in comparison with vertex networks[19], since the change of any input to the Exclusive-OR function will always propagate a change through to the function output, unlike vertex gates which require specific input pattern change to sensitise a path to output.

c) The Exclusive-OR implementation of digital circuits may be more economical than the conventional OR/AND/NOT implementation either in the number of gates or in the number of gate interconnections[23].

This advantage may be justified by using the Entropy definition of OR and Exclusive-OR logic. The entropy $E(y)$ of a binary function y may be defined as follows[23]

$$E(y) = -P_0 \log_2 p_0 - p_1 \log_2 p_1$$

where p_0 and p_1 are respectively the probability of being "0" and "1" for the function output. For a logic OR function $y_0 = x_0 \vee x_1$, if the independent variable x_0 and x_1 each, has an equal probability of being "0" and "1", then the logic OR function has probability 0.25 of being "0" and 0.75 of being "1". The entropy of logical OR function y_o is

$$E(y_o) = -0.25 \log_2 0.25 - 0.75 \log_2 0.75 = 0.8113$$

Similarly, the entropy of logical Exclusive-OR function $y_e = x_0 \oplus x_1$ is as follows

$$E(y_e) = -0.5 \log_2 0.5 - 0.5 \log_2 0.5 = 1$$

the latter is larger than the former.

Booth[2] first proposed an analytical representation of signals in linear sequential networks. Gill and Jacob[6] presented a mapping polynomial for Galois Fields, but only single-variable polynomial was discussed. Green and Taylor[9] developed the modular representation of multiple-valued logic systems. The mappings between polynomial and operational domains were presented, but they did not reveal the DFT property for computation of polynomial coefficients. Benjanthrit and Reed[13,15] presented the fundamental structure of Galois functions and a systematic method for calculating the coefficients of multiple-variable polynomials. This method calculates the constant term a_0 and the highest coefficient a_r using a separating formula from the other coefficients. This will be improved in this section. A number of authors are concerned with the Galois Field functions[12-18]. They only studied the case that the functions and the independent variables are in the same field. A finite state machine is a complex system and in general the states, inputs and outputs of machines are in different fields. In order to study the performance of systems, it is required to convert a multi-variable machine into a single-variable one and vice versa.

1.4 Outline of the Paper

Section 2 shows the representation of autonomous machines. It may be considered as a mapping from the present state s to the next state s' in the same field GF(2^M). The computation of polynomial coefficients may be converted into a DFT by rearranging the ordering of coefficients and field elements. Section 3 introduces the representation of output functions. For Moore Machine, this is a mapping from present state $s \in$ GF(2^M) to the output $y \in$ GF(2^K). The coefficients of the polynomial representation of the output function are in the common extension field GF(2^L) of GF(2^M) and GF(2^K) and possesses cyclic subgroup properties which can considerably reduce the memory size and amount of computation. The representation of next state function is a mapping from present state s and input u to the next state s' and will be discussed in Section 4. The coefficients of the polynomial representation of the next state functions may be computed in terms of a two-dimensional DFT and 2 one-dimensional DFTs. Any multiple-variable machine may be converted into a single-variable machine in terms of a variable transformation and vice versa. This is discussed in Section 5. Section 6 discusses the simplest, but the most important, Galois Field polynomial-Ring Sum expansion. The fast Reed-Muller transformation is also discussed in this section.

2. Representation of Autonomous Machines

Autonomous machines show the behaviour of systems under zero-input conditions. The next state of an autonomous machine is a function of only the present states. This is a simple, but important case, as the behaviour of IIR filters under zero-inputs can be described in terms of an autonomous machine. Zero-input limit cycle oscillations of IIR filters may be studied by means of the stabilities of autonomous machines. From eqn. (2), one-dimensional autonomous machines may be written as follows:

$$s' = f(s) \tag{5a}$$

$$y = g(s) \tag{5b}$$

where $s', s \in$ GF(2^M) are the next and present states, $y \in$ GF(2^K) is the output of the machine.

Eqn. (5a) is the next: state function of an autonomous machine. It is a mapping from GF(2^M) to GF(2^M), Eqn. (5b) is the output function of a FSM, It is a mapping from GF(2^M) to GF(2^M). The output function of the Moore Machine is dependent only on the present states, so there is no difference between the output functions of autonomous and non-autonomous Moore machines. The output mapping (5b) will be studied in the next section.

2.1 An Important Property of Galois Field GF(q)

We begin with the properties of Galois Fields. In 1978, Pradhan[14], Benjanthrit and Reed[15] introduced the following lemma.

Lemma 1

The sum of the ith power of all the nonzero elements over the field GF(q) is null for $0<i<q-1$ and -1 for $i=q-1$, mathematically,

$$\sum_{x\in GF'(q)} x^i = \begin{cases} 0 & \text{for } 0<i<q-1 \\ -1 & \text{for } i=q-1 \end{cases}$$

where $\sum_{x\in GF'(q)}$ denotes the sum for the nonzero elements of GF(q).

We improve upon the above lemma as follows.

Lemma 2

The sum of the $(r-j)$th power of all the elements over the field GF(q) is null for $0<j\leq q-1$ and -1 for $j=0$, mathematically,

$$\sum_{x\in GF(q)} x^{r-j} = -\delta(j) = \begin{cases} 0 & \text{for } 0<j\leq q-1 \\ -1 & \text{for } j=0 \end{cases} \tag{6}$$

where $r=q-1$; $\sum_{x\in GF(q)}$ denotes the sum for the whole field elements including null; δ is a Dirac symbol.

$$\delta(j) = \begin{cases} 0 & \text{for } j\neq 0, 0<j\leq q-1 \\ -1 & \text{for } j=0 \end{cases}$$

This lemma can be derived from Lemma 1. (Note: $0^0=1$, $0^r=0$.)

Three improvements are introduced in Lemma 2.

a) The sum $\sum_{x\in GF(q)}$ in Lemma 2 is over all the field elements including null, but the sum $\sum_{x\in GF'(q)}$ in Lemma 1 is for the non-zero elements of fields. This improvement leads to the next one.

b) The power of variables in Lemma 2 covers all the integers $0\leq(r-j)\leq q-1$, but Lemma 1 does not include the power $i=0$. This improvement avoids separately treating the coefficient a_{q-1} as in previous work[15].

c) The power $(r-j)$ is used in Lemma 2 instead of i in Lemma 1. This will reduce the complexity for further analysis, especially in multiple-variable machines.

In eqn. (6) $q=p^M$, here p is a prime. In this paper, we only deal with the case of $q=2^M$. However, all the results can be extended to GF(p^M).

2.2 The Mapping of GF(2^M)→GF(2^M)

The next state function of autonomous machines is a mapping from GF(2^M) to GF(2^M). It may be represented as a polynomial over GF(2^M) by means of the following Theorem.

Theorem 1

The next state mapping of autonomous machines may be represented as a polynomial of order $r=2^M-1$ over GF(2^M) as follows

$$s' = f(s) = \sum_{i=0}^{2^M-1} a_i s^i \tag{7}$$

where $s', s\in GF(2^M)$; $a_i\in GF(2^M)$, $i=0,1,2,\cdots,(2^M-1)$ and the coefficient a_j is determined by

$$\begin{cases} a_0 = f(0) \\ a_j = \sum_{s\in GF(2^M)} f(s)s^{r-j} \end{cases}$$

$$j = 1,2,\cdots,(2^M - 1) \quad r = 2^M - 1 \tag{8}$$

Proof

Substituting $f(s)$ of eqn. (8) by eqn. (7), then the right side of eqn. (8) is equal to

$$\sum_{s\in GF(2^M)} \left(\sum_{i=1}^{2^M-1} a_i s^i\right) s^{r-j} = \sum_{i=0}^{2^M-1} a_i \sum_{s\in GF(2^M)} s^{r-(j-i)}$$

Using eqn. (6) it follows

$$= \sum_{i=0}^{2^M-1} a_i \delta(j-i) = a_j$$

$$j = 1,2,\cdots,(2^M - 1)$$

This is the left-hand side of eqn. (8).

It is easy to prove from eqn. (7) that

$$a_0 = f(0)$$

Note:

i) $a_0 \neq \sum_{s\in GF(2^M)} f(s)s^{r-0} = a_0 + a_r$

ii) Because of $s^0 = 1$ for all the field elements including zero,

$$s^r = \begin{cases} 1 & \text{for } s\neq 0, s\in GF(2^M) \\ 0 & \text{for } s=0 \end{cases}$$

The terms a_0 and $a_r s^r$ play different roles in the polynomials.

Example 1

The state table of an autonomous machine is as follows.

Table 1 The state table of Example 1

s	$s' = f(s)$
0	0
1	0
2	1
3	2
−4	−2
−3	−2
−2	−1
−1	0

If α is a primitive root of the polynomial $(\alpha^3+\alpha+1)$ in $GF(2^3)$, then, from eqn. (3), the non-zero elements of $GF(2^3)$ may be represented as follows

$$\alpha^i = s_2\alpha^2 + s_1\alpha + s_0 \quad \mathrm{mod}\ (\alpha^3+\alpha+1)$$

$$i = 0,1,\cdots,6 \qquad s_2, s_1, s_0 \in GF(2)$$

Let $N(s) = s_2\alpha^2 + s_1\alpha + s_0$ be a decimal number ($N(0)=0$), then each field element code (s_2, s_1, s_0) and the number $N(s)$ are in one-to-one correspondence. This is named the positive-polarity assignment of field elements, as (s_2, s_1, s_0) are all in positive (non-complementary) polarities.

If the numbers in Table 1 are represented in two's complements and the positive-polarity assignment of field elements is used, then the above state table can be rewritten as follows

Table 2 The state table of Example 1 over GF(2^3)

s				$s'=f(s)$			
GF(2^3)	s_2	s_1	s_0	s'_2	s'_1	s'_0	GF(2^3)
0	0	0	0	0	0	0	0
1	0	0	1	0	0	0	0
α	0	1	0	0	0	1	1
α^3	0	1	1	0	1	0	α
α^2	1	0	0	1	1	0	α^4
α^6	1	0	1	1	1	0	α^4
α^4	1	1	0	1	1	1	α^5
α^5	1	1	1	0	0	0	0

From eqns. (7) and (8), the polynomial coefficients may be computed and the state table may be represented as the following polynomial

$$s'=f(s)=\alpha^3 s+s^2+\alpha^2 s^3+\alpha s^4+\alpha^4 s^5+\alpha^4 s^6+\alpha^2 s^7$$

$$s', s\in GF(2^3)$$

As a special case, any linear autonomous machine over GF(2^M) may be represented as a linear function over GF(2^M) as follows

$$s'=a_1 s+a_0$$

where $s', s\in GF(2^M)$.

a_1 and a_0 are elements of GF(2^M) and may be represented as powers of α which is a primitive element of GF(2^M). The constant term a_0 in the above equation may be eliminated by introducing a linear variable transformation. Therefore, a linear autonomous machine may be written as follows

$$s'=\alpha^n s \tag{9}$$

Note that if n and (2^M-1) are relatively prime, then the machine is a primitive linear machine having a state graph consisting of two cycles: a maximum cycle of length (2^M-1) and a cycle of length 1. There are $\phi(2^M-1)$ primitive linear finite state machines over GF(2^M). ϕ is Euler's function. The extension Galois Field representation of linear machines is much simpler than that of conventional matrix representations.

2.3 Discrete Fourier Transformation for Computing the Polynomial Coefficients

Theorem 1 presents the Galois Field representation of autonomous machines. The polynomial coefficients may be obtained by computing eqn. (8) over all the elements of GF(2^M). If the field elements and the polynomial coefficients $\{a_i\}$ are arranged according to the following new ordering

$$0,(\alpha^r=1),\alpha,\alpha^2,\cdots,\alpha^{r-1}$$

$$a_0,a_r,a_1,a_2,\cdots,a_{r-1} \tag{10}$$

where α is a primitive element of GF(2^M), $r=2^M-1$, then the next state mapping in eqn. (7) may be written as the following matrix form

$$\begin{vmatrix} f(0) \\ f(1) \\ f(\alpha) \\ f(\alpha^2) \\ \vdots \\ f(\alpha^{r-1}) \end{vmatrix} = \begin{vmatrix} 1 & 0 & 0 & 0 & \cdots & 0 \\ 1 & 1 & 1 & 1 & \cdots & 1 \\ 1 & 1 & \alpha & \alpha^2 & & \alpha^{r-1} \\ 1 & 1 & \alpha^2 & \alpha^4 & & \alpha^{2(r-1)} \\ \vdots & \vdots & & & & \vdots \\ 1 & 1 & \alpha^{r-1} & \alpha^{2(r-1)} & \cdots & \alpha^{(r-1)^2} \end{vmatrix} \begin{vmatrix} a_0 \\ a_r \\ a_1 \\ a_2 \\ \vdots \\ a_{r-1} \end{vmatrix}$$

Let

$$x(k)=f(\alpha^k)-a_0 \quad k=0,1,2,\cdots,(2^M-2)$$

$$\begin{cases} x(0)=a_r \\ x(i)=a_i \quad i=1,2,\cdots,(2^M-2) \end{cases} \tag{12}$$

then eqn. (7) may be represented as follows

$$\begin{cases} f(0)=a_0 & \text{(13a)} \\ x(k)=\sum_{i=0}^{2^M-2} x(i)\alpha^{ik} \quad k=0,1,2,\cdots,(2^M-2) & \text{(13b)} \end{cases}$$

Because α is a primitive unit root of order (2^M-1), so eqn. (13b) is a DFT of length (2^M-1).

Similarly, if the independent variable s and the polynomial coefficients a_i in eqn. (8) are reordered according to eqn. (10), then the computation of coefficients a_i in eqn. (8) may be written in the following matrix form

$$\begin{vmatrix} a_0 \\ a_r \\ a_1 \\ a_2 \\ \vdots \\ a_{r-1} \end{vmatrix} = \begin{vmatrix} 1 & 0 & 0 & 0 & \cdots & 0 \\ 1 & 1 & 1 & 1 & \cdots & 1 \\ 1 & 1 & \alpha^{-1} & \alpha^{-2} & & \alpha^{-(r-1)} \\ 1 & 1 & \alpha^{-2} & \alpha^{-4} & & \alpha^{-2(r-1)} \\ \vdots & \vdots & & & & \vdots \\ 1 & 1 & \alpha^{-r-1} & \alpha^{-2(r-1)} & \cdots & \alpha^{(r-1)^2} \end{vmatrix} \begin{vmatrix} f(0) \\ f(1) \\ f(\alpha) \\ f(\alpha^2) \\ \cdots \\ f(\alpha^{r-1}) \end{vmatrix}$$

Using the new variables in eqn. (12), the above matrix may be written as follows

$$\begin{cases} a_0=f(0) & \text{(15a)} \\ x(i)=\sum_{k=0}^{2^M-2} x(k)\alpha^{-ik} \quad i=0,1,2,\cdots,(2^M-2) & \text{(15b)} \end{cases}$$

It is obvious that eqn. (15b) is an inverse DFT of length (2^M-1).

For the example 1

$$a_0 = f(0) = 0$$

the other 7 coefficients may be computed in terms of an inverse DFT of length 7 as follows

$$\begin{vmatrix} a_7 \\ a_1 \\ a_2 \\ a_3 \\ a_4 \\ a_5 \\ a_6 \end{vmatrix} = \begin{vmatrix} 1 & 1 & 1 & 1 & 1 & 1 & 1 \\ 1 & \alpha^{-1} & \alpha^{-2} & \alpha^{-3} & \alpha^{-4} & \alpha^{-5} & \alpha^{-6} \\ 1 & \alpha^{-2} & \alpha^{-4} & \alpha^{-6} & \alpha^{-1} & \alpha^{-3} & \alpha^{-5} \\ 1 & \alpha^{-3} & \alpha^{-6} & \alpha^{-2} & \alpha^{-5} & \alpha^{-1} & \alpha^{-4} \\ 1 & \alpha^{-4} & \alpha^{-1} & \alpha^{-5} & \alpha^{-2} & \alpha^{-6} & \alpha^{-3} \\ 1 & \alpha^{-5} & \alpha^{-3} & \alpha^{-1} & \alpha^{-6} & \alpha^{-4} & \alpha^{-2} \\ 1 & \alpha^{-6} & \alpha^{-5} & \alpha^{-4} & \alpha^{-3} & \alpha^{-2} & \alpha^{-1} \end{vmatrix} \begin{vmatrix} 0 \\ 1 \\ \alpha^4 \\ \alpha \\ \alpha^5 \\ 0 \\ \alpha^4 \end{vmatrix} = \begin{vmatrix} \alpha^2 \\ \alpha^3 \\ 1 \\ \alpha^2 \\ \alpha \\ \alpha^4 \\ \alpha^{-1} \end{vmatrix}$$

3. Representation of Output Functions

3.1 The Output Mapping of $GF(2^M) \rightarrow GF(2^K)$

The output of the Moore Machine is a function of present states. In general, the state and output of a finite state machine are in different fields, e. g. the state s is in $GF(2^M)$, the output u is in $GF(2^K)$, Therefore the output function is a mapping from $GF(2^M)$ to $GF(2^K)$.

In order to represent the output mapping, an Extension Field $GF(2^L)$ is introduced such that both $GF(2^M)$ and $GF(2^K)$ are the subfields of $GF(2^L)$. Thus, the output function becomes a mapping from extension field $GF(2^L)$ to $GF(2^L)$ (exactly a mapping from a subfield to another subfield within the Extension Field $GF(2^L)$) as shown in Fig. 3.

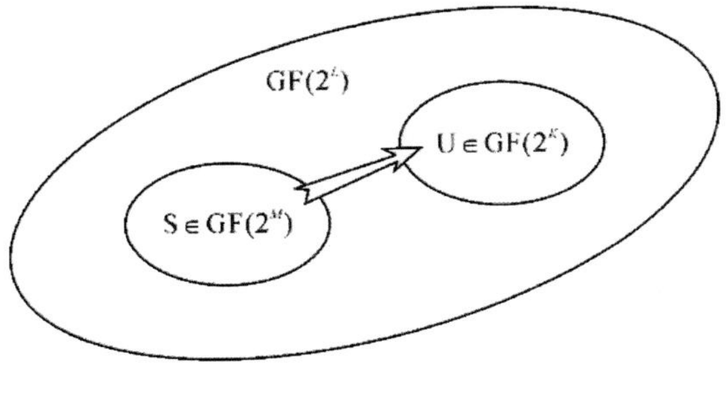

Fig. 3 The mapping between different subfields

Before giving the representation of output mappings, we introduce the following lemma[21].

Lemma 3

$GF(2^K)$ is a subfield of $GF(2^L)$, if and only if K divides L. If α be a primitive element of $GF(2^L)$, then the primitive element β of subfield $GF(2^K)$ is determined by

$$\beta = \alpha^{\eta}, \quad \eta = (2^L - 1)/(2^K - 1) \tag{16}$$

From the above lemma, if L is the Least Common Multiple (LCM) of K and M

$$L = \mathrm{LCM}(K, M)$$

then $\mathrm{GF}(2^K)$ and $\mathrm{GF}(2^M)$ both are the subfields of $\mathrm{GF}(2^L)$. $\mathrm{GF}(2^L)$ is the least common extension field of $\mathrm{GF}(2^K)$ and $\mathrm{GF}(2^K)$.

Using the above properties, the output mapping of machines may be obtained by the following theorem.

Theorem 2

The output function of a Moore Machine may be represented as a polynomial as follows:

$$y = g(s) = \sum_{i=0}^{2^M-1} b_i s^i \tag{17}$$

where

$$s \in \mathrm{GF}(2^M) \quad y \in \mathrm{GF}(2^K)$$

$$b_i \in \mathrm{GF}(2^L) \quad L = \mathrm{LCM}(K, M) \quad i = 0, 1, 2, \cdots, (2^M - 1)$$

The coefficients b_j are determined by

$$b_0 = g(0)$$

$$b_j = \sum_{s \in \mathrm{GF}(2^M)} g(s) s^{r-j}$$

$$j = 0, 1, 2, \cdots, (2^M - 1) \quad r = 2^M - 1 \tag{18}$$

Proof

See Appendix A2.

Theorem 2 is somewhat similar to Theorem 1, but the meaning of Theorem 2 is quite different from that of Theorem 1. The output y and state s in Theorem 2 are in different fields $\mathrm{GF}(2^M)$ and $\mathrm{GF}(2^M)$. The coefficients b_j of polynomial (17) may be neither in $\mathrm{GF}(2^K)$ nor in $\mathrm{GF}(2^M)$. In general, b_j is the element of Extension Field $\mathrm{GF}(2^L)$ such that the polynomial (17) maps $\mathrm{GF}(2^M)$ into $\mathrm{GF}(2^K)$.

Example 2

The output function table of a FSM is shown in Table 3.

Table 3 Truth table of Example 2

s	$y = g(s)$
0	0
1	0
2	1
3	1
−4	−2
−3	−1
−2	−2
−1	−1

Using the positive-polarity assignment of field elements and two's complement, the above table may be rewritten as follows:

Table 4 Truth table of Example 2 over GF(2^6)

s				$y=g(s)$		
GF(2^6)	s_2	s_1	s_0	y_1	y_0	GF(2^6)
$0=0$	0	0	0	0	0	$0=0$
$1=1$	0	0	1	0	0	$0=0$
$\beta=\alpha^9$	0	0	0	0	1	$1=1$
$\beta^3=\alpha^{27}$	0	1	1	0	1	$1=1$
$\beta^2=\alpha^{18}$	1	0	0	1	0	$\gamma=\alpha^{21}$
$\beta^6=\alpha^{54}$	1	0	1	1	1	$\gamma^2=\alpha^{42}$
$\beta^4=\alpha^{36}$	1	1	0	1	0	$\gamma=\alpha^{21}$
$\beta^5=\alpha^{45}$	1	1	1	1	1	$\gamma^2=\alpha^{42}$

In Table 4, β is a primitive element of GF(2^3), γ is a primitive element of GF(2^2), and α is a primitive element of GF(2^6). Because 6 is the Least Common Multiple of 3 and 2, from Lemma 3, GF(2^2) and GF(2^3) are subfields of GF(2^6) and it holds that

$$\beta=\alpha^{(2^6-1)/(2^3-1)}=\alpha^9$$
$$\gamma=\alpha^{(2^6-1)/(2^2-1)}=\alpha^{21}$$

The minimal polynomials of α,β,γ are $(\alpha^6+\alpha+1)$, $(\beta^3+\beta+1)$ and $(\gamma^2+\gamma+1)$ respectively. From Theorem 2, the above truth table may be represented as a polynomial over GF(2^6) as follow:

$$y=g(s)=\alpha^{36}s+\alpha^9s^2+\alpha^{38}s^3+\alpha^{18}s^4+\alpha^{26}s^5+\alpha^{41}s^6$$

where $b_3=\alpha^{38}$, $b_5=\alpha^{36}$ and $b_6=\alpha^{41}$ are not the elements of either GF(2^2) or GF(2^3), but the polynomial maps the above truth table from GF(2^3) into GF(2^2).

Theorem 2 may also be represented as a DFT. Let the state s and the coefficients $\{b_j\}$ be the order

$$0,1,\alpha^\eta,\alpha^{2\eta},\cdots,\alpha^{(r-1)\eta}$$
$$b_0,b_r,b_1,b_2,\cdots,b_{r-1} \tag{19}$$

where

$$y=(2^L-1)/(2^M-1)$$
$$r=2^M-1$$

then eqn. (17) and eqn. (18) may be written as matrix form as follows:

$$\begin{vmatrix} g(0) \\ g(1) \\ g(\alpha^\eta) \\ g(\alpha^{2\eta}) \\ \vdots \\ g(\alpha^{(r-1)\eta}) \end{vmatrix} = \begin{vmatrix} 1 & 0 & 0 & 0 & \cdots & 0 \\ 1 & 1 & 1 & 1 & \cdots & 1 \\ 1 & 1 & \alpha^\eta & \alpha^{2\eta} & & \alpha^{(r-1)\eta} \\ 1 & 1 & \alpha^{2\eta} & \alpha^{4\eta} & & \alpha^{2(r-1)\eta} \\ \vdots & \vdots & & & & \vdots \\ 1 & 1 & \alpha^{(r-1)\eta} & \alpha^{2(r-1)\eta} & \cdots & \alpha^{(r-1)2\eta} \end{vmatrix} \begin{vmatrix} b_0 \\ b_r \\ b_1 \\ b_2 \\ \\ b_{r-1} \end{vmatrix}$$

$$\begin{vmatrix} b_0 \\ b_r \\ b_1 \\ b_2 \\ \vdots \\ b_{r-1} \end{vmatrix} = \begin{vmatrix} 1 & 0 & 0 & 0 & \cdots & 0 \\ \hline 1 & 1 & 1 & 1 & \cdots & 1 \\ 0 & 1 & \alpha^{-\eta} & \alpha^{-2\eta} & & \alpha^{-(r-1)\eta} \\ 0 & 1 & \alpha^{-2\eta} & \alpha^{-4\eta} & & \alpha^{-2(r-1)\eta} \\ \vdots & \vdots & & & & \vdots \\ 0 & 1 & \alpha^{-(r-1)\eta} & \alpha^{-2(r-1)\eta} & \cdots & \alpha^{-(r-1)2\eta} \end{vmatrix} \begin{vmatrix} g(0) \\ g(1) \\ g(\alpha^{\eta}) \\ g(\alpha^{2\eta}) \\ \vdots \\ g(\alpha^{(r-1)\eta}) \end{vmatrix}$$

3.2 Cyclic Subgroup Properties of the Coefficients $\{b_j\}$

Equation (20) may be considered as a mapping from 2^M coefficients $\{b_j\}$ into the output vector $\{g(\alpha^{i\eta})\}$ of length 2^M. However, if $2^K < 2^M$ and $g(\alpha^{i\eta}) \in GF(2^K)$, then $g(\alpha^{i\eta})$ have at most 2^K different values, the other values are not independent. Therefore, the 2^M coefficients $\{b_j\}$ are also dependent. In general, the coefficients $\{b_j\}$ are elements of extension field $GF(2^L)$, but they are not arbitrary, some constraints exist such that the polynomial (17) is a mapping from $GF(2^M)$ to $GF(2^K)$.

Because $y = g(s)$ is an element of $GF(2^K)$, from eqn. (4) it holds

$$y^{2^K} = y \tag{22}$$

Applying the above constraint to eqn. (17) may give some cyclic properties of the coefficients $\{b_j\}$.

Before presenting these properties, a lemma is introduced[21].

Lemma 4

Let $g(x)$ be a polynomial over $GF(2^M)$ as follows

$$g(x) = \sum_{i=0}^{2^M-1} b_i x^i$$

where g, x and b_i are elements of any extension fields of $GF(2^M)$. Then it holds

$$[g(x)]^{2^n} = \sum_{i=0}^{2^M-1} b_i^{2^n} x^{i2^n} \tag{23}$$

where n is any integer.

Theorem 3

If the output function of a FSM is

$$y = g(s) = \sum_{i=1}^{2^M-1} b_i s^i$$

where $y \in GF(2^K)$, $s \in GF(2^M)$, and $b_j \in GF(2^L)$, $L = LCM(K, M)$, then the polynomial coefficients have the following cyclic properties

$$b_j = b_i^{2^k} \quad \text{for} \quad j = i * 2^K \mod (2^M - 1) \tag{24}$$

Proof

Because y is an element of $GF(2^K)$, from eqn. (22) it follows that

$$[g(s)]^{2^K} = g(s)$$

and using Lemma 4, the above equation may be written as

$$\sum_{i=0}^{2^M-1} b_i^{2^K} s^{i*2^K} = \sum_{j=0}^{2^M-1} b_j s^j$$

Compare the coefficients of the two polynomials in the above equation. If $s^j = s^{i*2^K \bmod (2^M-1)}$, then their coefficients are equal, i. e.

$$b_j = (b_i)^{2^K} \quad \text{for} \quad j = i * 2^K \quad \bmod (2^M - 1)$$

There are two special cases for the above result, they are

$$b_0 = b_0^{2^K} \text{ and } b_r = b_r^{2^K} \quad r = 2^M - 1 \tag{25}$$

as $0 = 0 * 2^K$ and $r = r * 2^K \bmod (2^M - 1)$.

For Example 2, $2^K = 4$, the polynomial coefficients have the following relations

$$(\alpha^{36} s)^4 = \alpha^{18} s^4, (\alpha^{18} s^4)^4 = \alpha^9 s^2, (\alpha^9 s^2)^4 = \alpha^{36} s$$

$$(\alpha^{38} s^3)^4 = \alpha^{26} s^5, (\alpha^{26} s^5)^4 = \alpha^{41} s^6, (\alpha^{41} s^6)^4 = \alpha^{38} s^3$$

The polynomial may be rewritten as the sum of two cyclic subgroups as follows

$$y = g(s) = (\alpha^{36} s + \alpha^{18} s^4 + \alpha^9 s^2) + (\alpha^{38} s^3 + \alpha^{26} s^5 + \alpha^{41} s^6)$$

From the above example, it is shown that only the coefficients b_1 and b_3 are independent, they should be computed by means of Theorem 3. The other coefficients can be obtained by using their cyclic subgroup properties as shown in Theorem 4. The cyclic subgroup properties of coefficients $\{b_j\}$ considerably reduce the size of computer memory and the amount of calculation.

4. Representation of Next State Functions

4.1 The Next State Mapping of $[GF(2^M), GF(2^N)] \rightarrow GF(2^M)$

The next state of FSMs is a function of present states and inputs. In general, the state s and input u are in different fields. From eqn. (1a), the next stats mapping of one-dimensional FSMs is a two variable function as follows

$$s' = f(s, u) \tag{26}$$

where

$$s', s \in GF(2^M), u \in GF(2^N)$$

This is a mapping from $GF(2^M)$ and $GF(2^N)$ into $GF(2^M)$.

In order to represent next state mappings, a common extension field $GF(2^L)$ of subfields $GF(2^M)$ and $GF(2^N)$ is introduced, so the function (26) may be considered as a mapping from two subfields to one subfield within the extension field.

Thr next state mapping of multiple-dimensional FSMs is a multiple-variable function. For the sake of brevity, only the two-variable function is dealt with in this section. The multiple-variable function will be discussed in Appendix A1.

Theorem 4

The next state mapping of one-dimensional FSMs may be represented as a two-variable polynomial with coefficients in a common extension field as follows

$$s' = f(s, u) = \sum_{i=0}^{2^M-1} \sum_{j=0}^{2^N-1} a_{ij} s^i u^j \tag{27}$$

where

$$s', s \in GF(2^M), u \in GF(2^N)$$
$$a_{ij} \in GF(2^L), L = LCM(M, N)$$
$$i = 0, 1, 2, \cdots, (2^M - 1), j = 0, 1, 2, \cdots, (2^N - 1)$$

The polynomial coefficients are determined by

$$a_{00} = f(0,0) \tag{28b}$$

$$a_{i0} = \sum_{s \in GF(2^M)} f(s,0) s^{r_1 - i} \qquad r_1 = 2^M - 1 \tag{28b}$$

$$a_{0j} = \sum_{u \in GF(2^N)} f(0,u) u^{r_2 - j} \qquad r_2 = 2^N - 1 \tag{28c}$$

$$a_{ij} = \sum_{s \in GF(2^M)} \sum_{u \in GF(2^N)} f(s,u) s^{r_1 - i} u^{r_2 - j} \tag{28d}$$

where $\sum$ denotes the sum over all the field elements including zero.

Proof

See Appendix A2.

If α, β and γ are primitive elements of $GF(2^L)$, $GF(2^M)$ and $GF(2^N)$ respectively, then from Lemma 3 it follows

$$\beta = \alpha^\eta \qquad \eta = (2^L - 1)/(2^M - 1)$$
$$\gamma = \alpha^\xi \qquad \xi = (2^L - 1)/(2^N - 1) \tag{29}$$

s and u in eqn. (28) may be represented as powers of α in $GF(2^L)$ as follows

$$s = 0, 1, \alpha^\eta, \alpha^{2\eta}, \cdots, \alpha^{(2^M - 2)\eta}$$
$$u = 0, 1, \alpha^\xi, \alpha^{2\xi}, \cdots, \alpha^{(2^N - 2)\xi} \tag{30}$$

Example 3

The state table of a FSM is given by

Table 5 The state table of Example 3

s	u			
	0	1	−2	1
0	0	1	−2	−1
1	1	2	−1	0
2	1	2	−1	0
3	2	3	0	1
−4	−2	−1	−4	−3
−3	−2	−1	−4	−3
−2	−1	0	−3	−2
−1	−1	0	−3	−2

Let the state s be in $GF(2^3)$, the input u in $GF(2^2)$, then the polynomial coefficients a_{ij} will be in the common extension field $GF(2^6)$, as $6 = LCM(2,3)$. If the primitive elements of $GF(2^6)$, $GF(2^3)$ and $GF(2^6)$ are α, β and γ respectively. From Lemma 3 it holds that $\beta = \alpha^9$, $\gamma = \alpha^{21}$. Using two's complement and positive-polarity assignment of field elements, then the above state table may be written in the following form.

Table 6 The state table of Example 3 over GF(2^6)

	0	1	α^{21}	α^{42}
0	0	1	α^{36}	α^{45}
1	1	α^9	α^{45}	0
α^9	1	α^9	α^{45}	0
α^{27}	α^9	α^{27}	0	1
α^{18}	α^{36}	α^{45}	α^{18}	α^{54}
α^{54}	α^{36}	α^{45}	α^{18}	α^{54}
α^{36}	α^{45}	0	α^{54}	α^{36}
α^{45}	α^{45}	0	α^{54}	α^{36}

From Theorem 4, the next state function of the above example may be obtained as follows

$$\begin{aligned} s' &= f(s,u) = \sum_{i=0}^{7}\sum_{j=0}^{3} a_{ij}s^i u^j \\ &= (\alpha^{62}u + \alpha^{55}u^2 + \alpha^{27}u^3) + (\alpha^{27}u + \alpha^{27}u^2 + \alpha^{27}u^3)s \\ &\quad + (\alpha^9 + \alpha^{61}u + \alpha^{47}u^2 + \alpha^{36}u^3)s^2 + (\alpha^{18} + \alpha^{38}u + \alpha^{52}u^2 + \alpha^{45}u^3)s^3 \\ &\quad + (\alpha^{27} + \alpha^{33}u + \alpha^{12}u^2 + \alpha^{27}u^3)s^4 + (\alpha^{54} + \alpha^{41}u + \alpha^{13}u^2 + \alpha^{45}u^3)s^5 \\ &\quad + (\alpha^{54} + \alpha^{18}u + \alpha^{18}u^2)s^6 + (\alpha^9 + \alpha^{14}u + \alpha^{49}u^2 + \alpha^{45}u^3)s^7 \end{aligned}$$

where

$$s', s \in \mathrm{GF}(2^3) \quad u \in \mathrm{GF}(2^2)$$

4.2 Two-Dimensional DFT for Computing the Coefficients of Next State Polynomials

From Theorem 4, the polynomial coefficients of next state functions may be obtained by computing eqn. (28) over all the elements of GF(2^M) and GF(2^N). If the values of s and u are ordered as in eqn. (30) and the subscripts of the polynomial coefficients are rearranged in the following ordering

$$\begin{aligned} i &= 0, (2^M - 1), 1, 2\cdots, (2^M - 2) \\ j &= 0, (2^N - 1), 1, 2\cdots, (2^N - 2) \end{aligned} \tag{31}$$

then eqn. (28) may be written as follows

$$a_{00} = f(0,0) \tag{32a}$$

$$a_{i0} = f(0,0)\delta(r_1 - i) + \sum_{k=0}^{2^M-2} f(\beta^i, 0)\beta^{-ik} \tag{32b}$$

$$a_{0j} = f(0,0)\delta(r_2 - j) + \sum_{k=0}^{2^N-2} f(0, \gamma^i)\gamma^{-jk} \tag{32c}$$

$$a_{ij} = f(0,0)\delta(r_1 - i)(r_2 - j) + \sum_{k=0}^{2^M-2}\sum_{k=0}^{2^N-2} f(\beta^i, \gamma_1^i)\beta^{-ik}\gamma^{-jk} \tag{32d}$$

where

$r_1=2^M-1, r_2=2^N-1$

δ is Dirac function

$\beta=\alpha^\eta \quad \eta=(2^L-1)/(2^M-1)$

$\gamma=\alpha^\xi \quad \xi=(2^L-1)/(2^N-1) \quad L=LCM(N,M)$

$i=0,(2^M-1),1,2\cdots,(2^M-2)$

$j=0,(2^N-1),1,2\cdots,(2^N-2)$

It is obvious that eqn. (32b) is a one-dimensional DFT of length (2^M-1), eqn. (32c) is a one-dimensional DFT of length (2^N-1), and eqn. (32d) is a two-dimensional DFT of size $(2^M-1)\times(2^N-1)$.

For Example 3, using the ordering (31) and $\beta=\alpha^9$, $\gamma=\alpha^{21}$, the computation of polynomial coefficients may be represented in the following matrix form:

$$\underline{A}=\left|\begin{array}{c:ccccccc} 1 & 0 & 0 & 0 & 0 & 0 & 0 & 0 \\ \hdashline 1 & 1 & 1 & 1 & 1 & 1 & 1 & 1 \\ 0 & 1 & \alpha^{-9} & \alpha^{-18} & \alpha^{-27} & \alpha^{-36} & \alpha^{-45} & \alpha^{-54} \\ 0 & 1 & \alpha^{-18} & \alpha^{-36} & \alpha^{-54} & \alpha^{-9} & \alpha^{-27} & \alpha^{-45} \\ 0 & 1 & \alpha^{-27} & \alpha^{-54} & \alpha^{-18} & \alpha^{-45} & \alpha^{-9} & \alpha^{-36} \\ 0 & 1 & \alpha^{-36} & \alpha^{-9} & \alpha^{-45} & \alpha^{-18} & \alpha^{-54} & \alpha^{-27} \\ 0 & 1 & \alpha^{-45} & \alpha^{-27} & \alpha^{-9} & \alpha^{-54} & \alpha^{-36} & \alpha^{-18} \\ 0 & 1 & \alpha^{-54} & \alpha^{-45} & \alpha^{-36} & \alpha^{-27} & \alpha^{-18} & \alpha^{-9} \end{array}\right| \underline{F} \left|\begin{array}{c:ccc} 1 & 1 & 0 & 0 \\ \hdashline 0 & 1 & 1 & 0 \\ 0 & 1 & \alpha^{-21} & \alpha^{-42} \\ 0 & 1 & \alpha^{-42} & \alpha^{-21} \end{array}\right|$$

where $\underline{F}$ is the state table matrix reordering according to eqn. (30).

$$\underline{F}=\left|\begin{array}{cccc} 0 & 0 & \alpha^{36} & \alpha^{45} \\ 1 & \alpha^{9} & \alpha^{45} & 0 \\ 1 & \alpha^{9} & \alpha^{15} & 0 \\ \alpha^{36} & \alpha^{45} & \alpha^{18} & \alpha^{54} \\ \alpha^{9} & \alpha^{27} & 0 & 1 \\ \alpha^{45} & 0 & \alpha^{54} & \alpha^{36} \\ \alpha^{45} & 0 & \alpha^{54} & \alpha^{36} \\ \alpha^{36} & \alpha^{45} & \alpha^{18} & \alpha^{54} \end{array}\right|$$

The above computation contains a two-dimensional DFT of size (7×3), three one-dimensional DFTs of length 3 and seven one-dimensional DFTs of length 7. The coefficient matrix $\underline{A}$ of (8×4) can be obtained in the new ordering (31) as follows

$$\underline{A}=\left|\begin{array}{cccc} a_{0,0} & a_{0,3} & a_{0,1} & a_{0,2} \\ a_{7,0} & a_{7,3} & a_{7,1} & a_{7,2} \\ a_{1,0} & a_{1,3} & a_{1,1} & a_{1,2} \\ a_{2,0} & a_{2,3} & a_{2,1} & a_{2,2} \\ a_{3,0} & a_{3,3} & a_{3,1} & a_{3,2} \\ a_{4,0} & a_{4,3} & a_{4,1} & a_{4,2} \\ a_{5,0} & a_{5,3} & a_{5,1} & a_{5,2} \\ a_{6,0} & a_{6,3} & a_{6,1} & a_{6,2} \end{array}\right| = \left|\begin{array}{cccc} 0 & \alpha^{27} & \alpha^{62} & \alpha^{55} \\ \alpha^{9} & \alpha^{45} & \alpha^{14} & \alpha^{49} \\ 0 & \alpha^{27} & \alpha^{27} & \alpha^{27} \\ \alpha^{9} & \alpha^{36} & \alpha^{61} & \alpha^{47} \\ \alpha^{18} & \alpha^{45} & \alpha^{38} & \alpha^{52} \\ \alpha^{27} & \alpha^{27} & \alpha^{33} & \alpha^{12} \\ \alpha^{54} & \alpha^{15} & \alpha^{41} & \alpha^{13} \\ \alpha^{54} & 0 & \alpha^{18} & \alpha^{18} \end{array}\right|$$

5. Conversion Between Single-Variable and Multiple-Variable Machines

5.1 Converting a Multiple-Variable Machine into a Single-Variable Machine

A high-order IIR digital filter may be represented as a multiple-variable FSM with polynomial coefficients which can be computed by means of multiple-dimensional DFTs, Single-variable machines are well studied. From the theoretical viewpoint, the coefficient computation, minimization, linearisation and dynamic behaviour analysis of single-variable machines are much simpler than that of multiple-variable machines. The conversion between multiple-variable and single-variable machines is useful for further study.

We begin with two-variable machines. The next state of one-dimensional machines is a function of present states s and inputs u as shown in eqn. (27). In general, such a machine may be considered as a two-variable machine with states $s \in GF(2^M)$ and inputs $u \in GF(2^N)$. The polynomial coefficients are in extension field $GF(2^L)$. All the elements of subfields $GF(2^M)$ and $GF(2^N)$ are members of extension field $GF(2^N)$. If a mapping from the two subfields to the extension field is defined such that each pair of elements of subfields $GF(2^M)$ and $GF(2^N)$ corresponds with a member of extension field $GF(2^L)$, then the next state function may be mapped into a single-variable polynomial over extension field $GF(2^L)$. There are a number of these mappings, most of them are non-linear. One of the simplest mappings is the following linear mapping

$$x = u\alpha + s \tag{33}$$

where α is a primitive element of extension field $GF(2^M)$. $x \in GF(2^L)$ is an extension field variable, it is similar to the complex variable which consists of two real variables. In the complex field, the real and imaginary components can be obtained in terms of linear combinations of the complex variable and its conjugate. In a Galois field, the subfield variables s and u may be represented as functions of the extension field variable x. In general, they are non-linear and can be obtained by using the extension field properties $s^{2^M} = s$ and $u^{2^N} = u$. Let

$$R = \mathrm{LCM}[(2^M - 1), (2^N - 1)] + 1 \tag{34}$$

From Lemma 3, the power of order R of eqn. (33) is as follows

$$x^R = u\alpha^R + s \tag{35}$$

The variables s and u may be obtained by solving the linear algebraic equations (33) and (55) as follows

$$s = (\alpha^R x + \alpha x^R)/(\alpha^R + \alpha)$$
$$u = (x + x^R)/(\alpha^R + \alpha) \tag{36}$$

These are two mappings from extension field $GF(2^L)$ to subfields $GF(2^M)$ and $GF(2^N)$. Substituting eqn. (36) into the next state functions, the machine with single variable x over extension field $GF(2^L)$ is obtained.

This method can also be used in the case that the variables s and u are in the same fields, but the extension field is $GF(2^M)$ instead of $GF(2^L)$.

Example 4

The state table of two-variable machines is given by Table 7.

Table 7　State table of Example 4

u	s	$s'=f(s,u)$
0	0	0
0	1	0
0	−2	1
0	−1	1
1	0	1
1	1	0
1	−2	1
1	−1	1
−2	0	−1
−2	1	−2
−2	−2	0
−2	−1	−1
−1	0	−1
−1	1	−1
−1	−2	0
−1	−1	0

Let β be a primitive element of $GF(2^2)$. Using the positive-polarity assignment of field elements, the above table may be written as follows.

Table 7　State table of Example 4 over $GF(2^2)$

u	s	$s'=f(s,u)$
0	0	0
0	1	0
0	β	1
0	β^2	1
1	0	1
1	1	0
1	β	1
1	β^2	1
β	0	β^2

(To be continued)

Table 7

u	s	$s'=f(s,u)$
β	1	β
β	β	0
β	β^2	β^2
β^2	0	β^2
β^2	1	β^2
β^2	β	0
β^2	β^2	0

From Theorem 3 the above state table may be represented as a polynomial with two variables over GF(2^2) as follows

$$s'=f(s,u)=(s+s^2)+(1+\beta s+\beta^2 s^3)u+(\beta s+\beta^2 s^2)u^2+(\beta^2 s^2+\beta s^3)u^3$$

where $s',s,u\in \mathrm{GF}(2^2)$.

Let x be a variable of extension field GF(2^4) is defined by

$$x=u\alpha+s$$

where α is a primitive root of the primitive polynomial $(\alpha^4+\alpha+1)$ over extension field GF(2^4). From eqn. (36) the variables s and u may be represented as follows

$$s=\alpha^4 x+\alpha x^4$$

$$u=x+x^4$$

Substituting the above two equations into the function $f(s,u)$, and using $\beta=\alpha^5$ (see Lemma 2), the next state function is as follows

$$s'=\alpha^3x^3+x^5+\alpha^4x^6+\alpha^3x^7+\alpha x^9+\alpha^5x^{10}+\alpha^5x^{11}+\alpha^{12}x^{12}+\alpha^{12}x^{13}+\alpha^5x^{14}+\alpha^5x^{15}$$

where $s'\in \mathrm{GF}(2^2)$ and $x\in \mathrm{GF}(2^4)$.

From Theorem 3, the coefficients of the above polynomials have the following cyclic subgroup properties

$$(\alpha^3x^3)^4=\alpha^{12}x^{12},\ (\alpha^4x^6)^4=\alpha x^9,$$

$$(\alpha^3x^7)^4=\alpha^{12}x^{13},\ (\alpha^5x^{11})^4=\alpha^5x^{14},$$

$$(x^5)^4=x^5,\ (\alpha^5x^{10})^4=\alpha^5x^{10},\ (\alpha^5x^{15})^4=\alpha^5x^{15}$$

The above result for two-variable machines can be extended to multiple-variable cases. An m-variable autonomous machine is given by

$$s' = f(s_0,s_1,\cdots,s_{m-1}) = \sum_{i_0=0}^{r}\sum_{i_1=0}^{r}\cdots\sum_{i_{m-1}=0}^{r} a_{i_0 i_1\cdots i_{m-1}} s_0^{i_0} s_1^{i_1} s_{m-1}^{i_{m-1}} \quad (37)$$

where $s_0,s_1,\cdots,s_{m-1},a\in \mathrm{GF}(2^M), r=2^M-1$.

An extension field GF(2^{mM}) is introduced, α is its primitive element and a variable x over GF(2^{mM}) is defined by

$$x=s_0+s_1\alpha+\cdots+s_{m-1}\alpha^{m-1}$$

Using the Galois Field property in eqn. (4): $s_j^{n(q-1)+1}=s_j\ (q=2^M, n\geq 0)$ and Lemma 3, the m linear algebraic equations of subfield variables $(s_0,s_1,\cdots,s_{m-1})$ can be obtained as follows

$$x^q = s_0 + s_1\alpha^q + \cdots + s_{m-1}\alpha^{q(m-1)}$$
$$x^{2q-1} = s_0 + s_1\alpha^{2q-1} + \cdots + s_{m-1}\alpha^{(2q-1)(m-1)} \tag{38}$$
$$\vdots$$
$$x^{(q-1)(m-1)+1} = s_0 + s_1\alpha^{(q-1)(m-1)+1} + \cdots + s_{m-1}\alpha^{(q-1)(m-1)^2+(m-1)}$$

Eqns. (38) are m linear algebraic equations of variables $(s_0, s_1, \cdots, s_{m-1})$. Solving the linear equations and substituting their solutions into eqn. (37), a single-variable machine over extension field $GF(2^{mM})$ may be obtained.

5.2 Converting a Single-Variable Machine into Multiple-Variable Machines

A single-variable machine over extension field $GF(q^m)$ may be converted into a number of multiple-variable machines over subfields $GF(q)$ by the following variable transformation

$$s = s_{m-1}\alpha^{m-1} + \cdots + s_1\alpha + s_0 \tag{39}$$

where

$$s \in GF(q^m),\ \alpha \text{ is its primitive element}$$
$$s_i \in GF(q),\ i = 0, 1, \cdots, (m-1), q = 2^M$$

Substituting eqn. (39) into the single-variable machine

$$s' = \sum_{i=0}^{q^m-1} a_i s^i \tag{40}$$

where $a_i \in GF(q^m)$ can be represented as a power of α.

Let

$$s' = s'_{m-1}\alpha^{m-1} + \cdots + s_1\alpha + s_0 \tag{41}$$

Compare the coefficients of $\alpha^i, i = 0, 1, \cdots, (m-1)$, m machines with m variables over subfield $GF(q)$ are obtained.

For Example 1, let

$$s = s_2\alpha^2 + s_1\alpha + s_0$$
$$s = s'_2\alpha^2 + s'_1\alpha + s'_0$$

Substituting the above new variables into the extension field polynomial, three polynomials over subfields may be obtained as follows

$$s'_0 = s_1 + s_0 s_1$$
$$s'_1 = s_2 + s_0 s_1$$
$$s'_2 = s_2 + s_0 s_1 s_2$$

In general, a FSM with single-variable over extension field $GF(2^n)$ may be converted into a number of multiple-variable machines over $GF(2)$, which can be represented as n-variable polynomials over $GF(2)$. This polynomial is called the modulo-2 Ring-Sum expansion, it is of particular interest for practical implementation of machines.

6. Ring-Sum Representations of FSMs

6.1 The Mapping of $GF(2)^n \rightarrow GF(2)$

Ring-sum expansions are special forms of Galois Field polynomials for GF(2). This is a very important case, as the elements of GF(2) are only 0 and 1, the polynomial over GF(2) can be implemented in terms of Exclusive-OR and AND logical circuits. The Exclusive-OR implementation is easily tested[19].

The above section shows that any single-variable FSM over extension field $GF(2^n)$ may be mapped into n-variable FSMs over GF(2) which may be represented as a number of ring-sum expansions with n binary variables. The general form of a ring-sum expansion with n positive-polarity variables is as follows

$$f(x_0, x_1, \cdots, x_{n-1}) = a_0 \oplus a_1 x_0 \oplus a_2 x_0 x_1 \oplus \cdots \oplus a_r x_0 x_1 \cdots x_{n-1} \tag{42}$$

where $\oplus$ denotes modulo-2 addition

$$x_i \in GF(2),\ i = 0, 1, \cdots, (n-1)$$

$$a_j \in GF(2),\ j = 0, 1, \cdots, r,\ r = 2^n - 1$$

This is a mapping from $(GF(2))^n$ to GF(2). Its coefficient vector $\underline{A}$ can be computed from its function output vector $\underline{F}$ as follows[9]

$$\underline{A} = \underline{T}_n \underline{F} \tag{43}$$

where $\underline{A} = (a_0, a_1, \cdots, a_r)^t$; $\underline{F} = (f(0,0,\cdots,0), f(0,0,\cdots,1), \cdots, f(1,1,\cdots,1))^t$ is the function output vector; $\underline{T}_n$ is a $(2^n \times 2^n)$ matrix recursively defined by

$$\underline{T}_n = \begin{vmatrix} \underline{T}_{n-1} & \underline{0} \\ \underline{T}_{n-1} & \underline{T}_{n-1} \end{vmatrix}, \underline{T}_1 = \begin{vmatrix} 1 & 0 \\ 1 & 1 \end{vmatrix} \tag{44}$$

It is named the Reed-Muller transformation matrix.

For example, the coefficients of three-variable ring-sum expansion may be obtained in terms of the following Reed-Muller transformations.

$$\begin{vmatrix} a_0 \\ a_1 \\ a_2 \\ a_3 \\ a_4 \\ a_5 \\ a_6 \\ a_7 \end{vmatrix} = \begin{vmatrix} 1 & & & & & & & \\ 1 & 1 & & & & & & \\ 1 & 0 & 1 & & & & & \\ 1 & 1 & 1 & 1 & & & & \\ 1 & 0 & 0 & 0 & 1 & & & \\ 1 & 1 & 0 & 0 & 1 & 1 & & \\ 1 & 0 & 1 & 0 & 1 & 0 & 1 & \\ 1 & 1 & 1 & 1 & 1 & 1 & 1 & 1 \end{vmatrix} \begin{vmatrix} f(0,0,0) \\ f(0,0,1) \\ f(0,1,0) \\ f(0,1,1) \\ f(1,0,0) \\ f(1,0,1) \\ f(1,1,0) \\ f(1,1,1) \end{vmatrix}$$

For Example 1 the ring-sum expansion of next state functions may be computed from the above Reed-Muller transformation as follows

$$s_0' = s_1 \oplus s_0 s_1$$

$$s_1' = s_0 s_1 \oplus s_2$$

$$s'_2 = s_2 \oplus s_0 s_1 s_2$$

6.2 Fast Reed-Muller Transformations

The above example shows that the computation of a three-variable Reed-Muller transformation needs ($3^3 - 2^3$) modulo-2 additions. In general, a Reed-Muller transformation with n variables requires ($3^n - 2^n$) modulo-2 additions. However, the Reed-Muller transformation matrix $\underline{T}_n$ may be factored into Kronecker products as follows

$$\underline{T}_n = \underline{T}_1 \otimes \underline{T}_{-1} \otimes \cdots \otimes \underline{T}_1 \tag{45}$$

where $\otimes$ denotes Kronecker products. This factorisation may considerably reduce the number of modulo-2 additions for the coefficient computation.

For example, a three-variable Reed-Muller transformation matrix may be factored into

$$\underline{T}_3 = \left|\begin{array}{cccc:cccc} 1 &&&&&&& \\ 1 & 1 &&&&&& \\ && 1 &&&&& \\ && 1 & 1 &&&& \\ \hdashline &&&& 1 &&& \\ &&&& 1 & 1 && \\ &&&&&& 1 & \\ &&&&&& 1 & 1 \end{array}\right| \left|\begin{array}{cccc:cccc} 1 &&&&&&& \\ & 1 &&&&&& \\ 1 && 1 &&&&& \\ & 1 && 1 &&&& \\ \hdashline &&&& 1 &&& \\ &&&&& 1 && \\ &&&& 1 && 1 & \\ &&&&& 1 && 1 \end{array}\right| \left|\begin{array}{cccc:cccc} 1 &&&&&&& \\ & 1 &&&&&& \\ && 1 &&&&& \\ &&& 1 &&&& \\ \hdashline 1 &&&& 1 &&& \\ & 1 &&&& 1 && \\ && 1 &&&& 1 & \\ &&& 1 &&&& 1 \end{array}\right|$$

It only needs $3 * 2^2 = 12$ modulo-2 additions. The flow graph of the above example is shown in Fig. 4.

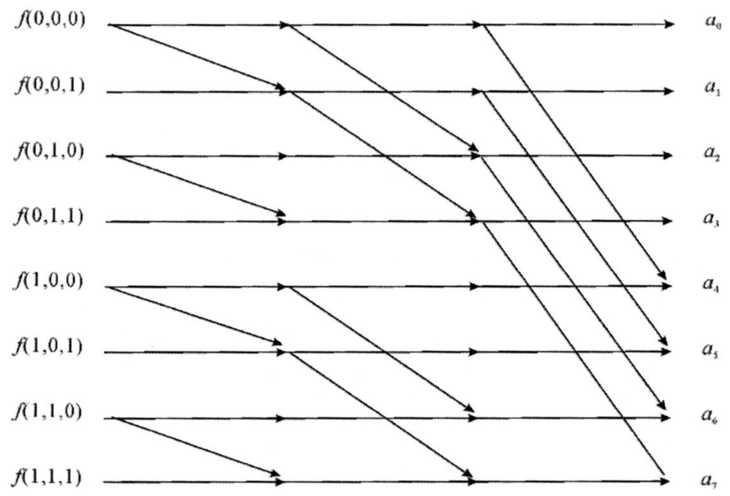

Fig. 4 Flow graph of fast Reed-Muller transformation

In general, using the Kronecker factorisation, the number of modulo-2 additions for computing a n-variable Reed-Muller transformation will reduce to

$$n * 2^{n-1} \tag{46}$$

This efficient algorithm is named the fast Reed-Muller Transformation. It is $2(1.5^n - 1)/n$ times faster than the Reed-Muller transformation. The comparison of two algorithms is shown in Fig. 5.

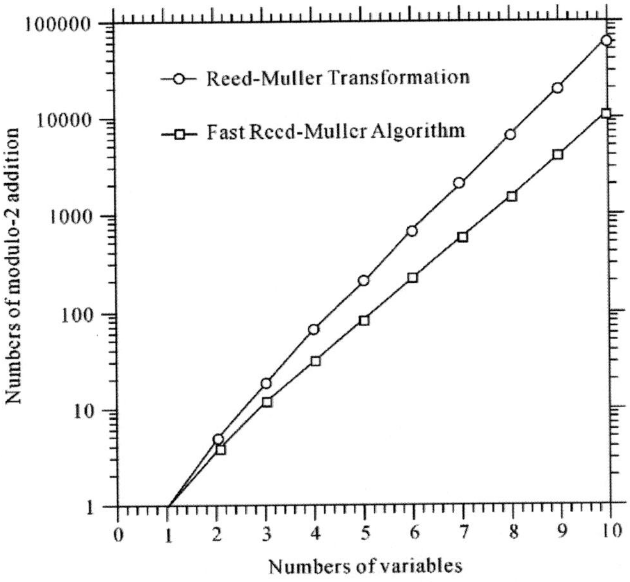

Fig. 5 The comparison of two algorithms

6.3 A Practical Method for Obtaining Ring-Sum Expansions

If the state table of FSMs is represented in the binary forms, then it may be considered as the truth table of Boolean functions. The truth table of logical functions may be easily represented as a number of Boolean OR/AND/NOT functions in the forms of sum of products as follows[22]

$$y = \bigvee_{i=0}^{2^n-1} m_i \tag{47}$$

where $\vee$ denotes logical OR operation; $m_i = \overset{*}{x}_0 \overset{*}{x}_1 \cdots \overset{*}{x}_{n-1}$ is minterm.

$$\overset{*}{x}_j = \begin{cases} x_j & \text{for } 1 \\ \bar{x}_j & \text{for } 0 \end{cases}$$

Using the following two logical relations[22]

$$m_i \vee m_j = m_i \oplus m_j \oplus m_i m_j \tag{48}$$

$$\bar{x}_i = x_i \oplus 1 \tag{49}$$

the above Boolean functions may be converted into modulo-2 ring-sum expansions. Fortunately, the minterms in the state table of FSMs are orthogonal, i. e.

$$m_i m_j = \begin{cases} m_i & \text{for } i = j \\ 0 & \text{for } i \neq j \end{cases} \tag{50}$$

Therefore, the ring-sum expansion with mixed-polarities can be directly obtained from the state table without any transforms as follows

$$y = \oplus \sum_{i=0}^{2^n-1} m_i \tag{51}$$

where $\oplus \sum$ denotes modulo-2 additions.

Using eqn. (49) the above functions may be written as positive-polarity.

For Example 2, the ring-sum expansion of output functions may be obtained directly from the state table as follows

$$y_0 = \bar{s}_0 \bar{s}_1 s_2 \oplus s_0 \bar{s}_1 s_2 \oplus \bar{s}_0 s_1 s_2 \oplus s_0 s_1 s_2 = s_2$$

$$y_1 = \bar{s}_0 s_1 \bar{s}_2 \oplus s_0 s_1 \bar{s}_2 \oplus s_0 \bar{s}_1 s_2 \oplus s_0 s_1 s_2 = 1 \oplus s_0 s_2 \oplus s_1 s_2$$

This practical method can be used in any machines with inputs, outputs and states which are in different fields, it is particularly useful when the outputs of state tables have very few "ones" or "zeros".

7. Conclusion

A systematic study for the representation of finite state machines as Galois Field polynomials is presented in this chapter. The polynomial coefficients can be computed from the state table of machines. A number of authors[12-15] have been concerned with the structure of Galois Field functions. The contributions of this paper are:

a) An improvement in the computation of polynomial coefficients is introduced in that the sum in Lemma 2 is for all the field elements including null element, and s^{r-i} is used instead of s^{-i}. This improvement avoids separately treating the coefficient a_{q-1} and considerably simplifies the representation of coefficient computations.

b) A DFT of length $(2^M - 1)$ for computing the polynomial coefficients is introduced by rearranging the ordering of coefficients and function output vectors.

c) The mapping between different fields $GF(2^M)$ and $GF(2^K)$ is developed in terms of a polynomial with coefficients which are in the common extension field $GF(2^L)$. These mapping polynomials are very useful for representation of the machines with states, inputs and outputs which are in different fields.

d) The cyclic subgroup property of the common extension polynomials is revealed. This property remarkably reduces the size of memory and amount of computations.

e) It is proved that a two-dimensional machine may be represented as a two-variable polynomial and the coefficients may be computed in terms of a two-dimensional DFT of size $(2^M - 1) \times (2^N - 1)$, and 2 one-dimensional DFTs of length $(2^M - 1)$ and $(2^N - 1)$ respectively by rearranging the order of coefficient and function output matrices.

f) A method for converting a multiple-variable machine into single-variable machines is given by means of a non-linear transformation which maps the extension field variable into several subfield variables. These non-linear transformations can be obtained by-solving a set of linear equations.

g) Ring-sum representations of finite state machines and fast Reed-Muller transformations for computing polynomial coefficients are discussed. The number of modulo-2 additions for computing a set of Reed-Muller polynomial coefficients with n-variables is reduced to $n * 2^{n-1}$ by using the fast Reed-Muller transformation.

Acknowledgements

The author would like to thank Dr P. J. W. Rayner for reading and correcting the manuscript, and Trinity College, Cambridge, for financial support during this period.

Appendix A1: Representations of Multiple-Variable Finite State Machines

A finite state machine with n inputs, k outputs and m states may be represented as a set of polynomials as follows

$$s'_i = f_i(s_0, s_1, \cdots, s_{m-1}, u_0, u_1, \cdots, u_{n-1})$$

$$= \sum_{i_0=0}^{r_1} \cdots \sum_{i_{m-1}=0}^{r_1} \sum_{j_0=0}^{r_2} \cdots \sum_{j_{n-1}=0}^{r_2} a^{(i)}_{i_0 \cdots i_{m-1} j_0 \cdots j_{n-1}} s_0^{i_0} s_{m-1}^{i_{m-1}} u_0^{j_0} \cdots u_{n-1}^{j_{n-1}}$$

$$i = 0, 1, \cdots, (m-1)$$

$$y_j = g_i(s_0, s_1, \cdots, s_{m-1})$$

$$= \sum_{i_0=0}^{r_1} \sum_{i_1=0}^{r_1} \cdots \sum_{i_{m-1}=0}^{r_1} b^{(j)}_{i_0, i_1 \cdots i_{m-1}} s_0^{i_0} s_1^{i_1} \cdots s_{m-1}^{i_{m-1}}$$

$$j = 0, 1, \cdots, (k-1)$$

where

$$s_0, s_1, \cdots s_{m-1} \in GF(2^M) \quad r_1 = 2^M - 1$$

$$u_0, u_1, \cdots u_{n-1} \in GF(2^N) \quad r_2 = 2^N - 1$$

$$y_0, y_1, \cdots y_{k-1} \in GF(2^K)$$

The coefficients

$$a^{(i)}_{i_0 \cdots i_{m-1} j_0 \cdots j_{n-1}} \in GF(2^{L_1}) \qquad L_1 = LCM(M, N)$$

$$b^{(j)}_{i_0, i_1 \cdots i_{m-1}} \in GF(2^{L_2}) \qquad L_2 = LCM(K, M, N)$$

$$0 \leq i_0, i_1, \cdots, i_{m-1} \leq r_1$$

$$0 \leq j_0, j_1, \cdots, j_{n-1} \leq r_2$$

The coefficients of polynomials are determined by

$$a^{(i)}_{00 \cdots 0} = f_i(0, 0, \cdots, 0)$$

$$a^{(i)}_{i_0 0 \cdots 0} = \sum_{s_0} f_i(s_0, 0, \cdots, 0) s_0^{r_1 - i_0}$$

$$\vdots$$

$$a^{(i)}_{0 \cdots 0 j_{n-1}} = \sum_{u_{n-1}} f_i(0, \cdots, 0, u_{n-1}) u_{n-1}^{r_2 - j_{n-1}}$$

$$a^{(i)}_{i_0 i_1 0 \cdots 0} = \sum_{s_0} \sum_{s_1} f_i(s_0, s_1, \cdots, 0) s_0^{r_1 - i_0} s_1^{r_1 - i_1}$$

$$\vdots$$

$$a^{(i)}_{0 \cdots 0 i_{m-1} j_0 \cdots 0} = \sum_{s_{m-1}} \sum_{u_0} f_i(0, \cdots, 0, s_{m-1}, u_0, 0, \cdots, 0) s_{m-1}^{r_1 - i_{m-1}} u_0^{r_2 - j_0}$$

$$\vdots$$

$$a^{(i)}_{0\cdots 0 j_{n-2} j_{n-1}} = \sum_{u_{n-2}} \sum_{u_{n-1}} f_i(0,\cdots,0,u_{n-2},u_{n-1}) u_{n-2}^{r_2 - j_{n-2}} u_{n-1}^{r_2 - j_{n-1}}$$

$$\vdots$$

$$a^{(i)}_{i_0 \cdots i_{m-1} j_0 \cdots j_{n-1}} = \sum_{s_0} \cdots \sum_{s_{m-1}} \sum_{u_0} \cdots \sum_{u_{n-1}} f_i(s_0,\cdots,s_{m-1},u_0,u_{n-1}) s_0^{r_1 - i_0} \cdots s_{m-1}^{r_1 - i_{m-1}} u_0^{r_2 - j_0} u_{n-1}^{r_2 - j_{n-1}}$$

$$i = 0,1,\cdots,(m-1)$$

$$b^{(j)}_{00\cdots 0} = g_j(0,0,\cdots,0)$$

$$b^{(j)}_{i_0 0\cdots 0} = \sum_{s_0} g_j(s_0,0,\cdots,0) s_0^{r_1 - i_0}$$

$$\vdots$$

$$b^{(j)}_{0\cdots 0 i_{n-1}} = \sum_{s_{n-1}} g_j(0,\cdots,0,s_{n-1}) s_{n-1}^{r_1 - i_{n-1}}$$

$$b^{(j)}_{i_0 i_1 0\cdots 0} = \sum_{s_0} \sum_{s_1} g_j(s_0,s_1,\cdots,0) s_0^{r_1 - i_0} s_1^{r_1 - i_1}$$

$$\vdots$$

$$b^{(j)}_{0\cdots 0 i_{m-2} i_{m-1}} = \sum_{s_{m-2}} \sum_{s_{m-1}} g_j(0,\cdots,0,s_{m-2},s_{m-1}) s_{m-2}^{r_1 - i_{m-2}} s_{m-1}^{r_1 - i_{m-1}}$$

$$\vdots$$

$$b^{(j)}_{i_0 i_1 \cdots i_{m-1}} = \sum_{s_0} \sum_{s_1} \cdots \sum_{s_{m-1}} g_j(s_0,s_1,\cdots,s_{m-1}) s_0^{r_1 - i_0} s_1^{r_1 - i_1} \cdots s_{m-1}^{r_1 - i_{m-1}}$$

$$j = 0,1,\cdots,(k-1)$$

$$0 \le i_0, i_1, \cdots, i_{m-1} \le r_1, j = 0,1,\cdots,(k-1)$$

$$0 \le j_0, j_1, \cdots, j_{n-1} \le r_2$$

$\sum_x$ denotes the sum for all the elements of variable x

Appendix A2 will show the proof for two-variable polynomials. It is easily extended to the general case.

If β_1, β_2 and β_3 are respectively the primitive elements of GF(2^M), GF(2^N) and GF(2^K), let the states, inputs and outputs of machines respectively be

$$0,1,\beta_1,\beta_1^2,\cdots,\beta_1^{2^M-2} \quad \text{for states } s_0, s_1, \cdots, s_{m-1}$$

$$0,1,\beta_2,\beta_2^2,\cdots,\beta_2^{2^N-2} \quad \text{for inputs } u_0, i_1, \cdots, u_{n-1}$$

$$0,1,\beta_3,\beta_3^2,\cdots,\beta_3^{2^K-2} \quad \text{for output } y_0, y_1, \cdots, y_{k-1}$$

and let the subscripts of polynomial coefficients and powers of variables in the above equations be reordered as follows

$$0,(2^M-1),1,2\cdots,(2^M-2) \quad \text{for } i_0, i_1, \cdots, i_{m-1}$$

$$0,(2^N-1),1,2\cdots,(2^N-2) \quad \text{for } j_0, j_1, \cdots, j_{n-1}$$

Then the above equations may be written in DFT form.

Appendix A2: Mathematical Proof of Theorems

Proof of Theorem 2

Since L is Least Common Multiple of N and K, from Lemma 3 GF(2^N) and GF(2^K) are both the subfields of extension field GF(2^L). Therefore, $s \in$ GF(2^M) $\subset$ GF(2^L) and $f \in$ GF(2^K) $\subset$ GF(2^L), eqn. (17) and eqn. (18) are arithmetic operations within extension field GF(2^L).

Substituting $g(s)$ of eqn. (17) into eqn. (18), the right side of eqn. (18)

$$\sum_s \left(\sum_{i=0}^{2^M-1} b_i s^i\right) s^{r-j} = \sum_{i=0}^{2^M-1} b_i \left(\sum_s s^{r(i-j)}\right)$$

Using Lemma 2, it follows

$$= \sum_{i=0}^{2^M-1} b_i \delta(i-j) = b_j \quad j = 1,2,\cdots,2^M-1$$

and is equal to the left-hand side of sqn. (18).

It is easy to prove that

$$b_0 = g(0)$$

Proof of Theorem 4

Since both $GF(2^M)$ and $GF(2^N)$ are subfields of $GF(2^L)$, Theorem 4 may be considered as an arithmetic operation within extension field $GF(2^L)$.

Substituting $f(s,u)$ of eqn. (27) into the right-hand side of eqn. (28d), it follows

$$\sum_s \sum_u f(s,u) s^{r_1-i} s^{r_2-j}$$

$$= \sum_s \sum_u \left(\sum_{i'=0}^{2^M-1} \sum_{j'=0}^{2^N-1} a_{i',j'} s^{i'} u^{j'}\right) s^{r_1-i} s^{r_2-j}$$

$$= \sum_{i'=0}^{2^M-1} \sum_{j'=0}^{2^N-1} a_{i',j'} \left(\sum_s s^{r_1(i-i')}\right)\left(\sum_u u^{r_2(j-j')}\right)$$

Using Lemma 3, it follows

$$= \sum_{i'=0}^{2^M-1} \sum_{j'=0}^{2^N-1} a_{i',j'} \delta(i-i')\delta(j-j') = a_{ij}$$

It is equal to the left-hand side of eqn. (28d).

Similarly, using $0^0 = 1$, and substituting $f(0,u)$ of eqn. (27) into the right-hand side of eqn. (28c), it follows

$$\sum_u f(0,u) u^{r_2-j}$$

$$= \sum_u \left(\sum_{j'=0}^{2^N-1} a_{0j'} u^{j'}\right) u^{r_2-j}$$

$$= \sum_{j'=0}^{2^N-1} a_{0j'} \delta(j-j') = a_{0j}$$

This is eqn. (28c).

Substituting $f(s,0)$ of eqn. (27) into the right-hand side of eqn. (28b), it follows

$$\sum_s f(s,0) s^{r_1-i}$$

$$= \sum_s \left(\sum_{i'=0}^{2^M-1} a_{i'0}, s^{i'}\right) s^{r_1-i}$$

$$= \sum_{i'=0}^{2^M-1} a_{i'0}, \delta(i-i') = a_{i0}$$

This is the left-hand side of eqn. (28b).

It is easy to prove that

$$a_{00} = f(0,0)$$

References

[1] Reed, I. S. and Stewart, R. M., "Note on the existence of perfect maps", IRE Transactions on Information Theory, Vol. IT-8, No. 1, January 1962, pp. 10-12.

[2] Booth, T. L., "An analytic representation of signals in sequential networks", Proceedings of Symposium on the Mathematical Theory of Automata, Polytechnic Institute of Brooklyn, April 1962, pp. 301-340.

[3] Bartee, T. C. and Schneider, D. I., "Computation with Finite Fields", Information and Control, Vol. 6, 1963, pp. 79-98.

[4] Bollraan, D. A., "Some periodicity properties of transformations on vector spaces over residue class rings", J. Soc. Indust. Appl. Math., Vol. 13, No. 3, September 1965, pp. 902-912.

[5] Richalet, J., "Operational calculus for Finite Rings", IEEE Transactions on Circuit Theory, Vol. CT-12, No. 4, December 1965, pp. 558-570.

[6] Gill, A. and Jacob, J. P., "On a mapping polynomial for Galois Fields", Quart. Appl. Math., Vol. 24, No. 1, 1966, pp. 57-62.

[7] Berlekamp, E. R., "Factoring polynomials over Finite Fields", The Bell System Technical Journal, Vol. 46, No. 8, October 1967, pp. 1853-1859.

[8] Tannak, H., Kasahara, M., Tezuka, Y. and Kasahara, Y., "Computation over Galois Fields using shift registers", Information and Control, Vol. 13, 1968, pp. 75-84.

[9] Green, D. H. and Taylor, I. S., "Modular representation of multiple-valued Logic systems", IEE Proceedings, Vol. 121, No. 6, June 1974, pp. 409-418.

[10] Lempel, A., "Matrix factorization over GF(2) and trace-orthogonal bases over GF(2^n)", SIAM J. Comput., Vol. 4, No. 2, June 1975, pp. 175-186.

[11] Santag, E. D., "On linear systems and noncommutative Rings", Mathematical Systems Theory, Vol. 9, No. 4, pp. 527-344.

[12] Pradhan, D. K. and Patel, A. M., "Reed-Muller like canonic forms for Multivalued functions", IEEE Transactions on Computers, Vol. C-24, No. 2, February 1975, pp. 206-210.

[13] Benjauthrit, B. and Reed, I. S., "Galois switching functions and their applications", IEEE Transactions on Computers, Vol. C-25, No. 1, January 1976, pp. 78-86.

[14] Pradhan, D. K., "A theory of Galois switching functions", IEEE Transactions on Computers, Vol. C-27, No. 3, March 1978, pp. 239-248.

[15] Benjanthrit, B. and Reed, I. S., "On the fundamental structure of Galois switching functions", IEEE Transactions on Computers, Vol. C-27, No. 8, August 1978, pp. 757-762.

[16] Imamura, K., "A method for computing addition tables in GF(p^n)", IEEE Transactions on Information Theory, Vol. IT-26, No. 3, May 1980, pp. 367-369.

[17] English, W. R., "Synthesis of finite state algorithms in a Galois Fields GF(p^n)", IEEE Transactions on Computers, Vol. C-30, No. 3, March 1981, pp. 225-229.

[18] Fleisher, H., Tavel, M. and Yeager, J., "Exclusive-OR representation of Boolean functions", IBM J. Res. Develop., Vol. 27, No. 4, July 1983, pp. 412-416.

[19] Reddy, S. M. , "Easily testable realization for Logical functions", IEEE Transactions-on Computers, Vol. C-21, 1972, pp. 1183-1188.

[20] Rayner, P. J. W. , "The application of finite arithmetic structures to design of digital signal processing systems", Proceedings of International Conference on DSP, Florence, Italy, September 1978.

[21] Stone, H. S. , "Discrete mathematical structures and their applications", Science Research Associates, Chicago, 1973.

[22] Friedman, A. D. and Menon, P. R. , "Theory and design of switching circuits", Woodland-Hills, California, 1975.

[23] Hellerman, L. , "A measure of computational work", IEEE Trans. on Computers, Vol. C-21, No. 5, May 1972.

Optimum Coefficients for Recursively Computing DFTs*

ZHANG Yan-Zhong

Changcheng Institute of Measurement and Metrology, Beijing, China

Abstract: The recursive computation of prime length DFTs has very simple structure and needs only one coefficient. The signal-to-noise ratio for fixed-point implementation of this algorithm is discussed. The optimum coefficients with maximum S/N ratio are presented.

1. Introduction

Discrete Fourier Transformation (DFT) is a central operation in digital signal processing. Since Cooley and Tukey[1] proposed the Fast Fourier Transformation (FFT) algorithm, the number of multiplication operations has been considerably reduced. However, the DFT and FFT algorithms need $(N-1)$ complex coefficients for computing each frequency componerit. This causes complexity of the algorithm implementation.

The N point DFT may be written as follows

$$\begin{aligned} X(k) &= \sum_{n=0}^{N-1} x(n) W_N^{nk} \\ &= (\cdots(W_N^k x(N-1) + x(N-2)W_N^k + \cdots + x(1))W_N^k + x(0) \end{aligned} \tag{1}$$

The above equation may be expressed as the following recursive form:

$$y_k(n) = W_N^k y_k(n-1) + u(n) \tag{2}$$

$$y_k(0) = u(0), k = 0, 1, 2, \cdots, (N-1)$$

where

$$u(n) = x(N-1-n), n = 0, 1, 2, \cdots, (N-1)$$

The kth frequency component $X(k)$ of DFTs is the $(N-1)$th output of the recursive presented $y_k(n)$

$$X(k) = y_k(N-1), k = 0, 1, 2, \cdots, (N-1)$$

This is Goertzel algorithm[2]. The advantages of the Goertzel algorithm are that only one coefficient W_N^k is required to compute each frequency component $X(k)$, and the algorithm can be realized by using the simple recursive equation. However, this algorithm needs $(N-1)$ coefficients for computing the whole N frequency components.

---

* Published in MELECON'85/Vol. Ⅱ: Digital Signal Processing, Elsevier Science Publishers B. V/IEEE, 1985: 11-14.

Y. Zhang, *Footprints in Cambridge and Aviation Industries of China*,
https://doi.org/10.1007/978-981-16-3176-4_7

2. Recursive Computation of Prime Length DFTs

Curts[3] and Zhang[4] have proposed that when the length of DFTs $N=P$ is prime, only one coefficient is required to compute $(P-1)$ frequency components by permuting the input sequence $\{x(n)\}$. This can be described by the following example. Consider a five-point DFT

$$X(k) = \sum_{n=0}^{4} x(n)W^{nk} \quad k = 0,1,2,3,4$$

where

$$W=\exp(-j2\pi/5)$$

The non-zero frequency components may be written as the following matrix form:

$$\begin{vmatrix} X(1) \\ X(2) \\ X(3) \\ X(4) \end{vmatrix} = \begin{vmatrix} 1 & w^1 & w^2 & w^3 & w^4 \\ 1 & w^2 & w^4 & w^1 & w^3 \\ 1 & w^3 & w^1 & w^4 & w^2 \\ 1 & w^4 & w^3 & w^2 & w^1 \end{vmatrix} \begin{vmatrix} x(0) \\ x(1) \\ x(2) \\ x(3) \\ x(4) \end{vmatrix}$$

It is obvious that each row of the coefficient matrix is a permutation of the first row. Rearrange the ordering of the Input sequences $\{x(n)\}$, the above matrix can be rewritten as follows:

$$\begin{vmatrix} X(1) \\ X(2) \\ X(3) \\ X(4) \end{vmatrix} = \begin{vmatrix} x(0) & x(1) & x(2) & x(3) & x(4) \\ x(0) & x(3) & x(1) & x(4) & x(2) \\ x(0) & x(2) & x(4) & x(1) & x(3) \\ x(0) & x(4) & x(3) & x(2) & x(1) \end{vmatrix} \begin{vmatrix} 1 \\ W^1 \\ W^2 \\ W^3 \\ W^4 \end{vmatrix}$$

One coefficient can be employed in recursively computing all the 4 non-zero frequency components by permuting the input data. For prime length DFTs, the recursive equation (2) may be changed to

$$y_k(n)=W_P^K y_k(n-1)+u_k(n) \tag{3}$$

$$y_k(0)=u_k(0),\ k=0,1,2,\cdots,(p-1)$$

where

$$u_k(p-1-n)=x(\langle nk^{-1}D\rangle P) \tag{4}$$

$$n=0,1,2,\cdots,(p-1)$$

k^{-1} is the inverse element of k in GF(p), $k^{-1}k=1 \bmod(p)$

D is an integer, $1\leq D<P$

The above recursive equation may be viewed as a first order IIR filter with a complex coefficient. Its z-transform function is as follows:

$$H(z)=\frac{1}{1-W_P^D z^{-1}} \tag{5}$$

It may be expressed as a real coefficient function by

$$\begin{aligned} H(z)&=\frac{1}{1-W_P^D z^{-1}}\frac{1-W_P^{-D}z^{-1}}{1-W_P^{-D}z^{-1}} \\ &=\frac{1-2\cos(2\pi D/p)z^{-1}-j\sin(2\pi D/p)z^{-1}}{1-2\cos(2\pi D/p)z^{-1}+z^{-2}} \end{aligned} \tag{6}$$

Eqn. (6) may be considered as a 2nd order IIR filter as shown in Fig. 1.

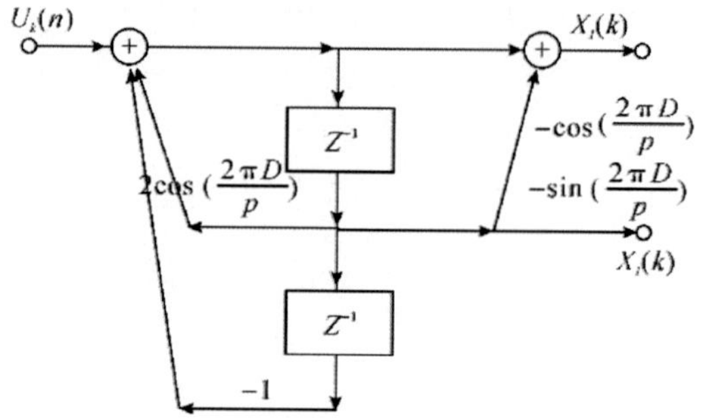

Fig. 1 2nd order recursive form of DFTs

3. Dynamic Range and Signal-to-Noise Ratio

As shown in eqns. (3) and (5), the recursive calculation of DFTs may be viewed as a process of finding the spectrum of input signals at a fixed point A on the unit cycle in z-plane (see Fig. 2). For a prime p, all the $(p-1)$ frequency components can be computed at the same point $A(=W_p)$ in the z-plane by permuting the input data. Since p is prime, any one of the $(p-1)$ pole points $B(=W_p^D$, $1\leq D<p)$ on unit circle can be used as the fixed-point A for recursively computing the DFT.

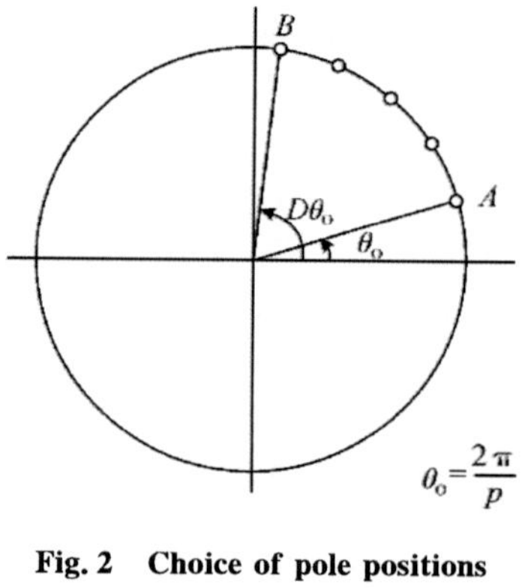

Fig. 2 Choice of pole positions

For the floating-point implementation of the recursive equation (3), any coefficients W_p^D $(1\leq D<p)$ can be used in the system, and they have no effect on the final results. However, in the fixed-point implementation, the coefficient W_p^D will affect the signal-to-noise ratio and dynamic range of systems.

In order to prevent overflows in fixed-point implementation, a scaling factor A_D should be introduced in the input of the second order all-pole filter (6), such that

$$A_D<\frac{1}{\sum_{n=0}^{p-1}|h_D(n)|} \quad 0<A_D<1 \tag{7}$$

where $h_D(n)$ is the impulse response of the all-pole filter (6).

The signal-to-noise ratio of systems is proportional to the scaling factor A_D. For the all-pole filter, its impulse $h_D(n)$ and scaling factor are as follows

$$h_D(n)=\frac{\sin\dfrac{2\pi(n+1)D}{p}}{\sin\dfrac{2\pi D}{p}} \tag{8}$$

$$A_D=\frac{\left|\sin\dfrac{2\pi D}{p}\right|}{\sum_{n=0}^{p-1}\left|\sin\dfrac{2\pi(n+1)D}{p}\right|} \tag{9}$$

Since p is prime, $\sum_{n=0}^{p-1}\left|\sin\frac{2\pi(n+1)D}{p}\right|$ is independent on the factor D, namely,

$$\sum_{n=0}^{p-1}\left|\sin\frac{2\pi(n+1)D}{p}\right|=c_p<p \tag{10}$$

where c_p is a constant and $c_p<p$.

Using the above equation, the scaling factor A_D can be written as

$$A_D=\left|\sin\frac{2\pi D}{p}\right|/c_p \tag{11}$$

It is obvious that when $\frac{2\pi D}{p}$ approches $\pm\pi/2$, A_D has the maximum value, and the system possesses the optimum signal-to-noise ratio without overflows.

4. Optimum Coefficients

Table 1 gives the optimum D and the recursive coefficient $\cos(\frac{2\pi D}{p})$ for all $p<100$. Some results for $p>100$ are shown in Table 2.

Table 1 Optimum coefficients ($p<100$)

p	D	$\cos(\frac{2\pi D}{p})$
3	1	−0.5
5	1	0.309017
7	2	−0.222521
11	3	−0.142315
13	3	0.120537
17	4	0.092268
19	5	−0.825793
23	6	−0.068242
29	7	0.054139
31	8	−0.050649

(To be continued)

Table 1

p	D	$\cos(\frac{2\pi D}{p})$
37	9	0.042441
41	10	0.038303
43	11	−0.036522
47	12	−0.033415
53	13	0.029633
59	15	−0.026621
61	15	0.025748
67	17	−0.023443
71	18	−0.022122
73	18	0.021516
79	20	−0.019882
83	21	−0.018924
97	24	0.016193

Table 2 Optimum coefficients ($p>100$)

p	D	$\cos(\frac{2\pi D}{p})$
127	32	-1.5831×2^{-7}
257	65	1.5647×2^{-8}
509	127	1.5801×2^{-9}
1021	255	1.5754×2^{-10}
2039	510	-1.5777×2^{-11}
4093	1023	1.5720×2^{-12}
8191	2048	-1.5710×2^{-13}

5. Conclusion

The recursive computation of prime length DFTs needs only one coefficient for all $(p-1)$ frequency components. The relationship between the coefficient and the signal-to-noise ratio of the algorithm has been discussed. The optimum coefficients with maximum S/N ratio have been n presented.

References

[1] Cooley, J. W. and Tukey, J. W. , "An algorithm for machine calculation of complex Fourier series", Math. Comut. Vol. 19, No. 90, April 1965, pp. 297-301.

[2] Gold, B. and Rader, C. , "Digtal Processing of signals", McGraw-Hill, New York, 1969.

[3] Curts, T. E. and Wickenden, J. D. , "Hard-ware-based Fourier transform: algorithm and architectures", IEE Proceedings, Vol. 130, Pt. F, No. 4, 1983, pp. 423-432.

[4] Zhang, Y. Z. , "Digital signal processing systems with finite state machine realisation", Ph. D. Dissertation, Cambridge University, England, 1984.

Linearization of Nonlinear Systems Over GF(2^M)*

Yan-Zhong Zhang
Ministry of Aero-Space Industry, Beijing, China

Abstract: A method for linearing any non-linear systems over GF(2^M) is presented in terms of increasing its dimensions. The linearization matrix can be obtained from a similarity transform of connection matrix. The similarity transform matrix $\underline{P}$ contains a DFT of length (2^M-1) over GF(2^M). The connection matrix $\underline{C}$ can be obtained from the truth table of systems. A computer method is proposed for studying of a non-linear system. The method can be used to detect the amplitude and period of oscillations in the systems.

1. Introduction

Exclusive-OR logic circuits possess many advantages of less number of gates, simpler connections and easier tests in comparison with Boolean Logic circuits. The presentation and minimization of Exclusive-OR logic circuits have been solved[1-2]. The Galois field representation of circuits is a very strong tool for studying the stability of systems.

Zhang[2] has proposed that any logic circuits or systems with M-bit inputs and outputs can be represented as a polynomial over Galois field GF(2^M) as follows

$$y = f(x) = \sum_{i=0}^{r} b_i x^i \tag{1}$$

where

$y, x, b_i \in \mathrm{GF}(2^M)$, $r = 2^M - 1$

$y = y_{M-1}a^{M-1} + y_{M-2}a^{M-2} + \cdots + y_0$

$x = x_{M-1}a^{M-1} + x_{M-2}a^{M-2} + \cdots + x_0$

$y_i, x_i \in (0,1)$ $i = 0, 1, \cdots, (M-1)$

a is a unity element of GF(2^M)

In general, this polynomial is non-linear. If the polynomial has the following form

$$y = b_0 + b_1 x, \text{ or } y = b_1 x \tag{2}$$

then the system is linear. Linear systems has been studied extensively. But the stability of non-linear systems is difficult to deal with. If a non-linear system can be transformed into a linear system, then the nonlinear system may be studied by using the theory and method of linear systems.

---

* Published in Proceeding of Int. Conf. on Circuits & Systems, ICCAS'89, July 1989: 217-220.

Y. Zhang, *Footprints in Cambridge and Aviation Industries of China*,
https://doi.org/10.1007/978-981-16-3176-4_8

2. Linearization Method

Eqn. (1) is a single variable function. Its variable and coefficients are elements of GF(2^M). Since Galois field GF(2^M) is finite and possesses the following property

$$x^{nr+1}=x, \quad \begin{matrix} r=2^M-1 \\ n=0,1,\cdots \end{matrix} \tag{3}$$

therefore, power of the output y is still an element of the field and can be also represented as a polynomial of the variable x as follows

$$y^k=f^k(x)=\sum_{i=0}^{r} b_{ki}x^i \tag{4}$$

It should be noted that the coefficient b_{ki} in eqn. (4) is not independent, and is a function of the coefficients b_i in eqn. (1).

Let k & i in eqn. (4) be re-arranged in the following order

$$0,r,1,2,\cdots,(r-1)$$

Then eqn. (4) may be written in matrix form:

$$\begin{vmatrix} 1 \\ y^r \\ y^1 \\ y^2 \\ \vdots \\ y^{r-1} \end{vmatrix} = \begin{vmatrix} 1 & 0 & \cdots & 0 \\ b_{r0} & b_{rr} & \cdots & b_{r(r-1)} \\ b_{10} & b_{1r} & \cdots & b_{1(r-1)} \\ b_{20} & b_{2r} & \cdots & b_{2(r-1)} \\ \vdots & \vdots & \vdots & \vdots \\ b_{(r-1)0} & b_{(r-1)r} & \cdots & b_{(r-1)(r-1)} \end{vmatrix} \begin{vmatrix} 1 \\ x^r \\ x \\ x^2 \\ \vdots \\ x^{r-1} \end{vmatrix} \tag{6}$$

or

$$\underline{Y}=\underline{LX} \tag{7}$$

Eqn. (7) has a similar form to eqn. (2) of the linear systems. It can be viewed as a multidimensional linear system over GF(2^M). But there are some important differences. The vector $\underline{X}$ in eqn. (7) have r variables which are not independent. In spite of this, the form of eqn. (7) allows the non-linearsystem to be studied as a linear system in an extended state space. $\underline{L}$ is named as a linearization matrix. A major disadvantage of this method is that the elements of linearization matrix $\underline{L}$ have to be calculated for each power of $\underline{Y}$ by using eqn. (3). This calculation is rather complex. However, this computation may be simplified in the state transition matrix formulation by the following sinilarity transformation.

3. Similarity Transformation of Matrix $\underline{L}$

Let the elements of GF(2^M) be denoted by

$$B=0,1,a,a^2,\cdots,a^{r-1} \quad r=2^M-1 \tag{8}$$

where a is a primitive element of GF(2^M). If the variable x in eqn. (6) takes successively the value of each field element, then

$$\begin{vmatrix} 1 & 1 & 1 & \cdots & 1 \\ f^r(0) & f^r(1) & f^r(a) & \cdots & f^r(a^{r-1}) \\ f(0) & f(1) & f(a) & \cdots & f(a^{r-1}) \\ f^2(0) & f^2(1) & f^2(a) & \cdots & f^2(a^{r-1}) \\ \vdots & & & & \\ f^{r-1}(0) & f^{r-1}(1) & f^{r-1}(a) & \cdots & f^{r-1}(a^{r-1}) \end{vmatrix} = \underline{L} \begin{vmatrix} 1 & 1 & 1 & \cdots & 1 \\ 0 & 1 & 1 & \cdots & 1 \\ 0 & 1 & a & \cdots & a^{r-1} \\ 0 & 1 & a^2 & \cdots & a^{2(r-1)} \\ \vdots & & & & \\ 0 & 1 & a^{r-1} & \cdots & a^{(r-1)^2} \end{vmatrix} \tag{9}$$

Each column of the left matrix in eqn. (9) forms a vector as follows

$$\underline{f}(B) = |1 \quad f^r(B) \quad f(B) \quad \cdots \quad f^{r-1}(B)|^t \tag{10}$$

Since $f(B)$ GF(2^M), it is one of the set of $(0, 1, a, a^r, \cdots, a^{r-1})$. The power of $f(B)$ in vector $\underline{f}(B)$ has the same ordering as that of a^i shown in eqn. (5). Therefore, $f(B)$ will be identical to one of the columns of the post-multiplying matrix on the right-hand side of eqn. (9). The eqn. (9) may be written as follows

$$\begin{vmatrix} 1 & 1 & 1 & \cdots & 1 \\ 0 & 1 & 1 & \cdots & 1 \\ 0 & 1 & a & \cdots & a^{r-1} \\ 0 & 1 & a^2 & \cdots & a^{2(r-1)} \\ \vdots & & & \vdots & \\ 0 & 1 & a^{r-1} & & a^{(r-1)^2} \end{vmatrix} |\underline{c}_0 \quad \underline{c}_r \quad \underline{c}_1 \quad \cdots \quad \underline{c}_{r-1}| = \underline{L} \begin{vmatrix} 1 & 1 & 1 & \cdots & 1 \\ 0 & 1 & 1 & \cdots & 1 \\ 0 & 1 & a & \cdots & a^{r-1} \\ 0 & 1 & a^2 & \cdots & a^{2(r-1)} \\ \vdots & \vdots & & \vdots & \\ 0 & 1 & a^{r-1} & \cdots & a^{(r-1)^2} \end{vmatrix} \tag{11}$$

where $\underline{c}_i$ is a connection column vector having only one element equal to unity "1" and all other elements equal to "0". The position of the unity element is such that the vector $\underline{c}_i$ selects that vector from the pre-multiplying matrix in eqn. (9), which is identical to the corresponding $\underline{f}(B)$. The matrix $\underline{c}_L = |\underline{c}_0 \underline{c}_r \underline{c}_l \cdots \underline{c}_{r-1}|$ is named a connection matrix. Let

$$\underline{P} = \begin{vmatrix} 1 & 1 & 1 & \cdots & 1 \\ 0 & 1 & 1 & \cdots & 1 \\ 0 & 1 & a & \cdots & a^{r-1} \\ 0 & 1 & a^2 & \cdots & a^{2(r-1)} \\ \vdots & \vdots & \vdots & \cdots & \vdots \\ 0 & 1 & a^{r-1} & \cdots & a^{(r-1)^2} \end{vmatrix} \tag{12}$$

From the eqn. (12), the linearization matrix $\underline{L}$ may be represented as the following similarity transformation form

$$\underline{L} = \underline{P}\,\underline{C}_L\,\underline{P}^{-1} \tag{13}$$

Using the Galois field properties[2] the inverse matrix $\underline{P}^{-1}$ may be obtained by

$$\underline{P}^{-1} = \begin{vmatrix} 1 & 1 & 1 & \cdots & 1 \\ 0 & 1 & 1 & \cdots & 1 \\ 0 & 1 & a^{-1} & \cdots & a^{-(r-1)} \\ 0 & 1 & a^{-2} & \cdots & a^{-2(r-1)} \\ \vdots & \vdots & \vdots & \cdots & \vdots \\ 0 & 1 & a^{-(r-1)} & \cdots & a^{-(r-1)^2} \end{vmatrix} \tag{14}$$

From eqns. (13), (12) and (14), it is shown that the linearization matrix $\underline{L}$ of a nonlinear system can be expressed as a similarity transformation of its connect matrix $\underline{C}_L$ as shown in eqn. (13). Since a is a primitive element of GF(2^M), the transformation matrix $\overline{P}$ contains a DFT of length(2^M-1), and the inverse matrix $\overline{P}^{-1}$ contains an inverse DFT of length (2^M-1) as shown in eqns. (12) and (14).

4. Property of Connection Matrix $\underline{C}_L$

If the output $f(x)$ in eqn. (1) represents the next state of sequential circuits, then the connection matrix $\underline{C}_L$ can be obtained from the state transition table of system. From the definition of the connection matrix $\underline{C}_L$, it is shown that the state transition direction of a system is from the position of each diagonal element c_{ii} to the position of the element "1" in the same column $\underline{c}_i$. Using this property, the connection matrix $\underline{C}_L$ can be directly obtained without any computation.

For example, the state table of a sequential system and its field element assignment 1 are given in Table 1.

Table 1 State table of example

x					$f(x)$				
	x_2	x_1	x_0	a^i		y_2	y_1	y_0	a^i
0	0	0	0	0	1	0	0	1	1
1	0	0	1	1	0	0	0	0	0
2	0	1	0	a	-1	1	1	1	a^5
3	0	1	1	a^3	-1	1	1	1	a^5
-4	1	0	0	a^2	3	0	1	1	a^3
-3	1	0	1	a^6	3	0	1	1	a^3
-2	1	1	0	a^4	3	0	1	1	a^3
-1	1	1	1	a^5	2	0	1	0	a

This is a non-linear system. Its output function is as follows

$$f(x)=1+a^5x^2+ax^3+a^2x^5$$

The connection matrix $\underline{C}_L$ can be obtained from the above property as follows

0	0	1	0	0	0	0	0	0
1	1	0	0	0	0	0	0	0
2	0	0	0	0	0	0	0	1
3	0	0	0	0	1	1	1	0
-4	0	0	0	0	0	0	0	0
-3	0	0	0	0	0	0	0	0
-2	0	0	0	0	0	0	0	0
-1	0	0	1	1	0	0	0	0
	$\underline{c}_0$	$\underline{c}_7$	$\underline{c}_1$	$\underline{c}_2$	$\underline{c}_3$	$\underline{c}_4$	$\underline{c}_5$	$\underline{c}_6$

The array denotes the state transition direction. The property of connection matrix may be used to study the stability of nonlinear systems.

The linearization matrix $\underline{L}$ can be easily obtained from the connection matrix $\underline{C_L}$ in terms of the similarity transform in eqn. (13). Since eqn. (13) is a similarity transformation, the characteristic polynomial of the linearization matrix $\underline{L}$ is equal to that of its connection matrix $\underline{C_L}$. The stability of a non-linear system can be specified by its connection matrix.

5. Detection of Oscillations in Systems

Oscillation in a non-linear system may be shown as cycle in the state graph of system. The states which are not on any cycles form numbers of branches. If all the states which are in braches can be removed from the state graph, the remaining states can clear show the behaviour of oscillations. This may be realized by cutting off all the states at the top of branches. These states have no predecessors and are not on any cycles. After cutting these top states, a new graph can be obtained which may have some new top states. Continue cutting the new top states, until all the states in every braches are cut off. Finally, only the states on cycles are left. This cutting can be realized in terms of the connection matrix.

From the connection matrix of systems, it is easy to find that every column of the matrix has only a single "1", but some rows have no "1". This rows represent the states which are on the tops of some branches. This rows and the corresponding columns can be deleted from the matrix without any effect on determining the oscillations. Continue the above procedure until every row of remaining matrix has a single "1". Finally, only the states which form pure cycles are left. For the above example,

0	0	1	0	0	0
1	1	0	0	0	0
2	0	0	0	0	1
3	0	0	0	0	0
−1	0	0	1	1	0

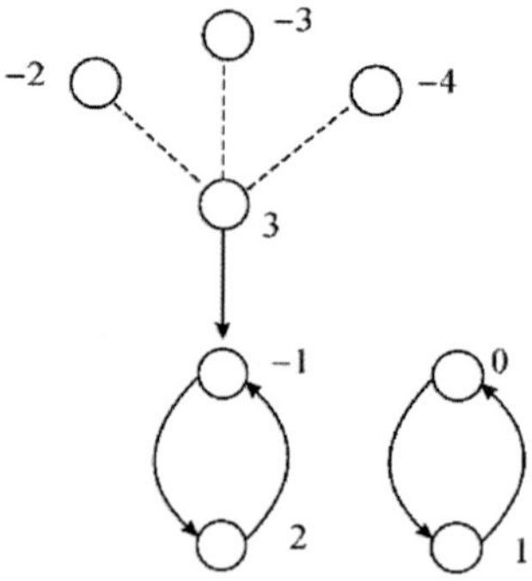

There are no "1" in the rows −4, −3 and −2. Delete the rows and the corresponding columns −4, −3 and −2 from the connection matrix. A 5×5 matrix is left as shown in the above Figure.

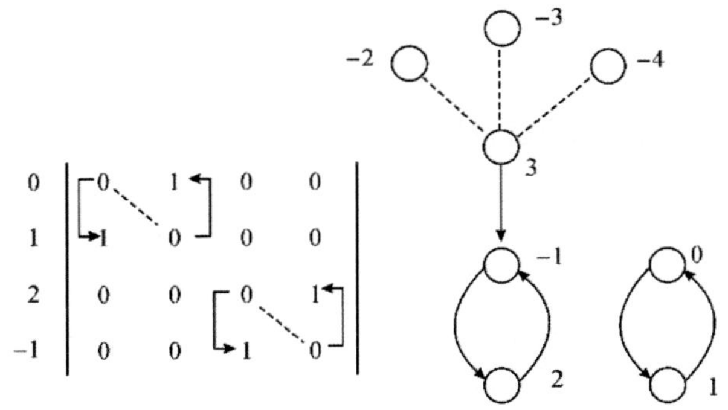

Check the new matrix of 5×5, there is no "1" in row 3. Delete the row 3 and the column 3. A 4×4 matrix remains. Every row has a "1" in this matrix. The final matrix can be represented as a block-diagonal matrix with 2×2 submatrices. The number of oscillation is equal to the number 2 of blocks. The dimension number of each submatrix denotes the period of oscillation. From the state values on cycles, the amplitudes of two oscillations are respectively 0 to 1 and -1 to 2 with period of 2.

The above method is called the connection matrix method and is easily programmed on computers. The connection matrix is a matrix of $2^M\times2^M$. However, it has only 2^M ones. Therefore, only 2^M memory locations are required. It saves the size of memory.

References

[1] Zhang Y. Z. and Rayner P. J. W. , "On Minimization of Reed-Muller Polynomials with fixed polarity", IEE Proceedings, Part E, Computers and Digital Techniques, Vol. 131, No. 5, 1984, pp. 176-186.

[2] Zhang Y. Z. , "Galois Field Representation of Finite state machines", Technical Research Reports of Cambridge University Engineering Department, CUED/B-Elect/TR70, February 1984, pp. 1-44.

Fast Implementation of Recursive DFTs*

Y. Z. Zhang and Y. F. Yao

Changcheng Institute of Measurement and Metrology, China

Abstract: A fast implementation of recursive DFTs is presented. It only needs $(N-1)/2$ real multiplications to compute all N frequency components. A factor R_T is introduced. If the ratio T_m/T_a of multiplier's and adder's period is greater than R_T, this scheme is faster than FFT. The error of systems is studied. A parallel scheme of adders is proposed. This scheme is much faster than the usual serial adder. A scheme for fast re-ordering the input data is proposed. This increases the reordering speed and saves the memory size.

1. Introduction

DFT is a central operation in digital signal processing. It can transform signal from time domain to frequency domain and vice versa. DFT is widely used to analyse and estimate spectrum of signals, to design and implement finite impulse response digital filters, and to compute convolution and correlation functions etc.

The direct evaluation of an N-point DFT needs about $4N^2$ real multiplications and additions. FFT for computing DFT of length $N=2^M$ can reduce the number of real multiplications to about $2N\log_2 N$. FFT requires to store or compute a large number of coefficients. This increases the memory size and the hardware complexity.

[1-2] proposed a recursive algorithm of DFTs with prime length. This algorithm uses looking-up table for reordering input data, therefore, it needs a lot of memory. The speed of adders is still a main problem for increasing the algorithm speed.

The principle and hardware scheme of Fast Recursive DFTs (FRDFT) is described in next section. Then the complexity of systems is studied. The error and S/N ratio of systems are presented. A fast adder with parallel structure is proposed. A scheme for fast reordering the inputs is presented.

---

* Published in IEEE Proceedings on ASSP, 27. D6. 3, ICCASSP'89, May 1989: 1071-1074.

Y. Zhang, *Footprints in Cambridge and Aviation Industries of China*,
https://doi.org/10.1007/978-981-16-3176-4_9

2. Principle of Fast Recursive DFTs

If the transform length $N=p$ is a prime, then the DFT may be computed by the following recursive form[1-2]

$$y_k(n)=W_P^D y_k(n-1)+u_k(n) \tag{1}$$
$$y_k(0)=u_k(0)$$
$$k=1,2,\cdots,(p-1)$$
$$n=0,1,\cdots,(p-1)$$

where

$$u_k(p-1-n)=x(\langle nk^{-1}D\rangle_p) \tag{2}$$

is a permutation of input sequences; D is any one of integers $(1\leqslant D<p)$; k^{-1} is an inverse element of k in GF(p); $W_P^D=\exp(-j2\pi D/p)$.

The DFT output is as follows

$$X(0)=\sum_{n=0}^{p-1}x(n) \tag{3}$$
$$X(k)=y_k(p-1),\ k=1,2,\cdots,(p-1)$$

The recursive equation (1) may be viewed as an IIR digital filter. Its z-transform function is as follows

$$H(z)=\frac{1}{1-W_P^D z^{-1}}=\frac{1-\cos(2\pi D/p)z^{-1}-j\sin(2\pi D/p)z^{-1}}{1-2\cos(2\pi D/p)z^{-1}+z^{-2}} \tag{4}$$

The above equation may be considered as a second order IIR digtal filter with real coefficients as shown in Fig. 1.

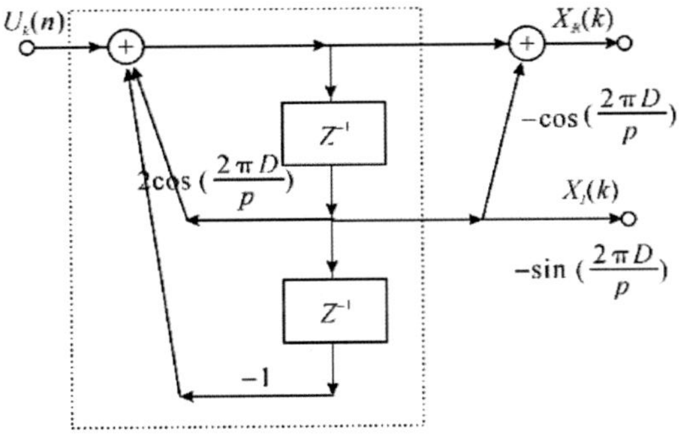

Fig. 1 2nd order recursive form of DFTs

In order to compute a frequency output, the section within the dotted line in Fig. 1 has to do $(p-1)$ real multiplications with the constant $2\cos(2\pi D/p)$, and the section beyond the dotted line only needs to do 2 real multiplications for computing the two frequency components $X_R(k)$ and $X_T(k)$. The coefficient $2\cos(2\pi D/p)$ is independent of the frequency index k. Only one coefficient can be used to compute all p frequency components. Because the figure D can be any one of the set $(1,2,\cdots,(p-1))$, the proper choice of D may make the coefficient have the approximate form of $\pm 2^{-m}$, then the multiplication with the constant $2\cos(2\pi D/p)$ can be replaced by shifting $(m-1)$

steps. The shift is much faster than multiplication. This implementation possesses a very high speed. Some coefficients with the form $\pm 2^{-m}$ are shown in Table 1.

Table 1 Some coefficients of FRDFTs

p	D	$\cos(2\pi D/p)$	$\pm 2^{-m} \pm e$	S/N (db)
3	1	−0.5	$-2^{-1}+0$	
7	2	−0.222521	$-2^{-2}+.8793\times 2^{-5}$	19
31	9	−0.250653	$-2^{-2}-.6682\times 2^{-10}$	45
127	33	−0.061803	$-2^{-4}+.7138\times 2^{-10}$	39
251	64	−0.031286	$-2^{-5}-.5842\times 2^{-14}$	61
503	127	−0.015614	$-2^{-6}+.7443\times 2^{-16}$	68
1063	223	0.250006	$-2^{-2}+.8430\times 2^{-17}$	70
2011	504	−0.003906	$-2^{-8}+.7858\times 2^{-20}$	86

3. Complexity

In Fig. 1, the multiplication of $\cos(\frac{2\pi D}{p})$ for computing $X_R(k)$ can be also replaced by shifting m steps. Only one multiplication with $\sin(2\pi D/p)$ is required to compute $X_I(k)$. Fig. 1 shows that $(p-1)$ recursive operations need $(2p-3)$ real additions. An extra real addition is required to compute $X_R(k)$. To sum up, $(2p-2)$ real additions and one real multiplication are needed for computing a pair of frequency components $X_R(k)$ and $X_I(k)$. For a real input, the DFT output has the following conjugate relationship

$$X(p-k)=X*(k), k=1,2,\cdots,(p-1)/2 \tag{5}$$

So only half of the frequency components are required to compute. The total real addition number A and multiplication number M for computing all p frequency components (including $(p-1)$ additions for $X(0)$) are

$$A=(2p-2)(p-1)/2+(p-1)=p(p-1)$$
$$M=(p-1)/2$$

For complex inputs (see Fig. 2), the two parallel filters are used. Only $2(p-1)$ additions are increased for the combination operations. So

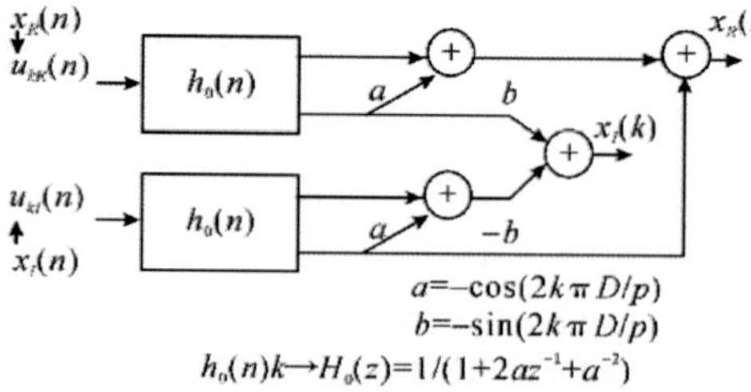

Fig. 2 Parallel scheme of complex FRDFT

$$A = p(p-1) + 2(p-1) = p^2 + p - 2 \qquad (6)$$
$$M = (p-1)/2$$

The real addition number A_F and real multiplication number M_F for FFT are about

$$A_F = 3N * \log_2 N \qquad (7)$$
$$M_F = 2N * \log_2 N$$

It is obviously that the multiplication number of FRDFT is less than that of FFT, but the addition number of FRDFT is more than that of FFT. The speed of algorithm depends on the speed of adder & multiplier. In general, the period of adder is less than that of multiplier. The shifting is fastest among the 3 operations. Ignore the shifting time, and consider the ratio of multiplier's period T_m to adder's period T_a, a criterion factor R_T is introduced by

$$R_T = \frac{p^2 + p - 2 - 3p * \log_2 p}{2p * \log_2 p - (p-1)/2} \qquad (8)$$

If $T_m/T_a R_T$, then the FRDFT is faster than FFT. Some R_T are shown in Table 2.

Table 2 Some criterion factor R_T

p	7	31	127	251	503	1063
R_T	0.13	1.82	7.94	14.8	27.3	52.7

4. Error and *S/N* Ratio of Systems

Because the coefficient $\cos(2\pi D/p)$ in recursive loop is replaced by 2^{-m}, some phase and amplitude errors are introduced in the DFT output. If coefficient error is

$$e = \cos\left(\frac{2\pi D}{p}\right) - 2^{-m} = \cos(Q_0) - \mathrm{Cos}(Q_0 + \Delta Q)$$

where $Q_0 = 2\pi D/p = Dq_0$, $Q = \cos^{-1}(2^{-m}) = Q_0 + \Delta Q$, then from the above 2 equations, it follows

$$\Delta Q \doteq e/\sin(Q_0)$$

The phase error Δq between 2 neighbour frequency components is as follows

$$\Delta q \doteq e/D\sin(Q_0) \qquad (9)$$

The maximum phase error $p\Delta q$ must satisfy the following condition

$$p\Delta q < q_0, q_0 = 2\pi/p$$

From eqn. (9), the necessary condition for the prime length DFTs is as follows

$$e < \frac{2\pi D}{p^2}\sin(Q_0) \qquad (10)$$

From Fig. 1, the recursive section within dotted line may be expressed as follows

$$w(n) = k\,w(n-1) - w(n-2) + u_k(n) \qquad (11)$$

where $k = 2^{-(m-1)} = k_0 + \Delta k$

$$k_0 = 2\cos(2\pi D/p), \Delta k = 2e$$

Consider the amplitude error caused by the coefficient error, let

$$w(n) = w_0(n) + E(n) \qquad (12)$$

where $w_0(n)$ is determined by error-free equation

$$w_0(n)=k_0 w_0(n-1)-w_0(n-2)+u_k(n) \tag{13}$$

$E(n)$ is the error of signal amplitudes. From eqns. (11), (12) and (13), neglecting second-order terms, $E(n)$ can be obtained by

$$E(n)=k_0 E(n-1)-E(n-2)+\Delta k w_0(n-1) \tag{14}$$

This is a 2nd order IIR filter with 2 poles and single zero. The RNS value of amplitude errors is the RMS output of this filter for a unity input. The impulse response of the filter (14) is

$$h(n)=\Delta k\,\frac{\sin(nQ_0)}{\sin(Q_0)} \tag{15}$$

If noise signal is uncorrelated, the output power of the filter for unity input is

$$\frac{E^2}{w^2}=\sum_{n=1}^{p}|h(n)|^2=\frac{p(\Delta k)^2}{2(\sin Q_0)^2} \tag{16}$$

The S/N ratio (Table 1) is obtained by

$$S/N=10\log(\sin^2 Q_0/2pe^2) \tag{17}$$

5. Parallel Fast Adder

The number of multiplication of FRDFTs is minimum, but its number of addition is rather large. The main problem for increasing its speed is how to decrease the tine of adders.

A general adder usually employed serial carry mode, is the higher bit begins to work after the lower bit finished its addition and generated a carry. This needs a long waiting time. For example, a 16-bit adder can consist of four 4-bit adders of 10 ns in serial mode. The period of the 16-bit adder is more than 40 ns. A fast parallel adder is presented in Fig. 3.

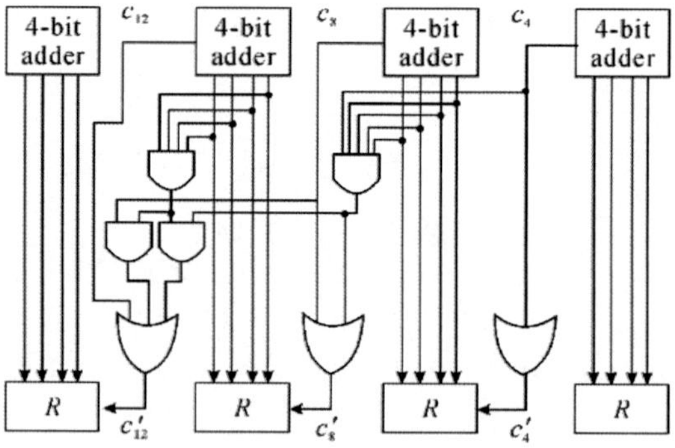

Fig. 3 Principal of fast parallel adder

In Fig. 3, the four 4-bit adders make the local addition at the same time without considering the carry from the lower stage. The carry effect on all the higher stages is considered then. The carry can be realized in terms of a combination circuit and binary counter as shown in Fig. 3. It is named as "parallel adder". A 16-bit adder can be "parallelly" implemented in terms of four 4-bit adders of 10 ns. The total time of the 16-bit addition is about 17 ns. It is much faster than the serial adder.

6. Fast Reordering of Input Data

FRDFT requires to reorder the input data. In general, the reorder of input data can use looking-up table or arithmetic method. The looking-up table of FRDFT needs about $(p-1)^2/2$ memory elements. For a sequence of length 1k, the table needs about 500k memory elements. It is rather large. The direct computation method spends very long time.

Eqn. (2) shows that the reordering of the input sequence has to do the computation of mod(p). In general, the module arithmatic spends rather long time. However, if the module $p=2^M-1$ is a Mersenne prime (3,7,31,127,…), the modulo-p computation can be implemented very fast.

The modulo-$N=2^M$ can be easily realized in terms of overflowing. The modulo (2^M-1) may be implemented by means of 1′ complement arithmetic.

Let $\overline{A}$ *and* $\overline{B}$ denote the 1′ complements of the number A and B, then

$$\overline{A}=2^M-1-A,\ \overline{B}=2^M-1-B$$

For example

No.	Binary	1′ Complement	
0	000	111	$\overline{0}$
1	001	110	$\overline{1}$
2	010	101	$\overline{2}$
3	011	100	$\overline{3}$
4	100	011	$\overline{4}$
5	101	010	$\overline{5}$
6	110	001	$\overline{6}$
7	111	000	$\overline{7}$

The sum of two 1′ complements can be obtained as follows

a) If $A+B\geq 2^M-1$(or $\overline{A}+\overline{B}<2^M$), then $\overline{A}+\overline{B}=2^M-1-((A+B)-(2^M-1))=\overline{\langle A+B\rangle_p}$, where $\langle\rangle_p$ denotes modulo $(p=2^M-1)$. It means that the sum of these 2 1′ complements equals a 1′ complement of their sum modulo(p). For example,

$$\overline{4}+\overline{5}=(0\ 1\ 1)+(0\ 1\ 0)=(1\ 0\ 1)=\overline{2}$$

$$(4+5)\ \mathrm{mod}\ (7)=2$$

b) if $A+B<2^M-1$(or $\overline{A}+\overline{B}\geq 2^M$), then $\overline{A}+\overline{B}=2^M+(2^M-1-(A+B))-1=2^M+\overline{(A+B)}-1$, where 2^M means overflow. The above equation can be written as

$$\overline{A}+\overline{B}+1=2^M+\overline{(A+B)}$$

It means that if $(\overline{A}+\overline{B})$ is overflowing, then the remainder plus one equals to the 1′ complement of the sum of two numbers.

For example,

$$(\overline{3}+\overline{2})+1=(1\ 0\ 0)+(1\ 0\ 1)+(0\ 0\ 1)=(10\ 0\ 0)+(0\ 1\ 0)=(1\ 0\ 0\ 0)+\overline{5}$$

Because the condition for plus one is overflow, the overflow can be fed to the sum as follows

$$
\begin{array}{cccc|c}
0 & 1 & 0 & 0 & \bar{3} \\
0 & 1 & 0 & 1 & \bar{2} \\
\hline
1 & 0 & 0 & 1 & \\
 & & & 1 & \\
\hline
0 & 0 & 1 & 0 & \bar{5}
\end{array}
$$

This operation can be implemented in terms of a loop feedback counter as shown in Fig. 4. The reordering address of input data can be obtained by computing $(nk^{-1}D)$ mod (p). Because $n=0, 1, \cdots, (p-1)$, the succeeding address can be computed by adding $k^{-1}D$ to previous address and mod(p). The scheme for fast reordering inputs is shown in Fig. 4. When the overflow is happened, the figure 1 is added to the sum. If overflow is not occurred, the loop counter doesn't feed back any number.

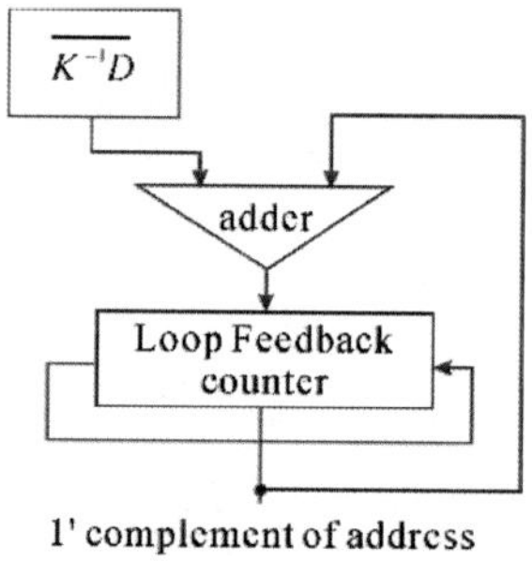

Fig. 4 Scheme of fast reordering

7. Conclusion

The advantage of FRDFT is that it has a very simple structure and needs only $(N-1)/2$ real multiplications for computing all N frequency components. If $T_m/T_a R_T$, it is faster than FFT. The error and S/N ratio is given. A fast parallel adder and a fast reordering scheme have been presented.

References

[1] Y. Z. Zhang, "Digital Signal Processing systems with Finite State Machine Realisation", Ph. D. Dissertation, Cambridge University, England, 1984.

[2] T. E. Curts and J. D. Wickenden, "Hardware-based Fourier Transform: algorithm and Architectures", IEE Proc. Pt. F, Vol. 130, No. 4, 1983, pp. 423-432.

[3] Y. Z. Zhang, "Optimum Recursive Computation of Prime Length DFTs", proceedings of 20th IREE Conference, Melbourne, Australia, October 1985, pp. 583-586.

Finite State Machine Realization of IIR Digital Filters*

Yan Zhong Zhang and Yu Fen Yao

Ministry of Aero-Space Industry, Beijing, China

Abstract: A method is proposed for representing an IIR digital filter as a Finite State Machine (FSM). This representation only has a maximum error of 1/2 bit from the final result of infinite accuracy arithmetic, and has not any increase in errors during the processing period. The realization avoids the effects of coefficient quantization and multiplication arithmetic errors caused by finite word length on the performance of DSP system. A graph theory method is presented for removing the oscillations from the system with a minimal increase in noise and without any increase in complexity of systems.

1. Introduction

A digital filter may be specified by a difference equation which is discrete in time and continuous in amplitude, or by a z-transform function as follows:

$$y(n) = \sum_{i=1}^{m} k_i y(n-i) + \sum_{i=0}^{m} L_i x(n-i) \tag{1}$$

$$H(z) = \frac{\sum_{i=1}^{m} L_i z^{-i}}{1 - \sum_{i=1}^{m} K_i z^{-i}} \tag{2}$$

where $x(n-i)$ and $y(n-i)$ are respectively the input and output at the $(n-i)$th clock instant which have infinite precision in amplitude, z is a complex variable, K_i and L_i are infinite precision coefficients.

The usual method of realising a digital filter is to approximate the infinite precision amplitude arithmetic by means of finite wordlength binary arithmetic. The roundoff and truncation errors are introduced. The error behaviour is dependent on the structures and properties of systems.

In some feedback systems, such as Infinite Impulse Response (IIR) filters, this approximation may lead to Limit Cycle Oscillation (LCO)[1].

Finite State Machine (FSM) is a finite arithmetic structure of sequential systems. It is free of amplitude arithmetic errors and can be used realize a finite storage system. Zhang[2] has proposed the application of FSMs to design a digital signal processing system. This paper is concerned with the

---

* Published in IREE Convention on Electronics, Sept. 1989: 387-390.

Y. Zhang, *Footprints in Cambridge and Aviation Industries of China*,
https://doi.org/10.1007/978-981-16-3176-4_10

FSM realization of IIR digital filters. This realization can successfully overcome the disadvantage of errors, noise and Limit Cycle Oscillation in systems.

The FSM is first introduced in Section 2, Section 3 describes the FSM representation of IIR digital filters. The optimum removal of Limit Cycle Oscillations is presented in Section 4.

2. Finite State Machine

Finite State Machine is a mathematical model of sequential systems. It comprises a finite input set $U \in (U_0, U_1, \cdots, U_{N_{u-1}})$, a finite output set $G \in (G_0, G_1, \cdots, G_{N_{g-1}})$, a finite state set $S \in (S_0, S_1, \cdots, S_{N_{s-1}})$, and two mappings $\underline{f}$ and $\underline{g}$ defined by

$$\underline{s}(n+1) = \underline{f}[\underline{s}(n), \underline{u}(n)] \tag{3}$$

$$\underline{g}(n) = \underline{g}[\underline{s}(n), \underline{u}(n)] \tag{4}$$

where $\underline{u}(n) \in U$, $\underline{g}(n) \in G$, and $\underline{s}(n) \in S$ are respectively the input, output and present state of the systems at the nth clock instant. $s(n+1) \in S$ denotes the state at the state at the $(n+1)$th clock instant and is named as the next state. The mapping $\underline{f}$ maps the present state $\underline{s}(n)$ and the input $\underline{u}(n)$ into the next state $\underline{s}(n+1)$, and is called as the next state mapping. $\underline{g}$ is referred to as the output mapping which produces the present output $\underline{g}(n)$, from the present state $s(n)$ and input $\underline{u}(n)$. This model is called the Mealy Model.

Finite state machine may be described by a state table or by a state transition graph. The state table has N_s rows and N_u columns. Each row corresponds to each member of the finite state set S, and each column corresponds to each member of the finite input set U. The entries of table denote the next states $s(n+1)$ and the machine's output $g(n)$.

The state transition graph is a directed graph which has N_s vertices and $N_s \times N_u$ arcs. Each vertex represents a number of the finite state set S and the arcs with arrays represent the state transitions from the present states to the next states with the corresponding inputs U_i and outputs G_i. The state transition graph can be used to show the dynamic behaviour and stability of machines.

The advantages of using finite state machines are:

The states, inputs and outputs of FSMs are elements of finite sets. The next state mapping and the output mapping may be defined over the finite sets. These mappings are infinite accuracy operations which are free of errors. Because the next state of an FSM depends on not only the present input but also the past states, FSM may simulate any finite storage system.

3. FSM Representation of IIR Filters

Finite state machine can be used to represent digital filters. If the continuous algebraic equation description of systems is given, one method of approach is to evaluate the system output for all possible combinations of finite inputs and system states using the continuous equation. The output calculated in this manner will not, of course, be in the same finite set as the input and state so that some form of approximation is needed to achieve this. However, the system output values determined in this way can be within 1/2 bit over the representable range of outputs. For example, consider a first order all pole digital filter described by the following difference equation

$$y(n)=K_1 y(n-1)+x(n) \tag{5}$$

where the present output is a function of the past output $y(n-1)$ and the present input $x(n)$. Compare eqn. (4) and (5), and choose the wordlength, give the quantization method assign to each output quantization level $\bar{y}(n)$ a member of the state set S, to each input quantization level $\bar{x}(n)$ a member of the input set U of a machine, then the 1st order filter may be represented as a finite state machine, and the next state may be obtained by quantizing the infinite accuracy computation result of eqn. (5)

$$\bar{y}(n)=Q[K_1\bar{y}(n-1)+\bar{x}(n)] \tag{6}$$

where Q denotes quantization operation.

If the coefficient $K_1=-0.8$, input $x(n)$ is 1.0 in eqn. (5), let output $\bar{y}(n)$ be represented in 3-bit word, then the state table may be set up by rounding the following equation for every possible $\bar{y}(n-1)$

$$y(n)=-0{,}8\ y(n-1)+1.0$$

$$\bar{y}(n)=Q[-0.8\ \bar{y}(n-1)+1.0]$$

Its state table for input $\bar{x}(n)=1.0$ is shown in the following table.

Table 1 State table of example

$\bar{y}(n-1)$	$y(n)$	$\bar{y}(n)$
0	1.0	1
1	0.2	0
2	−0.6	−1
3	−1.4	−1
−4	4.2	3
−3	3.4	3
−2	2.6	3
−1	1.8	2

This state table can be realized in terms of logic circuits. The state graph of the above machine is shown in Fig. 1.

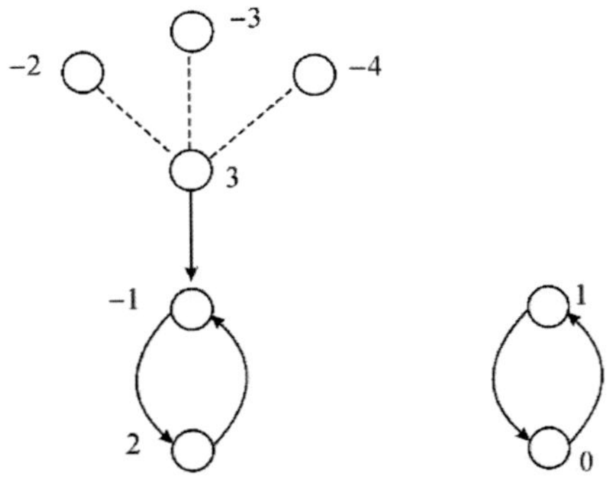

Fig. 1 State graph of example

This table may be realized by a one-dimensional FSM. The machine outputs for all combinations of inputs and past states have at most 1/2 bit error from the filter outputs with infinite accuracy. In this simple example, the obtained outputs are no different from those which would be obtained from the arithmetic realization. However in a system containing more than one multiplier the results obtained from the above method would more accurately match the results from an infinite precision filter.

For high order filters, from the transfer function of the mth order IIR digital filters, eqn. (2) may be re-written as follows:

$$H(z) = \frac{\sum_{i=0}^{m} L_i z^{-i}}{1 - \sum_{i=1}^{m} K_i z^{-i}} = \frac{-L_m}{K_m} + \frac{\sum_{i=0}^{m-1} (L_i - K_i \frac{L_m}{K_m}) z^{-i}}{1 - \sum_{i=1}^{m} K_i z^{-i}} \tag{7}$$

where $K_0 = -1, K_m \neq 0$.

The above equation may be considered as a parallel connection of two sections: a scaling factor $-L_m/K_m$ and a filter with m poles and $(m-1)$ zeros. The denominator of the second section represents an all-pole filter. Introducing a variable $w(n)$, the output $y(n)$ may be written as follows:

$$w(n) = x(n) + \sum_{i=1}^{m} K_i w(n-i) \tag{8}$$

$$y(n) \frac{-L_m}{K_m} x(n) + \sum_{i=0}^{m-1} (L_i - K_i \frac{L_m}{K_m}) w(n-i) \tag{9}$$

where $K_0 = -1, K_m \neq 0$.

This is a modified canonic form of an IIR filter. The modification is that its output is related to the input $x(n)$, but unrelated to the mth delayed variable $w(n-m)$. Let $\bar{x}(n)$, $\bar{y}(n)$ and $\bar{w}(n)$ denote the quantized discrete amplitude of $x(n)$, $y(n)$ and $w(n)$ respectively. From eqns. (8) and (9), $\bar{w}(n)$ and $\bar{y}(n)$ may be obtained by

$$\bar{w}(n) = Q[\bar{x}(n) + \sum_{i=1}^{m} K_i \bar{w}(n-i)] \tag{10}$$

$$\bar{y}(n) = Q[\frac{-L_m}{K_m} \bar{x}(n) + \sum_{i=0}^{m-1} (L_i - K_i \frac{L_m}{K_m}) \bar{w}(n-i)] \tag{11}$$

where Q denotes the quantization.

The discrete output determined in this way can be within 1/2 bit least significant bit over the representable range. If $\bar{x}(n)$ and $\bar{w}(n-i)$ in eqns. (10) and (11) are all possible values of finite sets U and S respectively, then a state table can be obtained. However, the state here is an m-dimensional vector defined by

$$\underline{s}(n) = | s_0(n), s_1(n), \cdots, s_{m-1}(n) |^{T}$$

where

$$s_0(n) = \bar{w}(n-m+1)$$
$$s_1(n) = \bar{w}(n-m+2)$$
$$\vdots$$
$$s_{m-1}(n) = \bar{w}(n) \tag{12}$$

Eqns. (10) and (11) may be represented as the following two mappings:

$$\underline{s}(n)=\underline{f}[\underline{s}(n-1),u(n)] \tag{13}$$

$$g(n)=g[\underline{s}(n),u(n)] \tag{14}$$

This is a Mealy Machine with an m-dimension next state mapping $\underline{f}$ and a one-dimensional output mapping g.

4. Stable Realization of IIR Filters

Fig. 1 shows that the system is not stable. It can not reach any stable output from an initial state. This behaviour is a constant input Limit Cycle Oscillation. The above example exhibits two limit cycles a limit cycle of period 2 with amplitude-1 to 2, another one of period 2 and amplitude 0 to 1. For a high order filter, limit cycles of period greater than 2 may exists.

Zhang[3] proposed a connection matrix method for determing the limit cycle in any system. The state graph of stable machine is a tree. The root state of trees can be determined and formed by using the methods in paper [3].

If the limit cycles of an IIR filter have been located and the root cycle has been formed, then the limit cycles can be broken and a stable system with a tree graph may be obtained. For the above example if the limit cycle $(-1,2)$ can be broken at point A and the state "2" is connected to the state "0", then a tree graph may be formed. This causes an increase in errors of system representations. The errors from a predecessor to a successor in the FSM representation of IIR filters may be defined by

$$e=[w(n)-s']^2 \tag{15}$$

where $w(n)$ is the infinite accuracy output of the system calculated from the filter eqn. (8); s' denotes the assignment of the given successor state.

The optimum successor of a state which has the minimum error among all the possible states of machines. The next state obtaining from the quantization equation(10) is an optimum successor. The optimum successor of each state of a cycle is also on the cycle, and the error can be calculated by

$$e_0=[w(n)-\overline{w}(n)]^2 \tag{16}$$

where $\overline{w}(n)$ denotes the quantization value of $w(n)$ in eqn. (10).

Breaking of a limit cycle requires reconnection of a state of the cycle to another state which is not on the cycle. A number of states in a machine are not on the cycle. One of them has the minimum error from the optimum successor on the cycle. This one is named a sub-optimum successor of that state on the cycle.

Sub-optimum means that this successor of the state of the cycle has the minimum error among all the successors which are not on the cycle, but its error is only larger than that of the optimum successor on the cycle. For the above example, if the state "2" on the cycle is considered as a predecessor, then all its possible successors and errors are shown in Fig. 2.

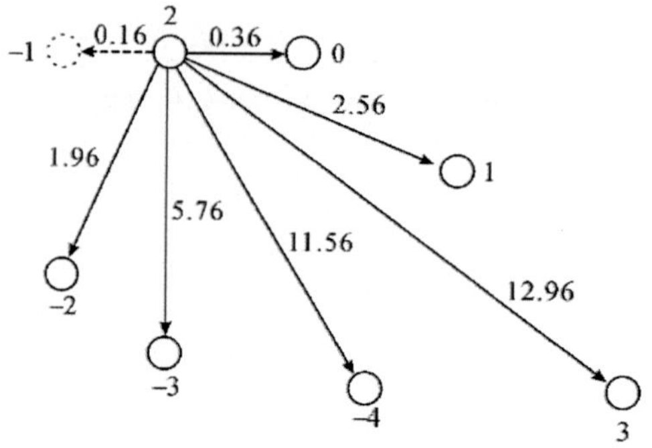

Fig. 2 All the successors of State "2"

In Fig. 2, the error is allocated to each edge. The error values may be calculated in terms of eqn. (15). For example, the system output $w(n)$ from $w(n-1)=2$ is calculated by the difference equation (for $x_c=1.0$) as follows:

$$w(n)=-0.8\,w(n-1)+1.0=-0.6$$

The error e in making the state transition "2" to "1" is:

$$e=(-0.6-1)^2=2.56$$

Fig. 2 shows that the state "−1" is the optimum successor of state "2". It has a minimum error 0.16, but it is on the cycle (−1,2). The state "0" is the sub-optimum successor which has an error 0.36, it is minimal among the successors not on cycle.

If the state "2" is connected to its sub-optimum successor "0", and the cycle is broken at point A as shown in Fig. 3, then the increase in error can be obtained by

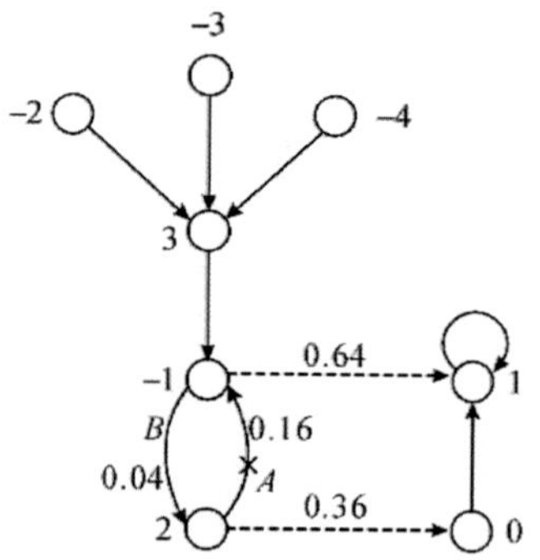

Fig. 3 Optimum break of limit cycles

$$\Delta e=0.36-0.16=0.20$$

The state "−1" is another state on the cycle. Similarly, its sub-optimum successor is the state "1" and with an error of 0.64 (its optimum successor is the state "2", and has error of 0.04). If the state "−1" is connected to its successor "1" and the cycle is broken at B, then an extra error is introduced as follows:

$$\Delta e=0.64-0.04=0.60$$

Compare the increases in errors between two breaks. It is obvious that the break at A is the optimum break of the cycle, only an increase in errors of 0. 20 is added. It is an optimum breaking.

In general, a limit cycle has N states on it. They have N sub-optimum successors. The breaking with minimal increase in noise is optimum breaking.

The above method can remove all the limit cycles with minimum increase in errors A stable machine with a tree graph can be obtained. The removal of limit cycle can be carried out either on the state table or in the polynomial by modifying the coefficient This is unlikely to increase the complexity of system implementations.

5. Conclusion

A method has been proposed for representing IIR filters as FSM. This representation has only 1/2 bit error from the infinite arithmetic results, and avoids the effect of coefficient quantization on pole-positions. As the state table is calculated from infinite accuracy coefficients and eqn. (8), and the state is realized by a finite algorithm structure which is free of errors. A graph method is presented for removing LCO from the systems. This method has a minimum increase in noise and hardly any increase in complexity of hardware implementation.

References

[1] Jackson, L. B., "An analysis of Limit Cycle due to Multiplication rounding in Recursive Digital (sub) Filters", Proceedings of 7th Allerton Conference on Circuits and System Theory, 1969, pp. 69-78.

[2] Zhang, Y. Z., "FSM Realisation of Digital Signal Processing Systems", Ph. D Thesis of Cambridge University, U. K., 1984, pp. 1-240.

[3] Zhang, Y. Z., "Linearization of Mon-Linear Systems Over GF(2^M)", Proceedings of the International Conference on Circuits and Systems, ICCAS-89, Nanjing, July 6-8, 1989.

Minimisation of Exclusive-OR Logic Functions*

ZHANG Yan-zhong
Ministry of Aero-Space Industry, Beijing, China

Abstract: An efficient computer algorithm for minimisation of exclusive-OR logic function is presented. The algorithm is $n/(1+(n-1)2^{-m})$ times faster than fast Reed-Muller transform algorithm for minimizing an exclusive-OR function with n input variables.

Keywords: minimisation; exclusive-OR logic functions

1. Introduction

The basic problem for designing a digital circuit is how to implement the logic function by using the least number of gates. A scheme with the least number of gates has many advantages such as low cast, high speed and great reliability. A digital circuit can be implemented either by means of OR, AND and NOT gates (named the Boolean circuit), or in terms of exclusive-OR, AND and NOT gates (called the exclusive-OR logic circuit). The former can be represented as a Boolean function, the latter can be described in the form of the Reed-Muller polynomial.

Exclusive-OR logic circuits possess the advantages of a smaller number of gates, simpler connections and easier tests in comparison with Boolean circuits. But it has not been widely used in engineering for the minimisation of exclusive-OR logic function is still an open problem.

Many authors[1-7] have been concerned about the minimisation of Reed-Muller polynomials. Saluja and Ong[7] proposed an exhaustive algorithm to obtain all the fixed-polarity modulo-2 expressions. This algorithm needs 2^n matrix multiplications and (2^n-1) permutations of output vectors. Mukhopadhyay and Sclimitz[2] gave the polarity functions of Reed-Muller polynomial coefficients. The best polarity for the minimum Reed-Muller polynomial can be obtained by finding the maximum clique of the polarity-compatibility graph. This graph possesses a very large number of vertices in most applications. Bioul, Davio and Deschamps[3] proposed a minimisation method for mixed-polarity ring-sum expansions with variable number $n \leqslant 5$. Papakonstantinou[6] developed a minimisation algorithm for mixed-polarity modulo-2 sum of product expressions with the number of terms $m<6$. Recently, an efficient computer method for single-output exclusive-OR logic design has been developed by Besslich[15], it does not deal with multiple-output functions. Robinson and Yeh[14]

* Published in Scientia Sinica, Series A, Vol. 33, No. 4, April 1990: 41-49.

Y. Zhang, *Footprints in Cambridge and Aviation Industries of China*,
https://doi.org/10.1007/978-981-16-3176-4_11

proposed a local minimisation procedure for mixed-polarity R-M polynomials using the minimisation polynomial coefficients with fixed polarities, but it does cot guarantee the global minimisation.

An efficient algorithm is presented for the minimisation of single-output R-M polynomials with fixed polarities.

2. Fast Reed-Muller Transformations

Any switching function of n-variables may be represented as a Reed-Muller polynomial with fixed polarity as follows:

$$f(x_0, x_1, \cdots, x_{n-1}) = a_0 \oplus a_1 x_0^* \oplus a_2 x_1^* \oplus a_3 x_0^* x_1^* \oplus \cdots \oplus a_r x_0^* x_1^* \cdots x_{n-1}^* \quad (1)$$

where $\oplus$ denotes modulo-2 addition,

$$a_i \in (0,1), i = 0, 1, 2, \cdots, 2^n - 1$$

x_i^* represents either complemented $\bar{x}_i$ for polarity 1 or uncomplemented x_i for polarity 0, but not both.

If one knows all the possible function outputs $f_0, f_1, \cdots, f_r, r = 2^n - 1$, then the coefficients of the above polynomial may be obtained by[19]

$$\underline{A} = \underline{T}_n \underline{F} \quad (2)$$

where $\underline{A} = (a_0, a_1, \cdots, a_r)^t$, t denotes transpose, $\underline{F} = [f(0,0,\cdots,0), f(0,0,\cdots,1), \cdots, f(1,1,\cdots,1)]^t$ is the function output vector, $\underline{T}_n$ is a $(2^n \times 2^n)$ matrix recursively defined by

$$\underline{T}_n = \begin{vmatrix} \underline{T}_{n-1} & \underline{0} \\ \underline{T}_{n-1} & \underline{T}_{n-1} \end{vmatrix}, \underline{T}_1 = \begin{vmatrix} 1 & 0 \\ 1 & 1 \end{vmatrix}$$

It is named the Reed-Muller transformation matrix.

For example, the 3-variable Reed-Muller polynomial with positive polarity (x_2, x_1, x_0) may be written as

$$f(x_2, x_1, x_0) = a_0 \oplus a_1 x_0 \oplus a_2 x_1 \oplus a_3 x_0 x_1 \oplus a_4 x_2 \oplus a_5 x_0 x_2 \oplus a_6 x_1 x_2 \oplus a_7 x_0 x_1 x_2$$

Its coefficients can be obtained by

$$\begin{vmatrix} a_0 \\ a_1 \\ a_2 \\ a_3 \\ \hdashline a_4 \\ a_5 \\ a_6 \\ a_7 \end{vmatrix} = \begin{vmatrix} 1 & & & & & & & \\ 1 & 1 & & & & & & \\ 1 & 0 & 1 & & & & & \\ 1 & 1 & 1 & 1 & & & & \\ \hdashline 1 & 0 & 0 & 0 & 1 & & & \\ 1 & 1 & 0 & 0 & 1 & 1 & & \\ 1 & 0 & 1 & 0 & 1 & 0 & 1 & \\ 1 & 1 & 1 & 1 & 1 & 1 & 1 & 1 \end{vmatrix} \begin{vmatrix} f(0,0,0) \\ f(0,0,1) \\ f(0,1,0) \\ f(0,1,1) \\ \hdashline f(1,0,0) \\ f(1,0,1) \\ f(1,1,0) \\ f(1,1,1) \end{vmatrix}$$

The above example shows that the computation of a three-variable Reed-Muller transformation needs $(3^3 - 2^3)$ modulo-2 additions. In general, a Reed-Muller transformation with n variables requires $(3^n - 2^n)$ modulo-2 additions. However, the Reed-Muller transformation matrix $\underline{T}_n$ may be factored into Kronecker products as follows:

$$\underline{T}_n = \underline{T}_1 \otimes \underline{T}_1 \otimes \cdots \otimes \underline{T}_1 \quad (3)$$

where $\otimes$ denotes Kronecker products. This factorization may considerably reduce the number of modulo-2 additions for the coefficient computation.

For example, the 3-variable Reed-Muller transformation matrix may be factored into

$$T_3 = \left|\begin{array}{cccc:cccc} 1 &&&&&&& \\ 1 & 1 &&&&&& \\ && 1 &&&&& \\ && 1 & 1 &&&& \\ \hdashline &&&& 1 &&& \\ &&&& 1 & 1 && \\ &&&&&& 1 & \\ &&&&&& 1 & 1 \end{array}\right| \left|\begin{array}{cccc:cccc} 1 &&&&&&& \\ & 1 &&&&&& \\ 1 && 1 &&&&& \\ & 1 && 1 &&&& \\ \hdashline &&&& 1 &&& \\ &&&&& 1 && \\ &&&& 1 && 1 & \\ &&&&& 1 && 1 \end{array}\right| \left|\begin{array}{cccc:cccc} 1 &&&&&&& \\ & 1 &&&&&& \\ && 1 &&&&& \\ &&& 1 &&&& \\ \hdashline 1 &&&& 1 &&& \\ & 1 &&&& 1 && \\ && 1 &&&& 1 & \\ &&& 1 &&&& 1 \end{array}\right|$$

It only needs $3 * 2^2 = 12$ modulo-2 additions. The flow graph of the above example is shown in Fig. 1.

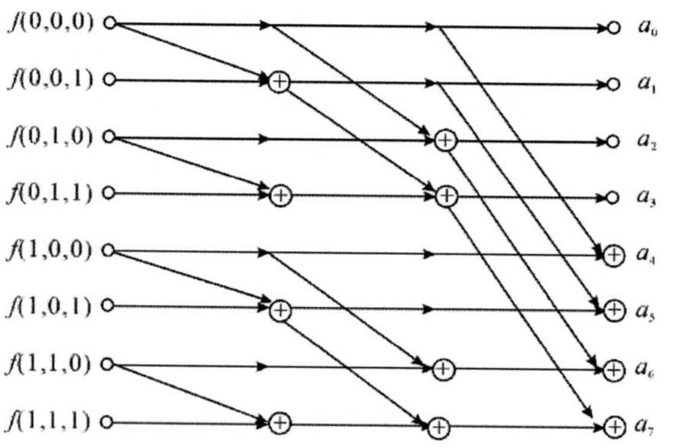

Fig. 1 Flow graph of the fast Reed-Muller transformation

In general, using the Kronecker factorisation, the number of modulo-2 additions for computing an n-variable Reed-Muller transformation will be reduced to

$$n * 2^{n-1} \tag{4}$$

This algorithm is named the fast Reed-Muller transformation (FRMT). It is $2(1.5^n - 1)/n$ times faster than the Reed-Muller transformation. The comparison of two algorithms is shown in Fig. 2.

There exist 2^n different polarities for n-variables. The minimisation of Reed-Muller polynomials is the process of finding the polarity from 2^n possible polarities such that the polynomial possesses the minimum number of modulo-2 additions.

Saluja and Ong[7] proposed that the function output vector for a new polarity is a permutation of another polarity function output vector. All the 2^n sets of polynomial coefficients may be obtained in terms of multiplying the successive modified output vectors by a Reed-Muller transformation matrix. One or more among them are minimum. If the FRMT algorithm is used, then the total number of modulo-2 additions for this exhaustive algorithm is

$$2^n \cdot n \cdot 2^{n-1} \tag{5}$$

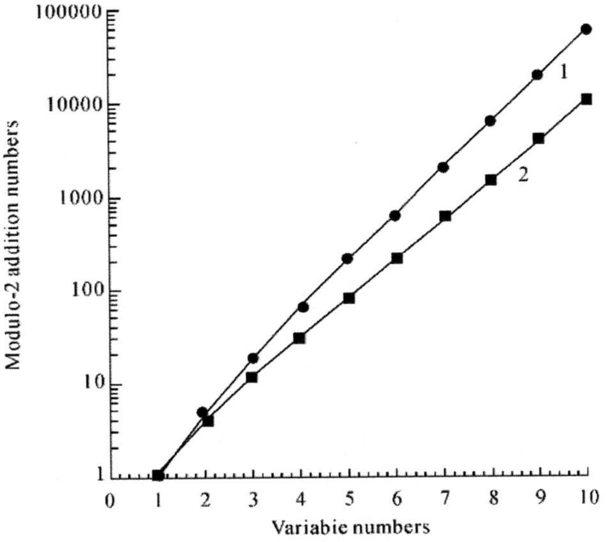

Fig. 2 Comparison between R-M and the fast R-M transformations

1—R-M transformation; 2—fast R-M transformation

3. Adjacent Polarity Mapping

It will now be shown that a set of polynomial coefficients with a new polarity may be obtained directly from another set of polynomial coefficients without permuting the output function vectors and computing Eq. (2). We begin with the adjacent polarity polynomials. The polarity of ($x_{n-1}^{*} \cdots x_{k}^{\bar{*}} \cdots x_{0}^{*}$) is said to be adjacent to the polarity of ($x_{n-1}^{*} \cdots x_{k}^{*} \cdots x_{0}^{*}$), as only one variable x_k has different polarities between the two sets of variables.

Theorem 1

The coefficients of a Reed-Muller polynomial with n-variables may be obtained directly from the coefficients of its adjacent polarity polynomial in terms of the map with 2^{n-1} modulo-2 additions. There exist n adjacent polarity maps for an n-variable polynomial.

Proof

From Eq. (1), the Reed-Muller polynomial with n-variables may be rewritten as follows:

$$f(x_0^*, x_1^*, \cdots, x_{n-1}^*) = a_{0\cdots00} \oplus a_{0\cdots01} x_0^* \oplus a_{0\cdots10} x_1^* \oplus a_{0\cdots11} x_0^* x_1^* \oplus \cdots \oplus a_{1\cdots11} x_0^* x_1^* \cdots x_{n-1}^* \quad (6)$$

Eq. (6) may be divided into two parts according to whether it contains x_0 or not.

$$f(x_0^*, x_1^*, \cdots, x_{n-1}^*) = f_0(x_1^*, x_2^*, \cdots, x_{n-1}^*) \oplus x_0^* f_1(x_1^*, x_2^*, \cdots, x_{n-1}^*) \quad (7)$$

where f_0 and f_1 are not functions of the variable x_0^*.

$$f_0(x_1^*, x_2^*, \cdots, x_{n-1}^*) = a_{0\cdots00} \oplus a_{0\cdots10} x_1^* \oplus \cdots \oplus a_{1\cdots10} x_1^* x_2^* \cdots x_{n-1}^*$$

$$f_1(x_1^*, x_2^*, \cdots, x_{n-1}^*) = a_{0\cdots01} \oplus a_{0\cdots11} x_1^* \oplus \cdots \oplus a_{1\cdots11} x_1^* x_1^* \cdots x_{n-1}^* \quad (8)$$

If one changes the polarity of x_0^* into $x_0^{\bar{*}}$ and uses

$$x_0^{\bar{*}} = 1 \oplus x_0^* \quad (9)$$

Eq. (7) may be changed into

$$f'(\overline{x_0^*}, x_1^*, \cdots, x_{n-1}^*) = f'_0(x_1^*, x_2^*, \cdots, x_{n-1}^*) \oplus \overline{x_0^*} f_1(x_1^*, x_2^*, \cdots, x_{n-1}^*) \quad (10)$$

where

$$f'_0(x_1^*, x_2^*, \cdots, x_{n-1}^*) = (a_{0\cdots 00} \oplus a_{0\cdots 01}) \oplus (a_{0\cdots 10} a_{0\cdots 11}) x_1^* \oplus \cdots \oplus (a_{1\cdots 10} \oplus a_{1\cdots 11}) x_1^* x_2^* \cdots x_{n-1}^* \quad (11)$$

$f_1(x_1^*, x_2^*, \cdots, x_{n-1}^*)$ is not changed. Here, $f'(x_0^*, x_1^*, \cdots, x_{n-1}^*)$ is the adjacent polarity polynomial of $f_1(x_0^*, x_1^*, \cdots, x_{n-1}^*)$ and can be represented as follows:

$$f'(\overline{x_0^*}, x_1^*, \cdots, x_{n-1}^*) = a'_{0\cdots 00} \oplus a'_{0\cdots 01} \overline{x_0^*} \oplus a'_{0\cdots 10} x_1^* \oplus a'_{0\cdots 11} \overline{x_0^*} x_1^* \oplus \cdots \oplus a'_{1\cdots 11} \overline{x_0^*} x_1^* \cdots x_{n-1}^* \quad (12)$$

Compare Eqs. (10), (11) and (12), the coefficient of $f'(\overline{x_0^*}, x_1^*, \cdots, x_{n-1}^*)$ is as follows:

$$a'_{xx\cdots x1} = a_{xx\cdots x1}$$

$$a'_{xx\cdots x0} = a_{xx\cdots x0} \oplus a_{xx\cdots x1} \quad (13)$$

where $XX\cdots X$ denotes $(00\cdots 0), (00\cdots 1), \cdots, (11\cdots 1)$.

This is the map of Reed-Muller polynomial coefficients from the polarity into the polarity $(x_0^* x_1^* \cdots x_{n-1}^*)$ into the polarity $(\overline{x_0^*} x_1^* \cdots x_{n-1}^*)$. The map needs 2^{n-1} modulo-2 additions.

Similarly the polarity x_1^* may be changed. So, there are n adjacent polarity maps for n variables.

For example, the coefficients of 3-variable R-M polynomials can be mapped from the polarity $(x_2^* x_1^* x_0^*)$ into the polarity $(x_2^* x_1^* \overline{x_0^*})$ by the following adjacent polarity map:

$$\begin{vmatrix} a_0(1) \\ a_1(1) \\ a_2(1) \\ a_3(1) \\ \hline a_4(1) \\ a_5(1) \\ a_6(1) \\ a_7(1) \end{vmatrix} = \begin{vmatrix} 1 & 1 & & & & & & \\ & 1 & & & & & & \\ & & 1 & 1 & & & & \\ & & & 1 & & & & \\ \hline & & & & 1 & & & \\ & & & & & 1 & 1 & \\ & & & & & & 1 & 1 \\ & & & & & & & 1 \end{vmatrix} \begin{vmatrix} a_0(0) \\ a_1(0) \\ a_2(0) \\ a_3(0) \\ \hline a_4(0) \\ a_5(0) \\ a_6(0) \\ a_7(0) \end{vmatrix}$$

where $a_i(1)$ denotes the coefficient with the polarity $(x_2^* x_1^* \overline{x_0^*})$, $a_i(0)$ denotes the coefficient with the polarity $(x_2^* x_1^* x_0^*)$, $i = 0, 1, \cdots, 7$.

Similarly, the coefficients with the polarities $(x_2^* \overline{x_1^*} x_0^*)$ and $(\overline{x_2^*} x_1^* x_0^*)$ can be mapped from the coefficient with the polarity $(x_2^* x_1^* x_0^*)$ by the following adjacent maps respectively:

$$\begin{vmatrix} a_0(2) \\ a_1(2) \\ a_2(2) \\ a_3(2) \\ \hline a_4(2) \\ a_5(2) \\ a_6(2) \\ a_7(2) \end{vmatrix} = \begin{vmatrix} 1 & & 1 & & & & & \\ & 1 & & 1 & & & & \\ & & 1 & & & & & \\ & & & 1 & & & & \\ \hline & & & & 1 & & 1 & \\ & & & & & 1 & & 1 \\ & & & & & & 1 & \\ & & & & & & & 1 \end{vmatrix} \begin{vmatrix} a_0(0) \\ a_1(0) \\ a_2(0) \\ a_3(0) \\ \hline a_4(0) \\ a_5(0) \\ a_6(0) \\ a_7(0) \end{vmatrix}$$

$$
\begin{bmatrix} a_0(4) \\ a_1(4) \\ a_2(4) \\ a_3(4) \\ \hdashline a_4(4) \\ a_5(4) \\ a_6(4) \\ a_7(4) \end{bmatrix}
=
\left[\begin{array}{cccc:cccc}
1 & & & & 1 & & & \\
 & 1 & & & & 1 & & \\
 & & 1 & & & & 1 & \\
 & & & 1 & & & & 1 \\ \hdashline
 & & & & 1 & & & \\
 & & & & & 1 & & \\
 & & & & & & 1 & \\
 & & & & & & & 1
\end{array}\right]
\begin{bmatrix} a_0(0) \\ a_1(0) \\ a_2(0) \\ a_3(0) \\ \hdashline a_4(0) \\ a_5(0) \\ a_6(0) \\ a_7(0) \end{bmatrix}
$$

where $a_i(2)$ and $a_i(4)$ denote the coefficients of polarities

$$(x_2^* \, \overline{x_1^*} \, x_0^*) \text{ and } (\overline{x_2^*} \, x_1^* \, x_0^*), \; i=0,1,\cdots,7$$

There exist 3 adjacent polarity maps for 3-variable Reed-Muller polynomials. Each map needs 4 modulo-2 additions.

4. Gray Code Ordering and Efficient Algorithms

Gray code is a reflective binary code[8], i. e. , in changing from one value to the next increment only one bit is changed at a time. If all the polarities of n variables are arranged according to Gray codes, then each polarity is adjacent to the next one. Therefore, all 2^n-sets of the Reed-Muller polynomial coefficients may be mapped by adjacent polarity maps based on the Gray code ordering. An efficient algorithm for exhaustive search of the minimum polynomial coefficients can be obtained in terms of (2^n-1) adjacent polarity maps.

For example, all the possible-polarity coefficients of a 3-variable Reed-Muller polynomial may be obtained in terms of 7 adjacent polarity maps according to the following Gray code ordering:

0 0 0	0 0 1	0 1 1	0 1 0	1 1 0	1 1 1	1 0 1	1 0 0
$x_2x_1x_0$	$x_2x_1\bar{x}_0$	$x_2\bar{x}_1\bar{x}_0$	$x_2\bar{x}_1x_0$	$\bar{x}_2\bar{x}_1x_0$	$\bar{x}_2\bar{x}_1\bar{x}_0$	$\bar{x}_2x_1\bar{x}_0$	$\bar{x}_2x_1x_0$

This efficient algorithm may be represented by Fig. 3.

5. Complexity Comparison

From Fig. 3, it is shown that $7\times 2^2=28$ modulo-2 additions are required to compute 7 sets of the polarity coefficients. The first set of the coefficients can be obtained from its output function vector by the FRMT algorithm, it needs 12 modulo-2 additions. The total amount of modulo-2 additions for computing all the polynomial coefficients is 40. For the case of variables, $(2^n-1)2^{n-1}$ modulo-2 additions are required to compute (2^n-1) sets of the polynomial polarity coefficients. $n\times 2^{n-1}$ modulo-2 additions are required to compute the first set of the coefficient by FRMT algorithm. The total amount of modulo-2 additions for computing all of the polynomial coefficients is

$$(2^n+n-1)\times 2^{n-1} \tag{14}$$

One or more among the polynomials are minimum. If only the minimal set of coefficients is required, then one can use the in-place algorithm. Only 2^n memory locations are required.

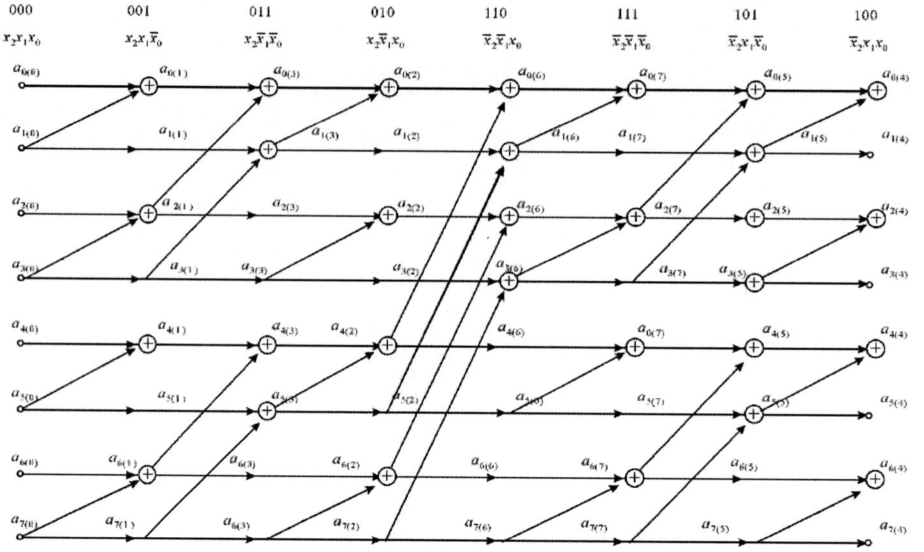

Fig. 3 Flow chart of the algorithm for 3 variables

From Eqs. (5) and (14), the amount of modulo-2 additions for this efficient algorithm is $n/(1+(n-1)2^{-n})$ times less than that of the FRMT algorithm. The comparison of two algorithms is shown in Fig. 4 and Table 1.

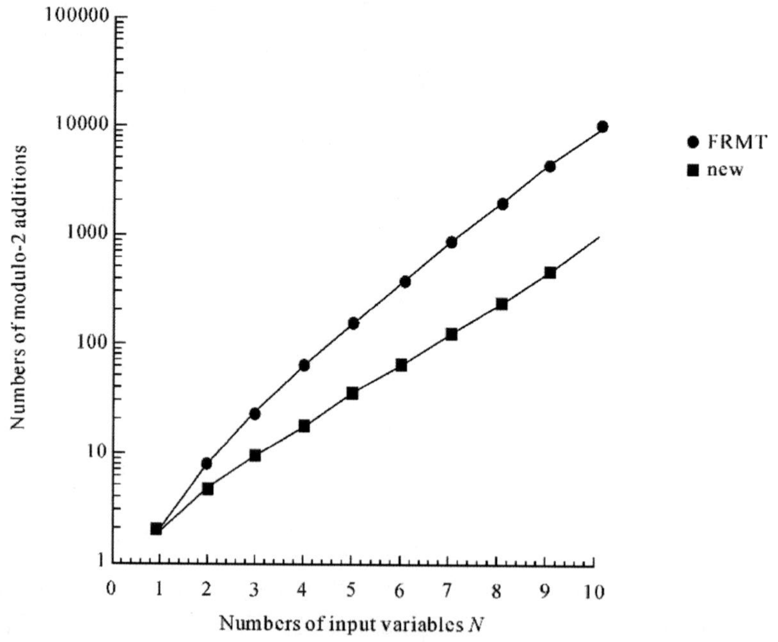

Fig. 4 Comparison between FRMT and new algorithm

Table 1 Comparison of the two algorithms

Variable Number n	No. of $\oplus$($*\ 2^{n-1}$)		Times
	New	FRMT	
2	5	8	1.600
3	10	24	2.400
4	19	64	3.368
5	36	160	4.444
6	69	384	5.565
7	134	896	6.686
8	263	2048	7.787
9	520	4608	8.861
10	1033	10240	9.912
12	4107	49152	11.967
16	65551	1048576	15.966

6. Conclusion

An efficient algorithm has been presented for minimizing exclusive-OR functions. The algorithm is $n/(1+(n-1)2^{-n})$ times faster than the fast Reed-Muller transformation for n-input logic function. This efficient algorithm possesses the advantages of minimum computation amount and high speed, and can be used for computer-aided logic circuit design.

References

[1] Even, S. , Kohavi, I. & Paz, A. , IEEE Trans. Electronic Computers, EC-16(1967), 671-674.
[2] Mukhopadkyay, A. & Schmitz, G. , IEEE Trans. on Computers, C-19(1970), 2:132-140.
[3] Bioul, G. , Davio, M. & Deschamps, J. P. , Philips Res. Repts. 28(1973), 17-36.
[4] Marincovic, S. B. & Tosic,Z. , IEEE Trans. on Computers, C-23(1974),1313-1315.
[5] Kodandapani, K. L. & Sethur, R. V. , ibid. , C-26(1977),310-313.
[6] Papakonstantinou, G. , ibid. , C-28(1979), 163-167.
[7] Saluja, K. K. & Ong, E. H. , ibid. , C-28(1979), 535-537.
[8] Wang, M. C. , IEEE Trans. on Electronic Computers, EC-18(1966), 659-660.
[9] Hellerman, L. IEEE Trans. on Computers, C-21(1972), 5: 439-446.
[10] Reddy, S. M. , ibid. , C-21(1972), 1183-1188.
[11] Kopandapani, K. L. , ibid. , C-23(1974), 332-333.
[12] Saluja, K. K. & Reddy, S. M. , ibid. , C-24(1975), 995-998.
[13] Wu, X. , Chen, X. & Hurst, S. L. , Proc. IEE, 129(1982), 1: 15-20.
[14] Robinson, J. P. & Yeh, C. -L. , IEEE Trans. on Computers, C-31(1982), 800-801.
[15] Besslich, Ph. W. , Proc. IEE, 130(1983), 6: 203-206.

Optimimization of Multiple-Output Exclusive-OR Logic Circuits*

ZHANG Yan-zhong
Ministry of Aero-Space Industry, Beijing, China

Abstract: A method is presented for optimizing multiple-output exclusive-OR logic circuits. The effect of common gates on the optimization of logic circuits is studied. The concepts of common function, residue function and remainder term are introduced. A computer method for searching the optimum polarity is proposed. This method can be used to design the logic circuit which needs the minimum number of exclusive-OR gates.

Keywords: excluftive-OR; logic circuits; optimization

1. Introduction

Many authors[1-3] studied optimization of exclusive-OR logic circuits. Zhang[3] proposed an efficient computer method for optimizing single-output exclusive-OR logic circuits in terms of adjacent polarity mapping and the Gray code ordering. But, they did not deal with optimization of multiple-output logic circuits. The effect of common gates on multiple-output circuits is complex and difficult to deal with.

First, the effect of common gates on multiple-output circuits is considered. Then, the concepts of common function, residue function and remainder are introduced, Finally, a computer method for optimizing multiple-output exclusive-OR logic circuits is presented.

2. Consideration of Common Gates

The optimization of single-output exclusive-OR logic circuits can be obtained by the efficient algorithm proposed in Ref. [3]. The minimum number $w(j)$ of exclusive-OR gates which are required to realize the circuits is determined by the non-zero coefficient number of the polynomial at the minimum polarity (j). Since

$$1 \oplus f = \overline{f} \tag{1}$$

the constant term of Reed-Muller polynomials may be realized in terms of a NOT-gate at an output terminal instead of an exclusive-OR gate in circuits. The number of exclusive-OR gates for realizing

* Published in Science in China, Series A, Vol. 33, No. 5, May 1990: 625-633.

Y. Zhang, *Footprints in Cambridge and Aviation Industries of China*,
https://doi.org/10.1007/978-981-16-3176-4_12

a polynomial is as follows:

$$w = w' - a_0 - 1 \tag{2}$$

where w' represents the total number of polynomial terms, $a_0 \in (0,1)$ is the constant term of the polynomial.

For the case of multiple outputs, the common terms must be considered. One common term among s output polynomials may be realized in terms of one exclusive-OR gate, $(s-1)$ exclusive-OR gates may be saved. Therefore, the optimum polarity of m output polynomials cannot be determined by minimizing the simple sum $w = \sum_{i=0}^{m-1} w_i(j)$. The saved number of gates must be counted.

For example, the truth table of 3 output functions and the coefficients of the polynomials with polarity (000) are shown in Table 1.

Table 1 The example of 3-output functions

Input			Output			Polynomial Coefficient		
x_2	x_1	x_0	f_2	f_1	f_0	c_i	b_i	a_i
0	0	0	1	0	0	1	0	0
0	0	1	0	0	0	1	0	0
0	1	0	1	1	1	0	1	1
0	1	1	0	0	0	0	1	1
1	0	0	1	0	1	0	0	1
1	0	1	1	1	1	1	1	0
1	1	0	1	1	0	0	0	0
1	1	1	0	0	0	1	1	1

Using the algorithm in Ref. [3], the coefficients of 3 polynomials for all possible polarities may be directly mapped from the coefficients $A(0)$, $B(0)$ and $C(0)$. The flow graph is shown in Fig. 1.

In Fig. 1, w'_i denotes the number of non-zero coefficients of the ith polynomial f_i, and it is referred to as the weight of the ith polynomial, w_i denotes the number of exclusive-OR gates for realisation of the ith polynomial.

Fig. 1 shows that the cum w has the minimum value 7 when the polarity is (100) or (110), say, 7 exclusive-OR gates are required. However, when the common terms are considered, the final number of the required exclusive-OR gates for the other polarity may be optimum. For example, the polarity (011) has $w=9$; its output polynomials are as follows:

$$f_0 = \bar{x}_0 \oplus \bar{x}_0\bar{x}_1 \oplus \bar{x}_0 x_2 \oplus \bar{x}_1 x_2 \oplus \bar{x}_0\bar{x}_1 x_2$$

$$f_1 = \bar{x}_0 \oplus \bar{x}_0\bar{x}_1 \oplus \bar{x}_0 x_2 \oplus \bar{x}_0\bar{x}_1 x_2$$

$$f_2 = \bar{x}_0 \oplus \bar{x}_1 x_2 \oplus \bar{x}_0\bar{x}_1 x_2$$

where $(\bar{x}_0 \oplus \bar{x}_1 x_2 \oplus \bar{x}_0\bar{x}_1 x_2)$ are the common terms for 3-output polynomials, 4 exclusive-OR gate may be saved, and $(\oplus \bar{x}_0\bar{x}_1)$ is the common term of the functions f_0 and f_1, one exclusive-OR gate may be saved. Only $9-(4+1)=4$ exclusive-OR gates are required to realise the 3 polynomials. Though w for the polarity (011) is larger than that for the polarity (100), the polarity (011) needs

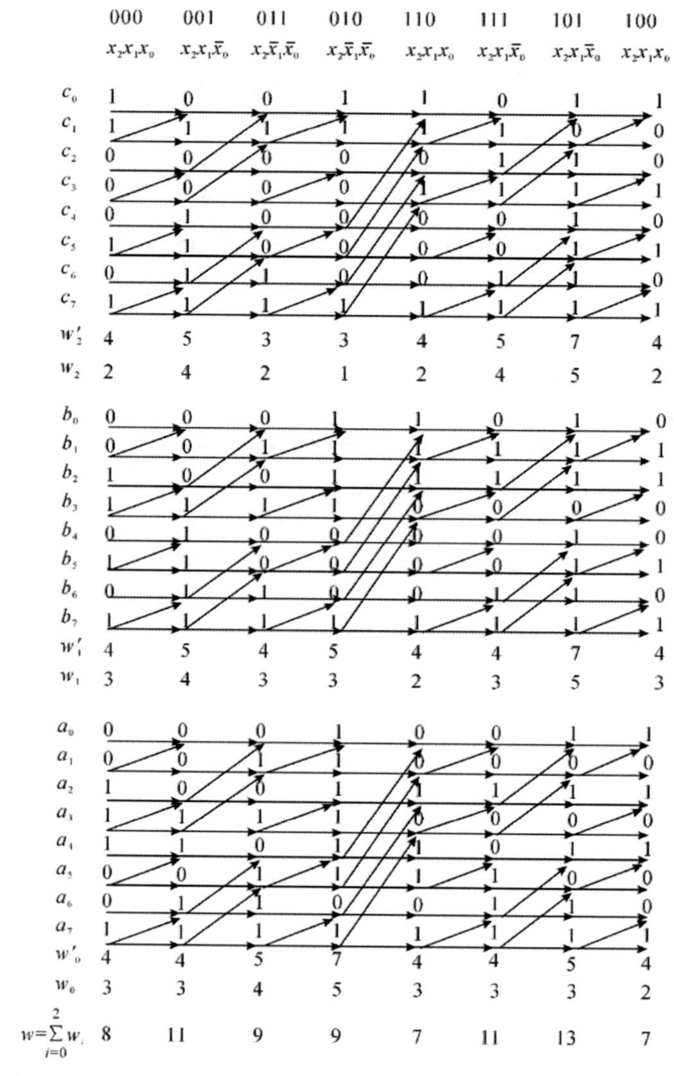

Fig. 1 Flow graph for multiple-output functions

fewer exclusive-OR gates as a result of the common terms.

From the above example, the optimum polarity of m output functions should be searched by considering the common terms. Theorem 1 will give a method for determining the common terms among the multiple-output polynomials.

3. Common and Residue Function Remainder Terms

Before giving the theorem, some definitions are introduced.

Definition 1

The function consisting of the common terms among s functions is called an s-common function.

The coefficients of common functions may be directly obtained by logical AND operations among the coefficients of the s functions

$$\underline{A}_{j_1 j_2 \cdots j_s} = \underline{A}_{j_1} \wedge \underline{A}_{j_2} \wedge \cdots \wedge \underline{A}_{j_s} \tag{3}$$

where $\underline{A}_{j_1 j_2 \cdots j_s}$ denotes the coefficient vector of the s-common function, $\underline{A}_{j_1}$, $\underline{A}_{j_2}$, $\cdots$, $\underline{A}_{j_s}$ are respectively the coefficient vectors of the output functions f_{j_1}, f_{j_2}, $\cdots$, f_{j_s}, $\wedge$ denotes logical AND operations.

Let $w'_{j_1 j_2 \cdots j_s}$ denote the weight of the s-common function, then the number $w_{j_1 j_2 \cdots j_s}$ of exclusive-OR gates may be obtained from Eq. (2).

Because $\underline{A}_{j_1}$, $\underline{A}_{j_2}$, $\cdots$, $\underline{A}_{j_s}$, are functions of the polarity, so the number $w_{j_1 j_2 \cdots j_s}$ of exclusive-OR gates of the common function is also a function of the polarity. The number of s-common functions among m output functions is

$$c_m^s, s = 2, 3, \cdots, m \tag{4}$$

The total number of common functions among m output functions is as follows:

$$c_m^2 + c_m^3 + \cdots + c_m^{m-1} + c_m^m = 2^m - (m+1) \tag{5}$$

In order to avoid repeated counting of the common exclusive-OR gates in different common functions, the residue function is introduced.

Definition 2

The residue function of a function f_i is the sub-function which consists of the terms of f_i without $(w_{j_1 j_2 \cdots j_s} - 1)$ common exclusive-OR terms of the s-common function, where $i = j_1, j_2, \cdots, j_s$.

The coefficients of residue functions may be obtained by module-2 additions of the coefficients of the function f_i, the s-common function $f_{j_1 j_2 \cdots j_s}$ and remainder term $\underline{C}_s$ as follows:

$$\underline{A}_i^{(s)} = \underline{A}_i \oplus \underline{A}_{j_1 j_2 \cdots j_s} \oplus \underline{C}_s \tag{6}$$

where $\oplus$ denotes modulo-2 additions, $\underline{A}_i$ is the coefficient vector of f_i, $\underline{A}_{j_1 j_2 \cdots j_s}$ is the coefficient vector of s-common vector $f_{j_1 j_2 \cdots j_s}$, $\underline{C}_s$ is a column vector, it has a single one remaining from $\underline{A}_{j_1 j_2 \cdots j_s}$ and all the other elements in $\underline{C}_s$ are zeros.

The existence of $\underline{C}_s$ means that one of the common function terms remains in the residue functions. It is named the remainder term.

As the common function possesses $w_{j_1 j_2 \cdots j_s}$ terms, so $\underline{C}_s$ has $w_{j_1 j_2 \cdots j_s}$ forms.

It is obvious that the choice of the remainder term and the form of $\underline{C}_s$ does not change the weight and number of exclusive-OR gates of the residue functions. Therefore, any term of the s-common function may be the remainder term. But, the remainder term of the s-common function must not be the remainder term of the $(s-1)$-common function, say

$$\underline{C}_s \neq \underline{C}_{s-1} \tag{7}$$

If $\underline{C}_s = \underline{C}_{s-1}$ then $\underline{C}_s \oplus \underline{C}_{s-1} = \underline{0}$, the incorrect common gates may be introduced.

After the common function is produced, the output function should be substituted by its residue function for further operations. The coefficient $\underline{A}_i$ is substituted by $\underline{A}_i^{(s)}$ in memory after $\underline{A}_{j_1 j_2 \cdots j_s}$ is produced.

For a 4-output function example, if the output functions are f_0, f_1, f_2 and f_3, then the 4-common function f_{0123} is

$$\underline{A}_{0123} = \underline{A}_0 \wedge \underline{A}_1 \wedge \underline{A}_2 \wedge \underline{A}_3$$

The 4-residue functions are

$$\underline{A}_i^{(s)} = \underline{A}_i \oplus \underline{A}_{0123} \oplus \underline{C}_4, i = 0, 1, 2, 3$$

The 3-common function f_{012} is

$$\underline{A}_{012}=\underline{A}_0^{(4)}\wedge\underline{A}_1^{(4)}\wedge\underline{A}_2^{(4)}$$

The residue functions are

$$\underline{A}_i^{012}=\underline{A}_i^{(4)}\oplus\underline{A}_{012}\oplus\underline{C}_3, i=0,1,2$$

But, the 3-common function f_{123} is

$$\underline{A}_{123}=\underline{A}_1^{012}\wedge\underline{A}_2^{012}\wedge\underline{A}_3^{(4)}$$

Here $\underline{A}_1^{012}$ and $\underline{A}_2^{012}$ are substituted for $\underline{A}_1^{(4)}$ and $\underline{A}_2^{(4)}$. The residue functions are

$$\underline{A}_1^{123}=\underline{A}_1^{012}\oplus\underline{A}_{123}\oplus\underline{C}'_3$$

$$\underline{A}_2^{123}=\underline{A}_1^{012}\oplus\underline{A}_{123}\oplus\underline{C}'_3$$

$$\underline{A}_3^{123}=\underline{A}_1^{(4)}\oplus\underline{A}_{123}\oplus\underline{C}'_3$$

Here $\underline{A}_1^{(4)}$ and $\underline{A}_2^{(4)}$ were substituted by $\underline{A}_1^{012}$ and $\underline{A}_2^{012}$ in the first two equations and $\underline{C}'_3\neq\underline{C}_3$. This substitution may be easily realised by an "in place" algorithm.

4. Minimum Number of Exclusive-OR Gates

After giving the definition and algorithm for finding common and residue functions, the number of gates which may be saved is given by the following theorem.

Theorem 1

The number of exclusive-OR gates that may be saved for m output functions is as follows:

$$N_s=(m-1)w^{(m)}+(m-2)w^{(m-1)}+\cdots+2w^{(3)}+w^{(2)} \tag{8}$$

where N_s denotes the number of gates that may be saved, $w^{(m)}$ is the number of common gates of m output functions, $w^{(m-1)}=w_{01\cdots(m-2)}+w_{01\cdots(m-3)(m-1)}+\cdots+w_{12\cdots(m-1)}$ denotes the number of common gates of C_m^1 the $(m-1)$-common functions, $w^{(3)}=\sum_{i\neq j\neq k}^{m-1}w_{ijk}$ denotes the gate number of all 3-common functions, $w^{(1)}=\sum_{i\neq j}^{m-1}w_{ij}$ denotes the gate number of all 2-common functions.

Corollary 1

The total number of exclusive-OR gates, w_i is the number of EOR gates for individual realisation of each single function f_i, N_s is the number of EOR gates that may be saved by common terms. Since w_i and N_s are functions of the polarities, so N_{cot} is also a function of polarities. The optimum polarity may be obtained by the minimum N_{eor}.

5. Algorithm for Finding Optimum Polarity

The optimum polarity and the number of EOR gates for m output functions may be obtained by the algorithm shown in Fig. 2.

Step 1

Compute the common gate number $w^{(m)}$ of m functions.

(1) Compute the m-common function coefficient $\underline{A}_{01\cdots m-1}$ based on Eq. (3) in terms of logical AND operations.

(2) Count the number of common gates $w^{(m)}$.

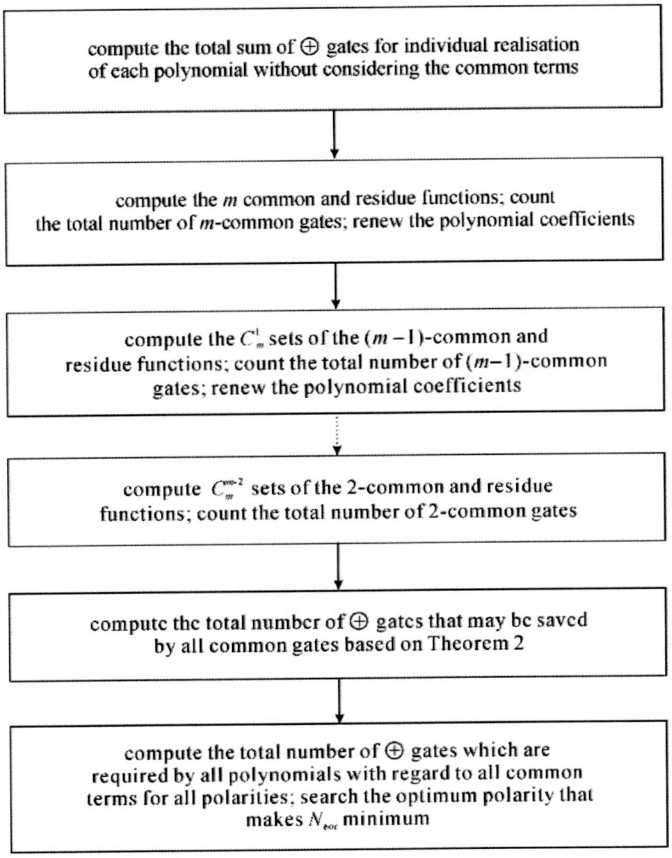

Fig. 2 Flow chart of the minimization algorithm

(3) Choose $\underline{C}_m$. Compute m residue function coefficients $\underline{A}_i^{(m)}$ $(i=0,1,\cdots,m-1)$ based on Eq. (6) in terms of modulo-2 additions.

(4) Substitute $\underline{A}_i^{(m)}$ for $\underline{A}_i$ in the computer memory $(i=1,0,\cdots,m-1)$.

Step 2

Compute the common gate number $w^{(m-1)}$ for all sets of $(m-1)$ output functions.

(1) Choose the first set of $(m-1)$ output functions, as Step 1. Compute the $(m-1)$-common function coefficients $\underline{A}_{01\cdots(m-2)}$ and the number of common gates $w_{01\cdots(m-2)}$. Choose $\underline{C}_{m-i} \neq \underline{C}_m$, compute $(m-1)$ residue function coefficients $\underline{A}_i^{(m-1)}$ $(i=0,1,\cdots,(m-2))$ and substitute $\underline{A}_i^{(m)}$ by $\underline{A}_i^{(m-1)}$ in the computer memory.

(2) Choose the second set of $(m-1)$ output functions, as step 2.1. Compute $\underline{A}_{01\cdots(m-3)(m-1)}$, $w_{01\cdots(m-3)(m-1)}$ and residue functions $\underline{A}'^{(m-1)}_i$ $(i=0,1,\cdots,(m-3),(m-1))$, and substitute $\underline{A}'^{(m-1)}_i$ for $A_i^{(m-1)}$ and $A_i^{(m)}$, but the second set of $(m-1)$ output function coefficients have been modified by the above step, and they will be modified by $A'^{(m-1)}_i$ $(i=0,1,\cdots,m-3,\ m-1)$ again in this step.

⋮

(m) Choose the mth set of $(m-1)$ output functions and compute $\underline{A}_{12\cdots m-1}$ and $w_{12\cdots m-1}$, etc., as the above step and modify the function coefficients.

(n)Compute the total number of all the ($m-1$)-common gates by

$$w^{(m-1)} = w_{01\cdots(m-2)} + w_{01\cdots(m-3)(m-1)} + \cdots + w_{12\cdots(m-1)}$$

Step 3

Compute the number of common gates $w^{(m-2)}, w^{(m-3)}, \cdots, w^{(2)}$ in the above steps.

Step 4

Compute the total number N_s of the EOR gates that may be saved based on Eq. (8). Compute the total gate number N_{eor} for one polarity by Eq. (9).

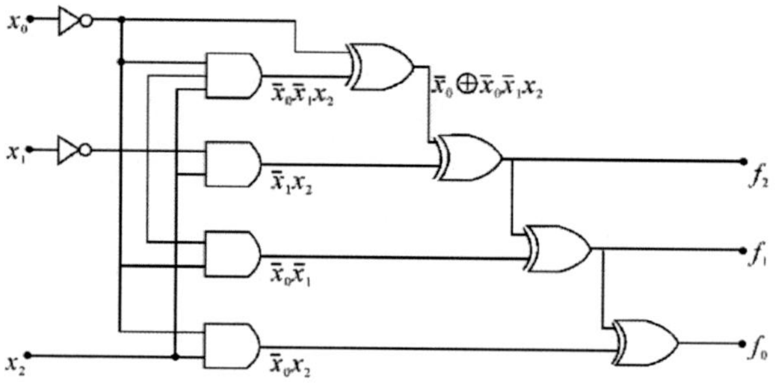

Fig. 3 Optimum logical circuit for the example

Step 5

Compute the total number N_{eor} of EOR gates for all possible polarities. The polarity that makes N_{eor} minimum is optimum.

For the above example, the total number of EOR gates for different polarities is shown in Table 2.

From Table 2 it may be seen that the polarity $x_2\bar{x}_1\bar{x}_0$ is the optimum one for realisation of the 3 output functions with regard to the common terms. Only 4 EOR gates are required to realise the circuit. The logical circuit of the example is shown in Fig. 3.

Table 2 The number of EOR gates for the example

Gray Code	000	001	011	010	110	111	101	100
Polarity	$x_2x_1x_0$	$x_2\bar{x}_1\bar{x}_0$	$x_2\bar{x}_1\bar{x}_0$	$x_2\bar{x}_1x_0$	$\bar{x}_2\bar{x}_1x_0$	$\bar{x}_2\bar{x}_1\bar{x}_0$	$\bar{x}_2x_1\bar{x}_0$	$\bar{x}_2x_1x_0$
w_0	2	4	2	1	2	4	5	2
w_1	3	4	3	3	2	3	5	3
w_2	3	3	4	5	3	3	3	2
$w = \sum_{i=0}^{2} w_i$	8	11	9	9	7	11	13	7
$w^{(3)}$	0	2	2	1	0	2	3	0
$w^{(2)}$	2	1	1	2	1	1	1	1
N_s	2	5	5	4	1	5	7	1
N_{eor}	6	6	4	5	6	6	6	6

6. Estimation of Total Number of AND and NOT Gates

The number N_{in} of input NOT gates depends on the input polarities. The number of output NOT gates and the number of AND gates are given by the following corollaries.

Corollary 2

The number N_{out} of output NOT gates is

$$N_{out} = \sum_{i=0}^{m-1} a_{i0}$$

where N_{out} denotes the number of NOT gates at output end, a_{i0} is the constant term of the ith output function. It is obvious that N_{out} is a function of the polarities.

Corollary 3

The number N_{and} of AND gates for realisation of m output functions is as follows:

$$N_{and} = w_d - \sum_{i=0}^{n-1} d_2 i - d_0 \qquad (11)$$

where w_d is the weight of the OR polynomial D for the m output functions,

$$D = A_0 \vee A_1 \vee \cdots \vee A_{m-1} \qquad (12)$$

$\vee$ denotes the logical OR operation, $d_2 i$ denotes the coefficients of single variable terms of the D polynomial, N_{and} denotes the number of AND gates.

For the example in Table 1, the number of input NOT gates, output NOT gates and EOR gates are shown in Table 3.

Table 3 Number of NOT, AND and EOR gates for the example

Gray Code	000	001	011	010	110	111	101	100
Polarity	$x_2 x_1 x_0$	$x_2 \bar{x}_1 \bar{x}_0$	$x_2 \bar{x}_1 \bar{x}_0$	$x_2 \bar{x}_1 x_0$	$\bar{x}_2 \bar{x}_1 x_0$	$\bar{x}_2 \bar{x}_1 \bar{x}_0$	$\bar{x}_2 x_1 \bar{x}_0$	$\bar{x}_2 x_1 x_0$
N_{in}	0	1	2	1	2	3	2	1
N_{out}	1	0	0	3	2	0	3	2
N_{not}	1	1	2	4	4	3	5	3
N_{and}	3	4	4	3	3	4	4	3
N_{eor}	6	6	4	5	6	6	6	6

If the price of NOT, AND and EOR gates are p_1, p_2 and p_3 respectively, then the total price of the circuits is

$$P = P_1 N_{not} + P_2 N_{and} + P_3 N_{eor} \qquad (13)$$

The optimum polarities should be obtained by minimum P.

7. Conclusion

A computer method has been proposed for optimizing multiple-output exclusive-OR logic circuits. The effect of common gates on multiple-output circuits is considered. The circuit designed by this method is optimum, and needs a minimal number of exclusive-OR gates. The theory, method and programme can be used in computer-aided design and optimization of logic circuits.

References

[1] Besslich, Ph. W., Efficient computer method for ExOR logic design, Proc. IEE., 130 (1983), 6:203-206.

[2] Zhang Y. Z., & Rayner P. J. W., Minimisation of Reed-Muller polynomials with fixed-polarity, IEE Proc., 131(1984), 5:177-186.

[3] Zhang Y. Z., Minimisation of Exclusive-OR logic functions, Science in China, Ser. A, 33(1990), 4: 477-485.

A Novel Interleaver Design Method for Turbo Codes*

F. Qi[a], Y. Z. Zhang[b], D. R. Shao[b] and M. Z. Wang[a]

[a] Department of Electronic and Information Engineering, The Hong Kong Polytechnic University, Hung Horn, Kowloon, Hong Kong, China

[b] Department of Electrical Engineering, Beijing University of Aeronautics and Astronautics, Beijing, China

Abstract: We introduce the concept of optimal period interleaver and present an interleaver design method based on optimal period interleavers. Simulation results are presented for turbo codes of memory length 3 using our interleavers. An order of magnitude improvement in bit error rate performance is obtained at high signal-to-noise ratio compared to turbo codes using other conventional interleavers.

1. Introduction

Recently, a class of parallel concatenated systematic codes termed turbo codes was introduced[1]. The performance of turbo codes using iterative decoding mainly depends on the type and depth of the interleaver used. This is because the interleaver structure affects the distance spectrum of the resulting turbo code. "Good" interleavers will result in spectrum "thinning", referring to the property that the number of codewords in a turbo code with low Hamming weight is reduced[2, 3].

Interleaver design for turbo codes has been commonly concentrated on the weight-two input sequences. In [4], Khandani designed interleavers which are in a sense optimum in this aspect. However, computer simulations show that the interleavers do not perform as well as expected in terms of bit error rate (BER) for the resultant turbo codes.

This paper is organized as follows. The following section introduces the concept of optimal period interleaver. Section 3 details our interleaver design method based on optimal period interleavers. Simulation results on bit error rate performance for rate 1/3 turbo codes using our interleaver are obtained and compared with those of turbo codes using other interleavers. Conclusions are drawn in the final section.

---

* Published in Proc. IEEE Wireless Communications & Network Conf, Sept. 1999: 476-479.

Y. Zhang, *Footprints in Cambridge and Aviation Industries of China*,
https://doi.org/10.1007/978-981-16-3176-4_13

2. Optimal Period Interleavers

Fig. 1 shows a turbo code using two constituent codes. We will use it to introduce terminology used later on. The code uses two identical recursive convolutional codes (RCCs) with memory length $m=3$. In the following, we use $a^n = a_0 a_1 \cdots a_{n-1}$ to denote a sequence of length n. The code shown is a rate $\frac{1}{3}$ turbo code. One of the three output streams is the uncoded information sequence $\underline{x}^n$. The other two output sequences $\underline{y_1}^{n+m}$ and $\underline{y_2}^{n+m}$ are parity sequences. The parity sequences are generated by the two RCCs. The two RCCs are in general identical denoted as $G(D)=\frac{F(D)}{B(D)}$, where $F(D)$ is the feedforward polynomial, $B(D)$ is the feedback polynomial, and D is an intermediate variable and can be considered as representing a delay element in the shift-register for the convolutional code. The period of the feedback polynomial (also loosely referred to as the period of the convolutional code) is given by the smallest r, such that the polynomial $Q(D)=1+D^r$ is a multiple of $B(D)$.

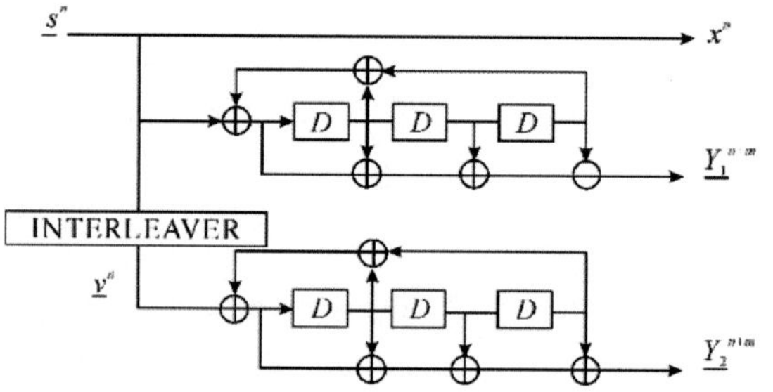

Fig. 1 A rate $\frac{1}{3}$ turbo encoder with feedback polynomial $1+D+D^3$ (13 in octal) and feedforward polynomial $1+D+D^2+D^3$ (17 in octal) (We also call this code a 13/17 turbo code, following a convention in the literature.)

For an information sequence of length n. $s^n = s_0 s_1 \cdots s_{n-1}$, we define its D-transformation as $S(D) = \sum_{i=0}^{n-1} s_i D^i$. We refer to $S(D)$ as information polynomial throughout the paper. $\underline{s}^n$ is then fed to an interleaver to obtain the interleaved sequenced $\underline{v}^n = v_0 v_1 \cdots v_{n-1}$. The interleaver is in essence a permutation matrix $P=[p_{ij}], 0 \leq i, j \leq n-1$, so that $\underline{v}=\underline{s} \cdot P$, where $\underline{s}=[s_0, s_1, \cdots, s_{n-1}]$ and $\underline{v}=[v_0, v_1, \cdots, v_{n-1}]$ are n-dimensional row vectors related to $\underline{s}^n$ and $\underline{v}^n$. For ease of reference, we denote the interleaver by an integer sequence $\rho^n = \rho_0 \rho_1 \cdots \rho_{n-1}$ where $\rho_j = \sum_{k=0}^{n-1} k P_{jk}, j = 0, 1, \cdots, n-1$.

For example, for $n=3$, $P=\begin{bmatrix} 0 & 1 & 0 \\ 0 & 0 & 1 \\ 1 & 0 & 0 \end{bmatrix}$ is a valid permutation matrix. The corresponding interleaver in sequence notation is $\underline{\rho}^n = 201$. The interleaver $\underline{\rho}^n = 201$ therefore permutes, for example, a sequence $s_0 s_1 s_2$ to $s_2 s_0 s_1$.

Similarly, the D-transformation of $\underline{v}^n$ is defined as $V(D) = \sum_{i=0}^{n-1} v_i D_i$. With a little abuse of terminology, we will refer $V(D)$ as the interleaved version of $S(D)$. The D-transformations of two output sequences $\underline{y_1}^{n+m}$ and $\underline{y_2}^{n+m}$ are defined as $Y_1(D) = \sum_{i=0}^{n+m-1} y_{1i} D_i$ and $Y_2(D) = \sum_{i=0}^{n+m-1} y_{2i} D_i$, respectively. It can be readily shown that

$$Y_1(D) = S(D)\frac{F(D)}{B(D)} \mod D^{n+m}$$

$$Y_2(D) = V(D)\frac{F(D)}{B(D)} \mod D^{n+m}$$

where $\frac{F(D)}{B(D)}$ denotes the quotient obtained when dividing $F(D)$ by $B(D)$ in a long-division manner, namely, $\frac{F(D)}{B(D)} = \sum_{i=0}^{\infty} F(D)(1+B(D))^i$.

Note that all the sequences, vectors and polynomials (D-transformations) are binary, i. e., over GF(2).

We have the following definition. For two polynomials in D, $S(D)$ and $B(D)$, if $S(D)=0$ mod $B(D)$, $S(D)$ is called a divisible polynomial by $B(D)$. The sequence $\underline{s}^n$ corresponding to $S(D)$ is called a divisible sequence with respect to $B(D)$. For example, let $B(D)=1+D+D^3$. If the binary information sequence is 0110100 and the information polynomial $S(D)=D+D^2+D^4$. then $B(D)$ divides $S(D)$. i. e.. $S(D)=0$ mod $B(D)$. When the last non-zero digit is presented to the encoder, the shift-register contains all zeros and remains in the all zero state. Thus, the encoder generates a sequence corresponding to $\frac{S(D)}{B(D)}F(D)$ in our example. It is clear that divisible information sequence generate a parity sequence with finite Hamming weight.

The role of the interleaver is to permute those low-weight input sequences which result in a low Hamming weight parity sequences from the first RCC, such that the the Hamming weight of the output of subsequent RCC is high. In other words, if a low weight information sequence is divisible, its interleaved counterpart, should not, divisible.

Let us consider an example. For an interleaver of length equal to 7, the feedback polynomial $B(D)=1+D+D^3$ is used. It is not a coincidence that the interleaver length is equal to the period of the feedback polynomial as we will see the reason in the next section. The information sequence length is consequently also 7. The weight of information sequences w varies from 1 to 7 (ignoring the all-zero information sequence). We also use the weight of a polynomial to refer the weight of the corresponding sequence.

The divisibility of $S(D)$ by $B(D)$ is listed in the following.

(1) $w=1,2$

Obviously, $B(D)$ can not divide $S(D)$.

(2) $w=7$

This is the all-one sequence, corresponding to $S_{(7)}(D)=1+D+D^2+D^3+D^4+D^5+D^6$, $B(D)$ always divides $S(D)$.

(3) $w=3$

All divisible $S(D)$ are listed below: $1+D+D^3$, $D+D^2+D^4$, $D^2+D^3+D^5$, $D^3+D^4+D^6$, $1+$

D^4+D^5, $D+D^5+D^6$, $1+D^2+D^6$.

For all divisible information polynomial $S(D)$ with weight 3, let $V(D)$ denote the interleaved version of $S(D)$. Interleavers can be found so that none of $V(D)$ is divisible by $B(D)$.

For instance, the interleaved version of the seven polynomials above for the interleaver 4156023 are: $D+D^4+D^6$, $1+D+D^5$, $D^2+D^5+D^6$, $1+D^3+D^6$, $1+D+D^2$, $D+D^2+D^3$, and $D+D^3+D^5$, none of which is divisible by $B(D)=1+D+D^3$.

(4) $w=4$

Let $S_{(3)}(D)$ be a polynomial in D of weight 3. $S_{(4)}(D)=S_{(3)}(D)+S_{(7)}$ is then a polynomial of weight 4 and can be thought of as a complement of $S_{(3)}(D)$ (or, $S_{(3)}(D)$ and $S_{(4)}(D)$ are complementary), where $S_{(7)}$ is an all-one polynomial. Since $S_{(7)}$ is always divisible by $B(D)$, the divisibility of $S_{(3)}(D)$ and its complement $S_{(4)}(D)$ is the same. In other words, if $S_{(3)}(D)$ is divisible by $B(D)$, so is its complement, a polynomial of weight 4; if $S_{(3)}(D)$ is not divisible by $B(D)$, nor is its complement.

Therefore, for each divisible polynomial of weight 3, there is a divisible polynomial of weight 4. Moreover, the interleaved versions of complementary polynomials are also complementary. This is so because the interleaver only permutes the digits in a sequence.

For example, $S_{(3)}=1+D+D^3$ and $S_{(4)}=D^2+D^4+D^5+D^6$ are complementary ($S_{(3)}+S_{(4)}=\sum_{i=0}^{6} D^i$. Both are divisible polynomials.

For the interleaver 4156023, the interleaved version of $S_{(3)}$ is $V_{(3)}=D^4+D^5+D^6$; the interleaved version of $S_{(4)}$ is $V_{(4)}=1+D+D^2+D^3+D^4$. $V_{(3)}$ and $V_{(4)}$ are complementary ($V_{(3)}+V_{(4)}=\sum_{i=0}^{6} D^i$). Both are not divisible polynomials.

Therefore there are 7 divisible polynomials of weight 4, corresponding to the 7 polynomials in case (3) previously discussed. If the interleaver 4156023 is used, none of the interleaved versions of the 7 divisible polynomials of weight 4 is divisible.

(5) $w=5,6$

These are complements of polynomials of weigh 1 and 2 in case (1). None is divisible by $B(D)=1+D+D^3$.

We now introduce the concept of the optimal period interleaver. Let L denote the period of the feedback polynomial in the RCCs in a turbo code. If an interleaver of length L converts (permutes) all divisible sequence of length L, except the all-zero $\underbrace{(00\cdots0)}_{L\text{ zeros}}$ and the all-one $\underbrace{(11\cdots1)}_{L\text{ ones}}$ sequences, to non-divisible sequences for the given feedback polynomial, the interleaver is an optimal period interleaver.

An interleaver of length in the order of the 7 is of course too short to be useful. As the interleaver gets longer, it becomes impossible to exhaustively search for the optimal interleaver (as defined previously) if one exists at all. We can however construct long interleavers based on a shorter interleaver as we demonstrate in the next section.

3. Interleaver Design and Simulation Results

We use an example to illustrate the interleaver design procedure. Let N denote the interleaver length. The binary RCCs is of constraint length $r=3$. The feedback polynomial is $B(D)=1+D+D^3$ which is primitive over GF(2). Therefore the period of $B(D)$ is $L=7$. Let $c_0 c_1 \cdots c_{L-1}$ denote the optimal period interleaver. For simplicity we assume that $N=m \cdot L$ (m a positive integer) is a multiple of L.

Let $\rho_0^N = 01 \cdots N-2\ N-1$ denote the identical interleaver (an interleaver that does nothing). Arrange the identical interleaver ρ_0^N into $m=N/L$ rows, each containing L elements. Let $x\rho_0^N$ denote the two dimensional representation of ρ_0^N in the manner $x_{\rho_0^N}[i,j]=\rho_{i,L+j}$, $0 \leq i < m$ and $0 \leq j < L$. Then the elements on the k-th row, $0 \leq A < N/L$, $x_{\rho_0^N}[k,0] x_{\rho_0^N}[k,1] \cdots x_{\rho_0^N}[k,L-1]$ are $k \cdot L$, $k \cdot L-1, \cdots, k \cdot L+(L-1)$, respectively. The procedure to obtain the interleaver of length N is as follows.

(a) Permute the elements in each row in ρ_0^N according to the optimal period interleaver to obtain the interleaver ρ_1^N namely, set $x_{\rho_1^N}[k,j]$ to $k \cdot L+c_j$, $0 \leq k < m$ and $0 \leq j < L$, where $x_{\rho_1^N}$ is the two dimensional representation of ρ_1^N.

(b) Permute elements in each column of $x_{\rho_1^N}$ separately, in a manner that elements of similar value are separated far apart.

The permutations applying to the columns in step (b) can be constructively chosen to further improve the bit error rate performance of the resultant turbo code. For example, the method presented in [5] can be used to permute the rows of $x_{\rho_0^N}$ (this is equivalent to using the same permutation for all columns) and then incrementally shift each column.

We have implemented the above algorithm for the 13/17 turbo-code as shown in Fig. 1. The constraint length of the RCC encoders is $r=3$. The feedback polynomial (13_{oct}) is primitive. Therefore, the period of the RCCs $L=2^r-1=7$. For interleaver (frame) length $N=105$, simulation results for bit error rate are obtained over an additive white Gaussian noise (AWGN) channel where 10 iterations are used in the iterative decoding algorithm. Fig. 2 shows the bit error rate of the turbo codes using random interleaver and the block interleaver against that of the turbo code using our interleaver.

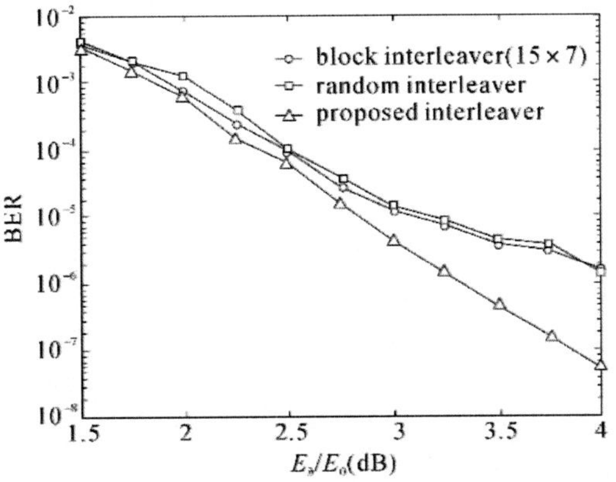

Fig. 2 BER of turbo codes, 13/17 code, frame length equal to 105

We have also simulated the BER performance for turbo codes using our interleaver with frame length $N=210$ and $N=420$. These are shown in Figs. 3 and 4.

It is evident that turbo codes using our interleavers perform better than turbo codes using other conventional interleavers, especially in high signal-to-noise range where the BER curves of turbo codes using pseudo-random and block interleavers start to flatten out.

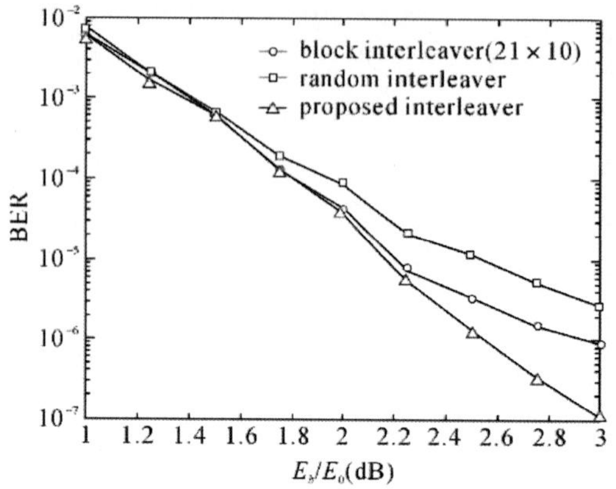

Fig. 3 BER of turbo codes, 13/17 code, frame length equal to 210

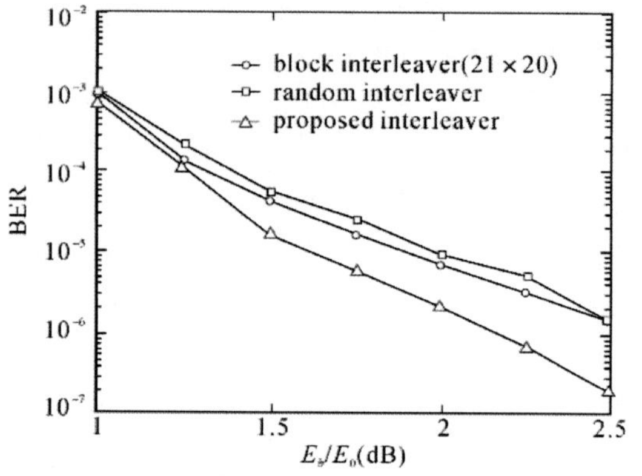

Fig. 4 BER of turbo codes, 13/17 code, frame length equal to 420

4. Conclusion

This paper presented a novel design scheme for interleavers used for turbo codes. The interleavers designed based on our method provide significant performance enhancement for turbo codes of short frames. This is particularly of interest to turbo coding applications in mobile communications where short frames are typical.

References

[1] C. Berrou, A. Glavieux, and P. Thitimajshima, "Near Shannon limit error-correcting coding and decoding Turbo-codes," in Proc. ICC'93, (Geneva, Switzerland), pp. 1064-1070, May 1993.

[2] S. Dolinar and D. Divsalar, "Weight distributions for turbo codes using random and nonrandom permutations," TDA progress rep. 42-122, pp. 56-65, Aug 1995.

[3] C. Perez and J. C. J. Seghers, "A distance spectrum interpertatin of turbo codes," IEEE Trans. Inform. Theory, vol. 42, pp. 1698-1709, Nov 1996.

[4] A. Khandani, "Design of turbo-code interleaver using hungarian method," Electron. Lett., vol. 34, pp. 63-65, Jan 1998.

[5] E. Dunscombe and F. C. Piper, "Optimal interleaving scheme for convolutional coding," Electron. Lett., vol. 25, pp. 1517-1518, Oct 1989.

A Turbo Code-Aided Adaptive Equalizer for Mobile Radio Communication*

F. Qi[a], M. Z. Wang[a], A. U. H. Sheikh[a] and Y. Z. Zhang[b]

[a] WISR Center, Department of Electronic and Information Engineering, The Hong Kong Polytechnic University, Hung Horn, Kowloon, Hong Kong, China

[b] Department of Electrical Engineering, Beijing University of Aeronautics and Astronautics, Beijing, China

Abstract: We propose to use turbo codes in a code-aided equalizer[1,2] for mobile radio communications. The turbo codes used take advantage of good short interleavers designed by an algorithm we developed earlier[3-6]. Bit error rate performance in a multipath environment for frame size equal to 210 bits will be presented.

1. Introduction

In radio channels, intersymbol interference (ISI) occurs due to multipath propagation. Equalization (signal processing operation that minimizes ISI) is required to mitigate the effects of intersymbol interference. The mobile radio channel is further impaired by Doppler spread because of the mobility of the transmitter and/or receiver. The equalizer must track the time varying characteristics of the mobile channel. A short preamble or training sequence is used to adjust the tap gains (the coefficients) of the equalizer at the start-up stage or periodically. The equalizer then enters into a so-called decision-directed mode, i. e. , the decisions (outputs) of the equalizer are used to adjust the tap gains of the equalizer. In normal operation, the decisions are correct often enough to allow the adaptive equalizer to maintain convergence. An occasional decision error from the equalizer, however, results in a burst of errors in the decision-directed mode. This usually happens when the equalizer is not able to track the fast varying fading channel, e. g. , when the channel goes into a deep fade.

Error correction coding improves mobile communication link performance by adding redundant bits in the transmitted messages. Error correction coding is a critical component in any modern communication system. In conventional receivers, equalizers and decoders of error-correcting codes are independently operated. Previous work shows that combining equalization and error correction coding provides performance advantage in terms of bit error rate. The joint adaptive equalizer and decoder structure described in [7,8] employs two sets of decision feedback equalizers (DFE's) and forward error control (FEC) decoders with exchange of channel state information and a reliability estimate of the received data which is then used to set the adaptation coefficient for the adaptation

* Published in Proc. IEEE Wireless Communications & Network Conf, Jan. 2000: 1703-1706.

Y. Zhang, *Footprints in Cambridge and Aviation Industries of China*,
https://doi.org/10.1007/978-981-16-3176-4_14

algorithm. In [9], a scheme called FEC-assisted adaptive equalizer was proposed that successfully combines adaptive equalization and coding. The FEC-assisted adaptive equalizer uses a short block code. It eliminates the need for periodic inclusion of a training sequence. For slow fading (e. g. , for the channel normalized fade rate, defined as the ratio of the Doppler frequency (shift) to symbol rate, smaller than 0. 001), the delay in the channel adaptation introduced by the code with the FEC-assisted adaptive equalizer does not affect the performance. As the the channel normalized fade rate increases, the advantage of the FEC-assisted adaptive equalizer diminishes[9, Fig. 8]. A code-aided adaptive equalizer was developed in [1, 2]. The code-aided adaptive equalizer employs a two-stage equalization. Our turbo code aided equalization scheme is based on the same two-stage structure.

The recently introduced turbo codes are a class of parallel concatenated systematic codes. Turbo codes achieve performance very close to the Shannon limit. One of the major disadvantages of a turbo code is its long delay, due to the large block lengths (in the range of ten thousand bits) in order to be effective. Recently, we successfully found effective short turbo codes (in the range of a few hundred bits)[3-6]. The short turbo codes use interleavers based on optimal period interleaver and provide considerable performance enhancement. This paper presents our investigation on these turbo codes used in a code-aided equalization scheme. The paper is organized as follows. The following section briefly describes the code-aided equalization scheme prsented in [1, 2], on which our turbo code-aided equalizer is based. Section 3 presents the simulation results for a turbo code-aided equalizer. Conclusion is drawn in the final section.

2. Code-Aided Equalization

A code-aided adaptive equalizer was developed in [1, 2]. The code-aided adaptive equalizer employs a two-stage equalization. This is shown in Fig. 1. Fig. 2 shows the details of the code-aided equalizer using a convolutional code and Viterbi decoder. The second-stage equalizer makes use of coding gain, hence code-aided. The code-aided adaptive equalizer consists of four blocks: the matched-filter front-end processor, first-stage equalizer, decision generation block, second-stage equalizer and decoding block. In the data transmission period (decision directed mode), the output of the first-stage equalizer is taken as the estimates of the transmitted data (codewords) and used to update the coefficients of the first-stage equalizer. Therefore, the first-stage equalizer is able to track the channel variation in decision-directed mode but lacks robustness in the presence of burst errors. The soft-output (no hard decision made) of the first-stage equalizer is fed to the second-stage equalizer as well as the decision generation block. D_1 denotes the delay shown in Fig. 1. The delay is a critical element in the scheme. It enables timing synchronization between the outputs of the decoder and the first stage equalizer.

The decision generation block consists of a hard-decision device, deinterleaver, Viterbi decoder and re-interleaver. The hard-decision device forms the estimates of transmitted codeword. These estimates will contain burst errors as a result of deep fades. The deinterleaver in this block successfully randomizes these error bursts into random errors as long as the burst length is covered by the interleaving depth. The Viterbi decoder is then used to estimate the most probable transmitted codeword.

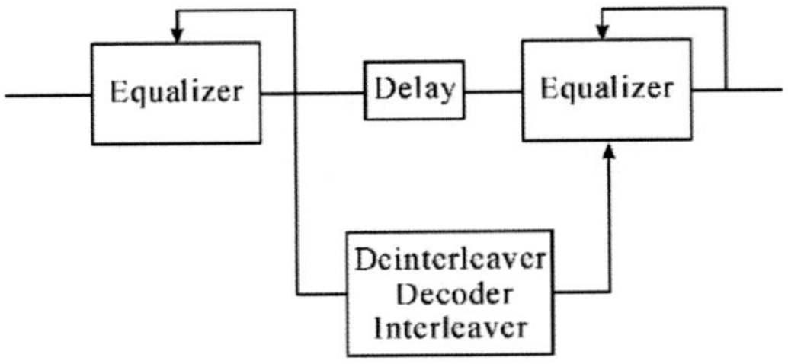

Fig. 1 A code-aided equalizer

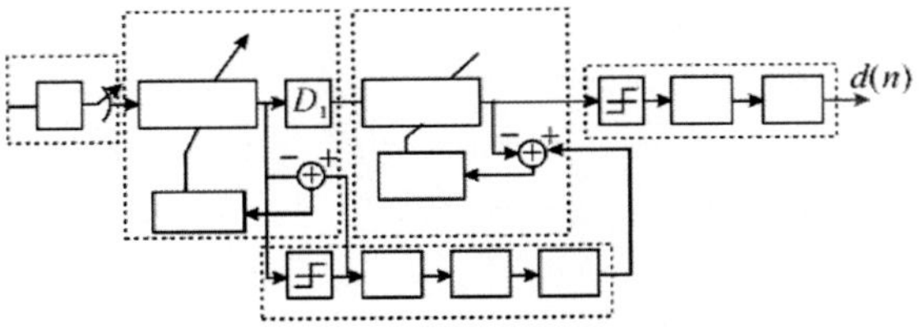

Fig. 2 A code-aided adaptive equalizer using a convolutional code

The code and, consequently, the decoder used in the code-aided adaptive equalizer can be selected by the designer. Turbo codes can therefore be used in the "deinterleaver/decoder/interleaver" box in Fig. 1. One requirement for turbo codes to be effective is, however, that the frame size (or interleaver depth) be large enough, say in the range of 10000 bits which is not suitable for mobile communications. Recently, we successfully found effective short turbo codes (in the range of a few hundred bits) for mobile communications[3-5]. In this paper, the (17/13)/3 turbo code in Fig. 3 is used. The frame size used in the simulation in this paper is 210. The number of coded bits (information bits, two parity check sequence bits and some trailing bits to drive the convolutional encoders into the all-zero state) is therefore 642. We simulated the performance of the code-aided equalizer using the turbo code over a two-ray Rayleigh fading channel as shown in the next section.

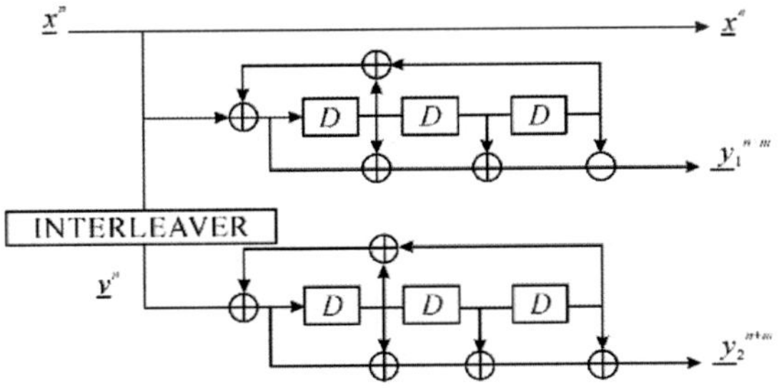

Fig. 3 A rate 1/3 turbo encoder with feedforward polynomial D^3+D^2+D+1 (17 in octal) and feedback polynomial D^3+D+1 (13 in octal) (We also call this code a 17/13 turbo code, following a convention in the literature.)

3. Simulation Results

We use a rate (17/13)/3 turbo code with interleaver length equal to 210. The interleavers discussed in [6] are used. We adopt the same two-ray fading channel model used in [1]: the relative delay between the rays is equal to one data symbol; average rms power of the two paths are 0.7 and 0.3, respectively. The fading simulators are synthesized by Jakes' model with 11 sinusoidals[10]. The two equalizers each use 11 taps. The normalized fade rate is equal to 0.005. The bit error rate (BER) curve is shown in Fig. 4. For comparison, the bit error rate performance for an uncoded system and a coded system with a convolutional code-aided equalizer (discussed in [1]) is also shown. As can be seen, the turbo code-aided equalizer is very effective. The improvement is mainly due to the soft-decision decoding nature of the turbo code used (the convolutional code-aided equalizer above used hard-decision Viterbi decoding) and due to the power of the turbo code itself.

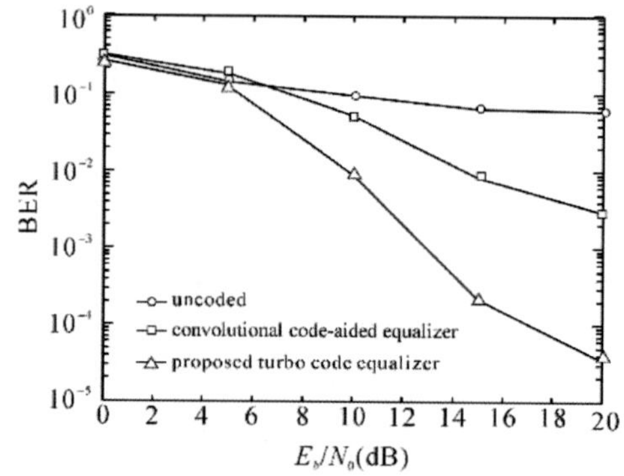

Fig. 4 BER of an uncoded system, a coded system with a convolutional code-aided equalizer and coded system with a turbo code-aided equalizer: normalized fade rate equal to 0.005, two-ray fading where the relative strength of the two paths equal to 0.7 and 0.3, respectively

4. Conclusion

A turbo code-aided equalizer is investigated in this paper and compared with an uncoded system and a convolutional code-aided equalizer in terms of bit error rate. Further improvement may be obtained in the following manner. Since iterative decoding is used for the turbo code, analog outputs can be obtained at each iteration. These analog outputs may be used to gradually adjust the tap coefficients of the equalizer.

Acknowledgements

This research is supported by Hong Kong UGC grant 357/042.

References

[1] S. Tantikovit, A. U. H. Sheikh, and M. Z. Wang, "Code-aided adaptive equalizer for cellular mobile radio systems," in *Proc. of Globecom*, 1998.

[2] S. Tantikovit, A. U. H. Sheikh, and M. Z. Wang, "Code-aided adaptive equalizer for mobile communication systems," *Electronics Letters*, vol. 34, pp. 1638-1640, Aug. 1998.

[3] M. Z. Wang, A. U. H. Sheikh, and R Qi, "Interleaver design for turbo codes based on divisibility," in *Proc. of IEEE Information Theory and Networking Workshop*, June 1999.

[4] M. Z. Wang, A. U. H. Sheikh, and F. Qi, "Interleaver design for short turbo codes," in *Proc. of Globecom*, Dec. 1999.

[5] F. Qi, Y. Z. Zhang, D. R. Shao, and M. Z. Wang, "A novel interleaver design method for turbo codes," in *Proc. of IEEE Wireless Communications and Networking Conference*, Sept. 1999.

[6] F. Qi, M. Z. Wang, A. U. H. Sheikh, and D. R. Shao, "A class of turbo code interleavers based on divisibility," *Electronics Letters*, vol. 36, Jan. 2000.

[7] R. Kohno, H. Imai, and M. Hatori, "Design of automatic equalizer including a decoder of error-correcting code," *IEEE Trans. on Commun.*, vol. 33, pp. 1142-1146, Oct. 1985.

[8] R. Kohno, S. Pasuparthy, H. Imai, and M. Hatori, "Combination of canceling intersymbol interference and decoding of error-correcting code," *IEE Proc. part F*, vol. 133, pp. 224-230, June 1986.

[9] R. Sharma, W. D. Grover, and W. A. Krzymien, "Forward error control (FEC)—assisted adaptive equalization for digital cellular mobile radio," *IEEE Trans, on Veh. Tech.*, vol. 42, pp. 94-102, Feb. 1993.

[10] W. C. Jakes, *Microwave Mobile Communications*. New York: IEEE Press, 1994.

Performance Analysis of a Digital Acquisition Circuit for DS/FH Spread Spectrum Signals*

Xiaojie Wen, Weining Song, Yanzhong Zhang, Dingrong Shao and Shujian Li

School of Electronic and Information Engineering, Beihang University, Beijing, China

Abstract: Fast acquisition is the major challenge for DS (direct sequence)/FH (frequency hopping) spread spectrum system. A DS/FH acquisition circuit which implements a digital noncoherent correlation scheme is studied extensively. The closed-form expressions of the mean acquisition time by single dwell decision and two-dwell decision are obtained with the state diagram respectively. The results show that, in the circumstance of low signal noise ratio (SNR), the approach of two-dwell decision performs better than the approach of single dwell decision.

Keywords: detection probability; false alarm probability; mean acquisition time; spread spectrum

1. Introduction

Hybrid direct sequence/frequency hopping (DS/FH) spread spectrum signals are widely used in the civil and military communications for its inherent properties such as low probability of intercept (LPI) and low probability of jamming (LPJ) capabilities[1-4]. Fast synchronization plays a critical role at the receiver. It is generally completed in two steps: first, acquisition is succeeded, and then tracking is continued in the whole transmission period. In order to increase the efficiency and decrease acquisition time a fast acquisition is required.

The acquisition problem has attracted considerable research in the recent past (see [1-7]). The most popular technique is based on a serial research [2-7], [9] followed by a control strategy, whose performance is expressed in terms of mean acquisition time. In [6], a digital acquisition scheme using an integrate-and-dump circuit is presented. It is less complex to implement, however, more time is needed for acquisition. In [7] a fast scanning and waiting method using the digital matched filter (DMF) is proposed. With this method, the mean acquisition time can be decreased significantly. There are two control strategies considered in the method, single dwell decision and two-dwell decision. The search strategy using single dwell decision can increase the detection probability while the one using two-dwell decision can decrease the false alarm probability. In this paper, the performance of the two search strategies is analyzed and the closed-form expressions of

---

* Published in Proc. IEEE Wireless Communication Conf. ,2009: 101-104.

Y. Zhang, *Footprints in Cambridge and Aviation Industries of China*,
https://doi.org/10.1007/978-981-16-3176-4_15

mean acquisition time are given.

This paper is organized as follows. The acquisition circuit and algorithm are proposed in Section 2. In section 3, the performance of two decision schemes is analyzed. In Section 4, the numerical results are given in terms of key system parameters in the circumstances of additive Gaussian noise.

2. Acquisition Circuit and Algorithm

2.1 Acquisition Circuit

The model of acquisition circuit is depicted in Fig. 1. The received DS/FH spread spectrum signal on one frequency hop can be expressed as

$$s(t) = \sqrt{2P}c(t-\zeta T_c)d(t)\cos(2\pi f_i t+\theta) \tag{1}$$

where

P is the transmitter signal power;

$c(t-\zeta T_c)$ is a ± 1 valued, M-chip-long PN code with chip interval T_c seconds, delayed by ζT_c with respect to an arbitrary time reference;

f_i is the frequency of hop i whose value is taken from the N-value set $\{f_1, f_2, f_3, \cdots, f_n\}$. N is the total number of hopping frequencies;

θ is random phase of hop i;

$d(t)$ is data sequence.

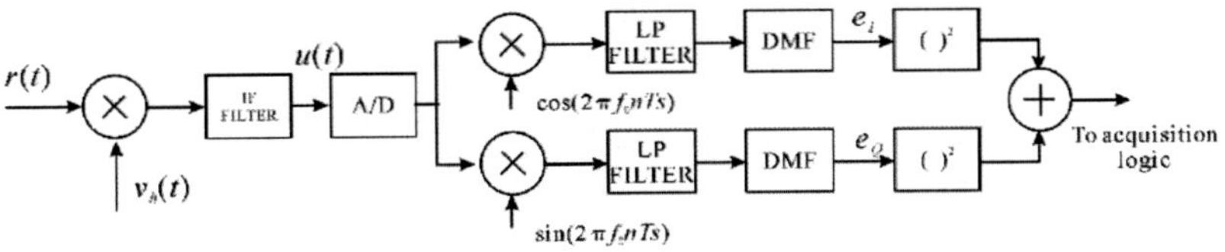

Fig. 1 Model of a digital acquisition circuit

For the whole acquisition time the data signal $d(t)$ is assumed to be $+1$, so that it can be ignored in (1).

At the receiver the signal $s(t)$ is corrupted by narrow-band Gaussian noise $n(t)$, whose single-side power spectral density is N_0. Then the received signal can be written as

$$r(t) = s(t) + n(t) \tag{2}$$

It is then multiplied by $v_h(t)$ to produce $u(t)$. $v_h(t)$ is generated by frequency synthesizer which is controlled to emit a certain frequency f'_j, namely

$$v_h(t) = \cos(2\pi f'_j t) \tag{3}$$

In (3) f'_j, $j=1,2,\cdots,N$ is taken from the N-value set $\{f'_1, f'_2, f'_3\cdots, f'_N\}$, where $f_i - f'_i = f_0$, $i=1,2,\cdots,N$. When the frequency of the incoming signal is f_i, and the local one is f'_j, $j\neq i$, there is no signal output which can be sent to the DMF circuit. If the signal with the correct frequency f'_i is got, the process of $u(t)$ is similar to DS signals and the correlation peak can be got after de-spreading, that is, the acquisition is achieved.

In order to synchronize the sender and receiver, the first task is to make both of them use the same hop frequency at the same time. In the acquisition process, H_1 denotes the hypothesis that the frequency is in synchronization and H_0 denotes the alternative hypothesis the frequency is asynchronous.

The $u(t)$ is changed to baseband signal through digital down converter. General expressions for the in-phase and quadrature components of the DMF output are

$$e_I = \sqrt{S}T_c\cos(\theta)\delta(i-j)\sum_{k=1}^{M} c_{p+k}c_k + \sum_{k=1}^{M} n_I(q+k)c_k = y\cos(\theta)\delta(i-j)+N_I \tag{4}$$

$$e_Q = \sqrt{S}T_c\sin(\theta)\delta(i-j)\sum_{k=1}^{M} c_{p+k}c_k + \sum_{k=1}^{M} n_Q(q+k)c_k = y\sin(\theta)\delta(i-j)+N_Q \tag{5}$$

where N_I and N_Q are noise components with zero mean value and variance σ_n^2 each, $\delta(x)$ is Kronecker delta function. The values of $\delta(i-j)$ corresponding to the H_1 and H_0 regions are thus given by

$$H_1:\delta(i-j)=1 \tag{6}$$

and

$$H_0:\delta(i-j)=0 \tag{7}$$

respectively.

The envelope can be obtained as

$$R=\sqrt{e_I^2+e_Q^2} \tag{8}$$

R is sampled at chip rate which yields M decision variables $\{R_m\}, m=1,2,\cdots,M$; every PN code period $T_d=MT_c$. Then $R_{\max}\triangleq\max\{R_m\}, m=1,2,\cdots,M$ will denote the largest decision variable. Comparing $R_{\max}$ with a preset threshold V_T allows for a decision to be made as to whether or not acquisition has been achieved.

2.2 Acquisition Algorithm

The algorithm for FH acquisition without any prior knowledge whose specific can be summarized as follows:

(1) set f'_j to frequency f'_1;

(2) scan f'_j 1 for LMT_c seconds;

(3) if acquisition is succeeded then stop and record $f'_i=f'_j$;

(4) if acquisition is succeeded and $j<N$, then increase j, go to step 2, when $j=N$, go to step 1.

An important factor to reduce the mean acquisition time is the acquisition strategy. There are two control strategies considered on Step 2: single dwell decision and two-dwell decision. Both of them make a decision during LMT_c seconds. $L-1$ noncoherent correlation peaks $R_{\max_l}$, $l=1,2,\cdots,L-1$ are got after MT_c seconds for the DMF initialization.

When the system adopts single dwell decision, once $R_{\max_l}>V_T$ a "hit" is declared and the acquisition algorithm goes to Step 3. On the other hand, the decision for acquisition is negative (H_0) when $R_{\max_l}<V_T, l=1,2,\cdots,L-1$ occurs and then the algorithm goes to Step 4.

When the system adopts two-dwell decision, once a "hit" is observed in the first dwell, it follows a second dwell for verification. During this period, a number of independent correlation tests sampled at MT_c intervals, named A, will be performed by the detector. If at least B of A tests indicate synchronization, it decides the acquisition is completed and algorithm goes to Step 3. Otherwise, it rejects the cell and search goes to Step 4.

3. Performance Analysis

The acquisition process of the proposed algorithm is modeled as the state diagram[8] shown in Fig. 2. In Fig. 2, State S is the state from which a transition to the acquisition state (ACQ) can occur, states $F_1, F_2, \cdots, F_{N-1}$ correspond to the cells under H_0, and the false alarm state (FA) is an intermediate state at which a transition form F_i to F_{i+1} may occur. The total number of states here is therefore $N+2$. Acquisition can be achieved only in S.

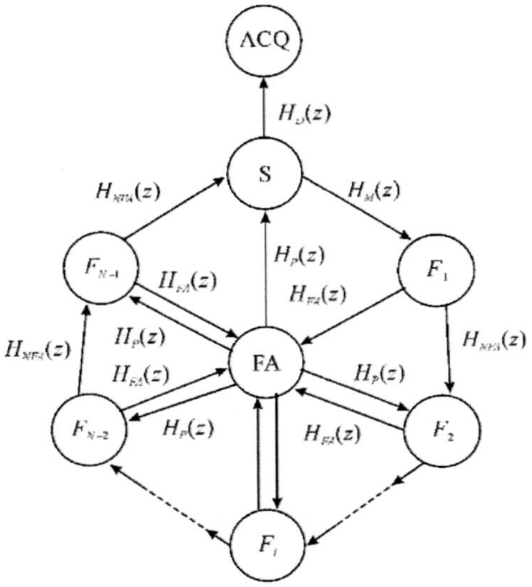

Fig. 2 A state diagram of the acquisition process

In Fig. 2, $H_D(z)$ is the gain of the branch leading from node S to node ACQ; $H_M(z)$ is the gain of the branch connecting S with F_1 while $H_0(z)$ is the gain of the branch connecting any other two successive nodes (F_i, F_{i+1}); $i=1,\cdots,N-1$. Furthermore, the process can move between any two successive node (F_i, F_{i+1}); $i=1,\cdots, N-1$ either without false alarm (associated gain $H_{NFA}(z)$) or by first reaching the FA state (branch gain $H_{FA}(z)$), then pass from the FA to node F_{i+1}[branch gain $H_P(z)$] so that the gain $H_0(z)$ is $H_{NFA}(z)+H_{FA}(z)H_P(z)$.

Since two control strategies are considered here, the details of acquisition process are analyzed respectively.

3.1 Single Dwell Decision

The details of state S for single dwell decision are shown in Fig. 3.

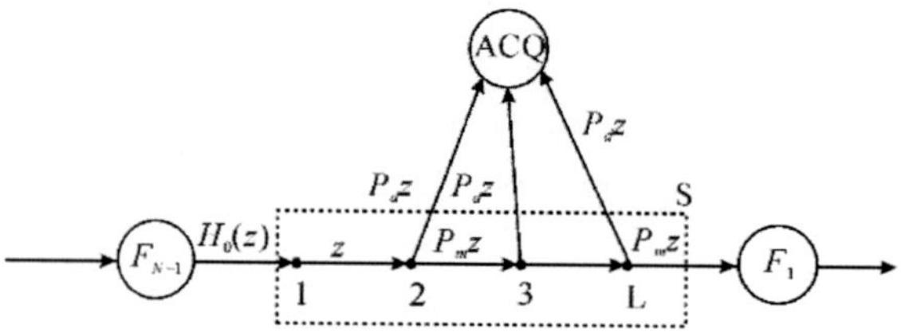

Fig. 3 The details of state S for single dwell decision

From Fig. 3, $H_D(z)$, $H_M(z)$, H_{NFA}, $H_{FA}(z)$, and $H_P(z)$ and be obtained as follows:

$$H_D(z) = z \cdot P_D z\{\sum_{i=0}^{L-2}[(1-P_D)z]^i\} \tag{9}$$

$$H_M(z) = z[(1-P_D)z]^{L-1} \tag{10}$$

$$H_{NFA}(z) = z \cdot (P_{NFA} z)^{L-1} \tag{11}$$

$$H_{FA}(z) = z \cdot P_{FA} z\{\sum_{i=0}^{L-2}[P_{NFA} z]^i\} \tag{12}$$

$$H_P(z) = z^K \tag{13}$$

and

$$H_0(z) = z \cdot (P_{NFA} z)^{L-1} + z \cdot P_{FA} z\{\sum_{i=0}^{L-2}[P_{NFA} z]^i\} z^K \tag{14}$$

The mean acquisition time can be obtained as

$$\overline{T_{acq}} = \frac{MT_c}{H_D(1)}[H_D(1) + H_M(1) + (N-1)H_0^1(1)(1-\frac{H_D(1)}{2})] \tag{15}$$

where

$$H_D(1) = 1-(1-P_D)^{L-1} \tag{16}$$

$$H'_D(1) = P_D \sum_{i=0}^{L-2}(i+2)(1-P_D)^i \tag{17}$$

$$H'_M(1) = L(1-P_D)^{L-1} \tag{18}$$

$$H'_0(1) = L(1-P_{FA})^{L-1} + P_{FA}\sum_{i=0}^{L-2}(i+2+K)(1-P_{FA})^i \tag{19}$$

3.2 Two-Dwell Decision

Fig. 4 shows the flow graph diagram of S for two-dwell decision system. Furthermore, (P_{d1}, P_{fa1}) and (P_{d2}, P_{fa2}) denote the detection and false alarm pairs for the first and second (verification) dwells, respectively.

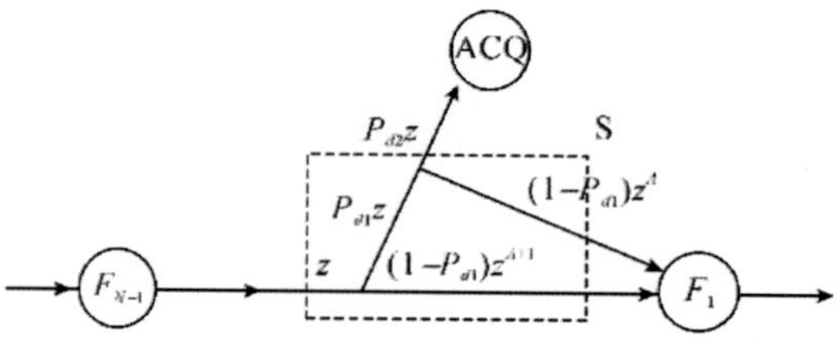

Fig. 4 The details of state S for two-dwell decision

From Fig. 4, $H_D(z)$, $H_M(z)$, $H_{NFA}(z)$, $H_{FA}(z)$, and $H_P(z)$ can be obtained as follows:

$$H_D(z)=z\cdot P_{d1}zP_{d2}z^A \tag{20}$$

$$H_M(z)=z[(1-P_{d1})z^{A+1}+P_{d1}z(1-P_{d2})z^A] \tag{21}$$

$$H_{NFA}(z)=z[(1-P_{fa1})z^{A+1}+P_{fa1}z(1-P_{fa2})z^A] \tag{22}$$

$$H_{FA}(z)=z\cdot P_{fa1}zP_{fa2}z^A \tag{23}$$

$$H_P(z)=z^K \tag{24}$$

and

$$H_0(z)=[(1-P_{fa1})+P_{fa1}(1-P_{fa2})+P_{fa1}P_{fa2}z^K]z^{A+2} \tag{25}$$

We define $P_{d,c}=P_{d1}P_{d2}$, $P_{fa,c}=P_{fa1}P_{fa2}$. After some algebra, the mean acquisition time of two-dwell system can be obtained as

$$\overline{T_{acq}}=\frac{MT_c}{P_{d,c}}[L+(N-1)(L+KP_{fa,c})(1-\frac{P_{d,c}}{2})] \tag{26}$$

3.3 Calculation of the Detection Probability and False Alarm Probability

To evaluate the mean acquisition time of the two decision systems, we need to have P_d and P_{fa}. Assume the detection threshold of the DMF is V_T, the detection probability P_d can be approximated by[9]

$$P_d=Q\left(\sqrt{\frac{2ST_cM}{N_0}},\sqrt{\frac{2V_T^2}{N_0MT_c}}\right) \tag{27}$$

And the false alarm probability is

$$P_{fa}=\exp\left(-\frac{V_T^2}{2\sigma_n^2}\right) \tag{28}$$

It is easily seen that the second dwell probability pair (P_{d_2}, P_{fa2}) can be expressed as

$$P_{d2}=\sum_{n=B}^{A}\binom{A}{n}P_d^n(1-P_d)^{A-n} \tag{29}$$

$$P_{fa2}=\sum_{n=B}^{A}\binom{A}{n}P_{fa}^n(1-P_{fa})^{A-n} \tag{30}$$

where $\binom{A}{n}$ is the binomial coefficient.

4. Numerical Results and Conclusions

In this section, the performance of the two control strategies is compared. The performance was evaluated with the following system parameters: the number of penalty PN code periods $K=100$ and the total number of hopping frequencies $N=16$. The threshold T_V was selected numerically to minimize the mean acquisition time at each value of SNR. Note that it was defined as $2PT_c/N_0$.

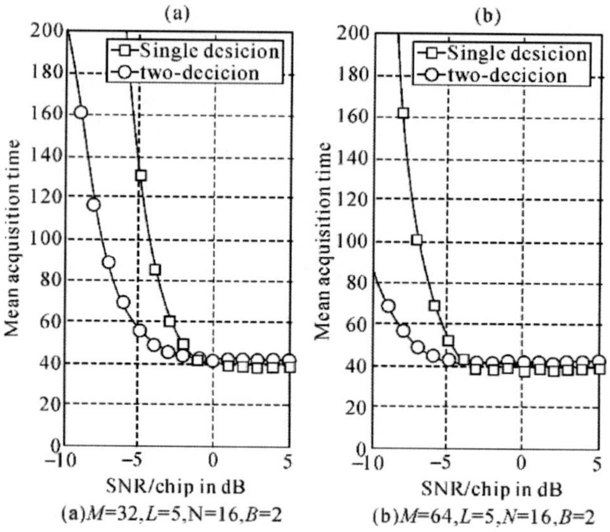

Fig. 5 Comparison of the mean acquisition time normalized the PN code period between the single dwell decision and two-dwell decision schemes

Fig. 5 shows the mean acquisition time, normalized to PN code period MT_c, of the single dwell decision and two-dwell decision schemes. From the figure, we can see that (1) when the SNR is high enough, the performance of two-dwell decision is similar to the performance of the single dwell decision; (2) under the condition of low SNR, the approach of the two-dwell decision performs better than the one using single dwell decision; (3) if SNR is invariable, the mean acquisition time is decreased as the PN code length increases, especially in the case of low SNR.

References

[1] T. Vanninen, H. Saarnisaari, M. Raustia and T. Koskela, "FH-code phase synchronization in a wireless multi-hop FH/DSSS adhoc network," Military Communications Conference, 2006, pp. 1-7.

[2] M. Song and S. Wigginton, "Frequency hopping pattern detection in wireless ad hoc networks," Proceedings of the International Conference on Information Technology: Coding and Computing, 2005, pp. 633-638.

[3] Minjian Zhao, Mingxia Xu, Jie Zhong, and Shiju Li, "Slot synchronization in ad hoc networks based on frequency hopping synchronization," Wireless Communications, Networking and Mobile Computing, 2006, pp. 1-4..

[4] A. Tabassam and S. Heiss, "Bluetooth clock recovery and hop sequence synchronization using software defined radios," Region 5 Conference, 2008, pp. 1-5..

[5] E. Sourour and A. Elezabi, "Robust acquisition of hybrid direct sequence-slow frequency hopping spread-spectrum under multi-tone and Gaussian interference in fading channels," Wireless Communications and Networking Conference, 2008, pp. 917-922.

[6] N. Benvenuto, G. Guidotti and S. Pupolin, "Performance of a digital acquisition circuit for hybrid FH-

DS spread spectrum systems," Military Communications Conference, 1988, pp. 971-975.

[7] Bo Zhang, Dingrong Shao and Shujian Li, "Study on fast synchronization of hybrid DS/FH system," Journal of Beijing University of Aeronautics and Astronautics, 2005,31(11),pp. 1226-1231. (in chinese).

[8] A. Polydoros and C. Weber, "A unified approach to serial search spread-spectrum code acquisition-part I: general theory," IEEE Transactions on Communications, 1984, 32(5), pp. 542-549.

[9] X. Tan,"Performance of acquisition in a digital matched-filter for DSSS," Microwave,Antenna, Propagation and EMC Technologies for Wireless Communications,2005, pp. 927-931.

Fine Timing and Frequency Synchronization for CMMB System*

SUN Yuming, ZHANG Yanzhong, SHAO Dingrong, LI Shujian
School of Electrical Engineering, Beihang University, Beijing, China

Abstract: A fine timing and frequency synchronization algorithm is presented for an orthogonal frequency division multiplexing (OFDM) system used in China mobile multimedia broadcasting (CMMB). The presented algorithm could accomplish the timing and frequency offset estimation using the perfect relation of training symbols. The simulation and practice results show the new algorithm could give out fine timing synchronization and accurate frequency offset estimation over AWGN channel and multipath channel.

Keywords: OFDM; CMMB; timing synchronization; frequency offset estimation

1. Introduction

In the broadband wireless digital communication system, a primary interference of the high speed information transmission is the frequency-selective fading. The main advantage of OFDM is that it allows transmission over highly frequency-selective channels at a low receiver implementation cost[1], but a well-known problem of OFDM is its vulnerability to synchronization errors. As a result, Synchronization at the receiver is one important step that must be performed in an OFDM system[2].

[3-6] proposed some synchronization methods using training sequence. In [3, 4], a symbol timing and frequency synchronization algorithm using repeated training sequence is presented. [5] proposed a specifically designed training sequence which consists of two segments of equal length where each segment is constructed from a different pseudo-noise (PN) sequence. [6] solved the synchronization acquisition problem of OFDM system in practice. The beacon in CMMB framing structure is a special training sequence[7].

In this paper, a synchronization algorithm about timing and carrier frequency estimation based on training symbols is proposed. In the CMMB beacon, two same OFDM training symbols generated by pseudo-random sequence are transmitted. So, the presented algorithm could accomplish the timing synchronization and frequency offset estimation using the perfect relation of training symbols. The simulation and practice results show the new algorithm could give out fine timing synchronization and

* Published in Proc. of 2009 Conf. on Communication Faculty: 498-501.

Y. Zhang, *Footprints in Cambridge and Aviation Industries of China*,
https://doi.org/10.1007/978-981-16-3176-4_16

accurate frequency offset estimation over additive white Gaussian noise (AWGN) channel and multipath channel.

2. System Model

Considering a frame of OFDM data which consists of M data symbols, the received OFDM signal with frequency and timing offset can be represented as

$$r[i,n]=\begin{cases} s[i,n-N_g],n\in[N_g,N-1+N_g]\cap i\in[1,M] \\ s[i,n+N-N_g],n\in[0,N_g-1]\cap i\in[1,M] \\ q[i,n],n\in[0,N_t+N_g]\cap i=0 \end{cases} \tag{1}$$

where i represents the OFDM symbols number, n represents the sampling number, N_t represents the synchronization signal length, N_g is the length of guard interval, N represents the OFDM symbol length, $q[i,n]$ is the received training sequence and $s[i,n]$ is the received data symbol.

The training sequence consists of two training symbols and one cyclic prefix. The received training sequence can be represented as (2), where θ denotes the timing offset, $N(n)$ is the additive noise, $PN(k)$ denotes the pseudo-random sequence, H_k is the k subcarrier channel frequency response, ξ represents the normalized sampling frequency offset, Δf is the normalized frequency offset.

$$\begin{cases} q[0,n]=\begin{cases} t[n-N_g], & n\in[N_g,N_t-1+N_g] \\ t[n+N_t-N_g], & n\in[0,N_g-1] \\ t[n-N_g-N_t], & n\in[N_t+N_g,2N_t+N_g-1] \end{cases} \\ t[n]=\frac{1}{\sqrt{N_t}}\sum_{k=0}^{N_t-1}PN(k)H_k\exp\{\frac{j2\pi kn(1+\xi)}{N_t}+\frac{j2\pi\Delta f(1+\xi)n}{N_t} \\ \quad +\frac{j2\pi k(\xi(i(N_t+N_g)+N_g)+\theta)}{N_t}+\frac{j2\pi\Delta f(i(N_t+N_g)+N_g)}{N_t}\}+N(n) \end{cases} \tag{2}$$

3. Synchronization Algorithm

Considering $PN(k)PN^*(k)=1$, the correlation between the demodulated data of training sequence and the known pseudo-random sequence could be written as

$$\lambda(\theta,\Delta f)=\sum_{k=0}^{N_t-1}R_p[k]PN^*[k]=\sum_{k=0}^{N_t-1}\{\frac{\sin(\pi\Delta f)}{N_t\sin(\frac{\pi\Delta f}{N_t})}\exp(-j2\pi k\theta/N_t-\frac{\pi(N_t-1)\Delta f}{N_t})\}+N_p \tag{3}$$

where N_p is the additive white Gaussian noise because of the pseudo-random characteristics. The fast fourier transform could be used to estimate θ. As a result, the estimation of θ and Δf could be written as[8]

$$\begin{cases} \hat{\theta}=\arg_\theta\max(\|FFT\{R_p[k]PN^*[k]\}\|) \\ \theta_c=\arg_\theta\max(\|\sum_{n=\hat{\theta}}^{n=N_t-1}r[0,n]r^*[0,n+N_t]\|) \\ f_I=\arg_{\Delta f}\max(\|\lambda(\hat{\theta},\Delta f)\|) \\ f_F=-\frac{1}{2\pi}\angle(\sum_{n=\hat{\theta}+x}^{n=N_t-1}r[0,n]r^*[0,n+N_t]) \end{cases} \tag{4}$$

where $\hat{\theta}$ is the fine estimation of θ, x represents the number of channel paths, f_I is the integral frequency offset estimation of Δf, f_F is the fractional frequency offset estimation of Δf, θ_c is the coarse timing estimation of θ, FFT{} denotes the fast fourier transform.

The new algorithm could implement as follows:

1) Estimate the coarse estimation of timing offset using (4).

2) Demodulate the second training symbol of training sequence by the fast fourier transform.

3) Frequency searching is applied to the fine timing and integral frequency capture, according to (4).

4) By means of (4), the fractional frequency offset estimation is realized.

5) Estimate the fine timing position use (4) and repeat (4) to track the OFDM signal.

4. Simulations

We use Monte Carlo simulations to evaluate the performance of the algorithm in CMMB system[8]. Suppose the simulation parameters as follow:

1) The 8 MHZ mode of CMMB would be used in the simulations. The baseband width is 10 MHz. In the beacon, pseudo-random sequence is modulated by the 2048 point inverse fast fourier transform. There are 1536 data sub-carriers and 512 virtual sub-carriers. As a result, the real baseband width is about 8 MHz. The guard interval length is 24.

2) The frequency offset is 3.24240625 KHz which is more than one sub-carrier interval about 801 Hz.

3) The pseudo-random sequence is m sequence which code length is 2047.

4) The signal is transmitted in AWGN, multipath A and B channel respectively. From Fig. 1, multipath A channel is consist of 5 paths. The delay of each path is 0, 3, 5, 7 and 11. The gain of each path could represent as $h_i = \exp\{-\tau_i/2\}$. where τ_i is the delay of the i path, h_i is the gain of the i path. The delay of multipath B channel path is the same as A. The gain of each path is 1, 0.891, 0.354, 0.316 and 0.1. It is obviously that the main paths of multipath B channel are more than A about 1 path, so the inter-symbol interference in multipath B channel is severer than A.

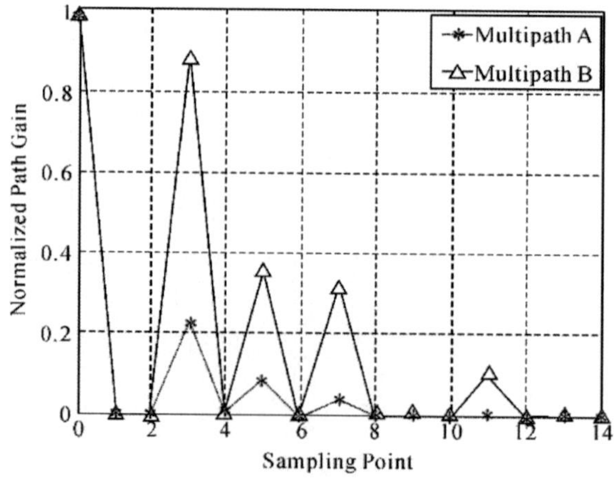

Fig. 1 Multipath channel

5) In each simulation, 1000 time slots are used. One slot includes 1 beacon and 53 OFDM data symbols.

The simulation results show the timing error is less than 1 sampling point and the frequency estimation error is less than 1 Hz. From Fig. 2, the new algorithm could estimate the amount of multipath. From Fig. 3, when signal-to-noise (SNR) ration is more than 10 dB, the mean-square error (MSE) of AWGN channel is almost the same as multipath channel.

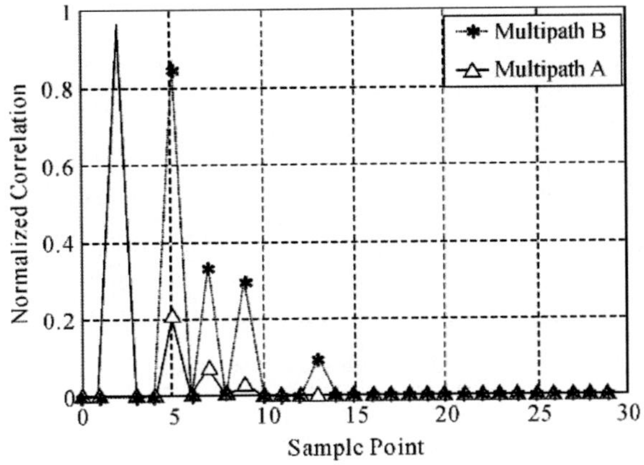

Fig. 2 Estimate the channel paths when SNR is 20 dB

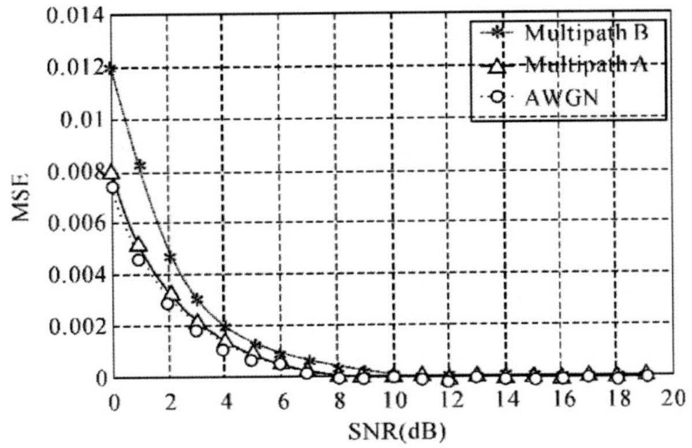

Fig. 3 The MSE of frequency estimation

5. Practice

The new algorithm had already been implemented in practice, from Fig. 4. Firstly, the received signal converted to baseband signal with frequency offset by the digital down conversion block. Secondly, estimate the approximate timing position using the coarse timing estimation block. Thirdly, compute the correlation by the FFT block which could use the same FFT block in FPGA because of the different working time. Finally, estimate the fine timing position and frequency offset.

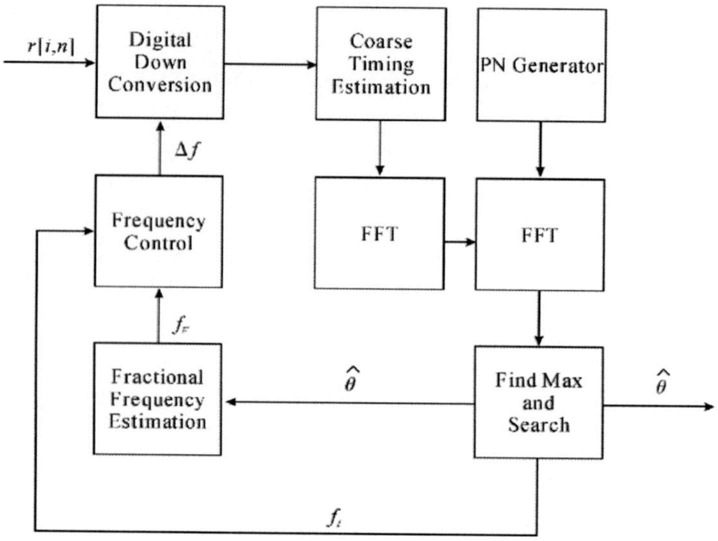

Fig. 4 Implement flow chart

To calculate the timing estimation mean-square error, a counter has been designed to record the amount of point between two correlation peaks. Standard signal has been generated by MATLAB which the signal-to-noise ration of is from 0 to 30 dB. The signal is sent by vector signal generator. Acquire the counter outcome in each signal-to-noise ration using ChipScope which is an online logic analyzer. The practice results show the timing mean-square error is less than 1 point which is the same as simulation and the frequency estimation error is less the 1 Hz.

6. Conclusions

In this paper, a fine timing and frequency synchronization algorithm is presented used in CMMB. The presented algorithm could accomplish the timing and frequency offset estimation over AWGN channel and multipath channel. The simulation and practice results show the new algorithm could give out fine timing synchronization within 1 sampling point and accurate frequency offset estimation within 1 Hz over AWGN channel and multipath channel.

References

[1] M. Speth, S. A. Fechtel, G. Fock et al., "Optimum receiver design for wireless broad-band systems using OFDM. I," Communications, IEEE Transactions on, Vol. 47, No. 11, pp. 1668-1677, 1999.
[2] T. M. Schmidl, and D. C. Cox, "Robust frequency and timing synchronization for OFDM," Communications, IEEE Transactions on, Vol. 45, No. 12, pp. 1613-1621, 1997.
[3] Z. H. Yu, C. Kai, H. Yu Mei et al., "OFDM Timing and Frequency Offset Estimation Based on Repeated Training Sequence." pp. 264-266.

[4] C. Yang, Y. Tan, S. Li et al., "A high performance frequency offset estimation method for OFDM systems." pp. 2236-2239 Vol. 3.

[5] A. Si, S. Leung, and C. Leung, "A Novel Synchronization Scheme for OFDM Over Fading Channels." pp. IV-IV.

[6] D. Liu, and J. Wang, "An efficient joint synchronization acquisition algorithm for OFDM." pp. 813-817.

[7] The State Administration of Radio Film and Television. GY/T 220. 1-2006. Mobile Multimedia Broadcasting Part 1: Framing Structure, Channel Modulation for Broadcasting Channel: 2006. (In Chinese)

[8] J. J. van de Beek, M. Sandell, and P. O. Borjesson, "ML estimation of time and frequency offset in OFDM systems," Signal Processing, IEEE Transactions on, Vol. 45, No. 7, pp. 1800-1805, 1997.

Study and Implementation of Receiver Based on High Data Rate M-ary Spread Spectrum with Non-Coherent Demodulation*

SONG Weining, ZHANG Yanzhong, SHAO Dingrong, LI Shujian, WEN Xiaojie

School of Electrical Engineering, Beihang University, Beijing, China

Abstract: Taking into account the request of high data rate and high spread spectrum processing gain, scheme of M-ary spread spectrum, fast PN-code acquisition and non-coherent demodulation is proposed. Experimental results show that with these methods, the system could reach 1 M data rate and 15 dB spread spectrum processing gain under 10^{-6} bit error rate.

Keywords: non-coherent demodulation; M-ary spread spectrum; high data rate

1. Introduction

Spread spectrum is widely used in the civil and military communications for its inherent properties such as low probability of intercept, low probability of jamming, and superior anti-jamming capabilities[1-4]. However, with high spread spectrum processing gain, it is very difficult to achieve high data rate[5].

In order to achieve high data rate with high spread spectrum processing gain, this paper proposes a project with M-ary spread spectrum using cyclic shift of PN-code or different PN-code to transmit data[6]. In this way, both high data rate and high spread spectrum processing gain could be reached.

This paper is organized as follows: the system model is proposed in Section 2; in Section 3, some key technologies are analyzed; in Section 4, simulation and experimental results are given out to verify system performance.

2. System Model

In this system, M-ary spread spectrum uses M cyclic shifts of PN-code to transmit M-ary data. Its modulation principle is as followings: serial data is converted to n bits parallel data $d[i]$. Every $d[i]$ is corresponded to a cyclic shift of M_2 by coding matrix, which is modulated in Q branch. I branch is modulated in normal spread spectrum way using M_1 for receiver synchronization. Then the

* Published in Proc. of 2009 Conf. on Communication Faculty: 502-504.

Y. Zhang, *Footprints in Cambridge and Aviation Industries of China*,
https://doi.org/10.1007/978-981-16-3176-4_17

received signal is

$$r(t)=M_1(t)\sqrt{2P}\cos(w_c t+\varphi)+M_2(t-dT_c)\sqrt{2P}\sin(w_c t+\varphi)+n(t)$$

where M_1, M_2 are I branch synchronous code and Q branch modulation code respectively; $n(t)$ is additive white noise; w_c is carrier frequency; φ is carrier phase.

3. Synchronization Algorithms

3.1 PN-code Fast Acquisition

The acquisition problem posses an extremely status in receiver. The most popular technique is based on the fast Fourier transform (FFT) and digital matched filter (DMF). FFT needs complicated calculations, while DMF needs only store units and adders, but longer the code is, more adders are needed. This paper presents folded digital matched filter (FDMF) to complete acquisition. DMF needs only store units and adders as DMF but less adders than DMF[7].

In fact, FDMF is a constant coefficient FIR filter. If its input is $x[n]$, coefficient sequence is $h[n]$, its output is $y[n]=\sum x[n-k]^* h[k]$, that is

$$y(n)\sum_{k=0}^{L-1} x(n-k)h(k) = [\cdots[[[\sum_{k=0}^{N-1} x(n-k)h(k)]+\sum_{k=N}^{2N-1} x(n-k)h(k)]$$

$$+\sum_{k=2N}^{3N-1} x(n-k)h(k)]+\cdots]+\sum_{k=(M-1)N}^{L-1} x(n-k)h(k)$$

where $L=N\times M$, M is folded times, L is code interval.

3.2 PN-Code Tracking

Digital delay-locked loop (DDLL) is invoked to tracking PN-code. Assume input PN-code is $r(t)=c(t-T_b)\cos\varphi$, in which T_b is delay of PN-code, φ is phase aberration.

Ignore noise, local two signals are

$$a_1(t)=c(t-\hat{T}_b+\frac{1}{2}T_c)$$

$$a_2(t)=c(t-\hat{T}_b-\frac{1}{2}T_c)$$

in which $\hat{T}_b$ is estimation of T_b, T_c is chip interval.

The outputs of correlator are

$$z_1=R_c[(\varepsilon+\frac{1}{2}T_c)]\cos\varphi$$

$$z_2=R_c[(\varepsilon-\frac{1}{2}T_c)]\cos\varphi$$

ε is deviation of $\hat{T}_b$ to T_b.

Then the output of DDLL is

$$z_1^2-z_2^2=A^2[R_c^2[(\varepsilon+\frac{1}{2}T_c)]-R_c^2[(\varepsilon-\frac{1}{2}T_c)]]=A^2D(\varepsilon)$$

where $A=\cos^2\varphi$,

$$D(\varepsilon)=\begin{cases} 0 & -N+\frac{3}{2}<\varepsilon\leq-\frac{3}{2} \\ -(\frac{3}{2}+\varepsilon)^2 & -\frac{3}{2}<\varepsilon\leq-\frac{1}{2} \\ 2\varepsilon & -\frac{1}{2}<\varepsilon\leq\frac{1}{2} \\ (\frac{3}{2}-\varepsilon)^2 & \frac{1}{2}<\varepsilon\leq\frac{3}{2} \\ 0 & \frac{3}{2}<\varepsilon\leq N-\frac{3}{2} \end{cases}$$

3.3 Non-Coherent Demodulation

As to BPSK signal, coherent demodulation should be used. Since in this system, Q branch modulates information data by means of cyclic shifts of PN-code, then non-coherent demodulation can be used. So even local carrier phase is not coherent to the input signal, data can be demodulated correctly.

Take the nth branch for example: the nth branch correlates with input data, the result is

$$IIN^n=\int_0^T M_1(t)\sqrt{2}P\cos(\omega t+\Delta\varphi)M_2(t-nT_c)\mathrm{d}t+\int_0^T M_2(t-dT_c)\sqrt{2}P\sin(\omega t+\Delta\varphi)M_2(t-nT_c)\mathrm{d}t$$

$$QIN^n=\int_0^T M_1(t)\sqrt{2}P\sin(\omega t+\Delta\varphi)M_2(t-nT_c)\mathrm{d}t+\int_0^T M_2(t-dT_c)\sqrt{2}P\cos(\omega t+\Delta\varphi)M_2(t-nT_c)\mathrm{d}t$$

After squarer and adder, the result is

$$r(t)=[\int_0^T M_2(t-dT_c)M_2(t-nT_c)\mathrm{d}t]^2\times 2P^2$$

That is, only when $d=n$, $r(t)$ will get its maximum value. In this way, d can be demodulated.

4. Simulations and Experimental

4.1 FDMF Simulation

Four folded matched filter is formed by ROMs, adders, delay units and feedback logic units. The work flow is like this: when the data comes, during the first clock cycle, code 0, 1, 2, ⋯, 7 give out DMF's coefficients, and 0 is added to the first adder. The final addition result is stored in the result register. During the second clock cycle, code 8, 9, ⋯, 15 give out DMF's coefficients, and the data in the result register is added to the first adder. The third and fourth clock is similar to the second clock. After the fourth clock, the data in the result register is the output of FDMF. Fig. 1 shows the simulation result of FDMF by

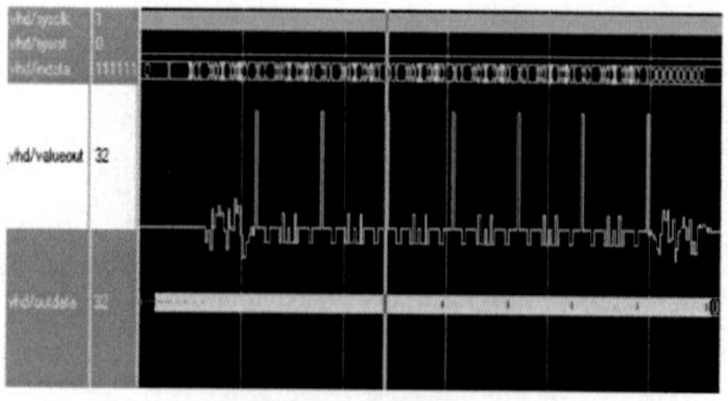

Fig. 1 FDMF simulation result

Modelsim.

It can be seen from Fig. 1 that, when input PN-code has the same phase as local PN-code, FDMF get maximum output value, that is, 32. Otherwise, the output value will be 0, +4 and −4.

4.2 Practical Results

To verify the system performance, comparing receiving data and emission data, result is showed in Fig. 2 and Fig. 3. The receiving data was got from logic analyzer and the test result was got by Matlab. Fig. 2 shows frame head compared result. It can be seen that there is no frame lost in the 37 frames data. And from Fig. 3 it can be found that there is no error in these 37 frames data. From net test (by SecureCRT), it can be seen that when receiving 10000 frames, there are 2 frames lost and 1 frame (3 bits) error.

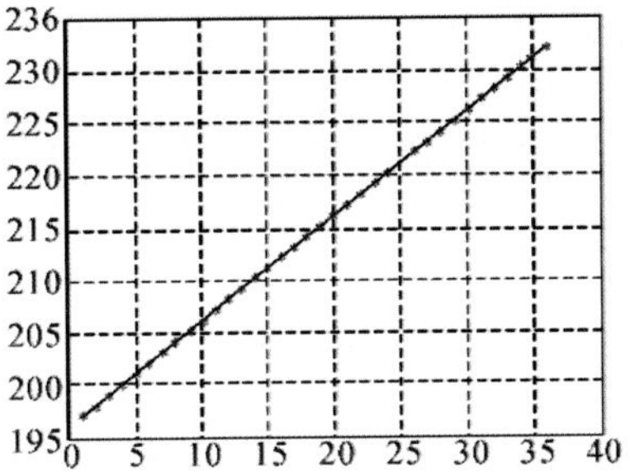

Fig. 2 System performance test Ⅰ

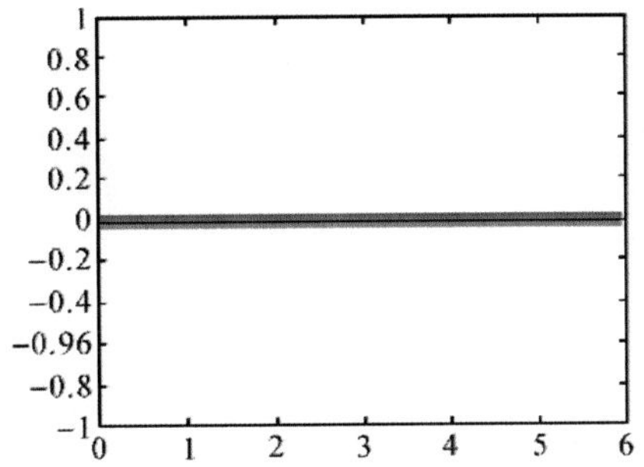

Fig. 3 System performance test Ⅱ

5. Conclusions

This system adopts M-ary spread spectrum, solving the problem between high data rate and high spread spectrum processing gain; using non-coherent demodulation, so that data can be correctly demodulated with residual phase error. Practice results show that in the case of 15 dB spread spectrum processing gain and 1 M data rate, system with the design above can reach 10^{-6} bit error rate and 0.01% frame loss rate, which meets the need of practical system.

References

[1] T. Vanninen. H. Saarnisaari, M. Raustia, and T. Koskela. FH-Code Phase Synchronization in a Wireless Multi-Hop FH/DSSS Adhoc Network[J]. Military Communications Conference, 2006: 1-7.

[2] M. Song, and S. Wigginton. Frequency Hopping Pattern Detection in Wireless Ad Hoc Networks [J]. Proceedings of the international Conference on information Technology: Coding and Computing, 2005: 633-638.

[3] Minjian Zhao, Mingxia Xu, Jie Zhong, and Shiju Li. Slot Synchronization in Ad Hoc Networks Based on Frequency Hopping Synchronization[J]. Wireless Communications, Networking and

Mobile Computing, 2006: 1-4.

[4] A. Tabassam, and S. Heiss. Bluetooth Clock Recovery and Hop Sequence Synchronization Using Software Defined Radios [J]. Region 5 Conference, 2008: 1-5.

[5] Bi Yanming, Li Wenxing, and Zhao Zhongyang. A New Method For High Data Transmission Rate of JTIDS[J]. Applied Science and Technology, 2007, 34(11):35-38.

[6] Jin Jian, Ying Guoping, and Kong Guoqiang. Application of M-ary Coding Spread Spectrum Technique in Wireless Communication System[J]. Avionics Technology, 2004, 39(4):16-20.

[7] Xu Feng, Shao Dingrong, and Li Shujian: Fast PN Code Acquisition for DSSS Receiver in High Dynamic Situation: Journal of Beijing University of Aeronautics and Astronautics[J]. 2007, 33(6):672-676.

On Channel Estimation in DTMB Standard Using Time Domain PN Sequences*

Zhang Zhan, Zhang Xiao-lin, Zhang Yan-zhong, Zhang Chao
School of Electronic and Information Engineering, BUAA, Beijing, China

Abstract: In some OFDM systems, time domain CP (cyclic prefix) and scattered pilots are used for synchronization and channel estimation. While in novel OFDM broadcasting systems, such as China Digital Terrestrial Multimedia Broadcasting Standard (DTMB), PN sequences rather than cyclic prefixes are inserted as guard interval between N-point IFFT blocks, which are also used as training symbols. This paper proposed a frequency domain transform method for channel estimation in time domain PN based OFDM systems. With simulation results, comparing to LS-DFT method that using scattered pilots for channel estimation, it is shown that our proposed method can provide satisfactory performance in time domain PN based terrestrial television broadcasting standard.

Keywords: channel estimation; time domain PN; LS-DFT; DTMB; OFDM

1. Introduction

As an effective technique of multi-carrier modulation, orthogonal frequency division multiplexing (OFDM) is widely used in wireless communication systems. In some OFDM systems, cyclic prefix (CP) is inserted at the beginning of every OFDM symbol, combating multipath fading. However, OFDM systems with CP do not provide accurate frame synchronization, and when the channel is complicated, more pilots need to be inserted in frame body which reduces channel efficiency. In Chinese DTV standard, a method of inserting pre-defined time domain PN sequences between OFDM symbols was proposed[1]. Although this approach caused more hardware complexities, it reduces transmission overhead, provides better channel spectrum efficiency and channel estimation performances[3].

This paper is organized as follows: In Section 2, the frame structure of DTMB is briefly overviewed. In Section 3, the details of channel estimation using time-domain PN and scattered pilots will be both discussed. In Section 4, simulation result is presented and performance is compared. Section 5 is the conclusion of this paper, a brief implementation result of channel estimation on FPGA will be introduced.

* Published in Proc. of IEEE Conf. on Communication, 2010: 1058-1061.

Y. Zhang, *Footprints in Cambridge and Aviation Industries of China*,
https://doi.org/10.1007/978-981-16-3176-4_18

2. Frame Structure of DTMB

The standard named "Framing Structure, Channel Coding and Modulation for Digital Television Terrestrial Broadcasting System" (referred to as DTMB) was published in August 2006[1]. DTMB standard supports 4.813 Mbps~32.486 Mbps net data rate in 8 MHz bandwidth and provides 330 kinds of transmission modes[4].

In DTMB standard, the signal frame is the basic element of the system frame structure. A signal frame consists of two parts, frame header and frame body. The baseband symbol rates for both frame header and frame body are the same, and defined as 7.56 Mbps. The frame structure of DTMB is shown in Fig. 1.

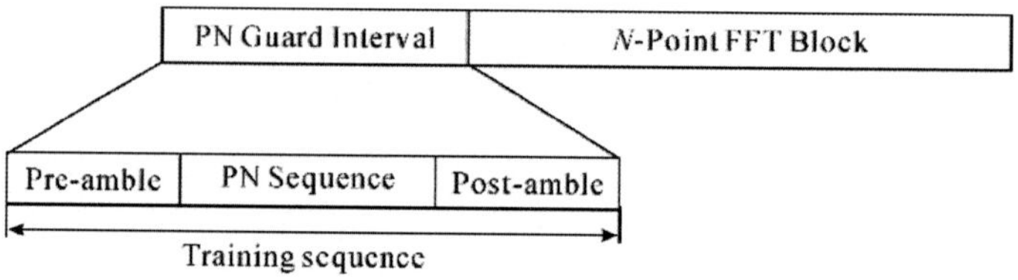

Fig. 1 The frame structure of DTMB

The frame header uses PN sequence and there are three options on the frame header lengths as shown in Fig. 2. Frame header uses 4QAM modulation scheme with the same I and Q components. Frame body consists of 36 system information symbols and 3744 data symbols, totally 3780 symbols. The duration time of Frame body is 500μs(3780 * 1/7.56μs)[1].

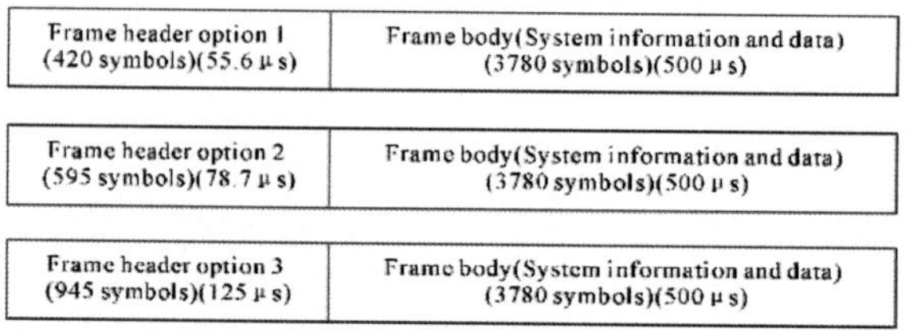

Fig. 2 Three oPtions for frame header

PN guard intervals can be used as training symbols for synchronization and estimation purpose so there is no additional pilot symbols needed, thus high spectrum efficiency is achieved. The self-correlation of PN sequence provides robustness to the system, the time domain correlative calculation period is 1~2 millisecond, it is much faster than other DTTB systems[2].

The cyclic structure of PN guard interval is deliberately designed. In Fig. 3(a), it is shown the slippage correlation result of three complete PN sequence, the one in the middle has circular correlation feature, so the correlating result is ideal, while the other 2 sequences aside introduced side lobe noise. Then for cyclic PN sequence composed of three same m sequences, as shown in Fig. 4, the Pre-amble is from the tail of the sequence, while the Post-amble is from the head. In Fig. 3(b), result shows the interference from the side lobe is reduced.

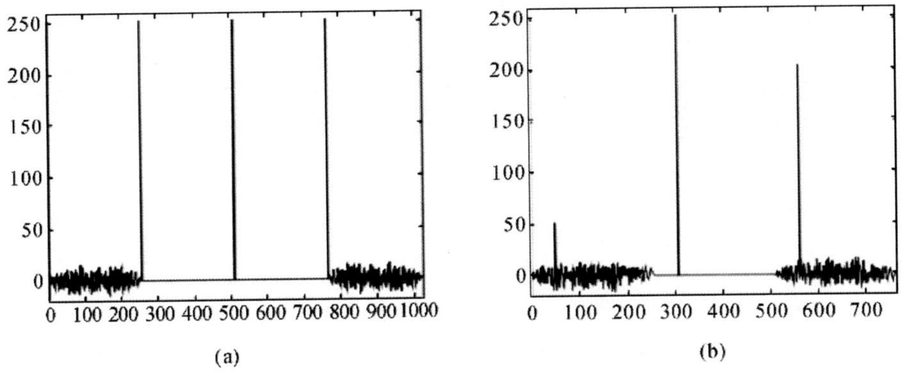

Fig. 3 (a) Result of non cyclic PN sequences
(b) Result of cyclic PN sequences

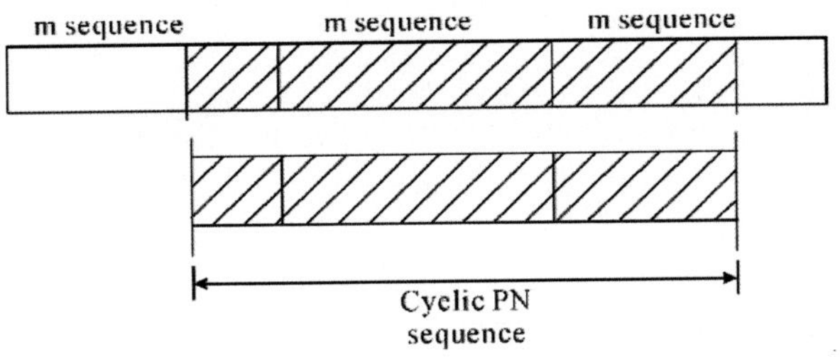

Fig. 4 Cyclic PN sequnce composed by 3 m sequences

3. Channel Estimation

The time varying multipath channel impulse response can be modeled as below:

$$h_m = [h_{m,l}]^T = [h_{m,1}, h_{m,2}, \cdots, h_{m,L}, 0, \cdots, 0]^T \quad (1)$$

In this expression, m is the number of symbols received; L is the maximum time delay spread. Let the length of training sequence $N_g > L$. Let the DFT length equal to N.

3.1 Channel Estimation Using Time-Domain *PN*

Assume the transmitted signal is:

$$x = [x_1, x_2, \cdots, x_{N_c}]^T \quad (2)$$

N_c is the length of m sequence in PN guard interval, if it is a cyclic PN sequence, $N_c < N_g$. So the received sequence will be the cyclic convolution of transmitted sequence and channel impulse response.

Without loss of generality, assume the length of Preamble and Post-amble is both N_c, then

$$r_n = h \otimes x_n \quad n = 1, 2, \cdots, N_c \quad (3)$$

r_n is the received signal, '$\otimes$' denotes the cyclic convolution of channel impulse response and transmitted signal. The received signal in matrix becomes:

$$r_m=\begin{bmatrix} r_{m,1} \\ r_{m,2} \\ \vdots \\ r_{m,N_c} \\ r_{m,N_c+1} \\ \vdots \\ r_{m,2N_c-1} \end{bmatrix}=\begin{bmatrix} x_1 & x_{N_c} & x_{N_c-1} & \cdots & x_2 \\ x_2 & x_1 & x_{N_c} & \cdots & x_3 \\ \vdots & \vdots & \vdots & \ddots & \vdots \\ x_{N_c-1} & x_{N_c-2} & x_{N_c-3} & \cdots & x_{N_c} \\ x_{N_c} & x_{N_c-1} & x_{N_c-2} & \cdots & x_1 \\ \vdots & \vdots & \vdots & \ddots & \vdots \\ x_{N_c-1} & x_{N_c-2} & x_{N_c-3} & \cdots & x_{N_c} \end{bmatrix}\cdot\begin{bmatrix} h_{m,1} \\ \vdots \\ h_{m,L} \\ \vdots \\ 0 \end{bmatrix}+w=X^{CYC}h_m+w \quad (4)$$

X^{CYC} is the cyclic convolution matrix composed of x. w is noise which cause estimation error.

Then the channel impulse response estimation value could be the liner convolution of r_m and X^H, where X^H is the matrix of local training sequence:

$$\hat{h}_m=\frac{1}{N_c}X^H r_m=\frac{1}{N_c}X^H(X^{CYC}h_m+w)=\frac{1}{N_c}X^H X^{CYC}h_m+\frac{1}{N_c}X^H w \quad (5)$$

where

$$X^H X^{CYC}=\begin{bmatrix} x_1^* & x_2^* & \cdots & x_{N_c}^* & 0 & \cdots & 0 \\ 0 & x_1^* & \cdots & x_{N_c-1}^* & x_{N_c}^* & \cdots & 0 \\ \vdots & \vdots & \ddots & \vdots & \vdots & \ddots & \vdots \\ 0 & 0 & \cdots & x_1^* & x_2^* & \cdots & x_{N_c}^* \end{bmatrix}\begin{bmatrix} x_1 & x_{N_c} & x_{N_c-1} & \cdots & x_2 \\ x_2 & x_1 & x_{N_c} & \cdots & x_3 \\ \vdots & \vdots & \vdots & \ddots & \vdots \\ x_{N_c-1} & x_{N_c-2} & x_{N_c-3} & \cdots & x_{N_c} \\ x_{N_c} & x_{N_c-1} & x_{N_c-2} & \cdots & x_1 \\ \vdots & \vdots & \vdots & \ddots & \vdots \\ x_{N_c-1} & x_{N_c-2} & x_{N_c-3} & \cdots & x_{N_c} \end{bmatrix}$$

$$=\begin{bmatrix} R(0) & R(1) & \cdots & R(N_c-1) \\ R(1) & R(0) & \cdots & R(N_c-2) \\ \vdots & \vdots & \ddots & \vdots \\ R(N_c-1) & R(N_c-2) & \cdots & R(0) \end{bmatrix}=R(0)I_{N_c}+R_p \quad (6)$$

I_{N_c} is N_c dimension identical matrix. R_p is the remaining matrix representing side lobe noise. $R(l)$ is cyclic self-correlation function, if the training sequence is m sequence, then:

$$R(l)=\begin{cases} N_c & l=0 \\ -1 & l=2,3,\cdots,N_c-1 \end{cases} \quad (7)$$

The remaining matrix is simply composed of -1, so the channel estimation result could be:

$$\hat{h}_m=h_m+\frac{1}{N}R_p h_m+\frac{1}{N}X^H w \quad (8)$$

where the first item is the real channel impulse response, the second item represents the side lobe noise, the third item represents the estimation error caused by noise.

The estimation mean squared error (MSE) is:

$$MSE=\frac{\mathrm{tr}\{E[(h_m-\hat{h}_m)(h_m-\hat{h}_m)^H]\}}{N}=\frac{\mathrm{tr}\{R_p E[h_m h_m^H](R_p)^H\}}{N^2}+\frac{\sigma_w^2}{N} \quad (9)$$

3.2 The Frequency Transform for Channel Estimation

Assume the first received N_c elements of r_m are:

$$r'_m=\begin{bmatrix} r_{m,1}\\ r_{m,2}\\ \vdots \\ r_{m,N_c}\end{bmatrix}=\begin{bmatrix} x_1 & x_{N_c} & x_{N_c-1} & \cdots & x_2\\ x_2 & x_1 & x_{N_c} & \cdots & x_3\\ \vdots & \vdots & \vdots & \ddots & \vdots\\ x_{N_c-1} & x_{N_c-2} & x_{N_c-3} & \cdots & x_{N_c}\\ x_{N_c} & x_{N_c-1} & x_{N_c-2} & \cdots & x_1\end{bmatrix}\cdot\begin{bmatrix} h_{m,1}\\ \vdots\\ h_{m,L}\\ \vdots\\ 0\end{bmatrix}+\eta=X_{N_c}h_m+\eta \tag{10}$$

The DFT transformation of r'_m is:

$$\begin{aligned} DFT(r'_m)&=W_{N_c}X_{N_c}h_m+W_{N_c}\eta=W_{N_c}X_{N_c}W_{N_c}^{H}H_m+W_{N_c}\eta\\ &=\text{diag}[DFT(X_{N_c})]\cdot H_m+W_{N_c}\eta \end{aligned} \tag{11}$$

where W_{N_c} is the DFT transform matrix:

$$W_{N_c}=\begin{bmatrix} W^0 & W^0 & W^0 & \cdots & W^0\\ W^0 & W^1 & W^2 & \cdots & W^{N_c-1}\\ W^0 & W^2 & W^4 & \cdots & W^{2(N_c-1)}\\ \vdots & \vdots & \vdots & \ddots & \vdots\\ W^0 & W^{N_c-1} & N_c^{2(N_c-1)} & \cdots & W^{(N_c-1)(N_c-1)}\end{bmatrix} \tag{12}$$

$$W=\frac{1}{/N_c}e^{-j(2\pi/N_c)} \tag{13}$$

Then we get the estimation result in frequency domain:

$$\widehat{H}_m=H_m+(\text{diag}[DFT(X_{N_c})])^{-1}DFT(r'_m) \tag{14}$$

where X_{N_c} is the matrix of local training sequence.

The estimation result in time domain could be:

$$\begin{aligned} \widehat{h}&=h_m+IDFT((\text{diag}[DFT(X_{N_c})])^{-1}W_{N_c}\eta)\\ &=h_m+W_{N_c}^{H}(\text{diag}[DFT(X_{N_c})])^{-1}W_{N_c}\eta=IDFT(\widehat{H}_m) \end{aligned} \tag{15}$$

We notice that with frequency domain transformation, through simple correlation and IDFT calculation, the channel response can be easily estimated.

3.3 Channel Estimation Using Scattered Pilots

Assume $X_m=(X_{m,1},X_{m,2},\cdots X_{m,N})^T$ is transmitted frequency domain signal, $H_m=(H_{m,1},H_{m,2},\cdots,H_{m,N})^T$ is the channel frequency domain response. And the channel response of subcarriers where pilots are inserted is:

$$H_m^P=(H_{m,1}^P,H_{m,2}^P\cdots H_{m,N_P}^P)^T=(H_{m,1},H_{m,M}\cdots H_{m,(N_P,M)})^T \tag{16}$$

Then the received signal is:

$$R_m^P=\text{diag}(X^P)H_m^P+W^P \tag{17}$$

The superscript P denotes the position of pilots. We then perform the least squares (LS) calculation:

$$\widehat{H}_m^{P,LS}=[\text{diag}(X^P)]^{-1}R_m^P=H_m^P+[\text{diag}(X^P)]^{-1}W^P=(\frac{R_{m,1}^P}{X_1^P}\frac{R_{m,2}^P}{X_2^P}\cdots\frac{R_{m,N_P}^P}{X_{N_P}^P})^T \tag{18}$$

which can be expressed as:

$$\widehat{H}_{m,l}^{P,LS}=H_{m,1}^P+\frac{W_{m,1}^P}{X_l^P}\quad l=1,2,\cdots,N_P \tag{19}$$

So the MSE is:

$$MSE^{LS}=E[(\widehat{H}_{m,l}^{P,LS}-\widehat{H}_{m,l}^{P})(\widehat{H}_{m,l}^{P,LS}-\widehat{H}_{m,l}^{P})^{*}]=E(\left|\frac{W_{m,l}^{P}}{X_{l}^{P}}\right|^{2})=\frac{\sigma_W^2}{|X_L^P|^2} \quad (20)$$

After performing LS calculation, full channel response can be achieved by DFT Interpolation process. As shown in Fig. 5, the time domain impulse response is calculated from the LS estimation result by performing IFFT with zero padding. After that, we fill the numbers outside the training sequence period with zero, and then we can get the full channel response by performing a FFT transform. We call it LS-DFT method in the following parts.

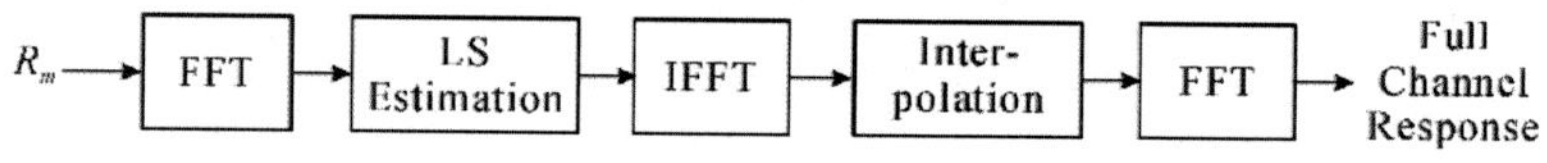

Fig. 5 DFT interpolation

4. Simulation Results

For wireless communication channel, we choose WSSUS model for simulation, the channel impulse response is[5]:

$$h(\tau,t) = a_0 e^{j2\pi f_{D_{LOS}} t}\delta(\tau) + a_1 e^{j\theta_0} e^{j2\pi f_{D_0} t}\delta(\tau-\tau_0) + \lim_{N\to\infty} a_2 \frac{1}{\sqrt{N}}\sum_{n=1}^{N} e^{j\theta_n} e^{j2\pi f_{D_0} t}\delta(\tau-\tau_n)\delta(\tau-\tau_0)a_1 \quad (21)$$

The first item is LOS (Line of Sight) component, the second item is reflection component and the third item is multipath component. a_0, a_1, a_2 is the strength of signal; θ_n, f_{Dn}, τ_n are random variables, which denote Phase of reflection signal, Doppler frequency offset and Delay of path (n) respectively; $1/\sqrt{N}$ is the normalization factor to ensure the summation power of n paths is 1.

The time and frequency domain of channel response we set is shown is Fig. 6. For stationary channel, we set the Doppler offset to 0, for dynamic channel, we set Doppler offset to 100 Hz and 300 Hz.

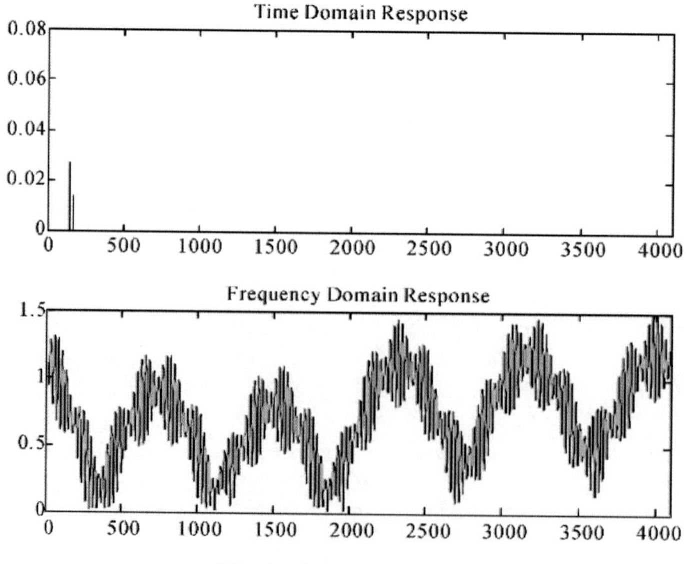

Fig. 6 Channel response

For convenience to compare the time domain and frequency domain estimation performance, we created a frame structure of 4k DFT body length with PN512 training sequence for guard interval, which is composed of complete PN255 with pre-amble and post-amble. We insert 512 scattered pilots in the frame body, then get both time and frequency domain estimation results in the same simulation circumstance.

For stationary channel simulation, the MSE of channel estimation under different signal to noise ratio (SNR) are shown in Fig. 7. When the SNR is under 15 dB, the MSE of time domain PN channel estimation is much lower than using scattered pilots method. When SNR is higher than 15 dB, because of the interference of side lobe noise, an error floor is introduced to the system.

The bit error rate (BER) of channel estimation under different SNR is shown in Fig. 8. Since equalization and demodulation may introduce processing noise, the difference of BER performance between time and frequency domain methods may not be as much as MSE performance difference. However, we notice that time domain estimation performance is better than LS-DFT method when the SNR is under 15 dB.

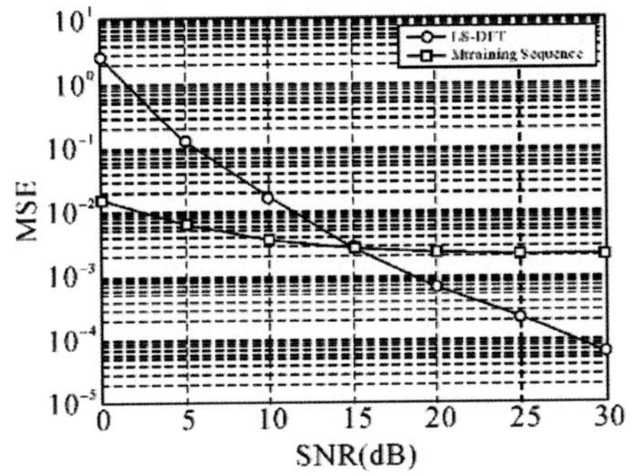

Fig. 7 MSE to SNR for channel estimation results

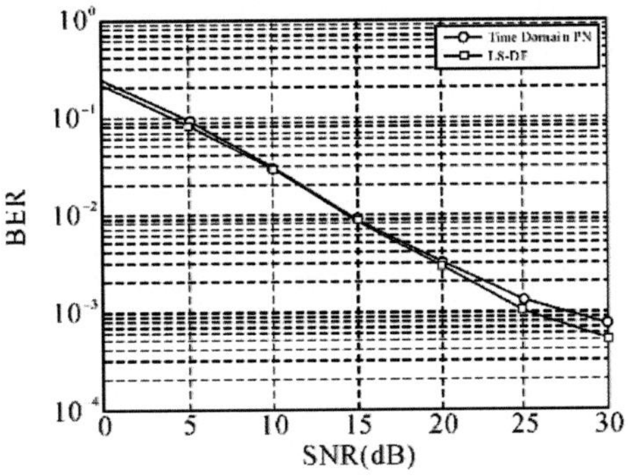

Fig. 8 BER to SNR for channel estimation results

For dynamic channel, the BER to SNR performance is shown in Fig. 9. Performance of LS-DFT method is better. In time domain, the PN guard interval is inserted at the beginning of the frame, there is no pilot in the frame body, so the channel estimation calculation can only be processed at the

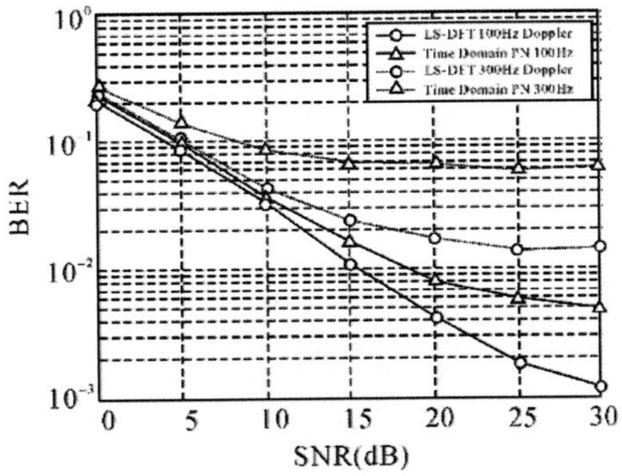

Fig. 9 BER to SNR under channel with Doppler offset

beginning of every frame. In frequency domain, the pilots are evenly inserted in the frame body, so the channel varieties in one frame time can be traced accurately, thus the performance of LS-DFT is better under dynamic channel.

5. Conclusion

In this paper, the frame structure of DTMB is introduced, the methods of time and frequency domain channel estimation methods are discussed. Then we propose a method that can easily get channel response from cyclic PN training sequence by frequency transformation. The simulation results are given for understanding performances of different estimation methods, we notice that for stationary channel, the time domain PN method is better, but under dynamic channel, using scattered pilots can improve performance, so we know that a mixed estimation method may introduce more performance under different channels. It is worth mentioning that the method we propose is already implemented in our FPGA development board based on Altera Stratix Ⅲ series. As part of DTMB standard based receiver, with help of powerful foreword error control encoding techniques such as LDPC code, our method can provide satisfactory system performances under more complicated channels.

References

[1] Framing Structure, Channel Coding and Modulation for Digital Television Terrestrial Broadcasting System (in Chinese), Chinese National Standard GB 20600-2006.

[2] Chinese DMB-TH Terrestrial Digital TV Transmission Standard White Paper. Digital Television Research Center of Tsinghua University, 2004.

[3] Xiao-lin Zhang. The principle of digital television design (in Chinese). Beijing: Higher Education Press, 2008.

[4] Bowei Song, Lin Gui, Yunfeng Guan and Wenjun Zhang, "On Channel Estimation and Equalization in TDS-OFDM based Terrestrial HDTV Broadcasting System", IEEE Transactions on Consumer Electronics, Vol. 51, No. 3, AUGUST 2005.

[5] Hoeher P. A statistical discrete-time model for the WSSUS multipath channel[J]. IEEE Trans Veh Techmol, 1992, 41(4): 461-468.

On Dual Timing Modulator for Chinese Digital Terrestrial Multimedia Broadcasting*

ZHANG Zhan, ZHANG Xiao-lin, ZHANG Yan-zhong
School of Electronic and Information Engineering, BUAA, Beijing, China

Abstract: A novel design of modulator for Chinese Digital Terrestrial Multimedia Broadcasting Standard is presented. Different from the traditional system of digital TV modulator, a simplified GPS/Compass2 dual timing receiver is embedded into the Single Frequency Network(SFN) block, which is removable and independent from the modulator. With this modification, the timing availability and the feasibility of switching between Multi-Frequency Network (MFN) and SFN mode is increased and meanwhile the costs are reduced. The lab tests are based on the prototype of this modulator, which is assembled with existing mature chips and modules, the results show the performance meets the Chinese National Standards on modulators and SFN applications.

Keywords: multimedia broadcasting; digital TV modulator; dual timing; terrestrial standard

1. Introduction

The digital terrestrial multimedia broadcasting standard (DTMB) named "Frame structure, channel coding and modulation for digital television terrestrial broadcasting system" was published in China in August 2006[1].

This standard describes the processing of transmission stream(TS) from 4.813 Mbps to 32.486 Mbps net data rate in 8 MHz bandwidth and provides 330 modulation modes for terrestrial DTV broadcasting. Following in 2008, the standard of digital television modulator and the standard of single frequency network adapter complementary to DTMB national standard are proposed[2, 3]. Since many modulators have been developed even massive produced, the State Administration of Radio, Film and Television (SARFT) announced that the forgoing institutions must not use any radio, film and TV equipments without a valid network access license. Relevant equipments to access broadcasting station, TV station, Coverage Network, and Monitoring Network must be certified of compliant with the national standards by SARFT. As the most important network of terrestrial multimedia broadcasting, single frequency network (SFN) strongly depends on the GPS timing system in the traditional modulators. In 2009, the USA government announced that GPS is becoming invalid due to technical issues. So another backup timing system such as Compass2 system becomes great need.

* Published in Proc. of IEEE Conf. on Information Technology & Computer Science 2010, 273-276.

Y. Zhang, *Footprints in Cambridge and Aviation Industries of China*,
https://doi.org/10.1007/978-981-16-3176-4_19

In this paper, a modulator embedded with GPS/Compass2 timing receiver is presented, which can benefit the SFN timing redundancy and MFN/SFN applications. It is also tested to meet the SARFT standards.

2. System Design and Structure

Shown in Fig. 1, the modulator is composed of 4 parts. The clock generator module generates synchronized system clocks for every working digital and RF parts. In the Base-band Module, the incoming Transport Stream (TS), which is transferred in Asynchronous Serial Interface (ASI) and Synchronous Parallel Interface (SPI) protocol, will be processed and reformed to Signal-Frames. In the IF/RF module, Base-band signal is interpolated and up-converted to IF/RF band. The controller module generates system commands from the input of user interface or remote interface. We will introduce the modules in following subtitles respectively.

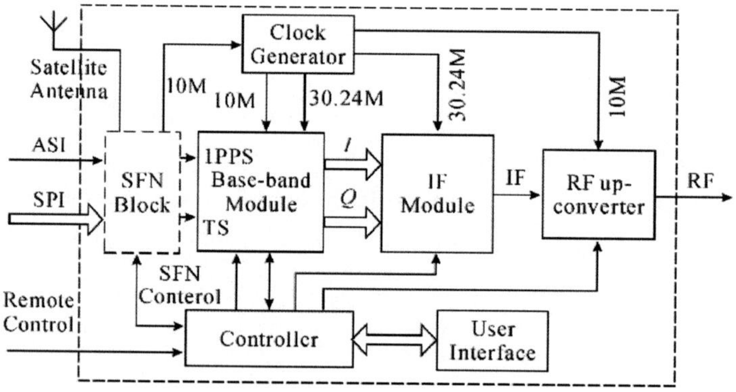

Fig. 1 Block diagram of the modulator

2.1 Base-Band Module

Shown in Fig. 2, Base-band module is the most important module in the modulator, since the standard mostly defines Base-band signals, nearly all signal processes are performed in this module.

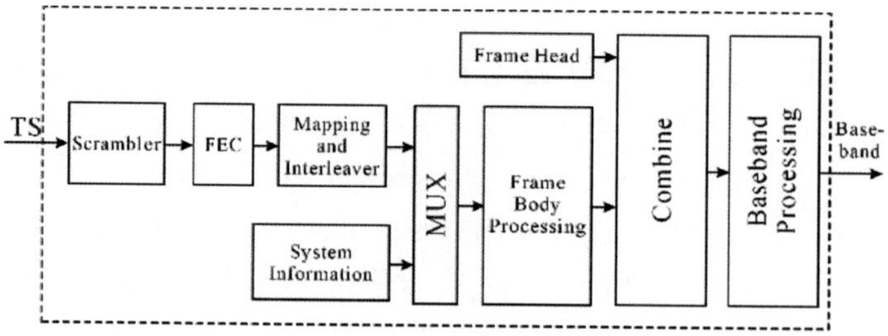

Fig. 2 Block diagram of base-band module

The input data is baseband transmission stream (TS) that compliant with standard GB/T 17975.1. Beside MPEG-2, the AVS/DRA standards that have been released recently are also taken into account of the common interface support. Anyway, the input of the modulator can only adopt the data format of TS standard, which may be transferred in ASI or SPI protocol. The length of data package is 188 bytes, one byte synchronous header 47 Hex with 187 bytes data. Therefore, any input bit stream meets the format can be transparently transmitted by the modulator[4].

To make the signal spectrum become more similar to the characteristics of the white noise, the base-band input data needs to be randomized by the scrambling code.

In the FEC block, the outer code is BCH (762, 752), which is the shorten code of BCH (1023, 1013). The BCH code is used to adapt the data rate. When modulator is working on various modes, the BCH code could adapt each signal frame length to the integer multiple of TS package length. The LDPC code is adopted as inner code. The total bit of the LDPC code word is 7493. Since there are 4, 6 and (or) 8 BCH code words are included in 1 LDPC codeword respectively, there are 3 modes of FEC coding rates, 0.4, 0.6 and 0.8, corresponding to the numbers of information bits 3048, 4572 and 6096. The first five parity bits of LDPC code word will be deleted, so the total bit of LDPC codeword is finally 7488.

Coded stream could be mapped to 5 modulation modes, namely 64QAM, 32QAM, 16QAM, 4QAM and 4QAM-NR. Then the symbols will be collected and reformed to Signal-Frames.

In Frame-Body Processing Block, if OFDM transmission mode is selected, 3780 points FFT will be performed on the Signal-Frames, if Single-Carrier mode is selected, the Signal-Frames will be sent to the shape-forming FIR filter directly.

In the Signal-Frame, a cyclic PN sequence is used as guard interval to perform fast synchronization, efficient channel estimation and equalization, which is widely discussed[5].

There are 2 clock domains in base-band module, one is the input TS stream clock domain, in which the stream is stabilized and stored in to a FIFO. The other is system working clock domain, which clocks are synthesized from 30.24 MHz clock. The FIFO is used to transfer the stream from slow TS clock domain to system working domain. All blocks are coded by Verilog language and implanted in FPGA device (Altera Stratix Ⅱ EP2S90). The utilization is 42% of available resources.

2.2 SFN Block

SFN is a significant function for DTV broadcasting, when the coverage range is much wider than one transmitter could reach, several transmitters should transmit the same bit at the same time on the same frequency[2], in order to broadcast the same TV programs in the same time without interfering each other. For accurate time synchronization, the 1 PPS from the timing receiver is needed as the base timing signal for every modulator in the network. To set the network parameters remotely, a Second-frame Initialization Packet (SIP) will be sent to modulators at every 1 PPS.

Shown in Fig. 3, at the raising edge of 1 PPS, a SIP is delivered from SFN adaptor to modulators in the network, and band then reaches a modulator at T_{rec}. To synchronize the transmit time of each modulator, the stream leading by SIP must be delayed to $T_{max\text{-}delay}$. So the local delay of the stream is calculated by $T_{delay} = T_{max\text{-}delay} - T_{rec}$. After the stream is delayed and stored, the SIP will be removed from the stream. T_{offset} is also defined in SIP, which is used to apply a deliberate offset of transmission time relative to the reference time $T_{max\text{-}delay}$. $T_{transmitted}$ indicates the actual transmission time of the stream.

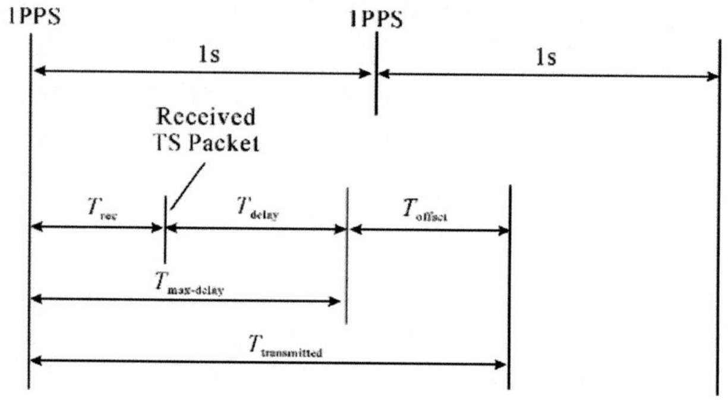

Fig. 3 Description of SFN synchronization

Normally, SFN block is supposed to be in the Base-band module in the traditional modulators. But according to our field experience, we found another way to implement the SFN block. We will discuss this topic in the next section.

2.3 Clock Generator Module

The clocks for each module are generated by the clock generator, which is composed of the internal source and Phase Locked Loop (PLL) circuits. The local source is an on-board 10 MHz oscillator. The reference signal for the PLL can be set as the internal or external 10 MHz source from the timing receiver. The 30.24 MHz clock is multiplied from the 10 MHz.

2.4 IF/RF Module

The IF chip is AD9857 which adjustment step of f_{IF} is 0.028Hz. To coordinate with the RF block, the f_{IF} range is defined within 36.5～37.5 MHz.

The RF Up-converter is UC-1852G, it is a low cost up-converter module and its output frequency range is in VHF and UHF band with 1 MHz adjustment step. In SFN mode, the external 10 MHz clock is required by RF up-converter as reference, to ensure the RF center frequency of all the modulators in SFN be completely identical and accurate. With the cascade of the IF and RF module, the center frequency of modulator output f_{OUT} can be expressed by $f_{OUT} = f_{RF} + f_{IF} - 37$ MHz, where f_{RF} is the selected frequency of up-converter, 37 MHz is the center frequency of IF output. The f_{RF} frequency can be adjusted from 48 MHz to 960 MHz with 1 MHz step. So if we use the user interface to change the output center frequency to 666.6 MHz, the number will be rounded up to 667 MHz and sent to RF module, then the residual number of −0.4 MHz will be sent to IF module, the frequency conversion will be performed in the 2 modules respectively, and then a mixed result of 666.6 MHz will be given to the output.

2.5 Controller and Interface

The controller receives the user instructions from the interface, and network parameters from SFN Block when working in SFN mode. Then the instructions will be translated into commands for other modules. The current state of the modulator is displayed on the interface at the same time. The instructions can also be given through the remote control port.

3. Implementation of SFN Block

3.1 Analysis on Dual Timing System

In field tests, we found that using traditional modulators to build a single frequency network needs many high-accuracy GPS timing receivers, which result in transportation, installation and costs problems, and for receiving the timing signal, the uncertainty of GPS system is also a problem. So that we design a SFN block that supports GPS/Compass2 dual timing system, which can be removed from the system when working in Multi-Frequency Network(MFN) mode. Usually at least 2 modulators are needed in a SFN broadcasting station, so for each modulator, one traditional timing receiver is saved, as well as for the MFN broadcasting stations, the SFN block costs can be saved.

The similarities and differences between GPS and Compass2 system can help us to optimize the SFN timing block. Firstly, Compass2 and GPS both work in the L-band and their signals have the same polarization, so the it is able to use the same RF antenna, the same pre-selector, low-noise preamplifier and filter on front-end of the receiver.

Secondly, they both use pseudo-code for ranging and spread spectrum modulation, so the same correlator can be used to de-spread the signal, restore the carrier, and to process acquisition and tracking in the same way as well as the pseudo range measurement. The ranging codes period is also the same, so that a local 1 millisecond signal can be used to read two systems' correlation cumulative data, pseudo-range measurements and Doppler measurements for navigation timing data solving.

Thirdly, the satellite identification mechanism is the same, both are code division multiple access identification system. So a Gold code generator can be shared.

Finally, though the system base coordinates are different, the 2 systems can be considered equivalent. GPS uses WGS-84 coordinate[6], Compass2 uses CGS2000 coordinate, while they only has little difference less than 10cm. In addition, according to the differences between 2 systems, the timing data can be processed as follows:

GPS system time is based on UTC, Compass2 system is based on the BD time, so there may have a few milliseconds difference, and this bias is a function of time. So we can increase the observation data, expand the observation matrix dimension, and resolve the synchronization error as an unknown variable[7, 8].

3.2 Structure of SFN Block

In this block, GPS/Compass2 dual system timing receiver's detailed structure is shown in Fig. 5. Analog front-end module includes the antenna module and RF module, which is for frequency shifting of satellite signals, power amplification and signal filtering, an A/D converter is inserted to convert analog IF signal to digital IF signal; Digital correlator is used to complete the digital IF signal correlation processing; Microprocessor is used to complete the pseudo-range measurements, navigation constellation selection, and timing. The TS process module is used to resolve SIP information and produce local TS time delay.

System availability directly depends on the number of visible satellites, or the number of ranging signals. With the number of satellites increases, dual timing receiver could obtain more constellation redundant information than single satellite timing receiver, so that the time accuracy will be improved while the system robustness enhanced, which is very important for DTV broadcasting safety and stability.

As shown in Table 1, at 10 degrees elevation angle constraint condition, the dual timing system can see satellites approximately twice as many as single system. Especially in the big constrain elevation, for example, 40 degrees, the visible satellites number for dual timing system is still 7 to 9, at which point the visible satellites number for single system is below 4.

Table 1 Availability of single system and combined system at elevation of different constraints

System	Elevation (Degree)					
	20	30	40	50	60	70
GPS	1	0.92	0.48	0.14	0	0
Compass2	1	0.90	0.30	0.02	0	0
GPS/Compass2	1	1	0.98	0.48	0.05	0

4. Test Results

According to the existing terrestrial DTV standards, the tests are taken in an ISO9001 certificated laboratory. The test system for SFN is shown in Fig. 4. A dual system antenna instead of 2 external timing receivers is connected to the modulators. The demodulation IC is 8G52 from Legend Silicon Co.

Spectrum of RF output in SFN mode is shown in Fig. 6. The working mode is 16QAM, 0.6 FEC code and OFDM. The output power of the modulators is at the same level, different delay time causes different "Bounce" in the spectrum, meaning the signals have the same content.

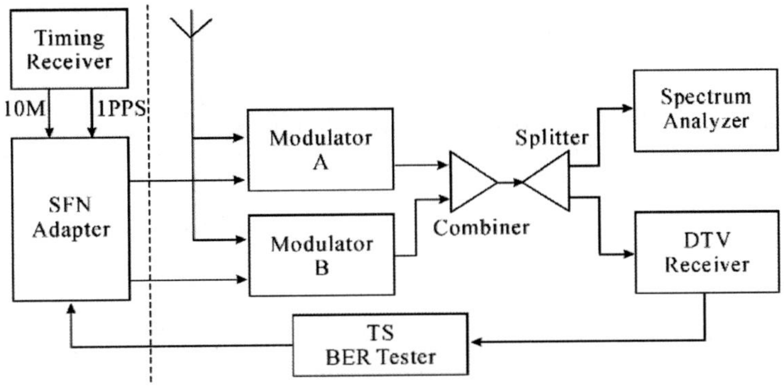

Fig. 4 Block diagram of laboratory test system

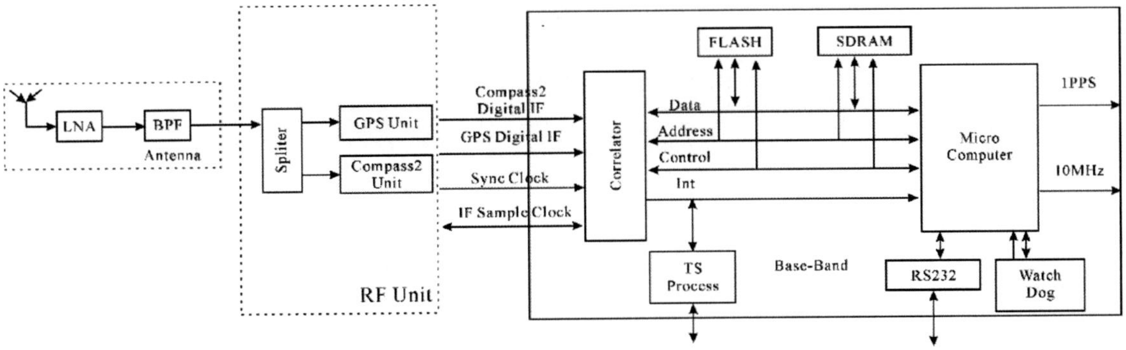

Fig. 5 Block diagram of GPS/Compass2 dual timing system

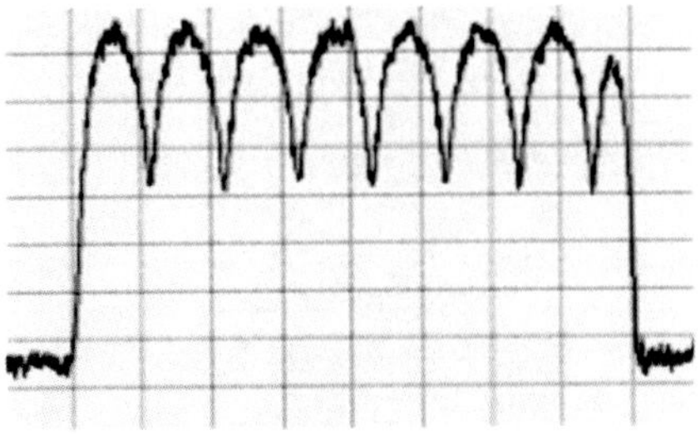

(a) 1μs delay

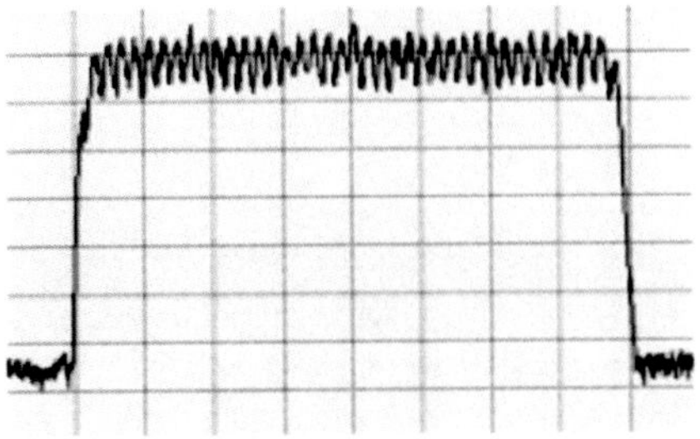

(b) 5μs delay

Fig. 6 Output spectrum of SFN mode

In Table 2, we present part measurement results of the modulator according to national modulator and SFN standard[2, 3].

Table 2 Test results of modulator

No.	Test Parameters	Test Results	Status
1	Frequency Range	48～960 MHz	Pass
2	SFN Frequency Step	1 Hz	Pass
3	Frequency Accuracy	±1 Hz	Pass
4	Output Power	0. 5 dBm	Pass
5	SFN Time Delay Step	100 ns	Pass
6	SFN Time Delay Range	0～999,9999 ms	Pass
7	RF Bandwidth	7. 56 MHz	Pass
8	Modulation Modes	330	Pass
9	Network Modes	MFN/SFN	Pass
10	Clock Reference	1 PPS/10 MHz	Pass

5. Conclusion

Based on the practical application experience, with new technology, the optimized modulator could provide economical efficiency and usability to both SFN and MFN broadcasting mode. For future terrestrial DTV broadcasting, the coverage will strongly depend on the scale of SFN, which leads to a new level demand on stability and accuracy. For the next step, GLONASS and Galileo will be also considered to merge into the system, as well as Compass2 system is gradually established, the modulator will have wide application prospects.

References

[1] GB20600-2006. Frame structure, channel coding and modulation for digital television terrestrial broadcasting transmission system[S]. Beijing: Standard press of China, 2006.

[2] GY/T 229. 1-2008. Single Frequency Network Adapter technical requirements and measurement methods for terrestrial digital television broadcasting [S]. Beijing: Standard press of China, 2008.

[3] GY/T 229. 2-2008. Exciter technical requirements and measurement methods for terrestrial digital television broadcasting[S]. Beijing: Standard press of China, 2008.

[4] LU Cheng, ZHANG Xiao-lin, ZHANG Zhan. Design of a modulator for China National Digital Television Terrestrial Broadcasting System. Proceeding 12th International Symposium on Broadcasting Technology, Beijing, Aug 2007: 390-394.

[5] Bowei Song, Lin Gui, Yunfeng Guan and Wenjun Zhang, "On Channel Estimation and

Equalization in TDS-OFDM based Terrestrial HDTV Broadcasting System", IEEE Transactions on Consumer Electronics, Vol. 51, No. 3, AUGUST 2005.

[6] Kaplan. Understanding GPS: principles and applications[M]. Second Edition. Beijing: Artech House. 2006: 213-242.

[7] Du Xue-tao, Li Nan, Liu Jie. Application of GPS/Compass dual timing system in TD-SCDMA [J]. Telecommunications Engineering Technology and Standardization, 2007 (7): 5-7 (In Chinese).

[8] Li Jian-hai, Yi Da-jiang, Wang Hao. Design and Implementation of Compass/GPS dual timing and navigation system processing module[J]. Computer knowledge and technology, Vol. 5, No. 9, 2009 (In Chinese).

Optimized Soft-Decision De-Mapping for Chinese Terrestrail DTV Standard*

ZHANG Zhan, ZHANG Yan-zhong, ZHANG Xiao-lin, FANG Lin-tang
School of Electronic and Information Engineering, Beihang University, Beijing, China

Abstract: An optimized soft-decision de-mapping method for Chinese Terrestrial DTV (Digital TV) Broadcasting Standard is proposed, which can be used on 64QAM, 32QAM, 16QAM and 4QAM de-mapping that compliant with the standard. Using the simplified de-mapping algorithm, the hardware resource usage and working speed are both optimized. Especially for 32QAM mode, comparing to traditional method, 75% hardware resource is saved, and the maximum working speed is 11% faster. This method is already implemented in the Chinese Terrestrial DTV receiver.

Keywords: soft-decision; de-mapping; LLR; Chinese terrestrial standard

1. Introduction

The digital terrestrial broadcasting standard named "Frame structure, channel coding and modulation for digital television terrestrial broadcasting system" was published in China in August 2006.

This standard describes the processing of transmission stream (TS) from 4.813 Mbps to 32.486 Mbps net data rate in 8 MHz bandwidth and provides 330 modulation modes for terrestrial DTV broadcasting, including 5 mapping modes which are 64QAM, 32QAM, 16QAM, 4QAM and 4QAM-NR[1]. Since the hardware utilization increases directly proportional to the modes and design complexity, the hardware implementation must be optimized on algorithm level.

In this paper, we propose an optimize method for de-mapper in Chinese terrestrial DTV receiver, which brings obvious benefits for hardware implementation.

2. Soft-Decision De-Mapping

In DTV receiver, the idea of soft-decision is to output the credibility value of the received signals, to give the channel decoder more information for decoding the correct information bits. Comparing to hard-decision, it could bring more system gain, and avoid much system noise and interfere. But the system complexity is one of its major problems[2].

---

* From 2010 IEEE 978-1-4244-3709-2/10.

Y. Zhang, *Footprints in Cambridge and Aviation Industries of China*,
https://doi.org/10.1007/978-981-16-3176-4_20

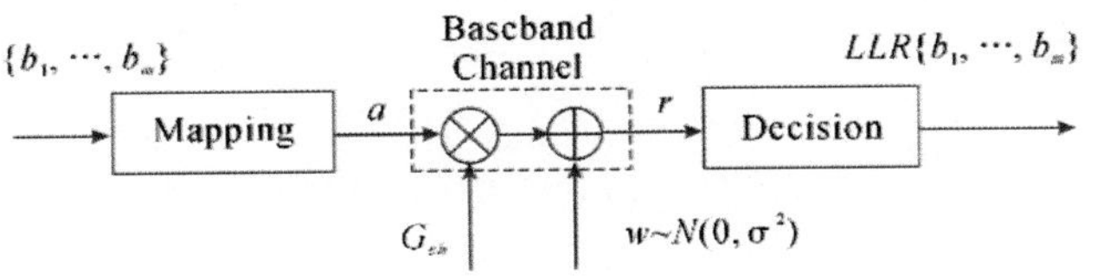

Fig. 1 Mapping signal decision model

Shown in Fig. 1, in the channel soft-decision model, $M=2^m$ is the symbol number of QAM constellation. Bit stream $\{b_1, \cdots, b_m\}$ through QAM are mapped to a, then the received signal is presented by:

$$r=G_{ch} \cdot a+w \tag{1}$$

where G_{ch} is the Channel Frequency Response(CFR), w is the Gaussian white noise.

Suppose the Channel estimation error can be ignored, the output of equalizer is[2]:

$$y=a+w/G_{ch}=a+w' \tag{2}$$

where w' is Gaussian white noise with variance $N_0/|G_{ch}|^2$.

According to [3] and the definition of log-likelihood ratio(LLR), the soft-decision equation of b_k is:

$$LLR(b_k) \triangleq \log \frac{P[b_k=1 \mid r]}{P[b_k=0 \mid r]} = \log \frac{\sum_{\alpha\in S_k^1} P[a=\alpha \mid r]}{\sum_{\alpha\in S_k^0} P[a=\alpha \mid r]} \tag{3}$$

where S_k^0 is the collection of stars for $\{b_1, \cdots, b_m\}$ when $b_k=0$, and S_k^1 is the collection of stars when $b_k=1$. α is the symbol in the corresponding position on the constellation map.

Suppose the transmitted signal is uniform distributed, according to Bayes formula and $\ln \sum_i e^{x_i} \approx \max_i\{x_i\}$, we have:

$$LLR(b_k)=\log \frac{\max_{\alpha\in S_k^1} P[r|a=\alpha]}{\max_{\alpha\in S_k^0} P[r|a=\alpha]} \tag{4}$$

Because the conditional probability density function of r is complex Gaussian process, so

$$P[r|a=\alpha]=\frac{1}{\sqrt{2\pi}\sigma}\exp\{-\frac{1}{2}\frac{|r-G_{ch}\alpha|^2}{\sigma^2}\} \tag{5}$$

Through normalization of (4) and (5), we obtain the final equation of b_k soft-decision result.

$$LLR(b_k)=\frac{|G_{ch}|^2}{4}\{\min_{\alpha\in S_k^0}|y-\alpha|^2-\min_{\alpha\in S_k^1}|y-\alpha|^2\}=|G_{ch}|^2 D_k \tag{6}$$

3. Optimization of Soft-Decision De-Mapping

The purpose of soft-decision is to maximum the Euclidian decision equation is: distance for the symbols in signal geometrical space. We notice the symmetrical characteristic of the constellation map from the space view, if the decision area is symmetrical vertical, then the LLR depends on the absolute value of the I channel; if the decision area is symmetrical horizontal, the LLR depends on the absolute value of the Q channel, so we can simplify the equations as follows:

Take 32QAM for example, it is the most complex mode in the standard. We will solve the map information D_k bit by bit. According to [2], the theory decision equation of 32QAM bit D_4 is:

$$D_4 = \begin{cases} 3(|y_I| - 4.5), y \in A \\ 1.5(|y_I| - 6), y \in \{B, C\} \\ 3(|y_Q| - 4.5), y \in D \\ 1.5(|y_Q| - 6), y \in \{E, F\} \end{cases} \tag{7}$$

The range A~F is shown in Fig. 2. The bold line is the decision border, where is the value 0 for soft-decision.

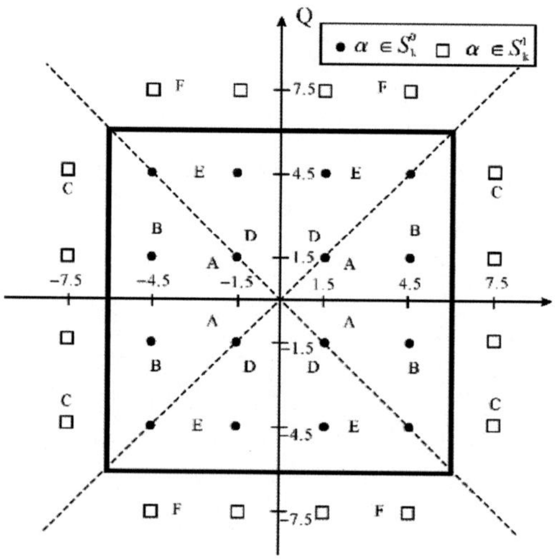

Fig. 2 Value range of 32QAM D_4

We can combine the equations that have near values:

$$D_4 = \begin{cases} |y_I| - 6), y \in \{A, B, C\} \\ |y_Q| - 6), y \in \{D, E, F\} \end{cases} \tag{8}$$

To maximum the Euclidian distance, the final equation for D_4 is:

$$D_4 = \max(|y_Q|, |y_I|) - 6 \tag{9}$$

The theory decision equation of 32QAM bit D_3 is:

$$D_3 = \begin{cases} 1.5y_I, |y_I| \leq 3 \\ 3(y_I - 1.5), 3 < y_I \leq 6 \\ 4.5(y_I - 3), y_I > 6 \\ 3(y_I + 1.5), -6 \leq y_I < -3 \\ 4.5(y_I + 3), y_I < -6 \end{cases} \tag{10}$$

which can be simplified to:

$$D_3 = y_I \tag{11}$$

The value range area of D_2 is shown in Fig. 3, its theory decision equation is:

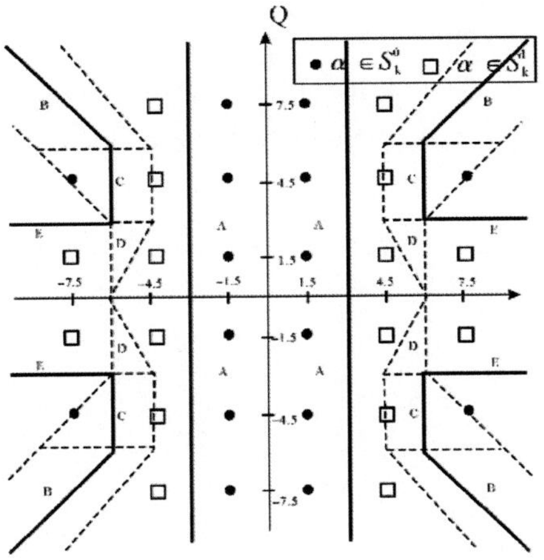

Fig. 3 Value range of 32QAM D_2

$$D_2=\begin{cases}1.5(|y_z|-3), y\in A\\ 1.5(|y_Q|-|y_I|), y\in B\\ -1.5(|y_1|-6), y\in C\\ -1.5(|y_Q|+|y_I|-9), y\in D\\ -1.5(|y_Q|-3), y\in E\end{cases} \quad (12)$$

The D_2 equation is complex, so we change the decision value through polygonal line approximation, but keep the decision border unchanged, it will be equivalent to original equations[4]:

$$D_2=\begin{cases}|y_I|-3, |y_I|<4.5\\ |y_Q|-|y_I|, |y_I|\geq 4.5\cap |y_Q|>6\\ \max[(3-|y_Q|),(6-|y_I|)], \text{ other}\end{cases} \quad (13)$$

For D_1:

$$D_1=\begin{cases}1.5y_Q, |y_Q|\leq 3\\ 3(y_Q-1.5), 3<y_Q\leq 6\\ 4.5(y_Q-3), y_Q>6\\ 3(y_Q+1.5), -6\leq y_Q<-3\\ 4.5(y_Q+3), y_Q<-6\end{cases} \quad (14)$$

Same as D_3:

$$D_1=y_Q \quad (15)$$

For D_0:

$$D_0=\begin{cases}1.5(|y_I|-3), y\in A\\ 3(|y_Q|-4.5), y\in B\\ 1.5(|y_Q|-|y_I|), y\in C\\ 1.5(|y_Q|-3), y\in D\\ -1.5(|y_I|-6), y\in E\\ 1.5(|y_Q|-|y_I|+3), y\in F\end{cases} \quad (16)$$

Same as D_2, we finally have:

$$D_0=\begin{cases}|y_Q|-|y_I|, |y_Q|>6\\ \min[(|y_Q|-3),(6-|y_I|)], \text{other}\end{cases} \tag{17}$$

Same as 32QAM, according to symmetrical and border simplify principle presented above, the simplified equations of 4QAM, 16QAM and 64QAM are respectively:

$$\begin{cases}D_1=y_Q\\ D_0=y_I\end{cases}\quad \begin{cases}D_3=y_Q\\ D_2=4-|y_Q|\\ D_1=y_I\\ D_0=4-|y_I|\end{cases}\quad \begin{cases}D_5=y_Q\\ D_4=4-|y_Q|\\ D_3=2-||y_Q|-4|\\ D_2=y_I\\ D_1=4-|y_I|\\ D_0=2-||y_I|-4|\end{cases} \tag{18}$$

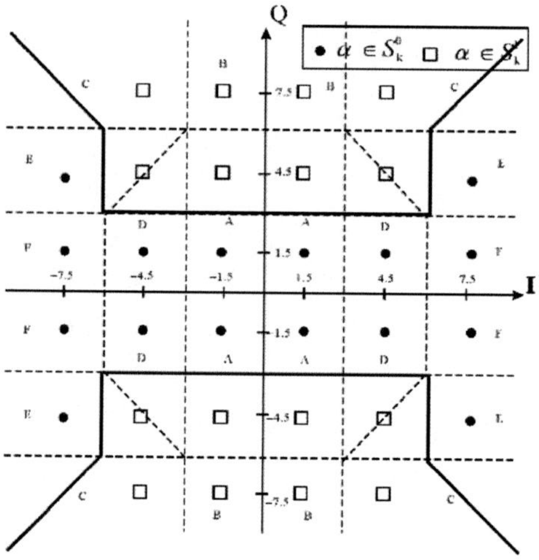

Fig. 4 Value range of 32QAM D_0

4. Test Results

The optimized method is written both in Matlab and Verilog HDL, tested in software simulation model and FPGA hardware. Shown in Table 1, the hardware look up table resource is saved, especially for 32QAM, only 25% resource is used than traditional theory method.

Table 1 Hardware resource usage

Look Up Table	4QAM	16QAM	32QAM	64QAM
Theory Method	2	4	21	8
Optimized Method	2	3	5	5

Shown in Table 2, in the FPGA synthesis software report, only 50% FPGA resource is used than traditional theory method, and the maximum working speed can be 11% faster. The FPGA device we used is Stratix Ⅱ GX series from Altera™.

Table 2 FPGA synthesis result*

	Hardware Resources	Max Frequency
Theory Method	38LUT+21FF	88.3 MHz
Optimized Method	19LUT+6FF	97.8 MHz

* From Synplicity™ Synplify pro.

Shown in Fig. 5, it is the BER vs E_b/N_0 performance of hard-decision, traditional soft-decision method and our method. It is tested under Chinese Broadcasting Channel Model 4, the Monte Carlo iterations are 3000. About 4~6 dB more system gain can be obtained by soft-decision than hard-decision, without performance loss, our method uses less hardware resource and can run faster. The complexity and the cost are reduced, which is very practical for saving chip area of ASIC and other industrial implementations.

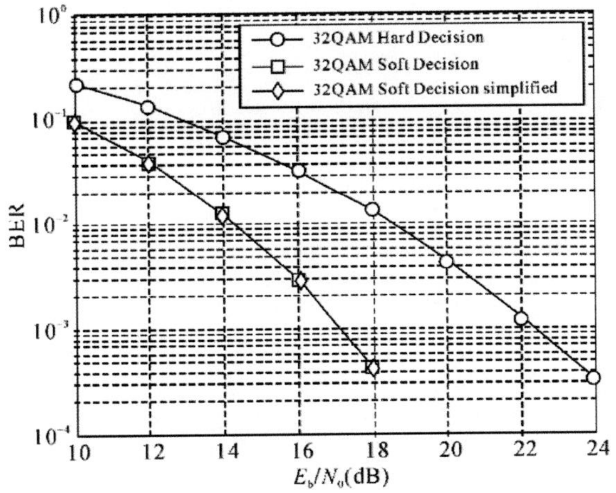

Fig. 5 BER of 32QAM decision

5. Conclusion

The LLR equations are simplified to reduce hardware cost and to optimize system performance. This method theoretically can be used on any M-QAM de-mapping. But for 4QAM-NR de-mapping, we have another way[5]. This method in the paper is already used and tested in our Chinese terrestrial DTV receiver.

References

[1] GB20600-2006. Frame structure, channel coding and modulation for digital television terrestrial broadcasting transmission system[S]. Beijing: Standard press of China, 2006.

[2] J. G. Proakis. Digital communications, 4th ed[M]. New York: McGraw Hill, 2001.

[3] Filippo Tosato and Paola Bisaglia. Simplified Soft-Output Demapper for Binary Interleaved COFDM with Application to HIPERLAN/2 [J], IEEE Trans. On Communications, 2002, vol. 2: pp. 664-668.

[4] Shannon C. A. Mathematical theory of communication[D]. Bell System Tech J. , 1948, 27: pp. 379-432, 623-656.

[5] Xiaolin Zhang, Lintang Fang, Zhan Zhang, et al. NR De-mapping Method and Device for Terrestail Digital TV Broadcasting System[P] (in Chinese). Chinese Patent, 2007.

Design and Implementation of Single-Carrier MIMO Transmission with Frequency Domain Equalization*

Liang Liu[a], Yong Liang Guan[b], Yanzhong Zhang[a], Dingrong Shao[a]

[a] Beihang University, Beijing, China

[b] Nanyang Technological University, Singapore

Abstract: A multiple-input multiple-output (MIMO) single-carrier frequency domain equalization (SC-FDE) testbed which can emulate a practical broadband system is described. A scheme combining space-time transmit diversity with SC-FDE was implemented and evaluated on the testbed. After the hardware configuration, software algorithms incorporating the space-time block coding and decoding, synchronization, channel estimation and frequency domain equalization are explained, followed by some field measurements which validating our design and implementation. It is showed that this software reconfigurable testbed is flexible and convenient to evaluate different system performance in kinds of wireless environments.

Keywords: single-carrier frequency domain equalization; multiple-input multiple-output; testbed; space-time block code

1. Introduction

Wireless broadband communication systems are characterized by severe time-dispersive channels. To deal with the detrimental effects of multipath fading, two block transmission schemes can be adopted: single-carrier modulation with frequency domain equalization (SC-FDE) and orthogonal frequency division multiplexing (OFDM)[1]. Although there are already tremendous applications of OFDM which range from DAB and DVB standards to wireless LAN standards, SC-FDE, which enjoys lower peak-to-average transmit power ratio and less sensitivity to carrier frequency offset than that of OFDM, has been elected as a promising technique for future broadband wireless networks[2-4].

On the other hand, digital communication using MIMO (multiple-input multiple-output), sometimes called a volume-to-volume wireless link, has recently emerged as one of the most significant technical breakthroughs in modern communications[5]. MIMO technology promises high data rate wireless communications and/or the potentiality of exploiting spatial transmit and receive diversity schemes[6]. Therefore, combining MIMO with SC-FDE has been identified as one key

* Published in the 5th Intl. Conf. con Computer Science & Education, Aug. 2010: 56-60.

Y. Zhang, *Footprints in Cambridge and Aviation Industries of China*,
https://doi.org/10.1007/978-981-16-3176-4_21

technique for supporting high data rates over frequency-selective fading channels. It has been shown in [7] that SC-FDE with Alamouti's space-time block codes (STBC) leads to significantly improved performance owing to the joint exploitation of spatial-frequency diversity in a rich scattering environment.

Although many papers have examined the remarkable performance of MIMO SC-FDE systems, most of their works still focus on the theoretical researches or computer simulations. In order to validate these theoretical conclusions, a testbed system, is emergent and essential to run in a real-world wireless environment and consider the presence of all analog hardware impairments. There have been a flourishing work on the evaluation MIMO-OFDM system through testbeds[8, 9]. However, to the best of the author's knowledge, there are few works reported the performance evaluation of MIMO SC-FDE under real-time or non-real-time testbed, which motivates us to address this issue.

In this paper, a MIMO SC-FDE testbed with WiMAX-like architecture is described, which is used to study practical issues in both RF and baseband models. Then a specific space-time coded SC-FDE was implemented on this testbed. It is worth emphasizing that this testbed is software reconfigurable and flexible to alter the system parameters such as carrier frequency, bandwidth, frame format, and transmission/reception antennas structures.

The remainder of the paper is organized as follows: In Section 2, the system model and architecture are described. The main baseband algorithms are introduced in Section 3. In Section 4, the implementation and field measurement results are given. Finally future relevant research works based on the MIMO SC-FDE testbed are introduced.

Notation: We denote matrices by bold uppercase letters and column vectors by bold lowercase letters; $(\cdot)_N$, $(\cdot)^*$, $(\cdot)^H$, and $(\cdot)^{-1}$ denote modulo-W operation, conjugate, conjugate transpose, and inverse, respectively; $D(h)$ represents a diagonal matrix with the diagonal coefficients from the vector h; The mark " $\sim$ " on top of a vector denotes the frequency domain representation of the time domain signal.

2. System Model and Architecture

In this section, we introduce the configuration of a specific space-time coded SC-FDE system based on a software-defined radio (SDR) testbed. The testbed consists of a transmitter and a receiver, each of which can be divided into hardware and software components. The National Instrument (NI) equipments construct the basic hardware architecture of the testbed. At the same time, the NI LabVIEW, embedding MATLAB, is programmed to control the hardware and perform the baseband signal generation, modulation, demodulation, equalization and decoding. All the baseband processing algorithms are achieved in the MATLAB environment.

The overall block diagram of the proposed space-time coded SC-FDE system is illustrated in Fig. 1. At the transmitter, the randomly generated bits are modulated according to MPSK or MQAM format, and subsequently space-time coded by Alamouti code[10]. Then the modulated signals are preceded with two preambles to constitute one frame and pushed into the LabVIEW to

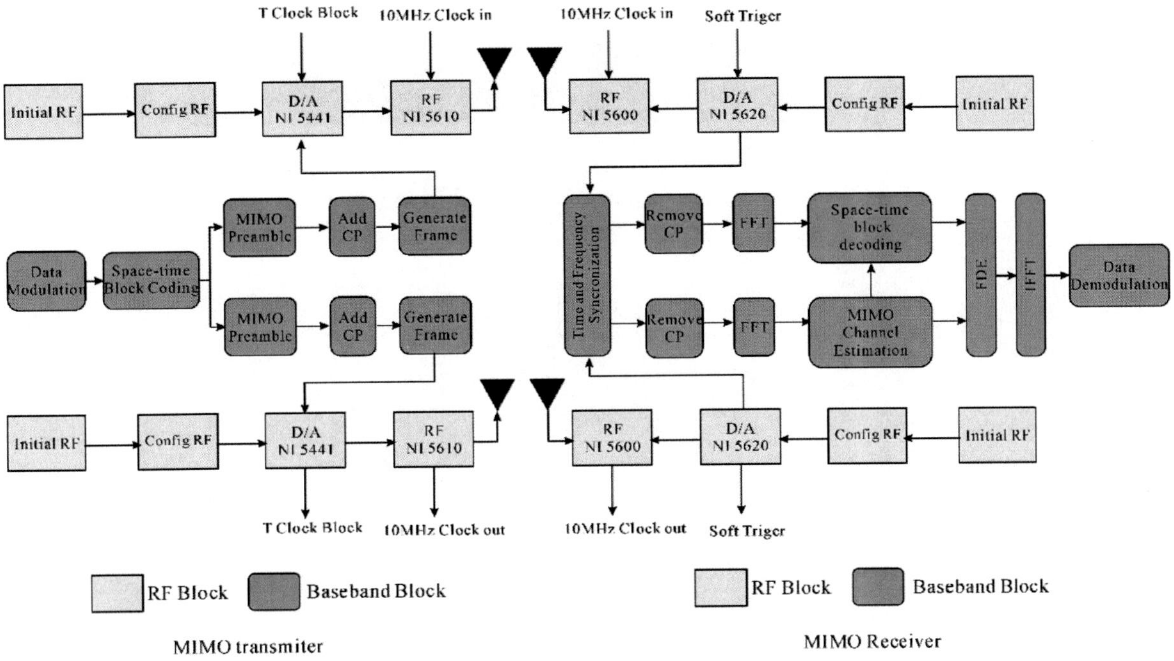

Fig. 1 Block diagram of the 2×2 MIMO SC-FDE testbed

drive the NI hardware to generate the analog waveform. At the receiver, the RF signals are received and downconverted by the RF hardware which are controlled by LabVIEW. The sampled data are then sent into MATLAB for baseband signal processing, including synchronization, fast Fourier transform (FFT), space-time decoding, FDE and inverse FFT (IFFT).

3. Algorithms of the Transceiver Functional Block

For the MIMO SC-FDE testbed, the RF blocks such as D/A, A/D and frequency converters are relatively mature. Thus in this section, we focus on the baseband signal processing such as frame generation, synchronization, channel estimation, space-time block decoding and FDE.

3.1 Frame Structure

The MIMO SC-FDE testbed supports a frame-by-frame transmission. Each frame starts with two preambles: a short preamble and a long preamble, where the short preamble is used for synchronization while the long preamble is constructed for the channel estimation. The preambles are followed by payload SC-FDE symbols. In our testbed, we set the payload length to be 4 to accommodate the WiMAX standard (termed it as WiMAX-like architecture). For 2 × 2 MIMO transmission case, the frame structure is shown in Fig. 2.

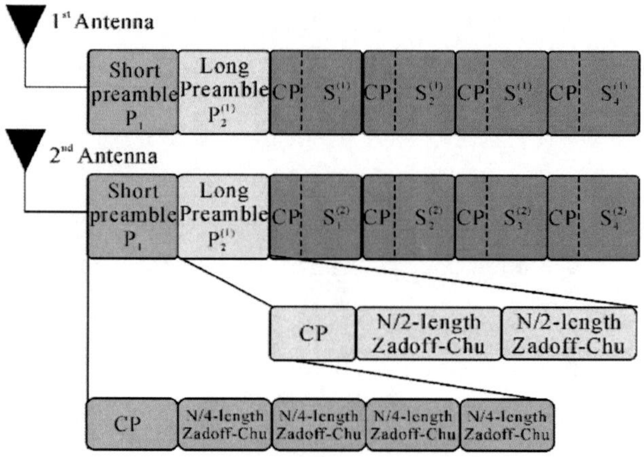

Fig. 2 Frame structure of the proposed 2×2 space-time coded SC-FDE system

It can be seen from Fig. 2 that the concatenated two preambles are structured as two consecutive SC-FDE blocks. The first SC-FDE block consists of four repetition of $N/4$-length Zadoff-Chu sequences[11], where N represents the number of modulated symbols in each SC-FDE block. A cyclic prefix (CP) which is a repetition of the last $N_g = N/4$ symbols in a SC-FDE block, is inserted at the beginning of each block. The second SC-FDE block are constructed by two identical $N/2$-length Zadoff-Chu sequences, preceded by CP. It should be noted that the preamble shall be transmitted from both transmit antennas simultaneously. The short preamble is set to be the same for both of the two antennas while the long preamble be different to guarantee that the channel estimation at each receiver antenna can be performed without interference from the other antenna. Specifically in the frequency domain, the long preamble transmitted from the first antenna uses only the even subcarriers, and the second antenna uses odd subcarriers.

Denote the nth modulated symbol of the k th SC-FDE block from antenna j by $s_k^{(j)}(n)$. In order to employ the Alamouti's STBC to achieve spatial diversity, the signals need to be coded as follows (see Fig. 2)

$$\begin{cases} s_{k+1}^{(1)}(n) = -\{s_k^{(2)}((-n)_N)\}^* \\ s_{k+1}^{(2)}(n) = \{s_k^{(1)}((-n)_N)\}^* \end{cases} \quad k=1 \text{ or } 3, n=0,1,\cdots,N-1 \tag{1}$$

Thus it satisfies the Alamouti architecture in the frequency domain as

$$\begin{cases} \tilde{s}_{k+1}^{(1)}(m) = -\{\tilde{s}_k^{(2)}(m)\}^* \\ \tilde{s}_{k+1}^{(2)}(m) = \{\tilde{s}_k^{(1)}(m)\}^* \end{cases} \quad k=1 \text{ or } 3, m=0,1,\cdots,N-1 \tag{2}$$

3.2 Time and Frequency Synchronization

It is well known that residual time and frequency offset due to non-ideal synchronization result in significant performance degradation, especially for block-based system. Hence, robust time and frequency synchronization is expected to be performed at the receiver.

1) Time synchronization: In order to accurately find the starting point of the frame in the presence of frequency offset and multipath, we resort to a preamble-aided cross-correlation method

given in (3)

$$C(m)=\frac{\left|\sum_{k=0}^{N/4-1} r(m+k)\beta_1^*(k)\right|^2}{\sum_{k=0}^{N/4-1}\left|r(m+k)\right|^2}\quad m=1,2,\cdots,N_{\text{samp}} \tag{3}$$

where $\beta_1(k)$ is the Zadoff-Chu training sequence with the length $N/4$ of the short preamble. $r(m+k)$ represents the samples of the received signal whose length is N_{samp}. Having obtained the cross-correlation $C(m)$, we may further calculate its partial auto-correlation $PA(m)$ as

$$PA(m)=\sum_{i=0}^{4} C(m+i\cdot N/4)\quad m=1,2,\cdots,N_{\text{samp}}-5 \tag{4}$$

Relying on (4), the starting point of the frame could be located by configuring adaptive threshold and searching the position of the pulse peak.

2) Frequency synchronization: Frequency synchronization can be divided into two steps: frequency offset estimation and compensation. After the frame detection, the fractional frequency offset $\hat{\varepsilon}_{\text{FFO}}$ can be readily estimated by exploiting the repeated feature of the short preamble

$$\hat{\varepsilon}_{\text{FFO}}=\frac{1}{\pi}\arg\{\sum_{m=0}^{L-1} r(m)r^*(m+N/2)\} \tag{5}$$

where L is set to $3N/4$. The maximal FFO that can be estimated is $\max\{\hat{\varepsilon}_{\text{FFO}}\}=\pm 1$ (normalized by virtual subcarrier spacing). Then the received preambles are compensated using the estimated $\hat{\varepsilon}_{\text{FFO}}$,

$$\hat{r}(n)=r(n)\exp\{-j2\pi\hat{\varepsilon}_{\text{FFO}}\frac{n}{N}\} \tag{6}$$

Once the fractional frequency offset is compensated, the FFT of the preambles will be circularly shifted by an even number of frequency subcarriers. The remaining integer frequency offset $\hat{\varepsilon}_{\text{IFO}}$ can be easily estimated by introducing the algorithms developed in [12]. Then the frequency synchronization of the payload part can be achieved using (6) by substituting $\hat{\varepsilon}_{\text{FFO}}$ with $\hat{\varepsilon}_{\text{FFO}}+\hat{\varepsilon}_{\text{IFO}}$.

3.3 Channel Estimation

As indicated in Section 1, we utilize the long preamble to perform channel estimation. In order to maintain orthogonality between two antennas, antenna 1 uses odd virtual subcarriers, and antenna 2 uses even virtual subcarriers to transmit training symbols.

Define $r_k^{(j)}$ denotes the received vector from the jth antenna of the kth SC-FDE block. Thus the received long preamble signal may be illustrated as

$$r_k^{(j)}=H_{j,1}p_2^{(1)}+H_{j,2}p_2^{(2)}+n^{(j)} \tag{7}$$

where $H_{j,t}$ represents the equivalent circulant channel matrix from transmit antenna t to the receive antenna j. Relying on the property that a circulant matrix can be decomposed into a diagonal matrix and applying FFT to $r_k^{(j)}$, we get

$$\tilde{r}_k^{(j)}=D(\tilde{h}_{j,1})\tilde{p}_2^{(1)}+D(\tilde{h}_{j,2})\tilde{p}_2^{(2)}+\tilde{n}^{(j)} \tag{8}$$

where $\tilde{h}_{j,1}$ and $\tilde{h}_{j,2}$ represent the channel frequency responses.

Estimation of the channel at the virtual subcarrier points can be done on the basis of least square (LS) and minimum mean square error (MMSE) criteria. Although MMSE performs better than LS, it needs knowledge of the channel statistics and the operating SNR. Thus we choose LS to tradeoff between implementation feasibility and system performance. The channel frequency responses $\tilde{h}_{j,t}(n), j=1,2$ are obtained by dividing $\tilde{r}_k^{(j)}(n)$ by the training symbols $\tilde{p}_2(n)$. In MIMO case, there have only half subcarriers can be used for each antenna. So linear interpolation is performed to obtain the channel frequency response $\tilde{h}_{j,t}$ for all the subcarriers corresponding to each transmission-reception antenna set.

3.4 Space-Time Decoding and FDE

Similar as (8), the received frequency domain payload signals are given by

$$\begin{cases}\tilde{r}_k^{(j)}=D(\tilde{h}_{j,1})\tilde{s}_k+D(\tilde{h}_{j,2})\tilde{s}_{k+1}+\tilde{n}_k^{(j)}\\ \tilde{r}_{k+1}^{(j)}=D(\tilde{h}_{j,1})(-\tilde{s}_k^*)+D(\tilde{h}_{j,2})\tilde{s}_{k+1}^*+\tilde{n}_{k+1}^{(j)}\end{cases}\quad j=1,2 \tag{9}$$

After some manipulations on (9), we get

$$\tilde{r}^{(j)}=\begin{bmatrix}\tilde{r}_k^{(j)}\\ (\tilde{r}_{k+1}^{(j)})^*\end{bmatrix}=\begin{bmatrix}D(\tilde{h}_{j,1}) & D(\tilde{h}_{j,2})\\ -D^*(\tilde{h}_{j,2}) & D^*(\tilde{h}_{j,1})\end{bmatrix}\begin{bmatrix}\tilde{s}_k\\ \tilde{s}_{k+1}\end{bmatrix}+\begin{bmatrix}\tilde{n}_k^{(j)}\\ (\tilde{n}_{k+1}^{(j)})^*\end{bmatrix} \tag{10}$$

Combining $\tilde{r}^{(1)}$ and $\tilde{r}^{(2)}$ in (10), we get

$$\tilde{r}=\begin{bmatrix}\tilde{r}_k^{(1)}\\ (\tilde{r}_{k+1}^{(1)})^*\\ \tilde{r}_k^{(2)}\\ (\tilde{r}_{k+1}^{(2)})^*\end{bmatrix}=\underbrace{\begin{bmatrix}D(\tilde{h}_{1,1}) & D(\tilde{h}_{1,2})\\ -D^*(\tilde{h}_{1,2}) & D^*(\tilde{h}_{1,1})\\ D(\tilde{h}_{2,1}) & D(\tilde{h}_{2,2})\\ -D^*(\tilde{h}_{2,2}) & D^*(\tilde{h}_{2,1})\end{bmatrix}}_{\Theta}\begin{bmatrix}\tilde{s}_k\\ \tilde{s}_{k+1}\end{bmatrix}+\tilde{n} \tag{11}$$

Since Θ is an orthogonal matrix, we can multiply both sides of (11) by Θ^H to decouple the two signals $\tilde{s}_k$ and $\tilde{s}_{k+1}$ resulting in

$$\hat{r}=\Theta^H\tilde{r}=\begin{bmatrix}\tilde{\Theta} & 0\\ 0 & \tilde{\Theta}\end{bmatrix}\begin{bmatrix}\tilde{s}_k\\ \tilde{s}_{k+1}\end{bmatrix}+\hat{n} \tag{12}$$

where $\tilde{\Theta}=|D(\tilde{h}_{1,1})|^2+|D(\tilde{h}_{1,1})|^2+|D(\tilde{h}_{1,1})|^2+|D(\tilde{h}_{1,1})|^2$ is an $N\times N$ diagonal matrix. Note that (12) signifies that the order-four spatial diversity gain can be achieved by this scheme. Applying the same FDE method as in SISO case, the frequency domain equalization matrix G maybe expressed as

$$G=\Theta^H(\Theta\Theta^H+\rho/\mathrm{SNR})^{-1} \tag{13}$$

where weight $\rho=1$ denotes MMSE-FDE, while $\rho=0$ represents the ZF-FDE. Then the estimated signal $\tilde{s}_k$ and $\tilde{s}_{k+1}$ could be easily obtained via IFFT processing.

4. Implementation and Field Measurements

One photo of our implemented testbed is given in Fig. 3. In this section, we present some of the measurement results to validate the system design and baseband signal processing algorithms. The system parameters are listed in Table 1. It is worth mentioning that the developed testbed is software reconfigurable, so all the parameters can be altered conveniently.

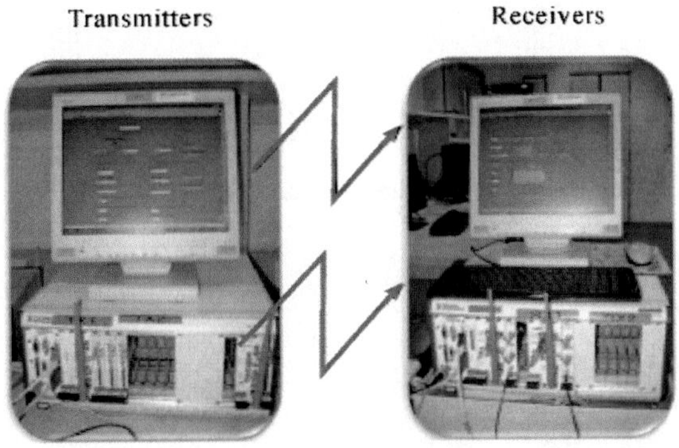

Fig. 3 MIMO SC-FDE testbed for fireld measurements

Table 1 System parameters

Carrier frequency	2.38 GHz
Bandwidth	8 MHz
FFT size	$N=256$
Cyclic Prefix ratio	$N_g/N=1/4$
SC-FDE block duration	32 μs+8 μs(Guard Interval)
Modulation	QPSK
Frame structure	2 preambles+4 data symbols
Frequency domain equalization	ZF-FDE

Before performing field trials, the synchronization of two hardware equipments has to be considered to accommodate the multiple antennas' transmission (see Fig. 1). Firstly, we need to synchronize the RF local oscillators (LO). At the TX, the first NI-5610 generates a 10 MHz clock signal and feed this clock signal to the second NI-5610 to achieve clock synchronization for the LO. At the RX, the first NI-5610 feeds its 10 MHz clock signal to the PXI chassis backplane and the second NI-5600 is then driven by the backplane clock. Next, the two sets of NI-5441 acting as D/A (NI-5620 acting as A/D) also have to be synchronized. At the TX, this is realized by the T-clock block. However at the RX, the NI hardware does not support the T-clock block and thus we resort to using a software trigger to achieve synchronization between the two NI-5620.

In order to validate the design, we perform the field measurements in the non-line-of-sight (NLOS) environment.

Fig. 4 shows the time synchronization results. It can be seen that the correlation calculation is performed in the first 5000 points (i. e. $N_{samp}=5000$). Since each frame has 320 * 6=1920 points, the frame detection can be guaranteed within the 5000 samples. The top figure shows the cross-correlation results $C(m)$, where five peaks could be detected due to the five repeated parts in the short preamble. The bottom figure plots the partial auto-correlation of $C(m)$ from which the peak corresponding to the starting point of a frame could be obtained.

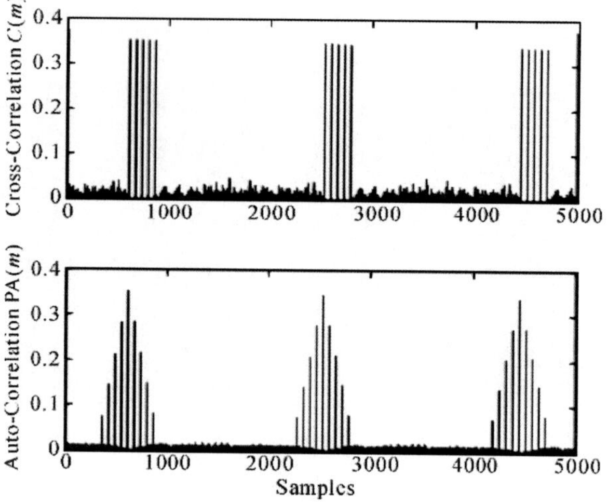

Fig. 4 Synchronization results using the short preamble

Fig. 5 shows the constellations and channel frequency response measurement results at the received RF power-40 dBm. The channel frequency response are obtained by averaging 20 frames on all 256 subcarriers. It can be observed that for different TX-RX antennas links, the channel fadings are different for each other. Some links are flat-fading like channel like Fig. 5 (f) and (g), while others may suffer from frequency-selective fading, like Fig. 5 (e) and (h). However, in the presence of fading, the detected signals are still within the correction region as shown in Fig. (a)(b)(c)(d), owing to the spatial-frequency diversity provided by space-time block codes and single-carrier frequency domain equalization. This measurement results validate our system design and implementation.

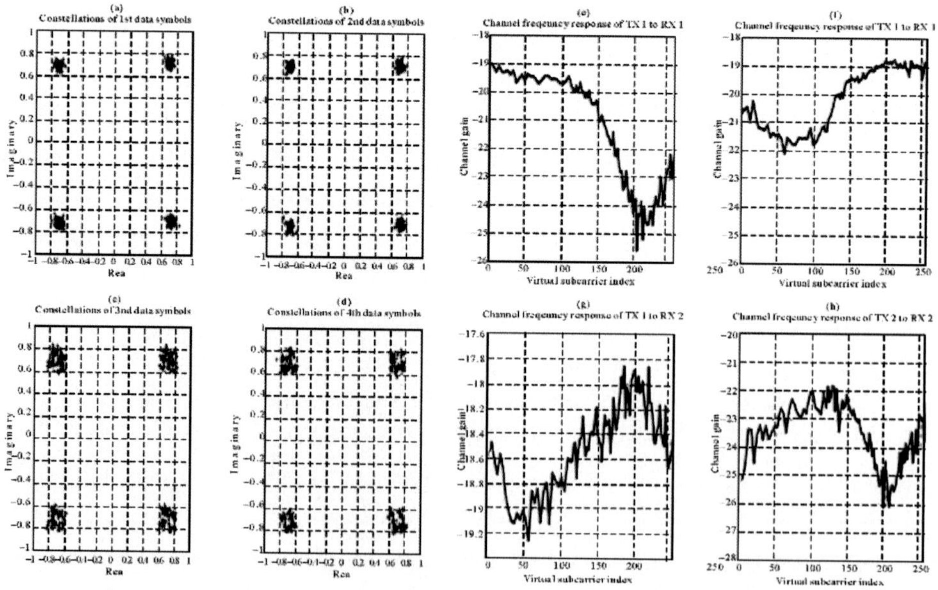

Fig. 5 Constellations and channel frequency response at received power-40 dBm over a NLOS environment

5. Conclusion

In this paper, we introduce a software defined MIMO SC-FDE testbed, based on which a specific space-time coded SC-FDE system was implemented. The developed testbed has some advantages. Firstly, it is software reconfigurable, which means we can easily change the kernel of the experiment system. Baseband processing are all achieved in the MATLAB environment. So we can just alter the MATLAB function to alter the system structure, such as frame fragment, modulation and coding methods, antenna selection, etc. Secondly, the testbed has a tunable RF front-end. Using NI equipments, the RF parameters including carrier frequency and bandwidth can be adjusted optionally.

However, only a preliminary work has been done on the testbed, our future work will focus on the design of SC-FDE system with spatial multiplexing and MIMO SC-FDE with multi-access (i. e. MIMO SC-FDMA)[13]. At the same time, any robust algorithms, such as channel estimation, frequency domain equalization over time-selective fading channels can be developed on this testbed.

Acknowledgment

The authors would like to thank Dr. D. W. Liu and Mr. Y. X. Yan from Positioning and Wireless Technology Center, Nanyang Technological University, for their technical reports and useful discussions. The first author also acknowledges the Chinese Scholarship Council (CSC) for providing scholarship to support this research.

References

[1] N. Benvenuto and S. Tomasin, "On the Comparison Between OFDM and Single Carrier Modulation With a DFE Using a Frequency-Domain Feedforward Filter.", IEEE Trans. Commun., vol. 50, no. 6, pp. 947-955, June 2002.

[2] H. Sari, G. Karam, and I. Jeanclaude, "Transmission Techniques for Digital Terrestrial TV Broadcasting." IEEE Commun. Mag., vol. 33, no. 2, pp. 100-109, Feb. 1995.

[3] D. Falconer, S. L. Ariyavisitakul, A. Benyamin-Seeyar, and B. Eidson, "Frequency domain equalization for single-carrier broadband wireless systems." IEEE Commun. Mag., vol. 40, no. 4, pp. 58-66, Apr. 2002.

[4] N. Benvenuto, R. Dinis, D. Falconer, and S. Tomasin, "Single carrier modulation with nonlinear frequency domain equalization: an idea whose time has come-again." IEEE Proceeding, vol. 98, no. 1, pp. 69-96, Jan. 2010.

[5] D. Gesbert, M. Shafi, D. S. Shiu, et al., "From theory to practice: An overview of MIMO space-time coded wireless systems." IEEE J. Select. Areas Commun., vol. 21, no. 3, pp. 281-302, Apr. 2003.

[6] A. J. Paulraj, D. Gore, R. U. Nabar, H. Btilcskei, "An Overview of MIMO Communications: A

Key to Gigabit Wireless." IEEE Proceeding, vol. 92, no. 2, pp. 198-218, Feb. 2004.
[7] S. Zhou, G. B. Giannakis, "Single-Carrier SpaceCTime Block-Coded Transmissions Over Frequency-Selective Fading Channels." IEEE Trans. Inform. Theory, vol. 49, no. 1, pp. 164-179, Jan. 2003.
[8] A. V. Zelst and T. C. W. Schenk, "Implementation of a MIMO-OFDM based wireless LAN system." IEEE Trans. Signal Processing, vol. 52, no. 2, Feb. 2004.
[9] R. M. Rao, S. Lang, and B. Daneshrad, "Field measurements with a 5.25 GHz broadband MIMO-OFDM communication system." IEEE Trans. Wireless Commun., vol. 6, no. 8, pp 2848-2858, Aug. 2007.
[10] S. M. Alamouti, "A simple transmit diversity technique for wireless communications." IEEE J. Select. Areas Commun., vol. 16, no. 8, pp. 1451-1458, Apr. 1998.
[11] D. C. Chu, "Polyphase codes with good periodic correlation properties." IEEE Trans. Inform. Theory, vol. 18, pp. 531-532, Jan. 1992.
[12] Y. Yan, Y. Gong, Y. L. Guan et al., "Joint timing and frequency synchronization for IEEE 802.16 OFDM system." IEEE WiMAX Symposium, Mar. 2007, pp. 27-31.
[13] Z. H. Lin, P. Xiao, B. Vucetic, et al., "Analysis of receiver algorithms for LTE SC-FDMA based uplink MIMO systems." IEEE Trans. Wireless Commun., vol. 9, no. 1, pp. 61-65, Jan. 2010.

Novel Fast Acquisition Algorithm for DS/FH System*

Weining Song, Yanzhong Zhang, Dingrong Shao, Shujian Li and Xiaojie Wen

School of Electronic and Information Engineering, Beihang University, Beijing, China

Abstract: Based on differential correlation technique, novel fast acquisition algorithm is developed for hybrid direct sequence and frequency hopping (DS/FH) system. Specifically, by using differential correlation for PN code phase and parallel searching for frequency, the fast acquisition is realized. Furthermore, in terms of acquisition time and hardware resources consumption, the proposed differential correlation acquisition method is compared with the conventional waiting self synchronization acquisition method and frequency parallel searching acquisition method. Analytical and simulation results reveal that the proposed algorithm can be used in DS/FH system.

Keywords: DS/FH; acquisition; differential correlation

1. Introduction

Hybrid direct sequence/frequency hopping (DS/FH) spread spectrum system is widely used in civil and military for its excellent properties such as low probability of intercept (LPI) and low probability of jamming (LPJ) capabilities[1-4]. In order to increase the efficiency and avoid long set-up time, fast acquisition is required.

In DS/FH system, acquisition includes two parts: frequency domain acquisition and PN code acquisition. DMF (Digital Matched Filter) is widely used in PN code acquisition. There are usually two methods for frequency domain acquisition: waiting self synchronization acquisition and frequency parallel searching acquisition. In waiting self synchronization acquisition, receiver's frequency synthesizer is controlled to emit a certain frequency f_j, waiting for the corresponding signals. In frequency parallel searching acquisition, M (M is the total number of hopping frequencies) frequency synthesizers are controlled to emit M frequencies. For waiting self synchronization acquisition, one DMF is needed, and acquisition time will be one hop duration. For frequency parallel searching acquisition, acquisition time is one hop, but M DMFs are needed, which costs too much source for practical system.

In this paper, a novel acquisition method based on differential correlation is proposed. By employing differential correlation to avoid frequency effects, PN code acquisition is accomplished rapidly and then frequency acquisition is carried out.

---

* Published in Proc. of Intl. Conf. on Communication, 2011: 460-462.

Y. Zhang, *Footprints in Cambridge and Aviation Industries of China*,
https://doi.org/10.1007/978-981-16-3176-4_22

This paper is organized as follows: the system model is proposed in Section 2; in Section 3, different acquisition algorithms are analyzed; in Section 4, simulations and analyses results are given out to verify algorithms' performance.

2. System Model

The received DS/FH spread spectrum signal (ignoring noise) can be expressed as

$$s(t)=Ac(t)d(t)\cos(2\pi f_i t+\theta_0) \tag{1}$$

- A is the signal amplitude.
- $c(t)$ is ± 1 valued, N-chip-long PN codes with chip interval T_c seconds.
- $d(t)$ is data sequence valued ± 1.
- f_i is the frequency of hop i whose value is taken from the frequency set $\{f_1, f_2, \cdots, f_M\}$. M is the total number of hopping frequencies.
- θ_0 is phase of carrier.

For the whole acquisition time, the data signal $d(t)$ is assumed to be one, so that it can be ignored in (1).

3. Acquisition Algorithms

3.1 Waiting Self Synchronization Acquisition

Waiting self synchronization acquisition algorithm block diagram is shown in Fig. 1.

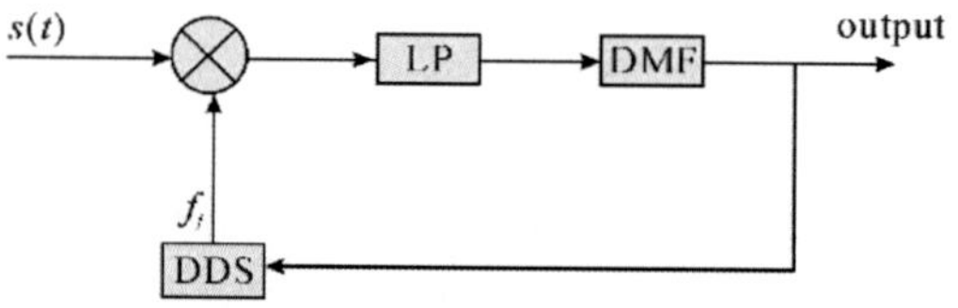

Fig. 1 Waiting self synchronization acquisition circuit

As shown in Fig. 1, the receiver waits at a certain frequency f_j ($j\in\{1,2,\cdots,M\}$), the received signal $s(t)$ is multiplied with $\cos(2\pi f_j t)$, the output will be:

$$\begin{aligned} s_{o1}(t) &= s(t)\cos(2\pi f_j t) \\ &= Ac(t)\cos(2\pi f_i t+\theta_0)\times\cos(2\pi f_j t) \\ &= \frac{Ac(t)}{2}\{\cos[2\pi(f_i+f_j)t+\theta_0]+\cos[2\pi(f_i-f_j)t+\theta_0]\} \end{aligned} \tag{2}$$

It can be got from Eq. (2) that, after low pass filter, if $f_i\neq f_j$, the input signal to DMF will be $\frac{Ac(t)}{2}\cos[2\pi(f_i-f_j)t+\theta_0]$, so that the correlation peak will be much smaller than threshold, then DDS will keep waiting f_j unless $f_i=f_j$, and then the frequency is changed according to frequency hopping pattern.

3.2 Frequency Parallel Searching Acquisition

Frequency parallel searching acquisition algorithm block diagram is shown in Fig. 2.

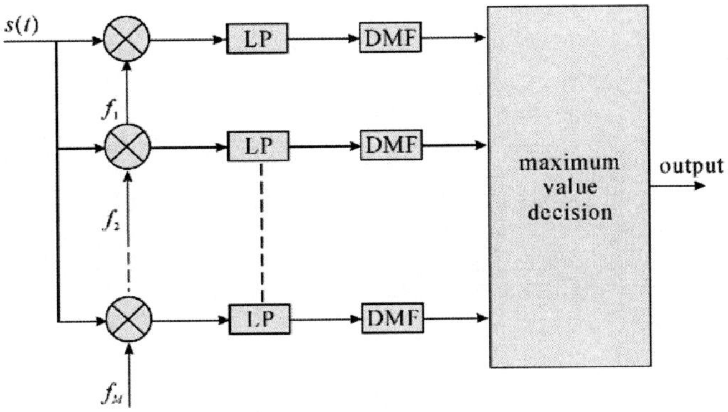

Fig. 2 Frequency parallel searching acquistion circuit

As shown in Fig. 2, there are M synthesizers. Take the j-th branch for example, the output will be:

$$\begin{aligned} s_{o2}(t) &= s(t)\cos(2\pi f_j t) \\ &= Ac(t)\cos(2\pi f_i t+\theta_0)\times\cos(2\pi f_j t) \\ &= \frac{Ac(t)}{2}\{\cos[2\pi(f_i+f_j)t+\theta_0]+\cos[2\pi(f_i-f_j)t+\theta_0]\} \end{aligned} \tag{3}$$

It can be got from Eq. (3) that, after low pass filter, corresponding to current received signal's frequency f_i, output of the i-th branch's DMF will be much larger than the others, and acquisition process is accomplished.

3.3 Differential Correlation Acquisition

Differential correlation acquisition algorithm block diagram is shown in Fig. 3.

The received signal $s(t)$ is multiplied with its delay $s(t-\tau)$ to find out PN code phase. Receiver waits at M hopping frequencies. Once the PN code phase is found out, M correlation results are obtained, and the maximum branch's frequency is current hopping frequency.

As shown in Fig. 3, the received signal $s(t)$ is multiplied with its delay $s(t-\tau)$, the output will be:

$$\begin{aligned} s_{o3}(t) &= s(t)s(t-\tau) \\ &= A^2q(t)\cos(2\pi f_i t+\theta_0)\times\cos[2\pi f_i(t-\tau)+\theta_0] \\ &= \frac{A^2q(t)}{2}\{\cos(4\pi f_i t+\psi)+\cos(2\pi f_i\tau)\} \end{aligned} \tag{4}$$

In which,

$$q(t)=c(t)c(t-\tau),\ \psi=-2\pi f_i\tau+2\theta_0$$

Eq. (4) contains two components: a low frequency term $\cos(2\pi f_i\tau)$ and a high frequency term. The low frequency term can be used to find PN code phase [5].

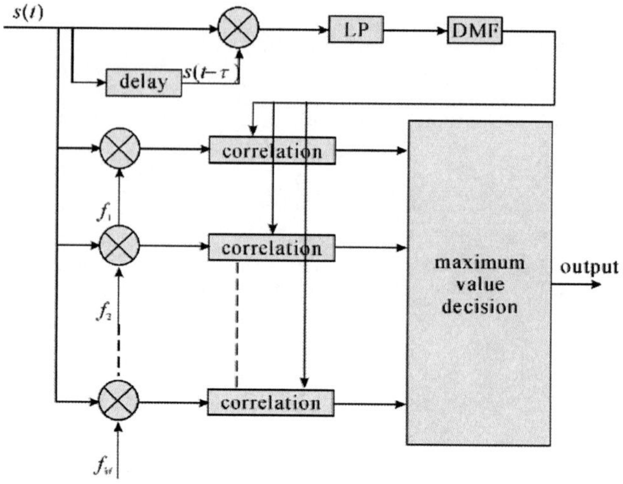

Fig. 3 Model of differential correlation acquisition circuit

Strictly speaking, it is difficult to use this to approach for acquisition. In order to make this approach function properly, one must make $\cos(2\pi f_i \tau)$ close to ± 1. Because the input frequency is variable, τ should be cautiously chosen to make sure that for all $f_i (i=1,2,\cdots,M)$, $\cos(2\pi f_i \tau)$ is closed to ± 1. For example, in a certain system, $M=1,\cdots,10$, $f_i=10$ MHz, $f_2=20$ MHz $\cdots f_{10}=$ 100 MHz, then τ can be chosen as $0.5\mu s$, so that for f_i:

$$\cos(2\pi f_i \tau)=\cos(2\pi f_i \times 0.5\mu s)=1 \tag{5}$$

It should be noticed that, during frequency changing period, Eq. (5) is not tenable. For DS/FH system, frequency stable time is much larger than changing time so differential correlation can be used in most time.

4. Performances Analysis

4.1 Displacement Sequence Self-Correlation

In differential correlation acquisition algorithm, a new PN code is formed by $c(t)c(t-\tau)$. As a certain example in (3.3), $c(t)$ is chosen as 63-chip-long m sequence, and T is one chip time ($0.5\mu s$). According to displacement additivity of m sequence, the new code is the displacement sequence of original m sequence, and its self-correlation is shown as Fig. 4.

4.2 Acquisition Performance Comparison

To analyze the acquisition performance, assuming that there is one PN code in one hop, besides, the number of hopping frequencies is M and PN code length is N. For ROM source is adequate relatively, hardware consumption analysis focuses on adder source.

If DMF is used for PN code acquisition in these three algorithms, it can be got that:

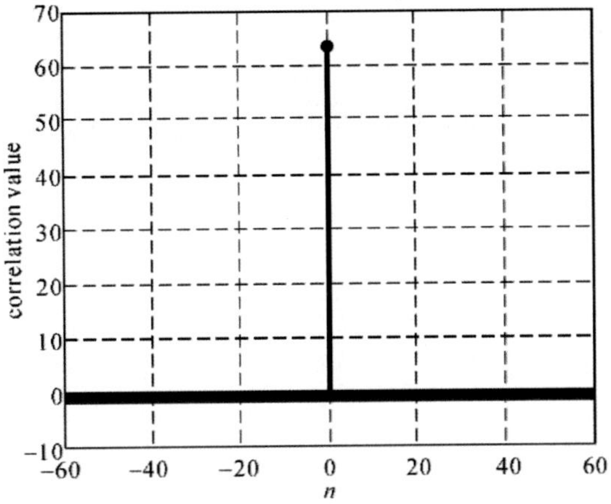

Fig. 4 Displacement sequence self-correlation

For waiting self synchronization acquisition, acquisition time is MT_h (T_h is one hop time) and the number of adders needed is N. For frequency parallel searching acquisition, acquisition time is T_h and the number of adders needed is NM.

For differential correlation acquisition, acquisition time is DMF's acquisition time (T_h) adding correlation time (T_h), that is, $2T_h$, and the number of adders needed is $N+M$, as shown in Table 1.

Table 1 Performance comparison

Algorithm	Acquisition Time	Adders Consumption
waiting self synchronization acquistion	MT_h	N
frequency parallel searching acquistion	T_h	NM
differential correlation acquisition	$2T_h$	$N+M$

5. Conclusions

In this paper, a fast acquisition algorithm is presented using differential correlation for DS/FH system. The simulation and analytical results show that the new algorithm could achieve fast acquisition with inconspicuous increase of adder's consumption comparing to conventional acquisition methods, which satisfies the fast requirement of practical system.

References

[1] T, Vanninen, H, Saarnisaari, M, Raustia, and T, Koskela, "FH-Code Phase Synchronization in a Wireless Multi-Hop FH/DSSS Adhoc Network," Military Communications Conference, 2006, pp. 1-7.

[2] M, Song, and S, Wigginton, "Frequency Hopping Pattern Detection in Wireless Ad Hoc Networks," Proceedings of the international Conference on information Technology: Coding and Computing, 2005, pp. 633-638.

[3] Minjian Zhao; Mingxia Xu; Jie Zhong; Shiju Li, "Slot Synchronization in Ad Hoc Networks Based on Frequency Hopping Synchronization," Wireless Communications, Networking and Mobile Computing, 2006, pp. 1-4.

[4] A, Tabassam, S, Heiss, "Bluetooth Clock Recovery and Hop Sequence Synchronization Using Software Defined Radios," Region 5 Conference, 2008, pp. 1-5.

[5] David M, Lin. James B. Y. Tsui. "Acquisition schemes for software GPS receiver," ION GPS 1998. 317-325.

A Frame-Related Approach for Performance Improvement of MPE-FEC in DVB-H*

Rongke Liu, Yan Wang, and Yanzhong Zhang
School of EIE, Beihang University, Beijing, China

Abstract: Digital Video Broadcasting-Handheld (DVB-H) is a new digital terrestrial broadcasting standard customized for the multimedia service on the mobile devices and the handheld terminals. It introduces a technique known as multi-protocol encapsulation forward error correction (MPE-FEC) for the IP-datagrams in the data link layer to improve the mobile performance. Every MPE-FEC frame is constructed independently in the DVB-H standard. In this paper, a frame-related coding approach for MPE-FEC frames is proposed. When a MPE-FEC frame is built during the data link layer encoding process, the rightmost n columns in the MPE part of the current MPE-FEC frame are filled with the data populated from the leftmost n columns of its previous MPE-FEC frame. After the encoding process, a correlation is set up between the adjacent frames. This correlation can be utilized by the proposed decoding algorithm to enhance the error correction capability. The theoretical analysis and simulations results show that the proposed approach can provide better error correction capability than the DVB-H standard as well as the conventional zero-padding method in random noise channels. As a more efficient decoding strategy, the proposed approach is a good candidate for the decoding of MPE-FEC frame.

Keywords: DVB-H; error correction capability; frame-related; multi-Protocol Encapsulation Forward Error Correction (MPE-FEC)

1. Introduction

DGITAL Video Broadcasting-Handheld (DVB-H)[1] is a customized standard for mobile devices and handheld terminals over two previous standards—Digital Video Broadcasting-Cable(DVB-C) and Digital Video Broadcasting—Terrestrial (DVB-T)[2, 3]. Compared with DVB-T, DVB-H provides low power consumption, flexible network planning, compatibility with IP network and good mobile performance. The reason is that several new techniques, such as multi-protocol encapsulation forward error correction (MPE-FEC) for IP-datagrams over the data link layer and time slicing, are introduced to improve the mobile performance. The MPE-FEC can increase the carrier noise ratio (C/N) and improve the receiving performance in mobile channels. It also has better tolerance to

* Published in IEEE Transactions on Broadcasting, Vol. 57, No. 4. Dec. 2011: 888-894.

Y. Zhang, *Footprints in Cambridge and Aviation Industries of China*,
https://doi.org/10.1007/978-981-16-3176-4_23

impulse interference[4, 5].

In DVB-H, every MPE-FEC frame is encoded and decoded independently. Although the DVB-H standard defines the MPE-FEC frame structure, the decoding methods are left open for each receiver manufacturer to decide[6]. In [6, 7], different decoding strategies for MPE-FEC were compared, including section erasure decoding, transport stream erasure decoding, section error and erasure decoding[8], transport stream error and erasure decoding[8] and non-erasure decoding. Section erasure decoding is the recommended approach in the DVB-H standard. In this paper, we focus on the section erasure decoding, in which the bytes in the sections are marked as "unreliable" or "reliable" depending on the decoding result of the CRC-32. A novel frame-related coding approach is proposed to enhance the error correction capability in a MPE-FEC frame with the complexity remaining the same order of magnitude as that of the conventional zero-padding method[9], in which the shorten RS codes are used. In the proposed algorithm when a frame-related MPE-FEC frame is built, the rightmost n columns in the so-called application data table of the current frame are padded with the data populated from the leftmost n columns of its previous frame. By this way an encoded MPE-FEC frame holds close relation with its previous and next neighbors. At the decoder side, the useful information can be easily obtained from the adjacent frames. The leftmost n columns in the previous frame and the rightmost n columns of the application data table in the next frame are able to benefit the decoding of the current frame; therefore the error correction capability of a frame-related MPE-FEC frame will be improved. In this paper, the mechanism to improve the MPE-FEC performance in the proposed frame-related solution is analyzed, and simulation results show the performance improvement.

The rest of this paper is organized as follows. In Section 2, an overview on the MPE-FEC technique as well as the conventional zero-padding method in the DVB-H standard are given, which is the basis of our innovation. In Section 3, our suggested algorithm is presented. The encoding and decoding algorithm of the frame-related MPE-FEC frame are detailed introduced. In Section 4, the error correcting capability of the frame-related MPE-FEC frame and the complexity of the proposed algorithm are analyzed in detail. In Section 5, the proposed algorithm is evaluated with the simulation results. Finally, a short conclusion is drawn in Section 6.

2. MPE-FEC Frame and Zero-Padding Method in DVB-H Standard

Compared with the DVB-T standard, the MPE-FEC method as well as other new technology elements, such as time-slicing, 4 K-mode and so on, are added to the data link layer and physical layer in the DVB-H standard in order to support the agile mobile devices better. The MPE-FEC method can increase the carrier noise ratio (C/N) and improve the Doppler performance in mobile channels, and it also provides a better tolerance to impulse interference[1, 4]. These new technologies make DVB-H standard better fulfill the demands of mobile and battery-operating handheld devices, such as multimedia cellular phone, mobile TV, personal digital assistant, and so on.

In the DVB-H system, Reed-Solomon (RS) codes are added on a MPE frame in the data link layer as forward error correction codes for the MPE sections. That is why it is called as MPE-FEC.

An MPE-FEC frame is organized as an array with 255 columns and a flexible number of rows, whose available numbers are 256, 512, 768 and 1024[1]. Each element in the array corresponds to one byte. As depicted in Fig. 1, the 191 leftmost columns of a MPE-FEC frame, called as the application data table, are dedicated for the network layer datagrams (e. g. IP datagrams)[10, 11], and may be partly padded by zero-valued bytes. The 64 rightmost columns of the MPE-FEC frame are dedicated for the parity symbols of the RS codes and are called the RS data table. After all the useful IP datagrams having been filled into the application data table, all the unfilled elements are padded with zero-valued bytes, so that the leftmost 191 columns are completely filled. The 64 parity bytes are generated from the 191 bytes in the application data table using RS (255, 191) code and are filled into the RS data table, which makes each row become an RS codeword.

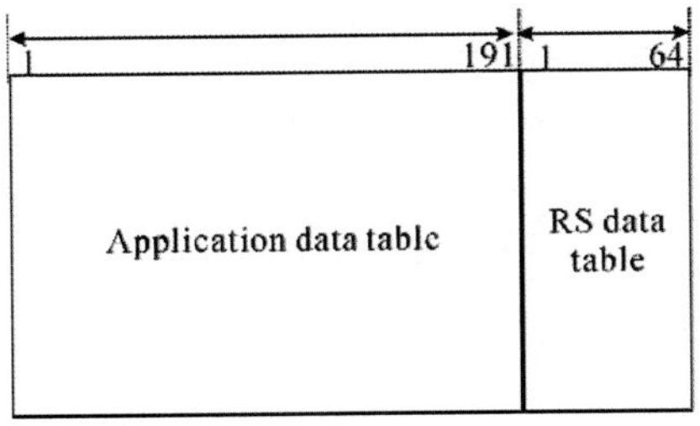

Fig. 1 Structure of an MPE-FEC frame in the DVB-H system

The columns in the application data table and RS data table are transmitted one by one. As depicted in Fig. 2, each IP datagram is encapsulated into one or more MPE sections which are in compliance with the DVB-H standard. The parity bytes are carried in MPE-FEC sections, and each section carries exactly one column in the RS data table. In each MPE or MPE-FEC section, there is a 12-byte header before the payload, and the whole section is protected by a 32 bits cyclic redundancy check (CRC-32) which is calculated by the payload as well as the header. All sections will be divided and filled into the payload of the MPEG-2 transport stream (TS) packages which are transmitted then[12].

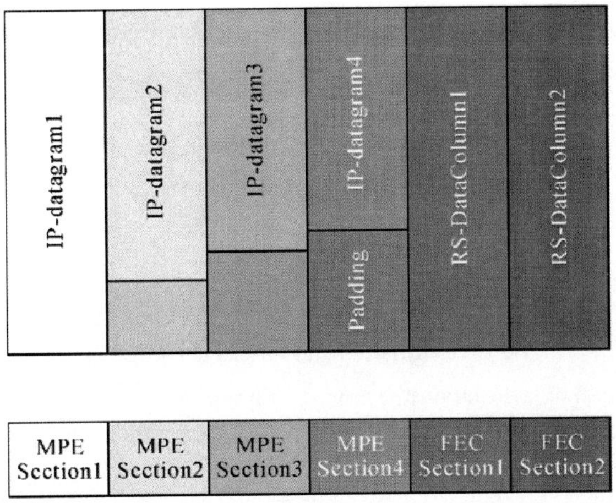

Fig. 2 Example of the transmitted sections in MPE-FEC frame

After receiving the TS packages, the receiver rebuilds all the MPE and MPE-FEC sections. The CRC-32 reliably detects all erroneous sections in the MPE and MPE-FEC sections. After this procedure, all the correct bytes and the padded zero-valued bytes are marked as "reliable", while all the error bytes are marked as "unreliable", which are treated by the RS decoder as erased symbols. All sections are used to reconstruct the MPE-FEC frame, so every element in the MPE-FEC frame is marked either as "reliable" or as "unreliable". With the marked reliability/erasure information, the RS decoder is able to correct up to 64 erased bytes in a 255-byte codeword by the erasure decoding algorithm[1, 13].

From the MPE-FEC method described above, it is obviously deduced that when there are more than 64 "unreliable" bytes in a row, the RS decoder will not be able to correct it. In spite of that, the RS decoder knows exactly the erroneous positions in the row. In order to improve the error correction capability, DVB-H standard pads zero-valued bytes in the rightmost n columns of the application data table, while the leftmost $191-n$ columns are still filled with the useful data. These padded zero-valued columns are employed to generate the parity bytes of RS code but not transmitted. At the decoder side, they are reintroduced and marked as "reliable"[14]. This method is the so-called zeropadding method[9].

Using the zero-padding method, the maximal error correction capability in a MPE-FEC frame can be increased at the cost of decreasing the code rate. Take the RS (255, 191) code for example. The RS (255, 191) code is able to correct at most 64 "unreliable" bytes in a codeword, and the maximal error correction capability can be denoted by 64/255. If n zero-valued bytes are padded in a RS codeword, namely a row in a MPE-FEC frame, the number of transmitted symbols in a codeword is decreased to $255-n$. In this case, the maximal error correction capability is increased from 64/255 to $64/(255-n)$, and the code rate is decreased from 191/255 to $(191-n)/(255-n)$.

Both the MPE-FEC frames and the frames in the zero-padding method in the original DVB-H standard are constructed independently. It is obviously to know that the successful decoding of a frame cannot benefit the decoding of its neighbor frames, because there is no correlation between them. The decoding of a frame can obtained benefit if the data in its adjacent frames can be made use of. In order to utilize the data in the contiguous frames, some kind of correlation should be introduced between them. That is why we propose the frame-related MPE-FEC coding approach which is described in detail in the following section.

3. Proposed Frame-Related MPE-FEC Method

Different from the zero-padding method in DVB-H standard, in the frame-related method proposed in this paper, the padding bytes used for the calculation of the RS parity bytes in a MPE-FEC frame are populated from its previous frame, other than zero-valued bytes. By this way, a correlation is set up between the adjacent frames. The padding bytes are not sent out in order to guarantee a high code rate, which is the same as the zero-padding method. The encoding process of the frame-related approach is as follows (see Fig. 3):

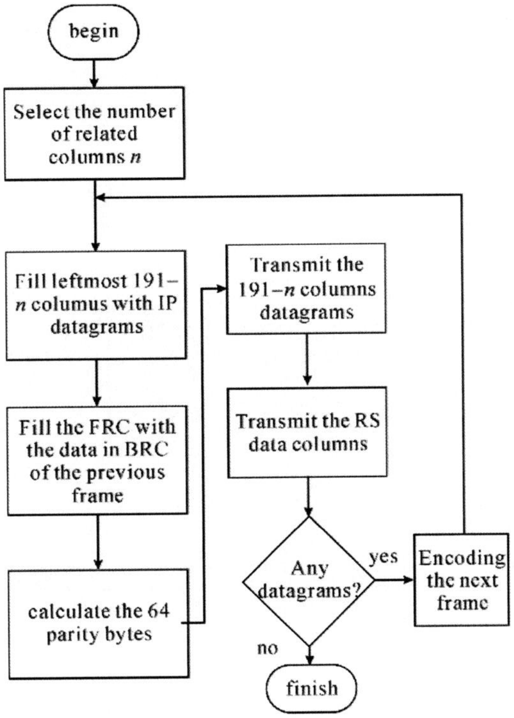

Fig. 3 Encoding flowchart of the frame-related approach

Step 1

Determine the value of n, which is the number of the related columns between the adjacent MPE-FEC frames. The rightmost n columns of the application data table in a frame-related MPE-FEC frame are defined as Forward Related Columns (FRC), and the leftmost n columns are defined as Backward Related Columns (BRC), which are depicted in Fig. 4.

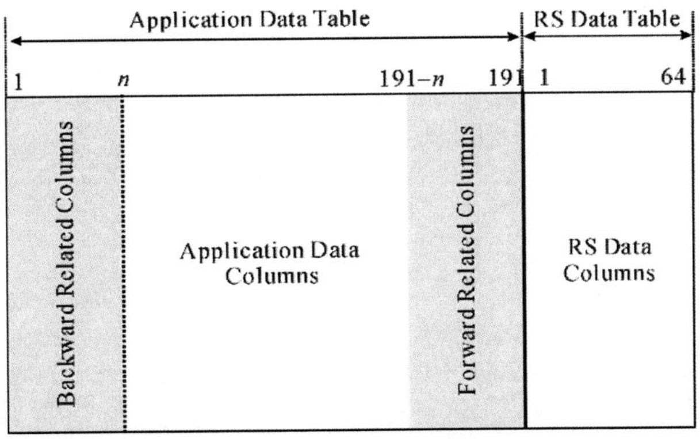

Fig. 4 Frame structure of a frame-related MPE-FEC frame

Step 2

Generate the frame-related MPE-FEC frame. The leftmost $191-n$, columns in the application data table are filled with the IP-datagrams. The first byte of the first datagram is filled into the upper left corner of the current frame. The filling process goes downwards the column direction and one datagram is filled immediately after the end of its previous datagram. The *n* FRC columns are filled with data in the BRC of the previous frame (see Fig. 5).

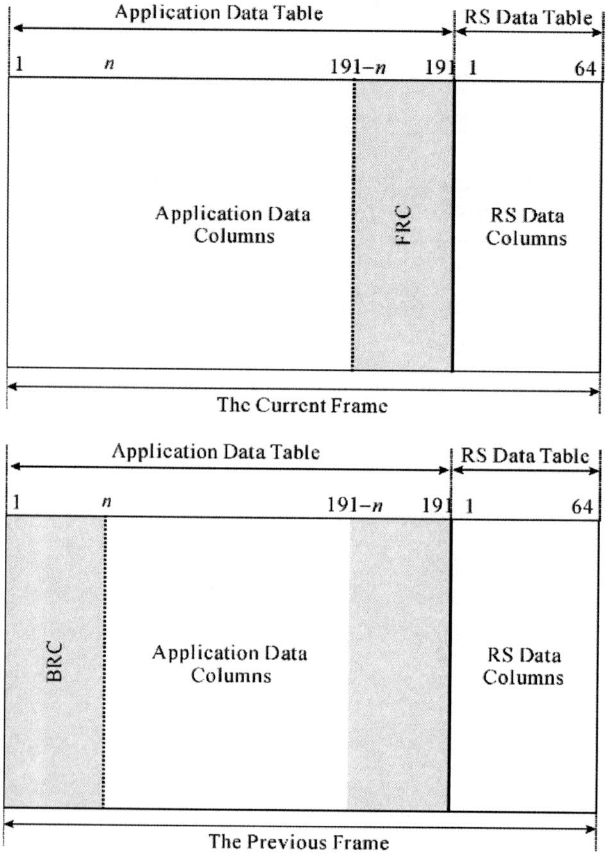

Fig. 5 Correlation between two adjacent frame-related MPE-FEC frames

Step 3

Use RS (255,191) code to generate the 64 parity bytes from the 191 bytes (including the FRC bytes) for each row, and fill the parity bytes into the RS data table.

Step 4

The leftmost $191-n$ columns in the application data table and the RS data table are transmitted. The data in the FRC are not transmitted.

Since the data in the FRC of a frame-related MPE-FEC frame are the copy of data in the BRC of its previous frame, the decoder can employ the correlation to correct more erased bytes in a frame-related MPE-FEC frame. As depicted in Fig. 6, the suggested decoding process is as follows:

Step 1

Use the CRC-32 to judge whether every received section is correct or not, and save the result of the judgment. For every section, the decoder finds the starting address of the payload in its header and fills the payload into the corresponding positions in the application data table or in the RS data table. Based on the result of the CRC-32 judgment, the decoder marks all the correct bytes in the

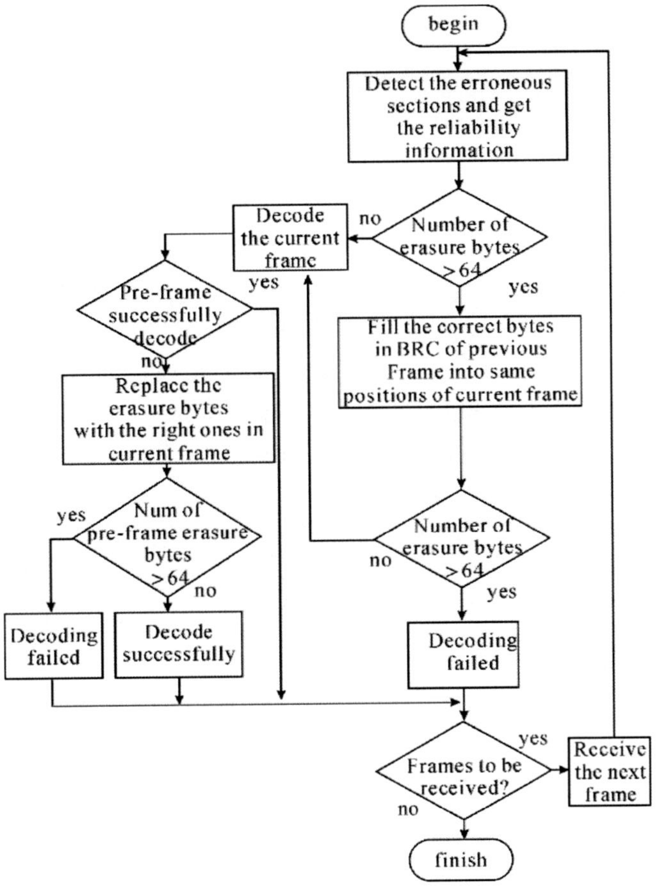

Fig. 6 Decoding flowchart of the frame-related approach

MPE-FEC frame as "reliable", and marks the bytes belonging to the error sections and the bytes in the FRC as "unreliable", which means they are erased bytes. If every row in the current frame has no more than 64 "unreliable" bytes, the decoder goes to Step 2; otherwise the decoder goes to Step 3.

Step 2

Decode the current frame by a RS decoder. Then judge whether the previous frame is successfully decoded or not. If the previous frame has been decoded successfully, the process turns to step 4. If the previous frame has not been decoded successfully, the decoder replaces all the BRC bytes in these rows which have not been successfully decoded in the previous frame with all the FRC bytes in the corresponding rows in the current frame. A corresponding row is a row which is in the same positions in another frame. If the numbers of the "unreliable" bytes in these rows are not more than 64 after this replacement, the decoder could decode these rows successfully. Or they can't be decoded. When all the above operations are finished, the process turns to Step 4.

Step 3

Replaced all the FRC bytes in the rows which have not been successfully decoded in the current frame with all the BRC bytes in the corresponding rows in the previous frame. Then the decoder judges whether the numbers of the "unreliable" bytes in these rows are more than 64 or not. If it is not more than 64, the decoder turns to Step 2; otherwise these rows can not be corrected temporarily, and the decoder stores the data and the reliability information of every positions in the current frame. Then the decoder goes to next step.

Step 4

If there are no frames to be decoded, the decoding process ends; otherwise the decoder receives the next frame and goes to Step 1.

4. Theoretical Performance Analysis of the Frame-Related MPE-FEC Method

The major difference between the proposed frame-related scheme and the zero-padding method in DVB-H standard is the way by which the rightmost n columns in the application data table are filled. In the frame-related method, the data filled in the rightmost n columns in the application data table in the current frame are not zero-valued bytes but the data in the BRC of the previous frame. The similarity of the two methods is that the data filled in the rightmost n columns in the application data table are not transmitted in order to guarantee a high code rate. If the number of the related columns n equals to the number of the zero-padding columns, the code rates in the two methods are the same.

Compared with the zero-padding method, the advantage of the proposed frame-related method, which will be proved later in this section, is that the maximal error correction capability is higher if the code rates in the two methods remain the same. In other words, if the error correction capabilities in the two methods remain the same, the code rate in the proposed frame-related method is higher. As depicted in Fig. 4, the columns in the application data table excluding the FRC are defined as application data columns (ADC). Without loss of generality, it is assumed that one column in a frame is encapsulated into just one transmitted section, which makes the following performance analysis more clear. Under this assumption, we assume that there are N "unreliable" columns in the ADC and the RS data table of the current frame, P "unreliable" columns in the BRC of the current frame and M "reliable" columns remaining in the BRC of the previous frame when the previous frame is not successfully decoded. From the structure of the frame-related MPE-FEC frame it is easy to get the following expressions:

$$0 \leq P \leq n \tag{1}$$

$$0 \leq M < n \tag{2}$$

P and M are both random variables. Eq. (1) shows that P is not bigger than the number of related columns n and Eq. (2) shows that M is smaller than the number of related columns n. According to the decoding algorithm of the proposed frame-related approach described above, it is easy to know that the current frame can be successfully decoded without the assistance of the adjacent frames when N is not bigger than $64-n$. Otherwise the decoding of the current frame needs the help of its adjacent frames. Whether it is successfully decoded or not, the previous frame can benefit the decoding of the current frame. If it isn't yet decoded successful with the assistant of its previous frame, the decoding of the current frame needs the help of its next frame. The next frame can help the decoding of the current frame only if it has been successfully decoded. Otherwise the next frame is not able to help the decoding of the current frame.

From the analysis above, it is easy to see that the decoding of the current frame can be divided into four cases when it is not successfully decoded. The first case is that both the previous frame and the next frame are successfully decoded. In this case, the previous frame and the next frames can both provide n columns of correct data to replace all the bytes in the FRC and the BRC of the current frame respectively. This replacement can correct all the n "unreliable" columns in the FRC and P "unreliable" columns in the BRC of the current frame. Therefore the maximum number of the "unreliable" columns that can be corrected in the current frame is $64+P$. The second case is that the previous frame is successfully decoded but the next frame is not. Then only the previous frame is helpful. All the n "unreliable" columns in the FRC of the current frame are replaced by the n columns of correct data in the BRC of the previous frame. So there are at most 64 "unreliable" columns that can be corrected in the current frame. The third case is that the next frame is decoded successfully but the previous frame is not. The previous frame provides only M "reliable" columns in its BRC, whereas the next frame provides n columns of correct data. P "unreliable" columns in the BRC and M "unreliable" columns in the FRC of the current frame are corrected after the replacement. Thus the maximum number of the "unreliable" columns that can be corrected is $64-n+M+P$. Finally, neither the previous frame nor the next frame is successfully decoded. In this case, only M "unreliable" columns in the FRC of the current frame can be corrected with the help of the columns in the BRC of the previous frame. So the current frame is able to correct $64-n+M$ "unreliable" columns at most.

Because the time interleaver[15] converts the burst noise to random noise, we only considered the performance of the proposed frame-related method in random noise channel, where the frame error rate is the same in each frame. Let FFER denote the frame error rate in data link layer. It is easy to see that the occurrence probability of the four cases discussed above are $(1-P_{FER})^2$, $(1-P_{FER}) * P_{FER}$, $(1-P_{FER}) * P_{FER}$ and P_{FER}^2 respectively. Let m denote the number of the "unreliable" columns that can be corrected in a frame-related frame in the proposed frame-related method, which can be expressed by the following:

$$\begin{aligned} m=&(64+P)\times(1-P_{FER})^2+64\times(1-P_{FER})\times P_{FER} \\ &+(64-n+M+P)\times(1-P_{FER})\times P_{FER}+(64-n+M)\times P_{FER}^2 \end{aligned} \tag{3}$$

P and M are random variables whose values vary in the ranges depicted in Eq. (1) and (2). Without loss of generality, it is supposed that P and M follow identical distribution. In order to get the expectation of m, the expectation of P and M, namely e, is used to replace P and M in (3). Therefore, the expectation of m, still denoted by m, can be represented by the follows:

$$\begin{aligned} m=&(64+e)\times(1-P_{FER})^2+64\times(1-P_{FER})\times P_{FER} \\ &+(64-n+2e)\times(1-P_{FER})\times P_{FER}+(64-n+e)\times P_{FER}^2 \end{aligned} \tag{4}$$

Eq. (4) can be simplified to the following expression:

$$m=64+(e-n\times P_{FER}) \tag{5}$$

The expectation e is in the ranges depicted in Eq. (1) and (2), so e can be denoted by the following expression:

$$e=\alpha\cdot n \tag{6}$$

Where α is the proportional factor, which is a positive real number between 0 and 1. For P it is not bigger than one and for M it is smaller than one. The slight difference of P and M has no influence on the performance analysis, so we omit it and assume that α is not bigger than one for both P and M. In fact the value of the proportional factor α is determined by the actual probability distribution of P and M, but in this paper we are not going to ravel this relationship out. We just use α to denote the relationship between e and n, which will make the analysis below more clear. Based on Eq. (6), Eq. (5) can be rewritten as:

$$m = 64 + n \cdot (\alpha - P_{FER}) \tag{7}$$

From Eq. (7) it is seen that the m is proportional to the number of the related columns n and is inverse proportional to the frame error rare P_{FER}. The reason is obvious. When n is bigger, e is bigger if α remains constant, which means that there are more "unreliable" columns that can be corrected in the BRC of the current frame and the previous frame can provide more "reliable" columns in its BRC for the replacement in the current frame. Therefore, the current frame is able to correct more "unreliable" columns when n is bigger. P_{FER} reflects the channel conditions. P_{FER} is big means that the noise in the channel is big. In this case, the decoding of frames are easier to failed, so the current frame may obtain less assistance from its adjacent frames and correct less "unreliable" columns.

The P_{FER} is determined by the noise in channel, which can not be changed by the designers. However, the related columns number n can be selected by the designers. Though big n brings high error correction capability, it is not appropriate to choose a very big n. The reason is that big n results in low code rate as mentioned in Section 2. Therefore the value of n should be a tradeoff between the code rate and the error correction capability.

In the conventional zero-padding method, the number of the "unreliable" columns that can be corrected in a frame is 64, no matter what the number of the zero-padding columns is. From Eq. (7), it is easy to see that when the frame-related method is employed, $n(\alpha - P_{FER})$ more "unreliable" columns can be corrected in a frame when the code rate remains the same as that of the zero-padding method. For example, when the P_{FER} reaches to 5%, the quality of the video in the DVB-H receiver will be obviously influenced[4]. In this case, if the proposed method is implemented, we can calculate that m is $64 + (\alpha - 0.05)n$. Different α result in different m. Take α as 1/3, 1/2, and 2/3 for example, the corresponding m is $64 + 0.28n$, $64 + 0.45n$ and $64 + 0.62n$ respectively, which means in the proposed frame-related method a frame can correct $0.28n$, $0.5n$ and $0.62n$ "unreliable" columns more than that in the zero-padding method.

There are some additional operations and an extra memory requirement for the decoder when the proposed frame-related method is employed. Besides the current frame, the decoder needs to store the data and the reliable information of every column in the previous frame and the next frame. Since the decoder originally has a buffer which stores several frames to ensure the quality of the play, this extra memory requirement is comparatively small for the decoder. The most complex operations in the decoding of the original DVB-H standard and the zero-padding method are the RS decoding operations[4]. The RS decoding algorithm is not actually modified in the proposed frame-related method. From the decoding algorithm in the frame-related approach mentioned above, it is easy to know that the number of the RS decoding operations in the current frame remains the same as that of the original DVB-H standard and that of the zero-padding method. Compared with the original DVB-H standard and the zero-padding method, the possible additional operations in the proposed frame-related approach are the data transfer from the BRC of the previous frame and the FRC of the next frame to the current frame. Take n as 16 for example, the maximum amounts of the possible transferred data are at least 4096 bytes and at most 16384 bytes, which depend on the number of the rows. These data are transferred twice at most. If the current frame can be successfully decoded without any help of its neighbor frames, even these data transfer operations can be avoided. Compared with the RS decoding, the complexity of the additional data transfer operations is small. Therefore, it can be concluded that the computational cost of the frame-related method has the same order of magnitude as that of the original DVB-H standard and that of the conventional zero-padding method.

5. Simulation Analysis

In this section, some simulation results of the proposed frame-related method are given. The setting of simulation parameters of channel coding in the physical layer is based on [2]. The following physical layer parameters of DVB-H are chosen for simulation: the hierarchical 64-QAM mode, 10 Hz Doppler frequency shift, 1/2 rate convolutional code, 8k transmission mode and 1/4 guard interval[16]. The number of rows in a MPE-FEC frame is 1024. The multipath channel model used in the simulation system is COST207 TU6, which is a six-tap multipath channel model corresponding to the typical urban propagation conditions[4, 17].

The data link layer performance of the proposed frame-related method compared with that of the zero-padding method as well as that of the original DVB-H standard is shown in Figs. 7～9. As mentioned above, when the FERs reach 5% or more, the video quality will be obviously influenced. Therefore, in these three figures the ranges of y-axes, which represents the FER, are chosen to be approximately from 10^{-2} to 10^{-1}. The x-axes represent the Bit Error Rate (BER) in the physical layer. These figures show that for a given FER, how high BER in the physical layer can be tolerated when different algorithms are employed. In other words, for the same BER in the physical layer, the lower the FER is, the bigger the error correction capability is. In Fig. 7 the number of the related columns in the frame-related method is 8, and the numbers of the zero-padding columns in the zero-padding method are 8 and 16 respectively. For a same BER, the FER of the frame-related method is lower than that of the 8 columns zero-padding method, which indicates a gain of error correction capability. In Fig. 8 the number of the related columns in the frame-related method is 16, and the numbers of the zero-padding columns in the zero-padding method are 16 and 32 respectively. From Fig. 8, it can be seen that when the FER is the same, the proposed algorithm can tolerant more bit error in physical layer. Therefore, it can get that the error correction capability of the frame-related method with 16 related columns is obviously more powerful than that of the 16 columns zero-padding method and that of the original DVB-H standard, and is comparable to that of the 32 columns zeropadding method. In Fig. 9, the number of the related columns in the frame-related method is 64, and the numbers of the zeropadding columns in zero-padding method are 64 and 80 respectively. For a same BER, the FER of the frame-related method with 64 related columns is lower than that of the 64 columns zero-padding method. Therefore, the performance of the frame-related method with 64 related columns is better than that of the 64 columns zero-padding method.

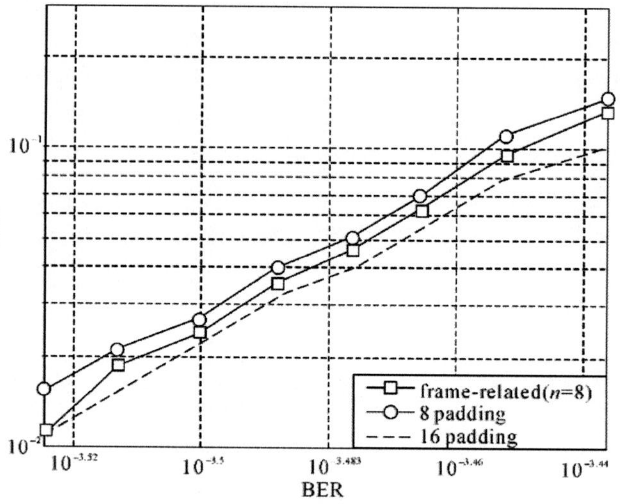

Fig. 7 FER simulation performance($n=8$)

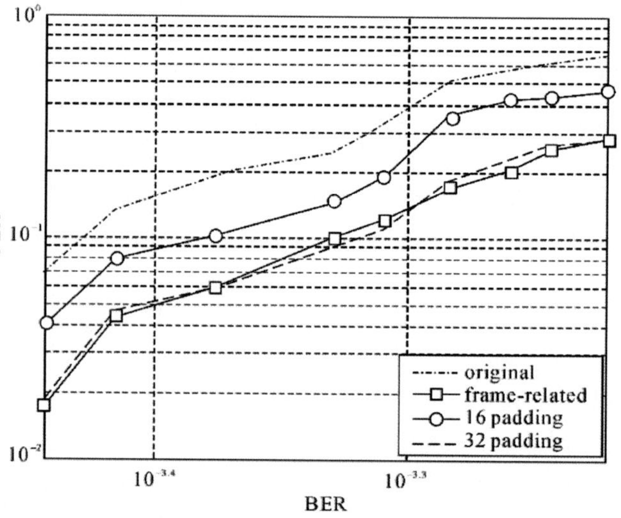

Fig. 8 FER simulation performance($n=16$)

The BERs and code rates of the original DVB-H standard, the zero-padding method and the proposed frame-related method when the FER is 0.1 are listed in Table 1. It can be seen that when the frame-related method with 16 related columns is adopted, the corresponding BER is 4.9×10^{-4}, which is more than that of the original DVB-H standard by 21.89%, and more than that of the 16 columns zero-padding method by 8.17%. It indicates that the proposed frame-related approach can tolerant more noise in the physical layer when the FER remains the same. It also demonstrates that the error correction capability of the frame-related method with 16 related columns is almost equal to that of the 32 columns zero-padding method. However, it should be noticed that the code rate of the frame-related method with 16 related columns, which remains the same as that of the 16 columns zero-padding one, is higher than that of the 32 columns zero-padding method by 2.70%.

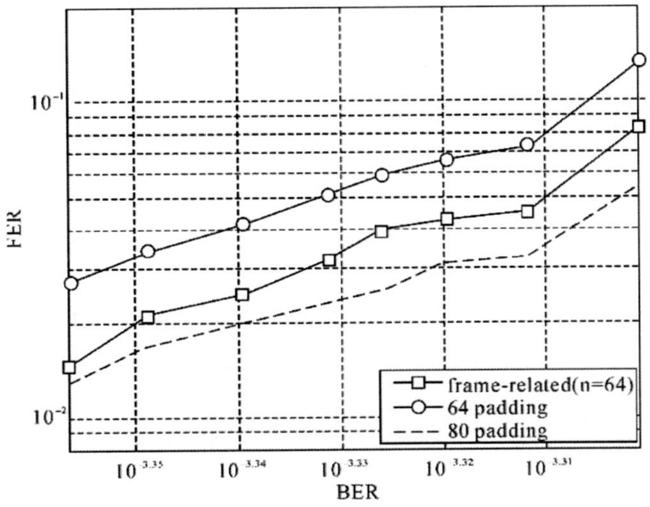

Fig. 9 FER simulation performance(n=64)

Table 1 Bit error rates and corresponding code rates when FER=0.1

Encoding Method	BER(e^{-4})	Code rate
Original DVB-H	4.02	191/255
16 zero-padding	4.53	175/239
16 frame-related	4.90	175/239
32 zero-padding	4.92	159/223

6. Conclusion

The contiguous MPE-FEC frames in the data link layer of DVB-H standard have no correlation with each other. In this paper, a frame-related coding scheme which establishes a correlation among the adjacent frames is proposed. In the frame-related method, the data in the FRC of the current frame are the copy of the data in the BRC of the previous frame, so the received adjacent frames can provide assistance to the decoding of the current frame. Both the analysis and the simulation results show that the error correction capability of the proposed frame-related approach is obviously higher than that of the zeropadding method and the original DVB-H standard. Although some data transfer operations are added, the complexity of the frame-related method remains the same order of magnitude as that of the zero-padding method. Future work includes considering the error correct capability in burst noise channels, and how to combine the proposed frame-related method with other decoding strategies.

References

[1] Digital Video Broadcasting (DVB): Transmission System for Handheld Terminals (DVB-H),

ETSIEN 302 304 V1. 1. 1, European Telecommunications Standards Institute, Nov. 2004.

[2] Digital Video Broadcasting (DVB): Framing Structure, Channel Coding, and Modulation for Digital Terrestrial Television, ETSI EN 300 744 V1. 5. 1, European Telecommunications Standards Institute, Nov. 2004.

[3] U. Reimers, "DVB—The family of international standards for Digital Video Broadcasting," Proc. IEEE, vol. 94, pp. 173-182, 2006.

[4] G. Faria, A. Jukka, and Henriksson, "DVB-H: Digital broadcast services to handheld devices," Proc. IEEE, vol. 94, no. 1, pp. 194-208, Jan. 2006.

[5] D. Plets, W. Joseph, L. Verloock, E. Tanghe, L. Martens, E. Deventer, and H. Gauderis, "Influence of reception condition, MPE-FEC rate and modulation scheme on performance of DVB-H," IEEE Trans. Broadcast., vol. 54, no. 3, pp. 590-598, Sep. 2008.

[6] J. Paavola, H. Himmanen, T. Jokela, J. Poikonen, and V. Ipatov, "The performance analysis of MPE-FEC decoding methods at the DVB-H link layer for efficient IP packet retrieval," IEEE Trans. Broadcast., vol. 53, no. 1, pp. 263-275, Mar. 2007.

[7] H. Himmanen, A. Hazmi, and J. Paavola, "Comparison of DVB-H link layer FEC decoding strategies in a mobile fading channel," in Proc. PIMRC, Helsinki, Finland, Sep. 2006.

[8] H. Joki and J. Paavola, "A novel algorithm for decapsulation and decoding of DVB-H link layer forward error correction," in Proc. ICC, Istanbul, Turkey, Jun. 2006.

[9] Digital Video Broadcasting (DVB): DVB-H Implementation Guidelines, ETSI TR 102 377 V1. 1. 1, Feb. 2005, European Telecommunications Standards Institute.

[10] M. Kornfeld and G. May, "DVB-H and IP Datacast—Broadcast to handheld devices," IEEE Trans. Broadcast., vol. 53, no. 1, pp. 161-170, Mar. 2007.

[11] Digital Video Broadcasting (DVB); IP Datacast Over DVB-H: Electronic Service Guide, ETSI TS 102 471 v1. 1. 1, European Telecommunications Standards Institute, Apr. 2006.

[12] Digital Video Broadcasting (DVB): DVB Specification for Data Broadcasting, ETSI EN 301 192 V1. 4. 1, European Telecommunications Standards Institute, Nov. 2004.

[13] R. E. Blahut, Algebraic Codes for Data Transmission. Cambridge, U. K.: Cambridge Univ. Press, 2003.

[14] H. Joki, J. Paavola, and V. Ipatov, "Analysis of Reed-Solomon coding combined with cyclic redundancy check in DVB-H link layer," in 2nd Int. Symp. Wireless Commun. Syst., Sep. 2005, pp. 313-317.

[15] M. Kornfeld, "Optimizing the DVB-H time interleaving scheme on the link layer for high quality mobile broadcasting reception," in Proc. 11th IEEE Int. Symp. Consum. Electron. (ISCE), 2007.

[16] L. Polak and T. Kratochvil, "Simulation of DVB-H transmission in Gaussian and fading channels," in ELMAR Proc., Sep. 2010, pp. 231-234.

[17] "Digital land mobile radio communications (Final report)," Commission of the European Communities, Directorate General Communications, Information Industries and Innovation, COST207, 1989.

Modified Block-Matching 3-D Filter in Laplacian Pyramid Domain for Speckle Reduction*

Donghai Wen[a], Yuesong Jiang[a], Yanzhong Zhang[a], Yuntao He[a], Houqiang Hua[a], Rong Yu[a], Xiaofang Wu[a], Qian Gao[b]

[a] Beihang University, Beijing, China

[b] Dalian Communication Sergeant School of Air Force, Dalian, China

Abstract: The Laplacian pyramid-based block-matching 3-D filtering (BM3D) is proposed (LPBM3D) for despeckling the speckle image. For BM3D in each pyramid layer, the criterion used to collect blocks in the 3-D groups to the actual data statistics is devised. An adaptive wavelet thresholding operator that depends on both noise level and signal characteristics is proposed. The performance of the proposed LPBM3D method has been compared with the state-of-the-art methods, including the recently proposed nonlocal mean (NLM) and BM3D method. Experimental results show that the visual quality and evaluation indexes outperform the other methods with no edge preservation. The proposed algorithm effectively realizes both despeckling and edge preservation.

Keywords: block-matching 3-D filter; Laplacian pyramid; speckle reduction

1. Introduction

The laser images are corrupted by speckles, which make images hard to interpret valuable information. So the speckle reduction is meaningful preprocessing for advance application such as segmentation and feature extraction.

Indeed, image despeckling has been an active field of research for almost 30 years, and many algorithms are proposed each year. A few reviews on speckle reduction approaches have been done[12]. Some of the classical methods have been proposed to reduce speckle noise. These filters use local pixel intensity statistics to adjust the amount of smoothing and noise removal in certain areas. In areas with large variance of pixel intensities, such as areas with high levels of detail or edges, the filter applies less smoothing in order to preserve those details. In homogeneous areas with little detail, a smoothing kernel is applied to remove the noise.

In the last decade, the Anisotropic Diffusion filter (AD)[10] and the Total Variation minimization scheme (TV)[13] have been developed for speckle reduction. These approaches are iterative and produce smooth images while preserving edges. Nevertheless, meaningful structural details are

* Published in Optics Communications, 322 (2014): 150-154.

Y. Zhang, *Footprints in Cambridge and Aviation Industries of China*,
https://doi.org/10.1007/978-981-16-3176-4_24

unfortunately removed during iterations.

Recently, speckle reduction approaches using wavelets[7], texture models[5], adaptive stack filters[2], Markov random field models[6], blind deconvolution methods[8] have been introduced for the speckle reduction.

These despeckling algorithms cannot perform well in both despeckling and preserving edges of speckle images simultaneously. Given these premises, the nonlocal approach[1], recently proposed for the additive white Gaussian noise (AWGN) denoising, looks like a potential breakthrough. The nonlocal approach principle has inspired several extensions, and the BM3D method[4] is one of them. The NLM and BM3D algorithms have been readily extended to speckle reduction[11] with suitable modifications aimed at taking into account the problem peculiarities.

Based on the conceptual path described earlier, and the related experimental evidence, in this work, we go one step further and propose a image-oriented version of BM3D. A speckle reduction method based on BM3D in the Laplacian pyramid domain is proposed to more effectively suppress speckle while preserving edges. A Canny operator is first utilized to detect and remove edges from the speckle image. Then, the pyramid transform and BM3D are used to decompose and despeckle the edge-removed image, respectively. For BM3D in each pyramid layer, a suitable threshold is automatically determined by an adaptive data-driven exponential operator. Finally, the removed edges are added to the reconstructed image.

This paper is organized as follows. The BM3D algorithm and its speckle image-oriented version is proposed in Section 2. In Section 3, the LPBM3D method is proposed. In Section 4, the experimental results demonstrating the effectiveness of the proposed method are given. Some concluding remarks are drawn in Section 5.

2. BM3D Algorithm and Its Polarization Image-Oriented Version

In this paper, the application of BM3D is extended to image despeckling. To this end, we introduce two major modifications.

First of all, we adapt the criterion used to collect blocks in the 3-D groups to the actual data statistics. The Euclidean distance is devised for AWGN. Once the noise statistics change, the Euclidean distance loses its significance. So the new block similarity measure is devised.

Second modification is an adaptive wavelet thresholding operator that depends on both noise level and signal characteristics is devised. Generally, in the BM3D method, the wavelet coefficients are passed through a threshold testing that requires replacing noisy coefficients below a fixed value with zeros, and keeping the others because they have most of the information. Then, the resulting coefficients are used to reconstruct the signal. The method depends on the choice of threshold value. Thresholding methods have the following two main drawbacks: (1) The choice of the threshold is made in an ad hoc manner; (2) the specific distribution of the signal and noise may not be well matched at different scales. So an adaptive wavelet thresholding operator is proposed.

2.1 New Block Similarity Measure

Mathematically, given two values, it results in

$$P[x(k),x(l) \mid y(k)=y(l)] = \int_R P[x(k) \mid y(k)=\delta]P[x(l) \mid y(l)=\delta]P\delta \mathrm{d}\delta \tag{1}$$

where $x(k)$, $x(l)$ are the noisy signal, $y(k)$, $y(l)$ are the corresponding values of the noise-free signal defined over the domain R, and $P(\cdot)$ indicates a probability density function. $x(k)$ and $x(l)$ are assumed to be conditionally independent given y. This expression further simplifies to

$$P[x(k),x(l) \mid y(k)=y(l)] \propto \int_R P[x(k) \mid y(k)=\delta]P[x(l) \mid y(l)=\delta]\mathrm{d}\delta \tag{2}$$

if we assume,lacking any prior knowledge, $P(\cdot)$ to be uniform over R, then (2) reads as

$$P[x(k),x(l) \mid y(k)=y(l)] \propto \int_0^{\infty} \frac{1}{\Gamma^2(\mu)}\left(\frac{\mu}{\delta}\right)^{2\mu} \mathrm{e}^{(\mu-1)[x(k)+x(l)]} \exp\left\{-\frac{\mu[\mathrm{e}^{x(k)}+\mathrm{e}^{x(l)}]}{\delta}\right\}\mathrm{d}\delta \tag{3}$$

with the integral equal to

$$\frac{\mu\Gamma(2\mu-1)\mathrm{e}^{(\mu-1)[x(k)+x(l)]}}{\Gamma^2(\mu)[\mathrm{e}^{x(k)}+\mathrm{e}^{x(l)}]^{2\mu-1}} \tag{4}$$

To rewrite this result into a manageable block similarity measure, we must rewrite Eq. (5) with vectors drawn from the blocks A_k and A_l in place of scalars and assume again the conditional independence of the observed values given the noise-free signal. Then, we define the block similarity measure as

$$\begin{aligned} d[x(A_x),x(A_l)] &= -\log\left\{\prod_q P[x(k+q),x(l+q) \mid y(k+q)=y(l+q)]\right\} \\ &= -\log\left\{\prod_q \frac{\mu\Gamma(2\mu-1)\mathrm{e}^{(\mu-1)[x(k+q)+x(l+q)]}}{\Gamma^2(\mu)[\mathrm{e}^{x(k+q)}+\mathrm{e}^{x(l+q)}]^{2\mu-1}}\right\} \end{aligned} \tag{5}$$

discarding the constant term

$$d_1[x(A_x),x(A_l)] = \sum_q \{(2\mu-1)\log[\mathrm{e}^{(k+q)}+\mathrm{e}^{(l+q)}] - (\mu-1)[x(k+q)+x(l+q)]\} \tag{6}$$

where the subscript 1 indicates that this measure is used in the first step. In the second step, in fact, the similarity measure must take into account the additional information provided by the first step, which is a coarse estimate of the noiseless signal. The similarity measure in the second step is defined as

$$\begin{aligned} d_2[x(A_x),x(A_l)] = \sum_q \{&(2\mu-1)\log[\mathrm{e}^{(k+q)}+\mathrm{e}^{(l+q)}] - (\mu-1) \\ &\times [x(k+q)+x(l+q)] + \beta\mu \frac{|\hat{y}(k+q)-\hat{y}(l+q)|^2}{\hat{y}(k+q)\hat{y}(l+q)}\} \end{aligned} \tag{7}$$

where β weighs the relative importance of the data and (loosely speaking) prior terms.

2.2 The Adaptive Wavelet Thresholding Operator

To overcome the shortcomings of thresholding methods, an adaptive wavelet thresholding operator is used. The following operator based on exponential function was defined as

$$et(m) = \begin{cases} m \cdot \mathrm{e}^{n_1 \cdot (|x|-T_{k_1})}, & |x|<T_{k_1} \\ m, & |x|\geq T_{k_1} \end{cases};\ T_{k_1}=k_1 \cdot T_{u_1} \tag{8}$$

where n_l is a real parameter identifying the fall degree of exponential function for l decomposition level, while k_l factor provides a modified version of l-level universal threshold. T_{u_1} is the VisuShrink threshold.

The modified BM3D is compared with the original BM3D algorithm. The main procedures of the original BM3D algorithm are illustrated in Fig. 1.

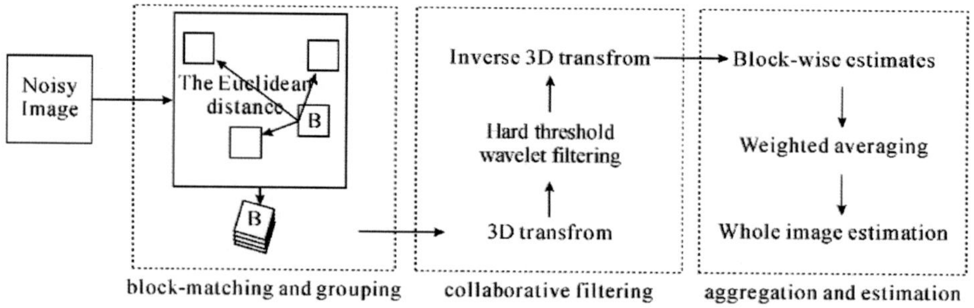

Fig. 1 Flow chart of original BM3D algorithm. The operations surrounded by dashed lines are repeated for each processed block (marked with B). The Euclidean distance is used for block-thresholding is used for collaborative filtering

The procedures of the polarization image-oriented BM3D algorithm are illustrated in Fig. 2. The differences with the original BM3D algorithm are the new block similarity measure and the adaptive wavelet thresholding operator.

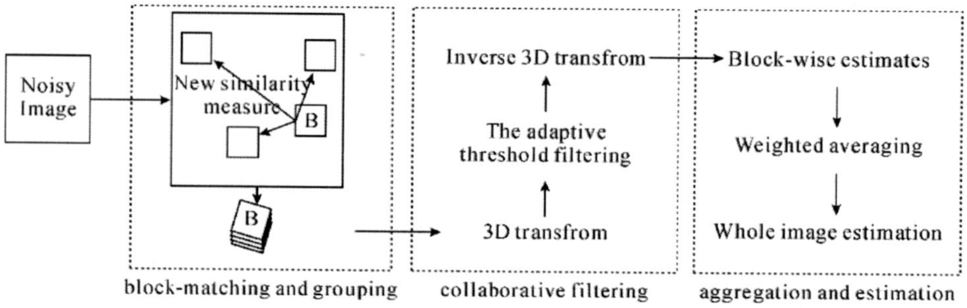

Fig. 2 Flow chart of modified BM3D algorithm. The operations surrounded by dashed lines are repeated for each processed block (marked with B). The differences with the original algorithm are the new block similarity measure and the adaptive wavelet thresholding. The new block similarity measure is used for block-matching. The adaptive wavelet thresholding is used for collaborative filtering

3. The Pyramid Transform and the Proposed Method

3.1 Pyramid Transform

A general structure of pyramid transforms consists of decomposition and reconstruction stages and can be described by approximation and interpolation filtering. In the decomposition stage, a signal is successively decomposed into a decimated approximation signal and a signal containing residual information. This residual signal is computed as the difference between the signal on a finer scale and the interpolated signal from a coarser scale. A finer scale corresponds to a lower pyramid layer. The lowest pyramid layer has the same size as the original image. In the Laplacian pyramid, two operators, REDUCE and EXPAND, are commonly used. The REDUCE operator performs a two-dimensional (2-D) lowpass filtering followed by a sub-sampling by a factor of two in both directions. The EXPAND operator enlarges an image to twice the size in both directions by up-sampling (i. e., insertion of zeros) and a lowpass filtering followed by a multiplication by a factor of four, which is necessary to maintain the average intensity being reduced by the insertion of zeros. For an input image x, let its Gaussian pyramid at layer l be G_l, and its Laplacian pyramid at layer l be L_l, where $l=0, 1, 2, \cdots, d-1$ and d is the total decomposition layer. Then, the Gaussian and the Laplacian pyramid can be defined as

$$G_0 = x \tag{9}$$

$$G_1 = \text{REDUCE}[G_{l-1}] \tag{10}$$

$$L_l = G_l - \text{EXPAND}[G_{l+1}] \tag{11}$$

3.2 The Proposed Method

The proposed LPBM3D method is shown in Fig. 3.

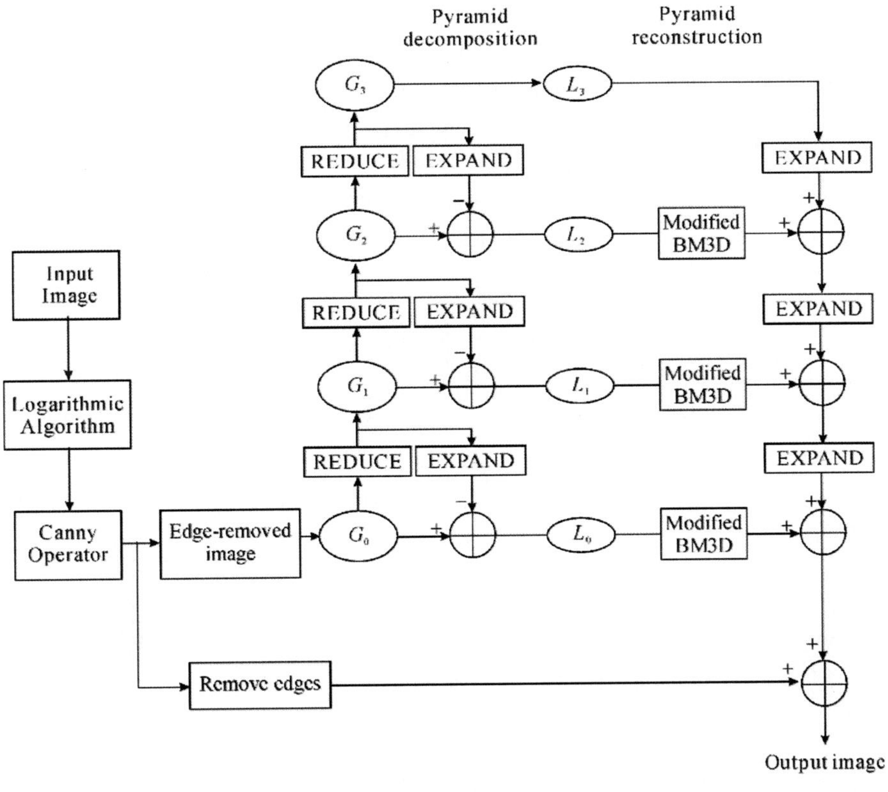

Fig. 3 The LPBM3D filter, including logarithmic algorithm, pyramid decomposition, modified BM3D, and pyramid reconstruction

The proposed method is described as follows:

(1) Change speckle into additive noise by logarithmic algorithm.

(2) Use the Canny operator to detect the edge, then the image is divided into two parts, one part is edge-removed image and the other is remove edges.

(3) The edge-removed image is handled by the Pyramid transform.

(4) Manipulation of pyramid coefficients by the modified BM3D method.

(5) Reconstruction of the image by adding the removed edges and the processed image.

4. Experimental Result

To validate the despeckling quality of the proposed algorithm, some speckle images are tested. The proposed algorithm was compared with NLM method[3], Parrilli et al. method[9]. For all these algorithms, if not stated otherwise, the free parameters are set as suggested in the reference papers. As for the proposed LPBM3D algorithm, in the first step of the modified BM3D algorithm, we use a Daubechies-8 undecimated discrete wavelet transform (UDWT) with a three-level decomposition, and fixed groups of dimension $8\times8\times16$. The search area is 39×39. Similar choices apply to the second step except for the transform, which is a spatial discrete cosine transform (DCT) followed by a Haar discrete wavelet transform (DWT) along the blocks with a maximum-level decomposition, and for the group dimensions that grow to $8\times8\times32$. Finally, the weight β in the similarity measure of formula (11) was set equal to one, which was found experimentally to guarantee a good performance. The well-known statistical estimators, mean, standard deviation, the equivalent number of looks (ENL), edge-preservation index (EPI), are used to quantitative analysis. The ENL is defined as

$$\mathrm{ENL}=\frac{\mu_{\mathrm{enl}}^2}{\delta_{\mathrm{enl}}^2} \tag{12}$$

where μ_{enl}^2 and δ_{enl}^2 are, respectively, the mean and the variance associated with a set of homogeneous regions in the image. The larger the ENL the better the speckle suppression. The EPI is expressed as

$$\mathrm{EPI}=\frac{\sum_{i=1}^{n}|\hat{x}_{i,1}-\hat{x}_{i,2}|}{\sum_{i=1}^{n}|x_{i,1}-x_{i,2}|} \tag{13}$$

where $x_{i,1}$, $x_{i,2}$, and $\hat{x}_{i,1}$, $\hat{x}_{i,2}$ are the values of the original and filtered images, respectively, observed on the one-pixel wide lines on both sides of the edge. Larger values correspond to a better edge retaining ability of the filter.

A grass picture is processed by these methods first. Fig. 4 shows the visual quality of the experiments. Table 1 gives the quantitative evaluation. Visually, from Fig. 4, the proposed method can get better balance in speckle removal and edge retention.

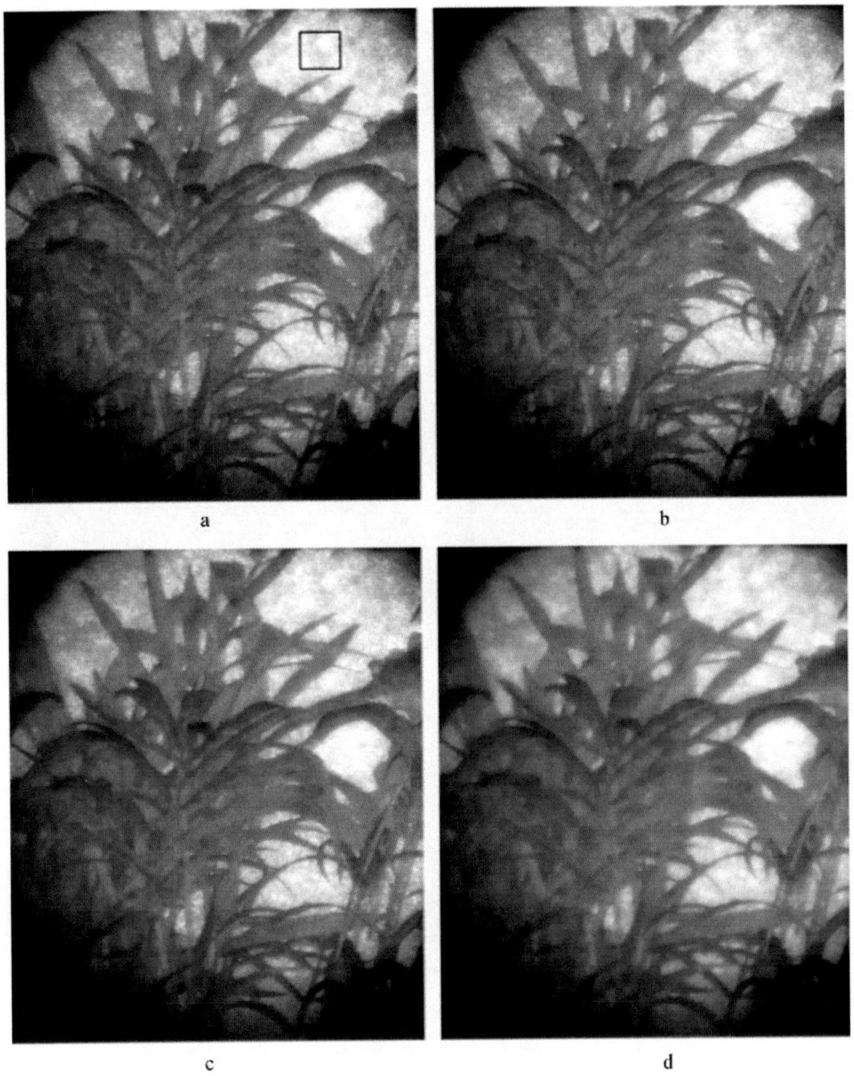

Fig. 4 The grass picture despeckling experiments

(a) Original image; (b) NLM; (c) Parrilli, S. method; (d) Proposed

Table 1 The quantitative evaluation for the grass picture

Filters	Mean	Standard deviation	ENL	EPI
NLM	175.2816	22.6481	70.5792	0.4729
Par.	170.5273	21.8476	72.1843	0.4035
LPBM3D	165.7268	19.5627	80.6251	0.5072

Fig. 5 shows despeckled brain images using the proposed and other comparison algorithms. The ENL and the EPI metric are shown in Table 2.

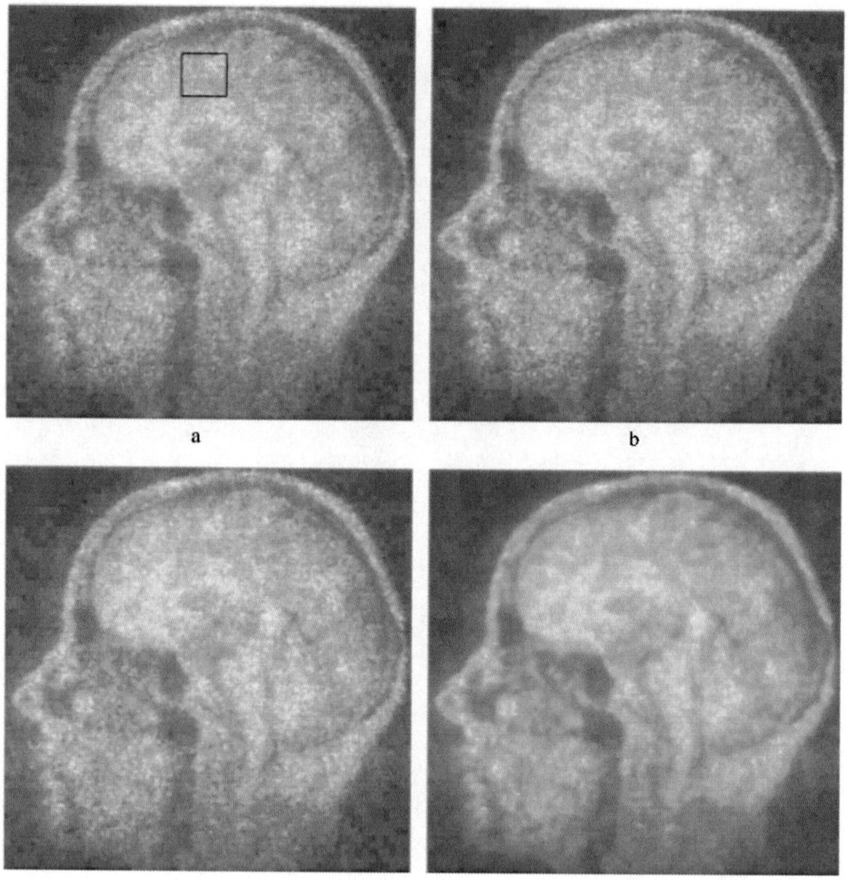

Fig. 5 The brain picture despeckling experiments
(a) Original image; (b) NLM; (c)Parrilli, S. method; (d)Proposed

Table 2 The quantitative evaluation for the brain picture

Filters	Mean	Standard deviation	ENL	EPI
NLM	61.2793	7.0351	62.5094	0.5709
Par.	60.8257	6.9546	65.0423	0.5918
LPBM3D	59.3254	6.4382	74.2681	0.6219

5. Conclusion

A novel algorithm for despeckling polarization images using BM3D and Laplacian pyramid domain was proposed. The noise statistics were examined and speckle noise distribution in a log-compressed image was computed. In the BM3D method, the new block similarity measure and the adaptive wavelet thresholding operator were proposed. Testing on real polarization images found that the LPBM3D algorithm shows strong despeckling performance when compared to state-of-the-art despeckling algorithms. Future work involves combining the proposed methodology directly with segmentation and classification algorithms to investigate the effectiveness of such combined systems for improving segmentation and classification of speckle images.

References

[1] A. Buades, B. Coll, J. M. Morel, Multiscale Model. Simul. 4 (2005) 490.

[2] Maria Elena Buemi, Alejandro C. Frery, Heitor S. Ramos, Pattern Recognit. Lett. 36 (2013) 281.

[3] P. Coupe, P. Hellier, C. Kervrann, IEEE Trans. Image Process. 18 (2009) 2221.

[4] K. Dabov, A. Foi, V. Katkovnik, IEEE Trans. Image Process. 16 (2007) 2080.

[5] J. Glaister, A. Wong, D. A. Clausi, IEEE Trans. Geosci. Remote Sens. 52 (2013) 1238.

[6] Xie Hua, L. E. Pierce, F. T. Ulaby, IEEE Trans. Geosci. Remote Sens. 40 (2002) 2196.

[7] Ma Jianwei, IEEE Trans. Image Process. 9 (2007) 2198.

[8] A. Lapini, T. Bianchi, F. Argenti, IEEE Trans. Geosci. Remote Sens. 52 (2013) 1044.

[9] S. Parrilli, M. Poderico, C. V. Angelino, IEEE Trans. Geosci. Remote Sens. 50 (2012) 606.

[10] P. Perona, J. Malik, IEEE Trans. Pattern Anal. Mach. Intell. 7 (1990) 629.

[11] Leonardo Torres Sidnei, J. S. Sant'Anna, Pattern Recogni. 47 (2014) 141.

[12] R. Touzi, IEEE Trans. Geosci. Remote Sens. 40 (1990) 2392.

[13] Hyenkyun Woo, Sangwoon Yun, IEEE Trans. Image Process. 21 (2012) 1701.

AERONAUTICAL ENGINEERING

Aviation Environmental Technology and Science*

Zhang Yanzhong
China Aviation Industry Corporation II, Beijing, China

1. Introduction

With the development of aviation undertaking, the effect caused by aircraft to the environment is increasingly growing. As aircraft are flying at high altitude, the pollution discharged as a result has more obvious effect on the environment than that discharged on the ground, and more easily leads to greenhouse effect and change of the global climate. As the environmental protection consciousness of people is intensified, people pay more and more attention to aircraft noise and pollution discharge. Some regulatory departments are formulating even stricter technical standard, so as to reduce aircraft noise and pollution discharge, thus bringing about a new challenge to the development of aviation technology. At present, people have used technology of aerodynamics, structures and engines, etc. to greatly reduce noise of the jet aircraft, improve flying efficiency and reduce pollution discharge.

In recent 100 years, aviation technology has grown out of nothing, and is continuously perfected. Aircraft are widely used in the areas of communication, transport, military affairs and trade, etc., creating enormous benefit to the society, and promoting the development of the global economy and social progress. According to the statistics by relevant authorities, main airlines across the world own totally 15,271 civil aircraft at present. Although affected by the "9 • 11" terrorist attack, in 2001, the passenger traffic volume of scheduled flights reached 1.62 billion person-times, and the volume of freight transport reached 29 million tons across the world. According to the forecast made by Boeing Company, in the future 20 years, the passenger traffic volume would increase at a rate of 4.9% annually, and the volume of freight transport at 6.4% in the world civil aviation industry. By 2021, main airlines of the world would own 32,495 aircraft in total.

However, with the continuous development of the aviation undertaking, the effect caused by aircraft to the environment is growing dramatically. According to the statistics by Intergovernmental Panel on Climate Change (IPCC) under the United Nations, the transport industry accounts for 20% of the fossil fuel consumed across the world, and the aviation transport accounts for 12% of the

* Published in Engineering Science, Vol. 6, No. 2, June 2008: 2-8.

Y. Zhang, *Footprints in Cambridge and Aviation Industries of China*,
https://doi.org/10.1007/978-981-16-3176-4_25

fossil fuel consumed by the transport industry across the world (See Fig. 1). That is to say, carbon dioxide generated by aviation transportation accounts for about 2.4% of the total volume of carbon dioxide generated through using fossil fuel by humankind. The aviation transport is one of the important factors that contribute to the increase of carbon dioxide content in the atmosphere.

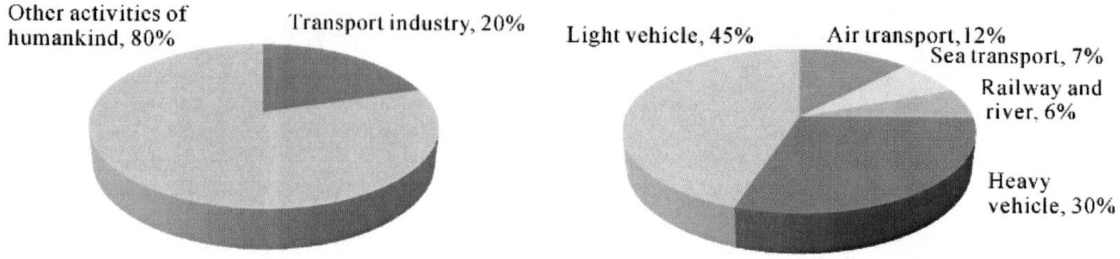

The proportion of the transport industry to the total volume of using fossil fuel by humankind

The proportion of carbon dioxide generated by the aviation transport to that by all kinds of transport

Fig. 1 Aviation transport is one of the important factors that contribute to the increase of carbon dioxide content in the atmosphere

With the increase of people's environmental protection consciousness, ground vehicles such as automobiles are gradually using electric power, natural gas and other green energies, and proportion of aviation fuel in the oil consumption across the world will increase every year. As aircraft pollutant is discharged at high altitude, its effect on the atmosphere becomes even greater. Therefore, people pay more and more attention to the environmental protection performance of aircraft. For this reason, International Civil Aviation Organization (ICAO) and some governmental organizations are gradually enhancing the regulation standards with respect to the environmental protection performance of aircraft. On Jan. 30th, 2002, EU Aviation Commission announced that from April 1st, civil aircraft of Tu-134, IL-62, IL-76 and IL-86 made by Russia would be prohibited to enter EU countries because their noise levels were not in conformity with the EU standard. This decision has caused a great loss to the Russian aviation manufacturers and transport industry.

In order to promote the implementation of sustainable development strategy and protect the common homestead of humankind, it is necessary to further improve the technological level of aircraft, so as to meet the environmental protection requirements of people.

2. Aircraft Effect on the Environment

Aircraft effect on the environment is mainly reflected by its exhaust pollution and noise. As aircraft are flying at high altitude, pollutant discharged thereof has more serious effect on the atmosphere than that discharged on the ground, and more easily leads to greenhouse effect and change of the global climate. According to calculations, total greenhouse effect created by humankind is about 1.5 times bigger than that created by carbon dioxide generated by humankind, while the total greenhouse effect created by aircraft is about 2～4 times bigger than that caused by carbon dioxide of equal magnitude on the ground. The main reason for this difference is that aircraft emissions such as hydrocarbon, nitric oxides and vapor at high altitude have more evident greenhouse effect than those on the ground.

2.1 Effect of Aircraft Emissions on the Environment

Fuel used by aircraft is aviation fuel. Its exhaust gas is mainly composed of carbon dioxide, nitric oxides, vapor, hydrocarbon, carbon monoxide, oxysulphide and small carbon particles. The main effect of these gases on the environment is to generate greenhouse effect of the atmosphere, which will in turn affect the change of the global climate.

Carbon dioxide: According to the molecular constitution of the aviation fuel, the aviation fuel of specific mass after it is burnt will produce about 3.15 units of carbon dioxide. The carbon dioxide discharged by aircraft in 1992 was about 514 million tons, which accounted for 2.4% of the total volume of carbon dioxide generated by all the fossil fuel consumed that year, or 2% of the carbon dioxide generated by humankind that year. By 2050, total volume of carbon dioxide generated by the aviation industry will reach 1468 million tons, which will account for 3% of the total volume of carbon dioxide generated by humankind. Carbon dioxide is the main gas that causes greenhouse effect.

Nitric oxides: Nitric oxides is a byproduct generated during burning. It is mainly composed of NO, NO_2 and N_2O, etc. NO is a colorless and scentless gas, and it has a stronger ability to integrate with haematoglobin than CO, and thus it more easily causes human body to run short of oxygen. NO_2 is a brown gas, and it can reach the depth of the lung during breathing, causing disease to the respiratory system. Furthermore, NO_2 can form acid rain, causing damages to the environment.

Vapor: Vapor generated by aircraft can form the wakes of cooled particles and high altitude cloud clusters. The wakes and cloud clusters, like greenhouse gases, can partially prevent solar energy reflected by the ground from radiating to the space, thus resulting in greenhouse effect. If there are also sulphate and carbon black generated during burning, the blocking radiation effect of vapor will be reinforced.

Hydrocarbon: Hydrocarbon is more capable of causing greenhouse effect than carbon dioxide. Especially because aircraft are normally flying over troposphere or in stratosphere, the greenhouse effect created by hydrocarbon discharged by them is more evident.

In order to control the effect of aircraft-discharged pollution, in May 1981, ICAO amended the content in Annex 16 of "International Civil Aviation Convention" by adding relevant regulations on pollutant discharged by civil aviation aircraft to the Annex, and required that engines produced after Jan. 1st, 1986 should meet these requirements.

These discharge standards mainly involve aircraft takeoff, landing, climbing, approaching and ground taxiing periods, and no specific regulations are set for aircraft cruising period. Since these standards are implemented, carbon monoxide and hydrocarbon discharged by aircraft have been greatly reduced, but nitric oxides discharged by aircraft have not been reduced much (See Fig. 2). The main reason for this phenomena is that temperature and pressure in the combustion chambers of engines have a tendency to increase gradually. To increase the combustion temperature of engines will be helpful for engines to improve their operating efficiency and promote complete fuel combustion, thus reducing the discharge of carbon monoxide and hydrocarbon. However, because of the higher combustion temperature, nitric oxides produced by engines have been increased.

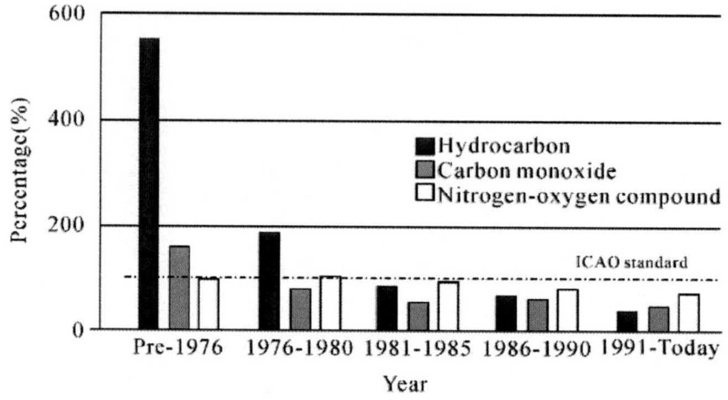

Fig. 2 Percentage of aeroengine's pollution discharge in relation to the ICAO requirements of 1996

In order to reduce aircraft discharge of nitric oxides, ICAO, based on the suggestions of its subordinate body, Committee on Aviation Environmental Protection (CAEP), revised the original aircraft discharge standards. The revised standards did not change the discharge standards for carbon monoxide, hydrocarbon or soot, but reduced the discharge volume of nitric oxides by 20%. In 1999, ICAO again raised the discharge standards for nitric oxides, and required that engines produced after Jan. 1, 2004 should further reduce their discharge volume of nitric oxides by 16.5% on the basis of the standards of 1993 (See Fig. 3). Meanwhile, although an engine of high compression ratio may increase the discharge of nitric oxides, it is contributing to the improvement of the engine overall performance and reduction of carbon dioxide discharge, hence it is permitted to relax the discharge standard of nitric oxides for engines with compression ratio of over 30.

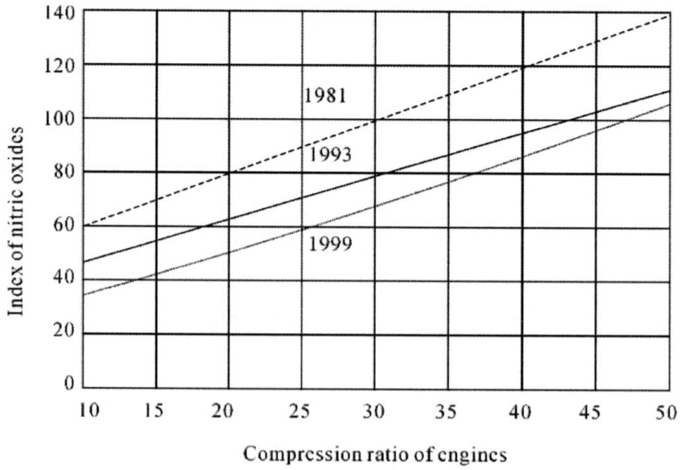

Fig. 3 Evolution of standard set by ICAO with regard to engine discharge amount of nitric oxides

2.2 Effect of Aircraft Noise on the Environment

Based on the different affected targets, aircraft noise can be classified as interior noise and exterior noise. Interior noise not only affects the comfort and health of passengers and flight crew on board, but also produces very strong noise load to the aircraft structure. When sound pressure level of the noise load exceeds 130 dB, it will result in fatigue damages to the structure. The higher the sound pressure level acts on the aircraft structure, and the longer the sound pressure level lasts, the greater damages it will cause to the structure.

Exterior noise mainly affects normal life of the residents near airports or air routes. Generally speaking, exterior noise can be categorized into low frequency noise and high frequency noise. High frequency noise brings more vexation to people than the low frequency noise, and yet most of noise generated by jet aircraft is exactly high frequency noise. Therefore, to control aircraft exterior noise has great significance for improving the living quality of people around the airports.

In addition to the ordinary noise, aircraft sometimes also produces an exterior noise called "sonic boom". "Sonic boom" takes place mainly because of the shock wave generated during the supersonic flying. The shock wave has very strong energy even after it reaches the ground, so it gives off thunder-like explosive sound. Sonic boom normally lasts a short period, about 350 ms for large supersonic aircraft, and about 100 ms for fighters. If sonic boom is transmitted into a room, it will last longer because of sound resonance caused by multiple reflections. Some high strength sonic boom even can damage the glasses of buildings, causing even greater damages to infrastructures on the ground.

In order to control aircraft noise, FAA formulated aircraft noise airworthiness regulation FAR 36 as early as 1969. In 1971, ICAO specified aircraft noise standards for application of airworthiness certificates before October, 1977 and after October 6, 1976 respectively in Chapters 2 and 3 of Annex 16 of "International Aviation Convention". Afterwards, ICAO revised and perfected these standards, and required all states to meet the requirements set forth in the standards. With the implementation of these standards and enhancement of the aviation technological level, aircraft noise has evidently taken a turn for better (See Fig. 4).

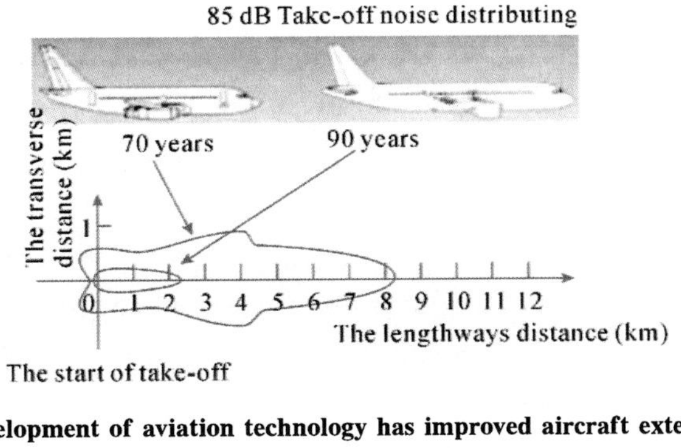

Fig. 4 Development of aviation technology has improved aircraft exterior noise

In October 2001, ICAO adopted a new proposal by CAEP to further improve aircraft environmental protection performance, and as a result, the requirements on aircraft environmental protection performance have been further enhanced. For example, according to the newly formulated standards in Chapter 4, Annex 16 of "International Civil Aviation Convention", for aircraft designed after Jan. 1, 2006, their noise level should be further reduced by 10 dB on the basis of the standard set forth in Chapter 3 of Annex 16.

3. Technology to Control Aircraft Discharge Pollution

In order to reduce aircraft discharge pollution, work can be done in many areas such as technology, management and tax policy. But the most important one is to constantly improve the technological level of aircraft to reduce pollution fundamentally. Among various technologies to reduce aircraft discharge pollution, technologies with respect of engines, aerodynamics, structure and air traffic control are the most important ones.

3.1 Engine

Since pollution discharged by aircraft is directly from engines, to improve the performance of engines and enhance combustion efficiency can greatly reduce aircraft fuel consumption and greenhouse gases discharge. From 1976 to 1994, with the improvement of aircraft structural design and engine technology, and the increase of aircraft seat occupancy, aircraft fuel consumption per seat-kilometer decreased by 50% (See Fig. 5), which meant that aircraft pollution discharge also decreased. The improvement of engine technology contributes a major half for this achievement.

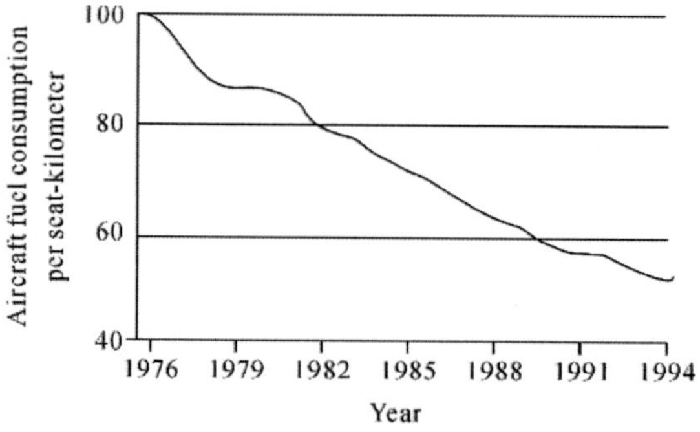

Fig. 5 Changes in fuel consumption of civil aviation aircraft

In the past forty years, technology of aeroengines has made great progress. From turbojet engines to the present turbofan engines of the third generation, fuel consumption of engines has been cumulatively reduced by 40%. By using new types of engines, fuel efficiency of aircraft can be evidently improved, and carbon dioxide discharge per seat-kilometer can be reduced. Especially for long-haul aircraft, the result is even more evident. As for the short-range route, the aircraft with advanced turbo propeller engines are more efficient than the aircraft powered by jet engines. The carbon dioxide discharge per seat-kilometer of the former is lower than that of the latter by 20%.

At present, a new generation of engines with high bypass ratios is being developed, whose fuel consumption is expected to be further lower than the present engines by 12%~20% (See Fig. 6). However, there is an important technological challenge in developing new aeroengines, that is the discharge of nitric oxides. This is because in order to improve the engine's operating efficiency, it is necessary to increase the engine's compression ratio and the combustion chambers' temperature; however, with pressure and temperature increases in combustion chambers, it is inevitable to lead to increase of nitric oxides (See Fig. 7).

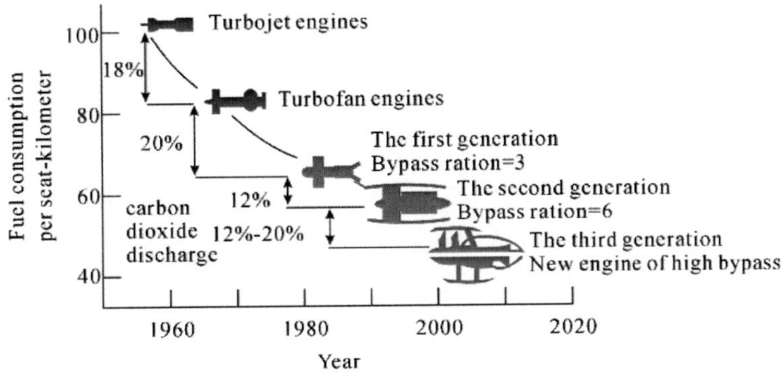

Fig. 6 Change in fuel consumption and carbon dioxide discharge of engines of different ages

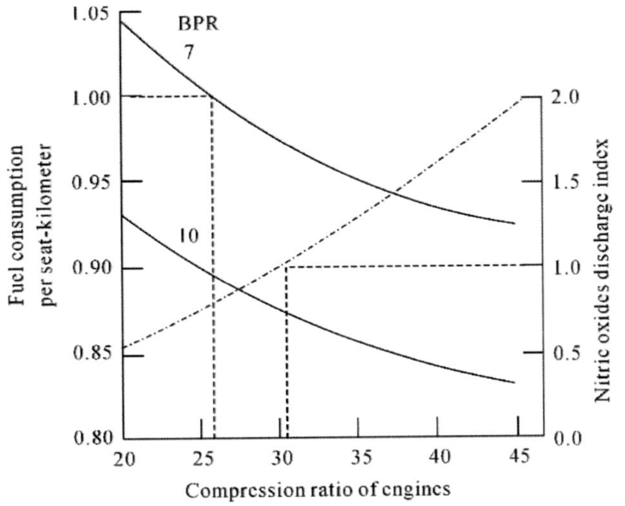

Fig. 7 Relationship between engine fuel consumption and nitric oxides discharge amount and changes of compression ratio

Researches show that the production of nitric oxides is related to combustion chambers' temperature, pressure and the time period that fuel gas passes through the chambers. By improving the flow field of the combustion chamber and the fuel-spraying mode, it is possible to reduce nitric oxides discharge by 20%~40%. In addition, by rationally controlling engines' fuel/air ratio, it is possible to decrease flame temperature and increase airflow velocity, thus reducing the time that the fuel gas dwells in the chambers and the production of nitric oxides.

At present, the USA and Europe are all energetically developing highly efficient and clean engines. According to the NASA working plan, in the next 10 years, nitric oxides discharge amount of aircraft is expected to be reduced to 1/3 of the present value, and carbon dioxide discharge amount reduced to 75% of the present value; in the next 25 years, nitric oxides discharge amount shall be reduced to 1/5 of the present value, and carbon dioxide discharge amount reduced to 50% of the present value.

In order to resolve the aviation fuel problem after petroleum becomes depleted, people now have already started looking for substitute fuels (e. g. hydrogen) for aviation kerosene. By using these fuels, it will be helpful to solve the aircraft pollution discharge problem. However, there are also some insurmountable problems to use these fuels. For example, the unit volume calorific value of liquid hydrogen fuel is only 23% of aviation kerosene. For the aircraft of the same fuel consumption, its fuel tank will be three times as big as the kerosene tank. Moreover, it is necessary to make a completely new design of aircraft, and replace the present fuel supply, distribution and storage infrastructure fundamentally. Presently there is no practical alternative option to replace aviation kerosene, therefore, to improve aircraft fuel efficiency is still the key measure to reduce aviation pollution discharge.

3.2 Aerodynamic Contour

Advanced aerodynamic technology can reduce aircraft aerodynamic resistance, thus achieving the aim of saving fuel and reducing pollution discharge. At present, people are intensifying research work on laminar flow technology and aerodynamic layout of blended wing-body, so as to improve the aerodynamic performance of aircraft. In addition, in the area of aerodynamic contour, the technology conducive to the improvement of aircraft efficiency and reduction of aircraft pollution discharge also includes using the advanced passive flow field controller (e. g. vertex generator) to increase aircraft lift, adding the winglet to the wing tip, using the super critical airfoil to improve and optimize aircraft lift-drag ratio during cruising, using the design method of computational fluid dynamics to carry out aerodynamic design, and using the advanced manufacturing technology to improve smoothness of aircraft surfaces. According to NASA's estimation, the lift-drag ratio of subsonic civil aircraft in the 21st century will increase by one time than the present, reaching 40.

3.3 Aircraft Structure

To reduce aircraft structural weight is conducive to the reduction of aircraft fuel consumption, thus reducing aircraft pollution discharge. In future aircraft, more aluminum lithium alloy and composite materials will be used. And the aluminum lithium alloy is mainly used in wings and main force-bearing structures of fuselage, and composite materials are mainly used in some secondary structures of aircraft.

In addition to the adoption of new materials, to reduce some of not-much-required components of aircraft can also decrease the weight of aircraft. For example, some people proposed to remove thrust reversers from some aircraft. Although thrust reversers can improve takeoff and landing performance of aircraft, it is obviously superfluous for flight missions in well-built airports. If the device is removed, the maximum takeoff weight of aircraft will be reduced by approximately 0.3%~1%, and moreover removal of thrust reversers can also improve the flow field characteristics within the engine. Currently, this proposal is being studied.

Another way to reduce aircraft weight is at the expense of passengers' comfort and enjoyment, for example, to cancel aircraft portholes and recreational equipment, or to reduce the aisle's width and seat pitch, etc. This way of reducing weight may be feasible for short-haul aircraft, but it needs further study to see whether it will be satisfied and accepted by passengers.

In addition, if active noise reducing technology is successful, it is possible to remove some decorations and sound insulation devices from the inside of aircraft, thus reducing aircraft weight.

It is estimated that after these weight reducing measures are comprehensively adopted, the takeoff weight of a medium range wide-body aircraft can be reduced by 2000 kg. This is approximately equivalent to 1% increase of engine fuel efficiency.

3.4 Air Traffic Control

With traffic flow increasingly growing, many airports and air routs across the world have experienced congestion of different degrees. In addition to the insufficient airports and other infrastructures that contribute to the congestion, there is also another important reason, that is, air traffic control systems need further improvement. In Europe, airlines suffered loss of about US$ 5 billion in 1999 because of flight delays. Half of the delays were caused by air traffic control system.

At present, with rapid development and gradual application of satellite communication and "free flying" technology of navigation, it is possible to increase air route capacity, shorten aircraft flight range and reduce aircraft waiting time both on the ground and in the air. And as a result, aircraft fuel consumption is decreased.

Since improvement of air traffic control system involves airports, airlines and air traffic managing authorities in many countries, all states are required to work hand in hand. According to a report by ICAO, if governments of all the countries can formulate proper aviation management framework and improve air traffic control system, the present fleets for international flights will be able to reduce fuel consumption and carbon dioxide by 6%~12% for each flight.

Furthermore, to improve the operating modes of airlines can also reduce aircraft fuel consumption and decrease aviation impact on the environment. At present, the main methods can be applied by airlines are as follows: to reduce the unnecessary weight on board, to increase the aircraft full-load rate, to optimize the aircraft flying velocity, to limit the use of APU and to shorten the aircraft taxiing distance, etc. Generally speaking, these fuel saving measures can reduce aircraft fuel consumption and carbon dioxide discharge by 2%~6%.

4. Technology to Reduce Aircraft Noise

In order to reduce jet aircraft interior noise, it is necessary to reduce its engine noise and airframe noise. As for propeller aircraft and helicopters, it is also necessary to reduce propeller noise in addition to the aforesaid noises. The main method to reduce damages caused by aircraft exterior noise is to rationally plan the land around airports, to use even quieter aircraft and to adopt noise-reducing means in operation.

4.1 Reduction of Engine Noise

Engine noise mainly includes turbine noise, combustion chamber noise, fan noise and jet noise, etc. As for turbojet engine, the airflow exhausted from its jet nozzle has high velocity, so the nozzle noise is a main source of the engine noise. Since turbofan engine replaces turbojet engine, the airflow velocity in the exhaust nozzle has been greatly reduced. With the increase of engine bypass ratio, the nozzle noise has been gradually reduced. In order to reduce the compressor noise, people have adopted such methods as elimination of the guide vanes in the air inlet duct, reduction of the number of fan blades and of rotating speed, and improvement of aerodynamic design of blades, etc. And people also have properly selected the number of fan blades and stator vanes to control the frequency of the compressor noise, so that it will not easily be transmitted to the outside of the engine nacelle to form noise. Furthermore, noise liners are also used on engines, so that the engine noise has been greatly reduced.

In order to meet even higher environmental protection requirements, German Aerospace Institute (DLR) is making researches on active noise control technology (ANC). They are planning to use a system that can generate counter sound waves to reduce the sound waves of the main airflow produced by the fan propeller. DLR stated that this technology was expected to make the attenuation rate of short passage engines bigger than that of the conventional passive sound-absorbing device. The American company GE is making researches on a new type of jet nozzle, which, through continuous mixing of hot gas with the ambient atmosphere, can make the energy of high speed jet flow reduce by 50%, thus reducing noise by 3.5 dB.

4.2 Reduction of Airframe Noise

During flying, there is friction between the airframe and the air, changing the air movement. As a result, airframe noise is produced. The airframe noise level is related to aircraft's aerodynamic contour, surface roughness and flying velocity. Since 1970s, with the reduction of the engine noise, the airframe noise has gradually become the main source of aircraft noise. It is impossible to further reduce aircraft noise just by reducing the engine noise, and some people even regard the airframe noise as the lower limit of aircraft noise, stating that it is almost impossible to reduce the airframe noise.

For recent ten years, with the help of the powerful tool of computational fluid dynamics, people have had a clear picture about the flow field distribution and the change around the key structural components that produce aircraft noise. And people are therefore able to fairly systematically understand the generating and controlling mechanism of the airframe noise, so as to seek for solutions to reduce the airframe noise. At present, NASA has already used these research achievements to design low noise flaps, and achieved success in model experiment. At next stage, NASA plans to make researches on low noise undercarriages, followed by researches on methods to reduce overall airframe noise.

Furthermore, people are now also making researches on active or passive methods to reduce aircraft interior noise. The principle of reducing the noise by active methods is to place some loudspeakers in certain areas of the aircraft, then use computers to control these loudspeakers to generate sound whose phase is opposite to that of noise signal, thus offsetting the noise. Although this noise-reducing method has now made some progress, there are still some difficulties to overcome. The passive way to reduce the airframe noise mainly includes the use of sound absorbing materials and shock absorption, sound insulation devices and dynamic mufflers, etc.

4.3 Reduction of Propeller Noise

Propeller noise is an important source of aircraft noise for helicopters and propeller aircraft. At present, the main method to reduce the propeller noise includes increase of propeller blades, reduction of diameters of propeller blades and improvement of profile of propeller blades, etc. Moreover, by increasing the number of propellers, it is also possible to increase the frequency of propeller noise. Since high frequency noise attenuates fast in the air, this feature can be used to reduce the propeller noise. Fig. 8 shows the relationship between number and diameter of the propeller blades and the far field deci-sound pressure level FL2. From this Figure, we notice that under fixed power, with the increase of the number of propeller blades and the diameter of the blades, rotating speed of the propeller has been reduced, hence the deci-sound pressure level decreases rapidly. As for aircraft with multiple propellers, using synchronized technology of propellers can offset noise generated by each propeller, and avoid mutual overlapping, thus reducing overall noise of aircraft.

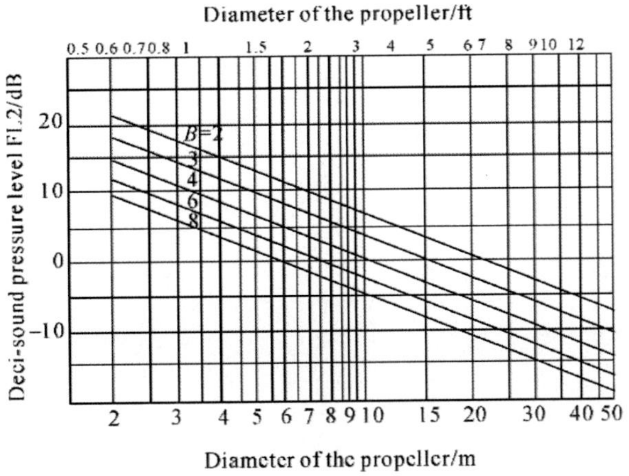

Fig. 8 Relationship between number and diameter of the propeller blades and the far field deci-sound pressure level FL2

5. Conclusion

With the social progress and economic development, aviation will play even more important role in the life of human society. Meanwhile, aviation will have bigger and bigger influences on the environment. States across the world have higher and higher requirements on the environment protective performance of aircraft, and ICAO is also formulating new standards on aircraft discharge and noise. This is an opportunity and a challenge as well for the development of our aviation undertakings.

As long as we make early efforts, increase investment in scientific researches, actively develop highly-efficient, environmental-protection and energy-saving new aviation technologies, and make use of the opportunity created by the entry of WTO to more actively carry out international cooperation, we will be able to reduce aircraft effect on environment to the minimum, promote the development of national economy, and improve people's living standards in China.

The Future Aircraft and Its Challenges on Aeronautical Technology*

Zhang Yanzhong
China Aviation Industry Corporation Ⅱ, Beijing, China

Abstract: The future special aircraft are presented. Different helicopters, mini-UAVs, ground effect aircraft, airships and tilt-rotors developing in China are described. The challenges of future aircraft on aeronautical technology are described. Supersonic flight, advanced aerodynamics, new aero-engines, smart structures, new materials, fly by light, free flight, new fuels, virtual reality technologies required by the 21st century aircraft are presented.

Keywords: special aircraft; aviation technology; 21st century

1. Introduction

The airplane is one of the greatest inventions of people in the 20th century. Before the 19th century, flying in the sky was only a dream of human beings. Since the Wright brothers invented the first airplane in 1903, aeronautical technology has experienced a leaping progress. Flying by plane has become a normal activity of people's daily life. The development of aviation has made a revolutionary influence on human society, living, economy, trade and military affairs.

During the approximate 100 years, the speed of airplanes has risen form dozes of kilometers per hour to more than Ma 3. The altitude of flying has grown from dozes of meters to more than 30,000 meters. The airplane has developed from single-seated to 400-seated. Prospecting the 21st century, we can see that the civil airplanes will become safer, cleaner, more economic, and more comfortable. Airliners having 500～1000 seats will enter service (see Fig. 1). These requirements will challenge the technology of civil aircraft. By the early part of the 21st century, it is forecasted that:

The drag of airplane and the structural weight will decrease by 20%;

The time and cost needed to develop an airplane will decline by 30%～35%;

The money needed to maintain an aircraft will be 40% less;

The emission of NO_x and the accident rate will drop by 80%.

The new generation of supersonic civil transport will be put into use in the 21st century.

---

* Published in International Conference on Engineering Sciences & Technology, ICEST 2000, Beijing, Oct. 2000: 141-151.

Y. Zhang, *Footprints in Cambridge and Aviation Industries of China*,
https://doi.org/10.1007/978-981-16-3176-4_26

Fig. 1 A future larger airliner capable of carrying 1000 passengers

The new generation fighters will be integrated with the abilities of stealth, super-cruise, post-stall maneuverable and short take-off and landing. They will also be equipped with more advanced avionics and armament.

The combat effectiveness of the new generation fighters will be 10 times that of the fighter today. The maintaining hours will be 70% lower.

Military hypersonic airplane of Ma 7～15 will also be developed in the 21st century. This will give an even harder challenge to aerodynamics, aero-engines, new materials, new structures, control technology, electronic equipment, sources of power and other aeronautical technologies.

Despite the rapid development of normal aircraft, special aircraft also encounters a good chance in the 21st century. Here, I would like to present the development trend of special aircraft in the 21st century and its development in China. The challenges of future aircraft on aeronautical technologies are also described.

2. Special Aircraft and Its Development in China

2.1 Helicopter

Since the first helicopter was successfully designed and manufactured in early 1940's, the helicopter technology has been developing rapidly. The helicopter has made a wide influence on human living, economy and military affairs, etc. About sixty thousand helicopters are used in the world. More than 230 models have been manufactured, the typical models include Z-8, S-70, AH-64, RAH-66, Boeing114/414, BA609, Mi-26, Ka-50, EH101, NH90, A129, Lynx, BO105, Tiger, BK117, EC120, etc.

Helicopter industries began developing in China from late 1950's. The developing phases of

China helicopter industries include introduction of foreign technologies, following design, self-design, and international cooperation, etc.

We have designed and manufactured eight models of helicopter, two models of unmanned helicopter and one model of super-light helicopter. They are Model 701, Yan'an-2, Z-5, Z-6, Z-8, Z-9, Z-11, EC-120, Soar Bird, Sea Gull and M16. Z-5, Z-8, Z-9, Z-11 and EC-120 were put into operation. AVIC Ⅱ is also responsible for design and manufacture of the vertical tail for the Sikorsky S-92 helicopter.

Z-8 is a multi-role military and civil helicopter which was designed and manufactured by AVIC Ⅱ. It is also the heaviest helicopter in China.

Z-9 is a twin-turbine general-purpose helicopter. It was license-production Eurocopter France AS 365N Dauphin 2 by AVIC Ⅱ.

Z-11 is a light utility and general-purpose helicopter (see Fig. 2). It is the first helicopter that was independently designed and manufactured by China. It is suitable for pilot training, police/security patrol, reconnaissance, coastguard, geological survey, rescue and forestry protection, etc.

EC-120 is a five-seated single-engined light helicopter. It is jointly designed and manufactured by Eurocopter France, CATIC (China National Aero-Technology Import & Export Corporation in the form of Harbin Aircraft Manufacturing Corporation) and STAe. China is responsible for the fuselage, the landing gear and the fuel system. China has 24% share and manufactured 110 sets last year.

Fig. 2 The Z-11 helicopter

Table 1 Helicopters in China

Model	Take-off weight(kg)	Velocity (km/h)	Service ceiling(m)	Range (km)	Endurance
Z-8	13000	315	6000	830	4 h 43 min
Z-9	3850	306	6000	1000	5 h
Z-11	2200	238	5240	600	3 h 42 min
EC-120	1770	232	6035	748	4 h 12 min

Soar Bird Unmanned Helicopter was designed by Nanjing University of Aeronautics and Astronautics (NUAA). It is a very practical air-platform with excellent performance and easy to take off and land. Upon the different payloads it fulfill the following tasks: reconnaissance of the enemy's situation, evaluation of situation on the battlefield, communication relay, electronic jamming, forest fireproof, fishing season search, etc.

Soar Bird has metal framework, glass fiber skin with honeycomb, composite rotor blade with advanced tip form, water cooling piston engine with two cylinders, digital flying control system, multiple special sensors, telemetering and telecontrol system with GPS and laser altimeter, and image-live display.

Table 2 Soar bird unmanned helicopter

Model	Weight	Payloads	Velocity	Endurance	Operating ceiling	Remote radius
Soar Bird	280 kg	30 kg	150 km/h	4 h	3000 m	150 km

Sea Gull is the first type of unmanned helicopter with coaxial contra-rotating rotors in China, designed by Beijing University of Aeronautics and Astronautics (BUAA). The first flight of the prototype was in September 29, 1995. Sea Gull weighs 300 kg. It has flying control and navigation systems, and telemetering and telecontrol systems.

M16 is a type of single-seated super-light helicopter with coaxial contra-rotating rotors, designed independently and manufactured by BUAA. The first flight of the prototype was in August 12, 1997.

Table 3 M16 Super-light helicopter

Model	Weight	Payloads	Velocity	Endurance
M16	350 kg	30 kg	132 km/h	1 h 42 min

2.2 Tilt-Rotor Aircraft

Tilt-rotor aircraft has the advantages of helicopters and turboprop aircraft. It has a faster velocity, a higher service ceiling and a longer range than helicopters, and can vertically take off, land and hover in the air (see Table 4).

Table 4 Comparison among tilt-rotor, helicopter and turboprop

Model	Type of take-off	Weight (kg)	Velocity (km/h)	Service ceiling (m)	Range (km)
V-22	Vertical take-off	24947	638	7925	3336
Bell 609	Vertical take-off	7265	509	7620	1400
Helicopter (S-70C)	Vertical take-off	9185	361	4360	550
Turboprop aircraft (Starship 1)	Run take-off	6591	641	10605	2629

Tilt-rotor aircraft will be an important means of transport in the future. Many countries are developing tilt-rotor technologies, designing and manufacturing tilt-rotor aircraft. Bell Helicopter Textron Inc and Boeing Helicopter Company has successfully designed and manufactured twin-engine tilt-rotor multi-mission V-22 Osprey (see Fig. 3) and nine-passenger tilt-rotor BA 609.

Fig. 3 The V-22 tilt-rotor

2.3 Unmanned Air Vehicle (UAV)

2.3.1 UAV trends

The UAVs developed very fast in the 1990's. Many advanced UAVs occurred during the last decade. It is possible that the high altitude endurance (HAE) UAVs will replace the manned reconnaissance aircraft, such as U-2R, after 2005. Unmanned combat air vehicles (UCAV) will become a new weapon platform. Micro air vehicles will get fast development and wide application. Some of the typical UAVs are as follows:

(1) Tier 2 predator

It is an middle-altitude-endurance UAV used for reconnaissance and surveillance. Its payload has optical-electro IR and all weather SAR. It was used in the Kosovo conflicts. Its specifications can be seen in Table 5.

(2) Tier 2+ "Global Hawk"

It's a big HAE UAV. It's one of the biggest UAVs developed in the world. It can carry a big payload. The sensors on it are very powerful. It can survey 137,320 km^2 everyday. Its details are listed in Table 5.

(3) Unmanned combat air vehicle (UCAV)

Boeing company will test-fly a full scale mockup of UCAV. It will enter service by 2010.

The overall weigh of the UCAV is 6800 kg, its engine is F124 turbofan. The thrust of the engine is 28 kN. Its details are listed in Table 5.

(4) Micro air vehicle (MAV)

The MAV is developing in the world. The first flight of "MicroStar" was in August, 1998. Its details can be seen in Table 5.

Table 5 Specifications of HAE, UCAV and MAV

Model	Tier 2	Tier 2+	UCAV	MicroStar
Span	14.85 m	35.42 m		15 cm
Length over all	8.13 m	13.53 m		
Height	2.21 m	4.63 m		
Empty weigh	350 kg	3469 kg		
Max T.O. weight	850 kg	10394 kg	6800 kg	85 g
Mission payload	204 kg	907 kg		18 g
Max level speed	202 km/h			
Cruise speed	130 km/h	635 km/h		4 km/h
Operation radius	805 km	5560 km		
Ceiling	7010 m	20500 m		
Endurance	24 h			20 min
Max endurance	40 h	42 h		

2.3.2 Chinese UAV

China has also developed some UAVs (see Table 6). These UAVs can be used as reconnaissance platform or target. Both ASN-104 and ASN-206 are developed by Xi'an ASN Technology Group. The ASN-104 is suitable for reconnaissance and the ASN-206 (see Fig. 4) is a multi-role UAV that is able to fulfill aerial reconnaissance, target positioning, artillery shooting adjustment, border patrol,

Fig. 4 The ASN-206 UAV

disaster survey. The Changhong-1 is developed by BUAA. It is also a multi-role UAV. The CK-1 is produced by NUAA. It is mainly used as a target.

Table 6 Specifications of Chinese UAV

Model	ASN-104	ASN-206	Changhong-1	CK-1
Span	4.3 m	6.0 m	9.76 m	7.5 m
Length over all	3.3 m	3.8 m	8.97 m	8.44 m
Height	0.9 m	1.4 m	2.18 m	2.955 m
Empty weigh				1537 kg
Max T.O. weight	140 kg	222 kg	1700 kg	2060 kg
Mission payload	30 kg	50 kg	65 kg	
Max level speed	205 km/h	210 km/h	800 km/h	920 km/h
Cruise speed	150 km/h			
Operation radius	60 km	150 km	17500 m	
Ceiling	3200 m	5000~6000 m		18000 m
Endurance	2 h	4~8 h	3 h	1 h 12 min

2.4 Ground effect Aircraft

Ground effect Aircraft (GEA) is a new kind of high-speed transport. The GEA has many advantages over high-speed ships and ordinary airplanes. It can fly as fast as an airplane, but has a larger payload. It is much safer and more comfortable than ships and has no jolt or weariness. Compared with the airplane, it also enjoys a much lower manufacture and operational cost. It can take off on water surface or plain ground without building the airport runway. So the GEA can be widely used for diversified purposes and it is recognized as one of the best means of high-speed transportation in the next century.

During the 1990's, China developed Ty-1 GEA (see Fig. 5). It was developed and manufactured by the

Fig. 5 The Ty-1 ground effect aircraft

GEA development center of Chinese academy of scientific & technological development which was founded in the beginning of 1995 by CASTA together with a special vehicle research institute of AVIC Ⅱ and Beijing institute of aerodynamics of CASC. The first flight of Ty-1 was in 1998, and it has been used as a touring vehicle in Taihu lake. Now, China is developing the DXF-200 GEA on the base of Ty-1. The Specifications of Ty-1 and DXF-200 are listed in Table 7 and Table 8.

Table 7 Specifications of Ty-1

Length over all	16 m
Width over all	11 m
Take-off weight	4800 kg
Accommodation	15 persons or load (1125 kg)
Engine	2 Lycoming IO-540-K1B5
Propeller	Two variable-pitch ducted propeller made of alum alloy
Total power	447 kW
Fuel consumption	140 kg/h
Maximum speed	200 km/h
Cruising speed	165 km/h
Flying height	0.6～1.2 m
Seaworthiness	level 3
Range	400km

Table 8 General Specifications of DXF200

Length over all	57.42 m
Width over all	42 m
Total height	14.639 m
Max. take-off weight	160 t
Load	27.2 t
Engine type	Д-30КП-2
Engine thrust	4×12000 kg
Full load range	2283 km
Maximum speed	600 km/h
Cruising height	0.5～3 m

2.5 Airships

China has also developed the FK series airships. Their details are listed in Table 9.

Table 9 Specifications of the FK series airships

Model	FK1	FK4	FK6	FK11	FK12	FK12B	FK100
Length overall(m)	10.00	39	14	10.4	11.80	11.8	42.8
Width overall (m)	3.34	12	4.5	3.33	3.333	3.333	12.28
Height overall(m)	3.77	14	5	3.7	3.866	3.866	14.71
Envelop volume(m^3)	35	2011	88	38	49	49	2580
Max t-o weight (kg)		1900					2650
Max payload (kg)		300					630
Max level speed (km/h)	35	73	40	60	65	65	99
Max flight altitude (m)	200	1800	300	300	300	300	1000
Engine power (kW)		2×34					2×73.5
Engine cylinder volume(cc)	2×10		2×40	2×40	2×40	2×52	
Endurance(h)	0.5	11	3	1	1.5	2	4
Flight radius (km)	1		1	1	1	1	
Advertising area(m^2)	2×5	2×108	2×11.1	2×6	2×13.5	2×13.5	2×200
Passenger number		4					9
Remark	passenger	passenger	Remote control	Remote control	Remote control	Remote control	passenger

3. Challenges to the Aeronautical Technology

Although the special aircraft has a fabulous future, there are still some challenges to the aeronautical technology. Unless we master these critical technologies, we can't develop those new generation of special aircraft.

3.1 Aerodynamics

During the 20th century, the aerodynamics was developing rapidly and this trend will remain in the 21st century (see Fig. 6).

At present, aerodynamic research is focused on unsteady flow. Such as more practical and accurate model of transit and turbulent flow, the model and the factors dominate the separation and laminar re-adherence of 3-D flow, the formation of vortex, the interaction of vortex, and the break down of vortex. With the development of computational fluid dynamics (CFD) and more powerful experimental facilities, it is likely that there will be a breakthrough in the theory of unsteady

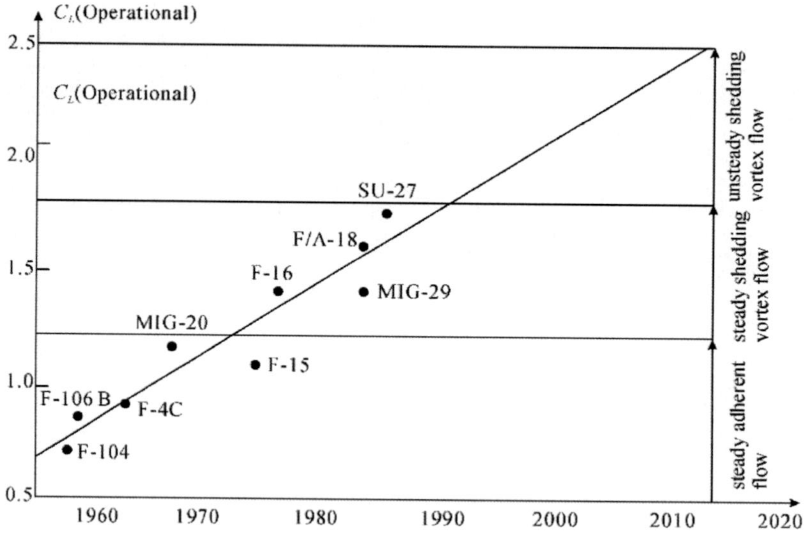

Fig. 6 The operational lift coefficient of fighters keeps growing

aerodynamics soon. This will make the theory of unsteady aerodynamics more accurate and practical. This will improve the aerodynamic efficiency of airplane greatly. Today, the maximum lift-drag ratio of civil airplane is 20, in the near future it will be doubled and become 40.

3.2 Aero-Engines

Different engines suit for different speeds (see Fig. 7). With the speed increasing, the efficiency of turbo engines declines. In order to meet the need of a higher speed, ramjet engine is adopted in some aircraft. As a ramjet cannot work at a low speed, the combined engine that incorporates ramjet and turbo engine appears. When the aircraft is taking off or landing, the combined engine works as a turbo engine. When the aircraft has accumulated certain speed, the combined engine works as a ramjet. Besides the combined engine, some new-concept engines are under development.

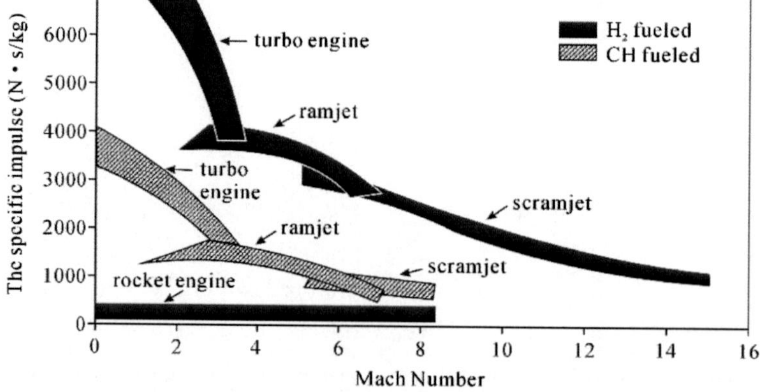

Fig. 7 The impetus ratio changes with the Mach number of several kinds aero-engines

One of these new concept engines is the scramjet. At the combustion chamber of normal engines, the speed of airflow is subsonic. If the Mach number of aircraft is more than 5 and the air speed at the chamber remains subsonic, the temperature there will be so high that nearly no material can stand at present. In a scramjet, the air speed at the chamber will be supersonic, thus make the temperature there reduced. This kind of engine suits for M6～M25, and it is also a combined engine to meet the requirement at low speed.

Another new concept of aero-engine is Pulse Detonation Engine (PDE). In this engine, it makes use of the detonation wave to generate thrust. The composition of PDE is much simpler than a turbo engine. It comprises of an air intake, an detonation chamber and an exhaust nozzle, and needn't a turbo. The pressure of the detonation is very high, which enables the PDE to have a greater thrust weight ratio. The PDE suits for a wide operating speed (from zero to M10), and it can be used for various aircraft, such as airplanes, UAVs, and aerospace planes.

Now, China has studied the theory of PDE for 6 years, and fulfilled a series of experiments. Many inspiring results have been reached and a solid foundation is set for the development of PDE.

3.3 Smart Structures

In the 21st century, smart materials and smart structures will be widely used in aircraft. If we implant sensors, control systems, radio emitters, or other actuator components in the structure of an aircraft, we will enable the structure to have more functions. Some smart structures can adapt to the environment automatically, and can change their shapes and stiffness according to the requirement of flight. If a crack happens in the structure, it will alert the maintainer automatically and change the distribution of loads on the structure to avoid the collapse. Some smart structures can emit and receive radio signals in place of the antennas, which will benefit to the stealthy of fighters greatly. The wide use of smart structures will improve the performance of future aircraft as well as bind aeronautical technology and electronic, computer, artificial intelligence, and automatic control technologies even more closely.

3.4 New Materials

According to the usage of different materials, aeronautic material can be divided into structural material, high temperature material and functional material. At present, the stress ratio and the specific modulus of the structural materials are becoming higher and higher. The importance of composite material is becoming higher and higher (see Table 11). It will be used extensively in the future aircraft. Its cost will decrease, and its maintenance will become more convenient.

In order to increase the thrust of an aero-engine, it is necessary to raise the temperature near the turbine. If the temperature is raised, the life of the engine will decrease greatly. The best way to solve this problem is to develop a new generation of high temperature materials. Now, many kinds of high temperature materials are being studied, such as polymer matrix composites, titanium alloys, and ceramic matrix composites, and carbon-carbon composites. It is estimated that the composite used in aero-engines will be sharply increased by 2010.

Table 11 The percentage of different materials used in fighters

	F-15	SU-27	EF2000	1.42	F-22
AL. Alloy	37.3%	60%	18%		15%
AL-Li Alloy		4%	25%	35%	
Ti Alloy	25.8%	18%	12%		41%
Composite Material	1.2%		43%	30%	24%
Steel	5.2%			3%	5%
Other	30.2%	18%	2%	5%	

Functional materials can change various physical signals into electrical signals, or change different forms of energy into another. By means of the function materials, we can sense convert and store different kinds of signals. In modern aviations, all the laser, infrared, and stealth technologies cannot exist without the help of functional materials. New functional materials will make the avionics more powerful, and make it possible to develop smart structures.

3.5 Fly by Light

Nowadays, fly-by-wire systems and hydraulic systems are widely used in the aircraft. But these systems have some inevitable shortcomings. The fly-by-wire system is vulnerable to thunder-stroke, electro-magnetic disturbance, and electro-magnetic pulse. The hydraulic system is relatively complex. If it is damaged, the survivability is very low. The sealing components of the hydraulic system are easy to become aging. If the hydraulic liquid leaks, it is easy to catch a fire.

In order to overcome these shortcomings, people are studying fly-by-light technology and electrical servo systems. After the maturity of these technologies, both the safety and the controllability of the airplanes will be improved.

3.6 Free Flight

The free flight technology is a new generation of air navigation technology based on the computer and satellite communication technology. With the help of this technology, pilots of the planes will have more freedom in choosing the routes so as to save time and fuel. This technology is also useful to solve the congestion in the airports. Although today's GPS and satellite communication systems cannot meet the safety and reliability requirements of free flight, it is possible that this technology will become perfect within 10~20 years. The free flight technology will be used in future air navigation system (FANS).

3.7 New Fuels

It is estimated that the reserves of natural gas all over the world can support our exploitation for 100~200 years, while the reserves of oil only can last 50 years. In order to cope with the resource shortage and environmental pollution, people are studying new fuels for aircraft. One hopeful fuel for future aircraft is hydrogen. Compared with gasoline, hydrogen is clean and can give off more heat at the same weight. As it is a gas at normal temperature, special containers and very low temperature are needed to liquidize it. Moreover, as the density of liquid hydrogen is lower than gasoline, it can give out only 23% heat of the same volume of gasoline. That means if we want the airplane bring the same amount of energy, the tanker of hydrogen-fueled airplane will be more than 4 times bigger than that of today's airplane. Many countries are studying this technology and have carried out a lot of experiments. But up to now, there are still some difficulties. It is possible that in 15~20 years, we will work out a practical hydrogen-fueled airplane.

3.8 Virtual Reality

Virtual reality is an artificial environment that is experienced through sensory stimuli provided by a computer and in which one's actions partially determine what happens in the environment. Computers, actuators, stereo videos and audio systems provide the audio and visual stimuli according to the programmed physical rules. Through the visualized and natural interaction between man and machine, we can feel the real world and try the real operation in the virtual environment.

During the development of an aircraft, we can use this technology to pre-demonstrate the aircraft to the designer and the customer, and verify the feasibility of the plan, technology, tactical performance, and configuration. During pilots' training, we can use the vivid environment produced by this technology to save the time and money of flying real planes, whereas the training quality remains the same or even better.

4. Conclusion

Now, we have started a new century. At the beginning of the 21st century, we are hopeful to the future of aviation. I believe that during the 21st century, aeronautical technology will get an even more brilliant outcome than that of the 20th century. However, to realize these dreams, it is not enough if we only depend on the struggle of the aviation world; it is still not enough if we only depend on a single nation's power. We must set out an even more extensive cooperation. Aeronautical technology should absorb more nutrition from electronic, material, computer, and other technologies. China should cooperate with more countries on the research of aeronautical technology.

In the near future, China will join the World Trade Organization. This will link China with the world more closely, and create an even better environment for China to cooperate with other countries on the research of aeronautical technology. I believe that with our diligent work together, the aeronautical technology of the 21st century will reach another splendid peak, and the development of China aviation will also be quickened.

References

[1] Y. Z. Zhang, Engineering Science, 1999, No. 1, 19.
[2] Y. L. Chen, Aeronautical material technology, 1999.
[3] L. B. Jiang, The advanced aerodynamic configurations of fighters, 1995.
[4] J. G. Liu, The engine of unmanned aerial vehicles, 1999.
[5] G. L. Zhang, The Helicopters, 1999.
[6] X. M. Ji, The Unmanned Aerial Vehicles, 1999.

Shock Response Spectrum Analysis with a Computer and a Fourier Analyser*

ZHANG YAN ZHONG

Changcheng Institute of Measurment and Metrology, Beijing, China

Abstract: In this paper the exact definition of shock response spectrum is given. The relation between Fourier spectrum of shock pulse and residual shock response spectrum is discussed with the following relationship: $S(f)=2\pi f|F(f)|$.

Finally, with paper introduces the programming of real-time analysing shock response spectrum with a mini-computer and a Fourier Analyser.

Shock may be defined as suddenly transmission of a large amount of kinetic energy to a mechanical system which takes place in a relatively short time compared with the natural period of the mechanical system. It can produce very large force and acceleration and cause severe mechanical damage. In order to study the mechanical damage effect of shock and find an effective method of shock control, it is necessary to introduce the concept of shock response spectrum.

1. Definition of Shock Response Spectrum

When shock force in question applies to a linear, undamped single degree-of-freedom system, the maximum response of the system as a function of system's natural frequency can be defined as shock response spectrum of shock pulse.

The differential equation of motion for a linear, undamped single-of-freedom system with mass M, stiffness K and shock force $mf(t)$ is formulated:

$$\ddot{X}(\omega t)+\omega^2 x(\omega t)=f(t) \qquad (1\text{-}1)$$

where $X(\omega t)$—displacement function;

$\omega=\sqrt{\frac{x}{m}}$—anguler natural frequency of system.

The acceleration of system is

$$A(\omega t)=\ddot{X}(\omega t) \qquad (1\text{-}2)$$

The maximum response acceleration of system i. e. shock response spectrum is

$$S(\omega)=|A(\omega)|_{\max} \qquad (1\text{-}3)$$

* Published in CIMM, Oct. 1979: 1-8.

Y. Zhang, *Footprints in Cambridge and Aviation Industries of China*,
https://doi.org/10.1007/978-981-16-3176-4_27

It is almost impossible to have zero damping for a practical mechanical system. The maximum response acceleration of undamped system is larger than that of damped system. So it represents the maximum possible motion of the system. The practical motion of the system is always less than the motion represented by shock response spectrum.

Usually, shock response spectrum can be divided into initial and residual shock spectrum. Initial shock spectrum stands for the shock response of system when the shock force is acting ($0<t<T$, T—duration of shock force). Residual shock spectrum deals with the shock response of system after the shock force has finished ($T<t<\infty$). Because the duration of shock pulse is very short, we tend mainly to research residual shock response spectrum[1].

Multi-galvanometer's analysis methods can be used in the experimental research of shock response spectrum[2], but its accuracy and resolution are rather poor and analysis speed very slow. It is possible to solve the equation (1-1) with computer, and get shock response spectrum, but it usually solves the question of some standard shock pulse, such as half-sine, rectangular and sawtooth pulse etc. Practical shock pulses are very complex. This method can not get real-time shock response spectrum.

In this paper a method of real-time analysing shock response spectrum is given. It is not necessary to solve the equation (1-1). The shock response spectrum can be obtained from the Fourier spectrum of shock pulse with a computer.

The Fourier spectrum of shock pulse is got with a Fourier Analyser. This method can very quickly analyse complex shock pulse.

It is very necessary to discuss the relation between Fourier spectrum of shock pulse and residual shock response spectrum of the system.

2. Relation Between Shock Response Spectrum and Fourier Spectrum

If the function of acceleration is $f(t)$, its Fourier Transformation is[6]

$$f(t) = \int_{-\infty}^{+\infty} B(f)\cos 2\pi f df + \int_{-\infty}^{+\infty} C(f)\sin 2\pi ft\, df$$

where $B(f) = \int_{-\infty}^{+\infty} f(t)\cos 2\pi ft\, dt$

$$C(f) = \int_{-\infty}^{+\infty} f(t)\sin 2\pi ft\, dt \tag{2-1}$$

Let $F(f)=B(f)-\lambda C(f)$, we have

$$f(t) = \int_{-\infty}^{+\infty} F(f)e^{i2\pi ft}\, df$$

$$F(f) = \int_{-\infty}^{+\infty} f(t)e^{-i2\pi ft}\, dt \tag{2-2}$$

$$|F(f)| = \sqrt{|B(f)|^2 + |C(f)|^2}$$

We consider the function $f(t)$ to be superposition of many step functions(Fig. 1).

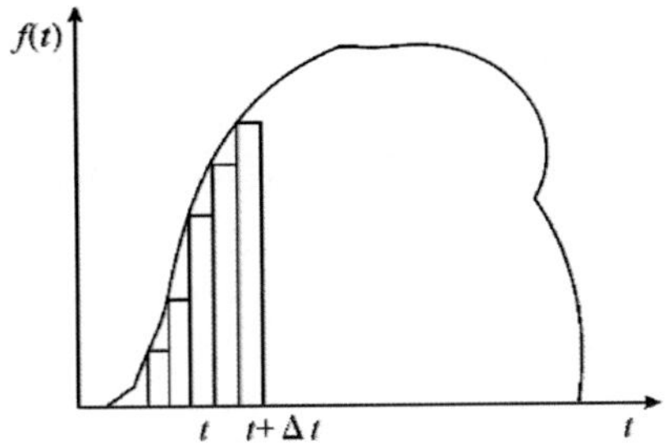

Fig. 1 Shock pulse of superposition in the time domain

If acceleration in duration $t—t+\Delta t$ is $f(t)$, with respect to *Dirac-δ* function following formular, can be obtained.

$$f(t)=\int_{-\infty}^{+\infty} f(t)\delta(t-\tau)\mathrm{d}\tau \tag{2-3}$$

According to the principle of superposition of linear system, the solution of equation (1-1) can be shown as well known Duhamel integration.

$$X(\omega t)=\int_{-\infty}^{+\infty} h(t-\tau)f(t)\mathrm{d}\tau \tag{2-4}$$

where $h(t-\tau)$ is the response of the same linear, undamped single-of-freedom system when unit-shock function $\delta(t-\tau)$ gets on it. $h(t)$ can be got from the following equation:

$$\ddot{h}t+w^2h(t)=\delta(t) \tag{2-5}$$

According to the definition of δ-function, equation (2-5) represents that acceleration acting on the system is

$$\delta(t)=\begin{cases}0 & \text{where} \quad t\neq 0\\ \infty & \text{where} \quad t=0\end{cases}$$

and $\int_{-\infty}^{+\infty}\delta(t)\mathrm{d}t=1$.

So the velocity of shock is "1". Equation (2-5) can be hanged into the following equation and initial conditions:

$$\begin{aligned}&\ddot{h}(t)+\omega h(t)=0\\&h(-0)=0\\&h(+0)=1\end{aligned} \tag{2-6}$$

The solution of equation (2-6) is

$$h(t)=\frac{1}{\omega}\sin\omega t \tag{2-7}$$

From formular (2-4) and (2-7), the solution of equation follows:

$$X(\omega t)=\int_{-\infty}^{+\infty}\frac{1}{\omega}\sin\omega(t-\tau)f(\tau)\mathrm{d}\tau$$

For computing the residual spectrum, let $\tau\rightarrow\infty$,

$$X(\omega t)=\int_{-\infty}^{+\infty}\frac{1}{\omega}\sin\omega(t-\tau)f(\tau)\mathrm{d}\tau \tag{2-8}$$

This is the displacement response of a linear, undamped single-of-freedom system to shock pulse $f(t)$, and its acceleration response is:

$$A(\omega t)=\ddot{X}=-\int_{-\infty}^{+\infty}\omega\sin\omega(t-\tau)\cdot f(\tau)\mathrm{d}\tau$$

$$=-\omega\sin\omega t\int_{-\infty}^{+\infty}f(\tau)\cos\omega\tau\,\mathrm{d}\tau+\omega\cos\omega t\int_{-\infty}^{+\infty}f(\tau)\sin\omega\tau\,\mathrm{d}\tau$$

According the definition formular (2-1) and $\omega=2\pi f$, we get:

$$A(ft)=-2\pi f\{B(f)\sin 2\pi ft-C(f)\cos 2\pi ft\}$$

$$=-2\pi f\mid F(f)\mid\sin(2\pi ft-\varphi) \tag{2-9}$$

where

$$|F(f)|=\sqrt{|B(f)|^2+|C(f)|^2}$$

$$\varphi=\mathrm{arctg}\frac{C(f)}{B(f)}$$

Shock response spectrum is the maximum response of acceleration in time domain:

$$S(f)=|A(ft)|_{\max}$$

$$=2\pi f|F(f)| \tag{2-10}$$

The shock response spectrum of system can be studied from the Fourier Spectrum of shock pulse without having to solve equation (1-1).

3. Real-Time Analysing Methods of Shock Response Spectrum

Upon the formular (2-10) we can analyse shock response spectrum with a mini-computer and Fourier Analyser. The Fourier spectrum $F(f)$ can be derived by a Fourier Analyser. Shock response spectrum can be computed from Fourier spectrum with a mini-computer.

We use type 3348 real-time Fourier analyser, the analysing time of which is 44.8 ms. Fourier spectrum consists of 400 lines. So analysing bandwidth is

$$B=f_h/400 \tag{3-1}$$

where f_h—the highest limited frequency of analyser.

The frequency of Nth line is

$$f_N=NB \tag{3-2}$$

where

$$N=1,2,3,\cdots,400$$

If the amplitude of N-th line is $F_{db}(N)$, the calibration factor of transducer is F_0 db, so the acceleration of N-th line is

$$F(f_N)=10\,\frac{F_{db}(N)-F_0}{20}$$

$$=e^{2.3026}\frac{F_{db}(N)-F_0}{20} \tag{3-3}$$

The shock response spectrum is

$$S(f_N)=2\pi BNe^{2.3026}\frac{F_{db}(N)-F_0}{20} \tag{3-4}$$

Depending on the formular (3-4), it is very to compute $S(f_N)$ from $F_{db}(N)$ with a mini-computer. F_{db} can be read out by Fourier analyser. The principle flow chart is shown in Fig. 2.

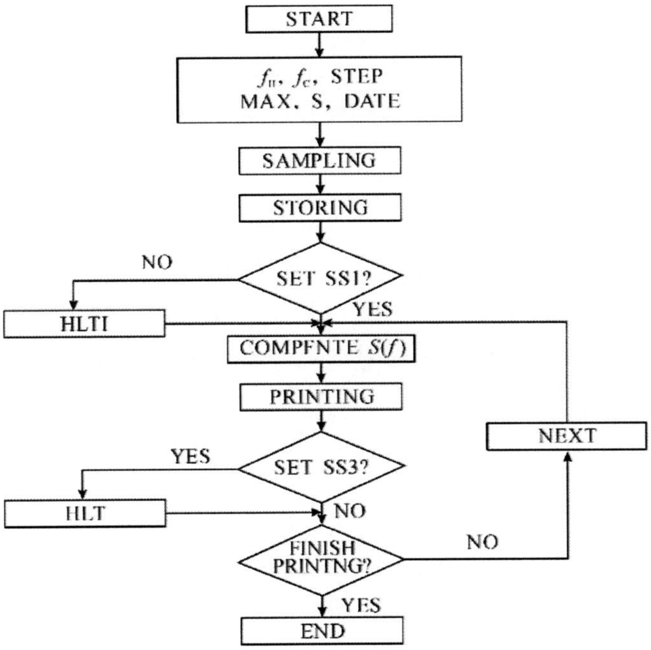

Fig. 2 Principal flow-chart of analying shock response spectrum

At first, the operater answers following questions one by one on the terminal: analyser's range, the lowest printed frequency, the highest printed frequency, step, max. number of spectrum to be printed, transducer's calibration facter, remark, etc. Then Fourier Analyser automatically captures and analyses shock pulse, on-line computer calculates shock response spectrum and prints $S(f_n)$.

Programme is designed in assember language. Its length is 600 words (16 bits). In order to increase the ability of storing spectrums, we combine two lines of a spectrum into the word of the computer. The principle is as follows: The dynamic range of Fourier analyser is 500 dB. So every line must be represented by 9 bits.

Because the accuracy of Fourier Analyser is that each line is divided by 2, the max. fixed-point numeral is 250. It can be represented by 8 bits ($2^8 = 256 > 250$) half a word of the computer (the word length of the computer is 16 bits). This method keeps the accuracy of analyser and doubles the ability of storing spectrums.

The analysing method has advantages of high speed, high resolution, good accuracy and avoiding complex calculation.

It has the ability of real-time analysing any practical shock pulse. The analysing time is 20～70 ms (it depends on line numbers to be saved).

The accuracy is ±0. 2 dB and resolution 1/400. Dynamic range is 50. 0 dB, frequency range is 0～10 kHz. Shock acceleration range is 0～9999 g.

References

[1] IEC RECOMMENDATION PUBLICATION. 1968-2-27. Basic Enviromental Testing Procedures Test Ea: shock.
[2] I. E. S Proceeding 1996.
[3] A. V Oppenheim. Digital Signal Processing, 1975.
[4] P. M Morse. Vibration and Sound. McGraw-Hill, 1948.
[5] J. T Broch. Mechanical Vibration and Shock Measurements, 1972.
[6] D. E Newland. Random Vibration and Spectral Analysis, 1975.

Optimum Removal of LCOs in Digital Systems*

Y. Z. Zhang, T. W. Cole and Y. F. Yao
School of Electrical Engineering, University of Sydney, Australia

Abstract: A Finite State Machine method is proposed for realizing IIR digital filters. This realization has a maximum error of 1/2 bit from the final result of infinite precision arithmetic, and has no growth of errors during the processing period. It avoids the effects of the filter coefficient quantization and multiplication arithmetic errors. A connection matrix method is presented for detecting and locating the limit cycle oscillations in digital filters. The concept of Sub-Optimum Successor is introduced based on the defined error function. A Graph Theory method is proposed for removing Limit Cycle Oscillations from digital filters. This method breaks the filter's cycles, connects the unstable state to its sub-optimum successor, and forms a tree graph of stable filters. It introduces a minimum increase in noise and hardly any increase in complexity of hardware.

1. Introduction

Any digital filter may be represented as a difference equation which is discrete in time, but continuous in amplitude as follows:

$$y(n) = \sum_{i=1}^{m} K_i y(n-i) + \sum_{i=0}^{m} L_i x(n-i) \tag{1}$$

where K_i and L_i are infinite accuracy coefficients, $x(n-i)$ and $y(n-i)$ are respectively the input and output at the $(n-i)$th clock instant which also have infinite precision amplitude.

The usual method for realizing digital filters is to approximate the infinite precision amplitudes and their arithmetic in terms of finite wordlength binary representation and arithmetic. Roundoff or truncation errors are introduced. The error behaviour depends on structures and properties of the systems. In some feedback systems, such as HR digital filters, this approximation may lead to Limit Cycle Oscillation (LCO)[1].

A Finite State Machine (FSM) is a mathematical model of sequential systems. It consists of three finite sets: input set U, output set G and state set S, and two mappings: $\underline{f}$ and $\underline{g}$ defined as follows:

$$\underline{s}(n+1) = \underline{f}[\underline{s}(n), u(n)] \tag{2}$$

$$\underline{g} = \underline{g}[\underline{s}(n), u(n)] \tag{3}$$

---

* Published in Proc. of IEEE Intl. Conf. on ASSP, ICASSP'90, April 1990: 1297-1300.

Y. Zhang, *Footprints in Cambridge and Aviation Industries of China*,
https://doi.org/10.1007/978-981-16-3176-4_28

where $\underline{s}(n) \in S$, $u(n) \in U$ and $\underline{g}(n) \in G$ are respectively the present state, input and output at the nth clock instant. $\underline{s}(n+1) \in S$ denotes the state at the $(n+1)$th clock instant and is named as the next state. The mapping $\underline{f}$ is called the next state mapping. $\underline{g}$ is referred to as the output mapping.

An FSM may be described by its state table and state graph. The State table is easy to establish on computer, and the state transition graph can show the dynamic behaviour and stability of machines. The advantages of FSM are: its state, input and output are elements of finite sets. The next state mapping and output mapping are defined over some finite set. They are infinite accuracy operations, and free of errors. Because the next state of an FSM depends on not only the present input but also the past states, FSM may be used to simulate any finite storage systems (e. g. IIR filter), and has been used in computer science, code theory and sequential circuits etc.

2. FSM Representation of Filters

Finite state machines can be used to represent digital filters, if the continuous algebraic equation description of the filter is given. One method of approach is to evaluate the system output for all possible combinations of finite input and system state using the continuous equation. The system output value determined in this way will be within 1/2 bit over the representable range of outputs. For example, consider a first order HR digital filter described by the following difference equation:

$$y(n)=K_1 y(n-1)+x(n) \tag{4}$$

where the present output $y(n)$ is a function of the past output $y(n-1)$ and the present input $x(n)$. Comparing eqns. (2) and (4), assign to each output quantization level $\bar{y}(n)$ a member of the state set S, to each input quantization level $\bar{x}(n)$ a member of the input set U of a machine, and then the 1st order filter may be represented as a finite state machine. Its next states may be obtained by quantizing the infinite accuracy arithmetic results of the above equation as follows

$$\bar{y}(n)=Q[K_1 \bar{y}(n-1)+\bar{x}(n)] \tag{5}$$

where Q denotes quantizing operation.

If the coefficient $K_1=-0.8$, input $x(n)$ is 1.0, and let the output be represented in a 3-bit word, then the state table may be set up by rounding the following equation for all possible $\bar{y}(n-1)$

$$\bar{y}(n)=Q[-0.8\bar{y}(n-1)+1.0]$$

Its state table for input $\bar{x}(n)=1.0$ is shown in Table 1. This state table can be realized in terms of logic circuits. The state graph of the machine is shown in Fig. 1.

This table may be realized in terms of a one dimensional FSM. The machined outputs for all combinations of inputs and past states have at most 1/2 bit error from infinite accuracy outputs of the filters. In this simple example, the obtained outputs are no different from those which would be obtained from the arithmetic realization. However, in a system containing more than one multiplier the results obtained from the above method would more accurately match the results of an infinite precision filter.

Table 1 State table of example

$\overline{y}(n-1)$	$y(n)$	$\overline{y}(n)$
0	1.0	1
1	0.2	0
2	−0.6	−1
3	−1.4	−1
−4	4.2	3
−3	3.4	3
−2	2.6	3
−1	1.8	2

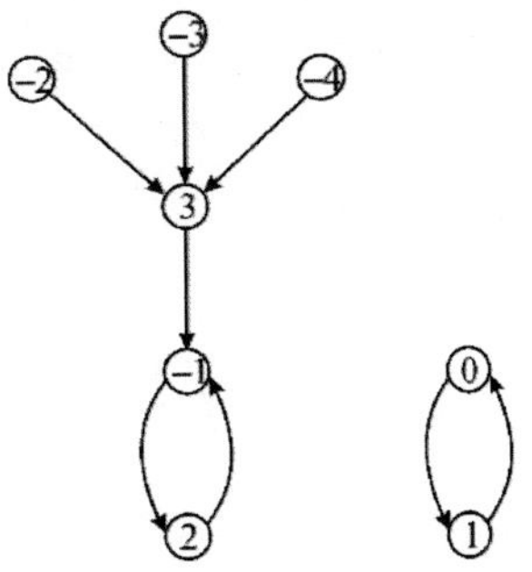

Fig. 1 State graph of example

For high order filters, introducing a variable $w(n)$, the equation (1) may be re-written as follows:

$$w(n) = x(n) + \sum_{i=1}^{m} K_i w(n-i) \tag{6}$$

$$y(n) = \frac{-L_m}{K_m} x(n) + \sum_{i=0}^{m-1} (L_i - K_i \frac{L_m}{K_m}) w(n-i) \tag{7}$$

where $K_0 = -1$, $K_m \neq 0$.

This is a modified canonic form of an HR filter. The modification is that its output is related to the input $x(n)$, but unrelated to the mth delayed variable $w(n-m)$. Let $\overline{x}(n)$, $\overline{y}(n)$ and $\overline{w}(n)$ denote the quantized discrete amplitude of $x(n)$, $y(n)$ and $w(n)$ respectively. $\overline{w}(n)$ and $\overline{y}(n)$ may be obtained by

$$\overline{w}(n) = Q[\overline{x}(n) + \sum_{i=1}^{m} K_i \overline{w}(n-i)] \tag{8}$$

$$\overline{y}(n) = Q[\frac{-L_m}{K_m} x(n) + \sum_{i=0}^{m-1} (L_i - K_i \frac{L_m}{K_m}) \overline{w}(n-i)] \tag{9}$$

where Q denotes quantization.

The discrete output determined in this way can be within 1/2 least significant bit over the representable range. If $\overline{x}(n)$ and $\overline{w}(n-i)$ in the above two equations are all possible values of finite sets U and S respectively, then a state table can be obtained. However, the state here is an m-dimensional vector defined by

$$\underline{s}(n) = | s_0(n), s_1(n), \cdots, s_{m-1}(n) |^t \tag{10}$$

$$\begin{aligned} s_0(n) &= \overline{w}(n-m+1) \\ s_1(n) &= \overline{w}(n-m+2) \\ &\vdots \\ s_{m-1}(n) &= \overline{w}(n) \end{aligned} \tag{11}$$

Using this vector the above quantization equations may be represented by the following two mappings:

$$\underline{s}(n) = \underline{f}[\underline{s}(n-1), u(n)] \tag{12}$$

$$g(n) = g[\underline{s}(n), u(n)] \tag{13}$$

This is a FSM with an m-dimensional next state mapping $\underline{f}$ and a one-dimensional output mapping. It can be used to express the mth order IIR filter.

3. Detection of LCOs

The LCOs in an IIR filter may be shown as cycles in the state graph of the FSM which represents the filter. However, for a practical filter (e. g. 10-bit wordlength), it is too complex to show the graph state. For detecting and locating the LCOs in a machine, a connection matrix method is introduced.

Connection matrix C is a $(2^M \times 2^M)$ matrix (M is the wordlength of the state variable) consisting of "0" and "1". Each column of a connection matrix has only one element equal to unity and all other elements equal to zero. The position of the unity element is such that the direction from the position of each diagonal element c_{ii} to the position of the unity element in the same column denotes the state transition direction. For the above example, its connection is as follows.

0	0	1	0	0	0	0	0	0
1	1	0	0	0	0	0	0	0
2	0	0	0	0	0	0	0	1
3	0	0	0	0	1	1	1	0
−4	0	0	0	0	0	0	0	0
−3	0	0	0	0	0	0	0	0
−2	0	0	0	0	0	0	0	0
−1	0	0	1	1	0	0	0	0

If all the states which are not on the cycles can be removed from cycles, then the remaining states show the behaviour of limit cycles. This may be realized by cutting off all the states at the top of branches step by step. The connection matrix can be used to cut these states by deleting the rows which have no "1" element, and the corresponding columns one by one. For the above example, the states −4, −3 and −2 are at the top of branches, there is no "one" in rows −4, −3 and −2. Delete the rows −4, −3 and −2, and the columns −4, −3 and −2. This procedure is shown as follows.

0	0	1	0	0	0
1	1	0	0	0	0
2	0	0	0	0	1
3	0	0	0	0	0
−1	0	0	1	1	0

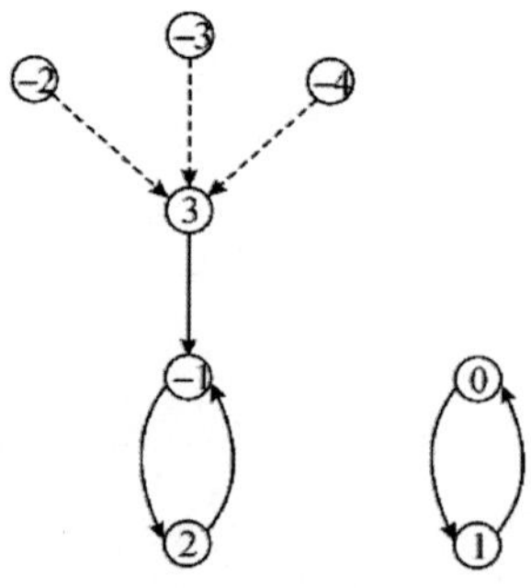

Continue the above procedure. As shown in the figure, the state forms a new top state, these is no "one" in row "3". Delete the row "3" and the corresponding column "3". A (4×4) matrix remains. This matrix represents pure cycle states in the state graph.

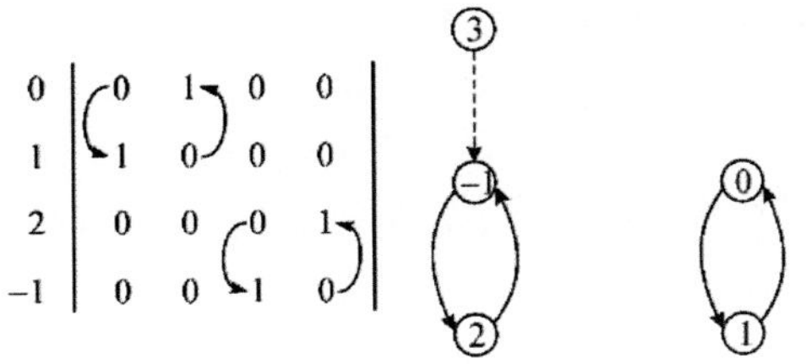

4. Stable Realization of Systems

If the limit cycles of an IIR filter have been detected and located, then the limit cycles can be broken by relocating the state table. For the example, if the state "2" can be connected to the state "0", then the limit cycle (−1,2) is broken at point "A". This can be realized by changing its next state from "−1" to "0". This change causes an increase in errors of system representation. The errors from a predecessor to a successor in the FSM representation of IIR filters may be defined by

$$e = [w(n) - s']^2 \tag{14}$$

where $w(n)$ is the infinite accuracy output of the system from the filter equation. s' denotes the assignment of the given successor.

The optimum successor of a state is that which has the minimal error among all the possible states of the machine. The next state obtaining from the quantization equation (8) is the optimum. The optimum successor of the states on a cycle is also on the cycle, and the error can be calculated by

$$e_0 = [w(n) - \overline{w}(n)]^2 \tag{15}$$

where $\overline{w}(n)$ denotes the quantization value of $w(n)$.

Breaking of a limit cycle requires re-connection of a state of the cycle to another state which is not on the cycle. A lot of states in a machine are not on the cycle, and can be the succcessor. Every successor has a different error. One of them has a minimal error among the states in branches. This state is named as the sub-optimum successor. Sub-optimum means that this successor of the state of the cycle has a minimal error among all the successors which are not on the cycle, but its error is only larger than that of the optimum successor on the cycle. For the above example, if the state "2" on cycle is considered as a predecessor, then all its possible successors and errors are shown in the following Fig. 2.

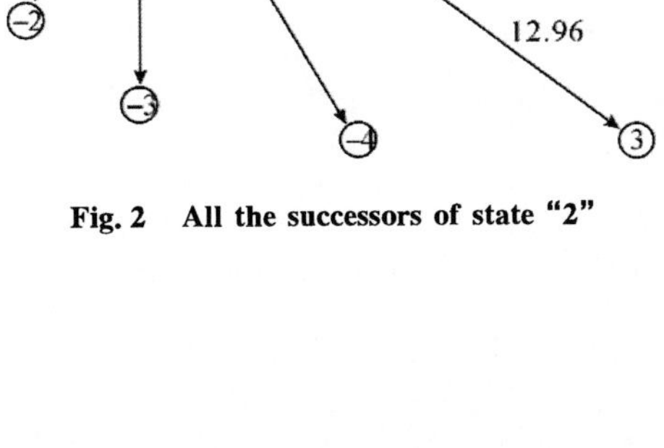

Fig. 2 All the successors of state "2"

In Fig. 2, the error is allocated to each edge. The error values may be calculated in terms of eqn. (14). For example, the system output $w(n)$ from $w(n-l)=2$ is calculated by the difference equation ($x_c=1.0$) as follows

$$w(n)=-0.8w(n-1)+1.0=-0.6$$

The error e in making the state transition "2" to "1" is

$$e=(-0.6-1.0)^2=2.56$$

Fig. 2 shows that the state "−1" is the optimum successor of state "2". It has the minimum error of 0.16, but it is on the cycle (−1, 2) and unstable. The state "0" is the sub-optimum successor which has an error of 0.36. It is minimal among the successors not on the cycle.

However, there are many states on a cycle. The breaking of a cycle at a distinct point may cause a distinct error. If a state is on a cycle, its optimum successor has the error e_o, and its sub-optimum successor possesses the error e_s. Then breaking the cycle at the state and connecting the state to its sub-optimum successor may cause an increase in error as follows

$$\Delta e=e_s-e_o \tag{16}$$

The distinct predecessor on a cycle for breaking the cycle can cause distinct increase in errors. The optimum breaking of a limit cycle can be obtained by minimizing the error increase in the above equation.

For example, if the state "2" is connected to its suboptimum successor "0", and the cycle is broken at point "A" as shown in Fig. 3, then the increase in error can be obtained by

$$\Delta e=0.36-0.16=0.20$$

Similarly, the state "−1" is another state on the cycle. Its optimum successor is the state "2" and has an error of 0.04. Its sub-optimum successor is the state "1" with an error of 0.64. If the cycle is broken at the point "B", the state "−1" is connected to its sub-optimum successor "1" and an extra error is introduced as follows

$$\Delta e=0.64-0.04=0.60$$

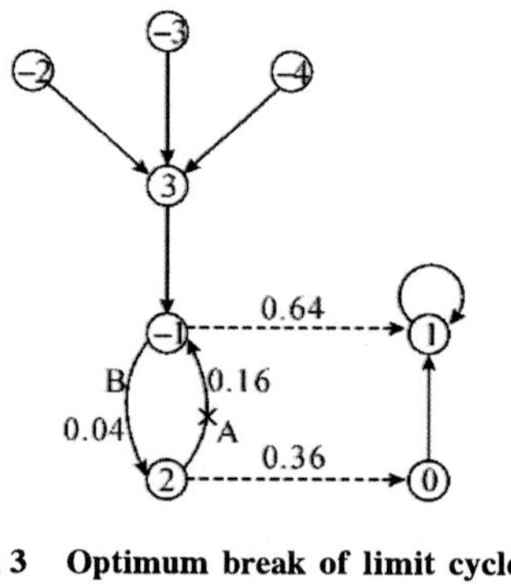

Fig. 3 Optimum break of limit cycles

Compare the increases in errors between these two breaks. It is obvious that the break at point "A" is the optimum breaking of this cycle. A minimal increase in error of 0.20 is added.

In general, a limit cycle has C states on it. They have C sub-optimum successors. The breaking with minimal increase in noise among the C breakings is optimum.

This method can remove all the limit cycles with minimum increase in errors. A stable machine can be obtained. The break of cycles is only a relocation of the state table. This is unlikely to increase the complexity of system implementation.

5. Conclusion

A method is proposed for representing an IIR digital filter as an FSM. This representation has only a maximum error of 1/2 bit from the final results of infinite accuracy arithmetic, and has no growth of errors during processing. A connection matrix is presented for detecting and locating the limit cycle oscillations in the machine. A graph theory method is proposed for removing all limit cycles from the system with a minimal increase in noise and hardly any increase in complexity of hardware.

References

[1] Jackson, L. B., "Analysis of Limit Cycle Due To Multiplication Rounding in Recursive Digital (Sub) Filters", Proc. of 7th Allerton Conference on Circuits and System Theory, 1969, pp. 69-78.

[2] Zhang, Y. Z., and Yao, Y. F., "Digital Signal Processing System and Finite State Machine", Proc. of 22nd International Electronics, IREECON' 89, Melbourne, Australia. September 1989. pp. 96-99.

[3] Zhang, Y. Z., "Linearization of Non-linear System over GF(2^M)", Proc. of International Conference on Circuits and Systems ICCAS-89, Nanjing, 1989.

Composite Attitude Control for Flexible Spacecraft with Simultaneous Disturbance Attenuation and Rejection Performance*

H Liu, L Guo, and Y Zhang

National Key Laboratory of Science and Technology on Aircraft Control,
Beihang University, Beijing, China

Abstract: In this paper, a composite attitude control approach for orbiting spacecraft with rigid central hubs and flexible appendages is presented. The established attitude control model consists of the vibration modes excited by the rigid body, the space environment disturbances, the measurement noises, and the model uncertainty. The model is formulated into a dynamic system with two types of disturbance inputs. A composite control law with the simultaneous disturbance attenuation and rejection performance is presented for the flexible spacecraft system subject to multiple disturbances. The disturbance-observer-based control is designed for feedforward compensation of the elastic vibration. The H_∞ state-feedback controller is designed to perform the robust attitude control in the presence of the space environment disturbances, measurement noises, and the model uncertainty. Numerical simulations show that the performance of the attitude control systems can be improved by combining the disturbance observer with H_∞ state-feedback control.

Keywords: flexible spacecraft; attitude control; disturbance-observer-based control; H_∞ control; multiple disturbance; vibration control; anti-disturbance control; composite hierarchical anti-disturbance control

1. Introduction

Flexible spacecraft have played an increasingly important role in space missions. To reduce launch costs, lightweight materials are applied to the spacecraft structure, which may lead to low-frequency elastic modes. The unwanted excitation of the flexible modes will deteriorate the pointing and stability performance of attitude control systems (ACSs). This situation arises because flexible spacecraft structures are usually coupled systems of elastically deformable bodies whose behaviour is characterized by non homogeneous equations with uncertain parameters. The dynamical model of spacecraft is non-linear and its inertia matrix and elastic modes are usually unknown exactly, so the parametric uncertainty should be taken into account. Moreover, space environmental disturbances

* Published in Proc. IMech Vol. 226, Part I: J System & Control Engineering, 2011: 154-161.

Y. Zhang, *Footprints in Cambridge and Aviation Industries of China*,
https://doi.org/10.1007/978-981-16-3176-4_29

also may degrade the pointing accuracy and stability of the spacecraft. Therefore, the fundamental attitude of the control design problem for flexible spacecraft is to overcome the influence of vibration, together with the model uncertainty, unmodelled dynamics, the space environment disturbances, and measurement noise.

In recent decades, the proportional-integral-derivative (PID) controller has been widely applied in ACS design, due to its simplicity and reliability[1, 2]. When model uncertainties exist or external disturbance varies, it is difficult to achieve satisfactory PID control performance. New design schemes are required for the simultaneous attitude control and vibration suppression. Optimal control of flexible spacecraft has been considered in references [3, 4], but in general this solution requires a high-order controller, which is not acceptable in the case of an onboard computer where memory is very limited. Variable structure control and sliding mode control, known to be efficient control strategies for systems with strong non-linearity and modelling uncertainty, have also been applied to ACS design by combining with active vibration control via piezoelectric actuators[5-7]. It is noted that a possible chattering phenomenon resulting from variable structure control and sliding mode control has an inevitable effect on the steady-state precision.

In modern control theory, disturbance attenuation and rejection is a fundamental problem since the model uncertainty, unmodelled dynamics, and even non-linearities can be described as an equivalent disturbance input, besides various noises and disturbances. In the presence of external disturbance and model uncertainty, robust control theory provides designers with a systematic approach for analysis and synthesis of the whole system. In paper [8], the robust control is investigated for the influence of the flexibility on the rigid motion, the presence of disturbances acting on the structure, and parameter variations. It is noted that H_∞ controllers have demonstrated their effectiveness in on-orbit attitude control experiments for Engineering Test Satellites ETS-VI and ETS-VIII[9, 10]. Furthermore, a feasible mixed H_2/H_∞ controller has been designed for rigid spacecraft[11]. In reference [12], an H_∞ multi-objective controller based on the linear matrix inequality (LMI) framework was designed for flexible spacecraft. H_∞ control can provide satisfactory disturbance attenuation performance, but it is of limited usefulness for attitude control and active vibration control problems since the characteristics of the disturbance are neglected.

To overcome the conservativeness of H_∞ control approaches, disturbance-observer-based control, or DOBC, has received extensive attention for many practical plants[13, 14]. A survey can be seen in paper [15] for non-linear DOBC. Compensation through feedforward for the modelling error or the exogenous disturbance has been considered as a robust control scheme when the error or disturbance can be estimated. In references [16, 17], DOBC for non-linear robotic manipulator was investigated in the presence of constant and harmonic disturbances. In paper [13], the DOBC approach in a state space framework was presented for a class of non-linear systems, where the disturbance was generated by a linear exogenous system. Other recent developments can be seen in references [18-20]. Although DOBC can provide elegant disturbance rejection performance, the obstacle to DOBC is that the dynamics of the disturbance estimation error rely on strict confinement of the disturbance and plant model.

Anti-disturbance control methodologies can be divided into two main types. One is the disturbance attenuation method (such as H_∞ control) where the influence of the disturbance can be decreased for the reference output. The other is the disturbance rejection method which can realize compensation of the disturbance with internal mode controllers or adaptive compensation controllers. However, these approaches only deal with one type of disturbance. In practice, together with the rapid development on sensor and data processing technologies, the disturbances or noise from different sources (e. g. sensor and actuator noise, friction, vibration, etc.) can be characterized by different mathematical models. Also, disturbance can represent the unmodelled dynamics and system uncertainties. For the case of multiple disturbances, a composite hierarchical anti-disturbance control was proposed[21, 22] to guarantee the simultaneous disturbance attenuation and rejection performance. However, the disturbances rejected in these two references are confined to be an exosystem with known parameters, which cannot be used for the attitude and vibration control problems studied in this paper.

In this paper, a composite attitude controller design approach is designed for flexible spacecraft based on DOBC and H_∞ state-feedback control. DOBC can reject the effect of vibrations from flexible appendages, and H_∞ state-feedback control can attenuate the influence of the norm bounded disturbances, and correspondingly guarantee the robust stability against other disturbances and model uncertainty. Simulations of flexible spacecraft show that the performance of ACSs can be improved with comparisons to H_∞ state-feedback control.

The remainder of this paper is organized as follows. In Section 2, the dynamic model of the flexible spacecraft is introduced. In Section 3, the stabilization of the system under the given controller is analysed, and the solution of the controller is resolved. In Section 4, the effectiveness of the proposed control algorithm is confirmed by numerical simulations. Conclusions are provided in Section 5. In the following, if not otherwise stated, matrices are assumed to have compatible dimensions. The identity and zero matrices are denoted by I and 0, respectively, with appropriate dimensions. For a square matrix M, $\mathrm{sym}(M)=M+M^T$ is denoted. For a symmetric matrix M, the notation $M>(\geqslant 0)$ is used to denote that it is positive definite (positive semi-definite). The case for $M<(\leqslant 0)$ follows similarly. The norms $\| \cdot \|$ of a real vector function and a matrix are defined as their Euclidean norms.

2. Problem Formulation

Similarly to references [5, 8], the single-axis model can be derived from the non-linear attitude dynamics of the flexible spacecraft. In this paper, the problem is simplified and only considers the single-axis rotational manoeuvre. It is assumed that this model includes one rigid body and one flexible appendage (see Fig. 1), and the relative elastic spacecraft model is described as

$$J\ddot{\theta}+F\ddot{\eta}=u+w \tag{1}$$

$$\ddot{\eta}+2\xi\omega\dot{\eta}+\omega^2\eta+F^T\ddot{\theta}=0 \tag{2}$$

where θ is the attitude angle, J is the spacecraft inertia, F is the modal participation matrix, u is the control torque, w represents the merged disturbance including the space environmental disturbances, moment-of-inertia uncertainty, and noises from sensors and actuators, η is the flexible modal coordinate, ξ is the damping ratio, and ω is the modal frequency. Combining (1) with

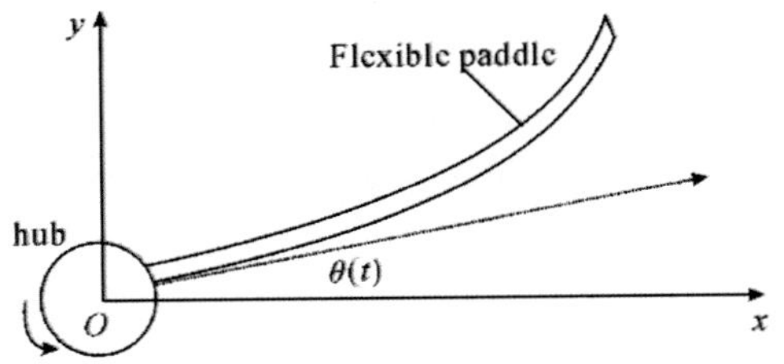

Fig. 1 Spacecraft with flexible appendages

(2) yields

$$(J-FF^T)\ddot{\theta}=F(2\xi\omega\dot{\eta}+\omega^2\eta)+u+w \tag{3}$$

In (3), $F(2\xi\omega\dot{\eta}+\omega^2\eta)$ is caused by the flexible appendages and can be modelled as the disturbance. Denoting

$$x(t)=[\theta(t),\dot{\theta}(t)]^T$$

then (3) can be transformed into

$$\dot{x}(t)=A_0x(t)+\{B_uu(t)+B_fd_0(t)+B_dd_1(t)\} \tag{4}$$

where the coefficient matrices are

$$A_0=\begin{bmatrix}0 & 1\\ 0 & 0\end{bmatrix};\ B_u=B_f=B_d=\begin{bmatrix}0\\ (J-FF^T)^{-1}\end{bmatrix}$$

In this paper, $d_0(t)=F(2\xi\omega\dot{\eta}+\omega^2\eta)$ is the disturbance to represent the flexible appendages; $d_1(t)$ represents the merged disturbance from space environmental disturbances, moment-of-inertia uncertainty, and noises from sensors and actuators.

According to the practical situations, it can be supposed that $\|\dot{d}_0(t)\|\leqslant W_0$ and $\|d_1(t)\|\leqslant W_1$, where W_0 and W_1 are known positive constants.

The reference output equation of the system for the H_∞ performance index is defined as $z(f)=C_1x(t)$. Then, the system model can be described by

$$\begin{cases}\dot{x}(t)=A_0x(t)+B_uu(t)+B_fd_0(t)+B_dd_1(t)\\ z(t)=C_1x(t)\end{cases} \tag{5}$$

where $u(t)$ and $z(t)$ are the control input and the output, respectively.

As state feedback controller has been widely applied in many practical systems, a direct application of state feedback control strategy would lead to

$$u_c(t)=Kx(t)$$

where K are control gains to be determined. Our aim is to design state feedback controllers for (5) such that the closed-loop system is asymptotically stable with the guaranteed generalized H_∞ performance, where the generalized H_∞ performance is defined as

$$J_\infty:=\|z(t)\|^2-\gamma^2\|d_1(t)\|^2$$

Different from previous works such as references [3, 5, 8], the dynamic model for the flexible spacecraft includes two types of disturbances which describe the vibration modes excited by the rigid body, the space environment disturbances, the measurement noise, and the model uncertainty respectively. Both DOBC and H_∞ control are used to enhance the anti-disturbance performance, compared with PID-type atlitude control laws[2]. In the following, the composite anti-disturbance control approach presented recently in papers [21, 22] will be generalized to the considered attitude control problem. To solve the attitude and vibration control problem, a different type of disturbance $d_0(t)$ which satisfies $\|d_0(t)\| \leqslant W_0$ is considered, while in papers [21, 22], the disturbance to be rejected is described by an exosystem with known parameters.

3. Composite Attitude Controller Design

3.1 Disturbance Observer Design

According to system (5), the disturbance observer is formulated as

$$\begin{cases} \dot{\tau}(t) = -N(x)B_f(\tau + p(x)) - N(x)(A_0 x(t) + B_u u(t)) \\ \hat{d}_0 = \tau + p(x) \end{cases} \tag{6}$$

where $N(x)$ is the gain of the observer, defined by

$$N(x) = \frac{\partial p(x)}{\partial x}$$

Here, $N(x)$ is a constant, and is abbreviated by N. The error of disturbance observer is defined as

$$e(t) = d_0(t) - \hat{d}_0(t)$$

Then

$$\dot{e}(t) = \dot{d}_0 - NB_f e(t) - NB_d d_1(t) \tag{7}$$

According to the practical situation of the flexible appendages, here, we should design an appropriate N to make $e(t) \to 0$.

In the DOBC scheme, the controller can be constructed as $u(t) = -\hat{d}_0(t) + u_c(t)$, which is directly described by Fig. 2 (where $\hat{d}_0(t)$ is the estimation of $d_0(t)$). It is shown that the composite attitude controller possesses the hierarchical architecture. In fact, it has two loops: the inner loop is used to estimate the vibration and compensate it, while the outer loop provides the H_∞ control to attenuate the norm bounded disturbances. After the above transformations, it is possible to deal with the parameter design by adopting the popular robust control techniques.

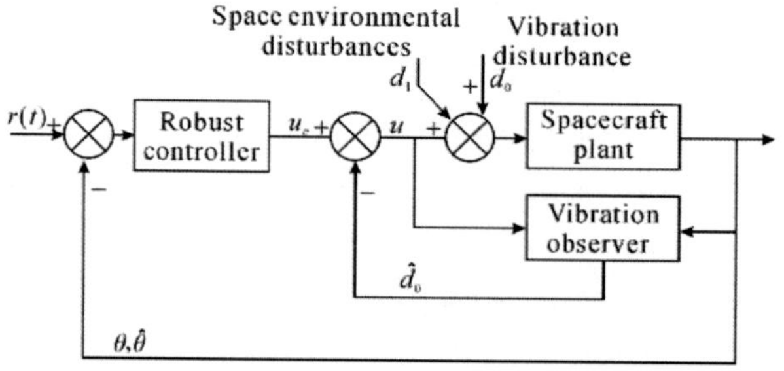

Fig. 2 Composite controller design

Substituting $u(t)$ to (5) and (7), it is possible to obtain the augmented system

$$\begin{bmatrix} \dot{x}(t) \\ \dot{e}(t) \end{bmatrix} = \begin{bmatrix} A_0 + B_u K & B_f \\ 0 & -NB_f \end{bmatrix} \begin{bmatrix} x(t) \\ e(t) \end{bmatrix} + \begin{bmatrix} 0 & B_d \\ I & -NB_d \end{bmatrix} \begin{bmatrix} \dot{d}_0(t) \\ d_1(t) \end{bmatrix} \tag{8}$$

Denoting $\bar{x} = \begin{bmatrix} x(t) \\ e(t) \end{bmatrix}$

$$\begin{cases} A = \begin{bmatrix} A_0 + B_u K & B_f \\ 0 & -NB_f \end{bmatrix}, & d(t) = \begin{bmatrix} \dot{d}_0(t) \\ d_1(t) \end{bmatrix} \\ B = \begin{bmatrix} 0 & B_d \\ I & -NB_d \end{bmatrix}, & C = [C_1 \quad C_2] \end{cases} \tag{9}$$

For $\dot{d}_0(t)$ and $d_1(t)$ are all bounded, then $d(t)$ is bounded also, thus, (8) and its reference output equation in the H_∞ performance index can be described as

$$\begin{cases} \dot{\bar{x}}(t) = A\bar{x}(t) + Bd(t) \\ z(t) = C\bar{x}(t) \end{cases} \tag{10}$$

3.2 Stability of the Composite Systems

In this section, the stability of the composite system is proved, and the parameters of the composite controller can be given via the solution of a class of linear matrix inequalities (LMIs).

Theorem 1

To system (8), for $\gamma > 0$, if there exists symmetrical matrix $Q_1 > 0$, $P_2. > 0$, R_1, and R_2 that satisfies

$$\begin{bmatrix} \Phi_{11} & Q_1 C_1^T & 0 & B_d & B_f \\ * & -\gamma I & 0 & 0 & C_2 \\ * & * & -\gamma I & 0 & P_2 \\ * & * & * & -\gamma I & -B_d^T R_2^T \\ * & * & * & * & \Phi_{55} \end{bmatrix} < 0 \tag{11}$$

where

$$\Phi_{11} = \mathrm{sym}(A_0 Q_1 + B_u R_1)$$
$$\Phi_{55} = \mathrm{sym}(-R_2 B_f)$$

then system (8) is asymptotically stable and satisfies H_∞ performance, and the controller gain is given by $K = R_1 Q_1^{-1}$ and the observer gain is given by $N = P_2^{-1} R_2$

Proof

Based on the bounded real lemma, the system (10) satisfies H_∞ performance $\|z\| < \gamma \|d(t)\|$, if there exists symmetrical matrix $P > 0$ and $\gamma > 0$ satisfying (see for example[13])

$$\begin{bmatrix} A^T P + PA & PB & C^T \\ B^T P & -\gamma I & 0 \\ C & 0 & -\gamma I \end{bmatrix} < 0 \tag{12}$$

Substituting (9) into (12), and denoting

$$P = \begin{bmatrix} P_1 & 0 \\ 0 & P_2 \end{bmatrix}$$

we can get

$$\begin{bmatrix} \Theta_{11} & P_1B_f & 0 & P_1B_d & C_1^T \\ * & \Theta_{22} & P_2 & -P_2NB_d & C_2^T \\ * & * & -\gamma I & 0 & 0 \\ * & * & * & -\gamma I & 0 \\ * & * & * & * & -\gamma I \end{bmatrix} < 0 \tag{13}$$

where

$$\Theta_{11} = \mathrm{sym}(P_1P_0 + P_1B_uK)$$
$$\Theta_{22} = \mathrm{sym}(-P_2NB_f)$$

From (13), by some matrix transformations, it is possible to obtain

$$\begin{bmatrix} \Theta_{11} & C_1^T & 0 & P_1B_d & P_1B_f \\ * & -\gamma I & 0 & 0 & C_2 \\ * & * & -\gamma I & 0 & P_2 \\ * & * & * & -\gamma I & -B_d^TN^TP_2 \\ * & * & * & * & 0 \\ * & * & * & * & \Theta_{22} \end{bmatrix} < 0 \tag{14}$$

Denoting

$$Q = \begin{bmatrix} Q_1 & 0 \\ 0 & Q_2 \end{bmatrix} = \begin{bmatrix} P_1^{-1} & 0 \\ 0 & P_2^{-1} \end{bmatrix} = P^{-1}$$
$$R_1 = KQ_1, R_2 = P_2N$$

then pre-multiplying diag$\{Q_1, I, I, I, I\}$ and post-multiplying diag$\{Q_1, I, I, I, I\}$ to the left and right side of the matrix of (14), it is possible to obtain the conclusion.

4. Simulations

4.1 Simulation Parameter

In order to demonstrate the effectiveness of the proposed control algorithm, numerical simulations will be performed in this section. The composite control scheme will be applied to a spacecraft with one flexible appendage. Since low-frequency modes are generally dominant in a flexible system, only the lowest two bending modes have been considered for the implemented spacecraft model.

We suppose that $\omega_1 = 1.27$ and $\omega_2 = 6.58$ with damping $\xi_1 = 0.004\ 705$ and $\xi_2 = 0.005\ 590$ respectively. As an example, this paper tries to control the attitude in the pitch channel, where $J=35.72\ \mathrm{kg \cdot m^2}$ (J is the nominal principal moment of inertia of pitch axis). In addition, 35 percent perturbation of the nominal moment of inertia will also be considered.

The flexible spacecraft is designed to move in a circular orbit with the altitude of 500 km, then the orbit rate $n=0.0011$ rad/s, The disturbance torques acting on the satellite are assumed as follows

$$\begin{cases} T_{dx} = 5.2 \times 10^{-5}(3\cos nt + 1) \\ T_{dy} = 5.2 \times 10^{-5}(3\cos nt + 1.5\sin nt) \\ T_{dz} = 5.2 \times 10^{-5}(3\sin nt + 1) \end{cases}$$

The initial pitch attitude of the spacecraft is $\theta=0.08$ rad, and $\dot{\theta}=0.0006$ rad/s. $F(2\xi\omega\dot{\eta}+\omega^2\eta)$ is defined as vibration torque which comes from the flexible appendages.

Selecting $\gamma=5.7$, it is possible to obtain the anticipated controller gain

$$K=[-67.85 \quad -65.5]$$

and the observer gain $N=[0\ 1000]$.

4.2 Simulation Analysis

Fig. 3(a) shows the elastic vibration, its estimation, and the estimation error respectively. The estimation error is amplified in Fig. 3(b), where one can see that satisfactory tracking performance can be achieved for the vibration from the flexible appendages. With the estimation, the effect of the elastic vibration can be rejected by feed-forward compensation. In Fig. 4(a), the attitude angle of the spacecraft is demonstrated and compared with the pure H_∞ attitude control. Fig. 4 (b) is the amplification of Fig. 4 (a), where one can see that the improved response performance can be guaranteed under the composite controller. Correspondingly, Fig. 5 (a) and (b) show that the attitude stabilization can also be improved with a composite controller in the presence of flexible vibration. In addition, + 35 per cent perturbation of the nominal moment of inertia is also considered. Fig. 6 shows that the proposed controller has improved robustness against model uncertainty, when compared with H_∞ control.

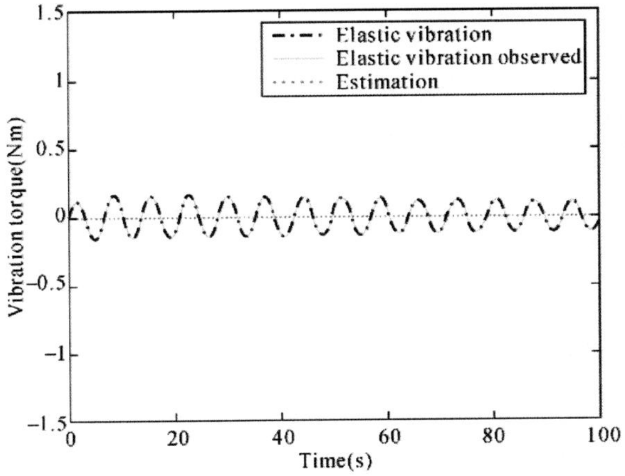

(a) Time responses of vibration and vibration observed

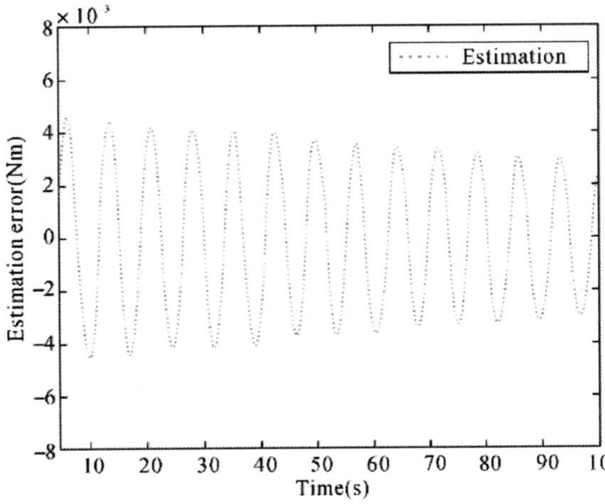

(b) Estimation error zoom of the above plot

Fig. 3 Vibration estimation error in disturbance observer

From the simulations, it can be seen that the composite controller based on DOBC and H_∞ control is capable of compensating the effect of the elastic vibration actively, and can improve the pointing precision and stabilization of the flexible spacecraft in the presence of the model uncertainty and space environmental disturbances.

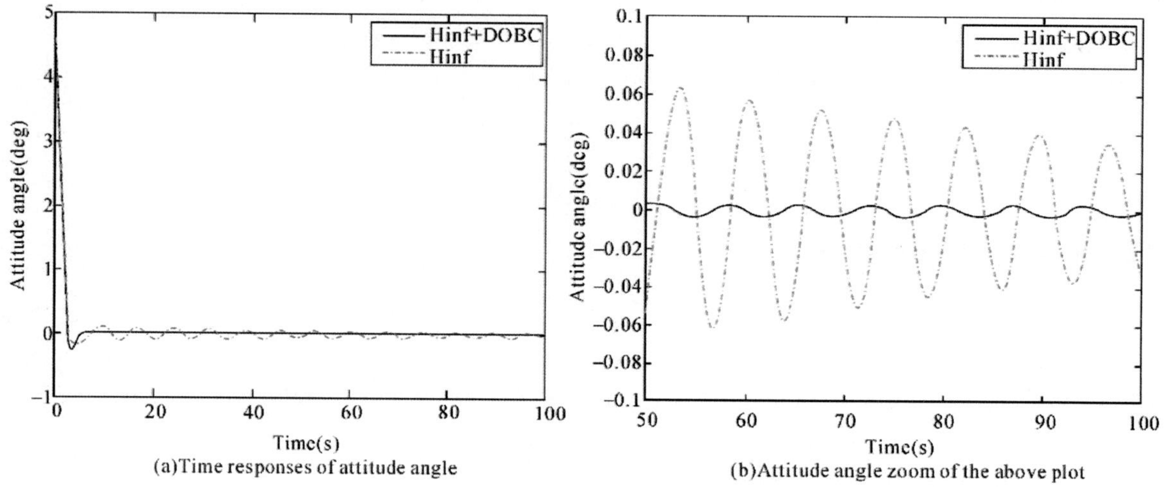

Fig. 4 Time responses of attitude angle

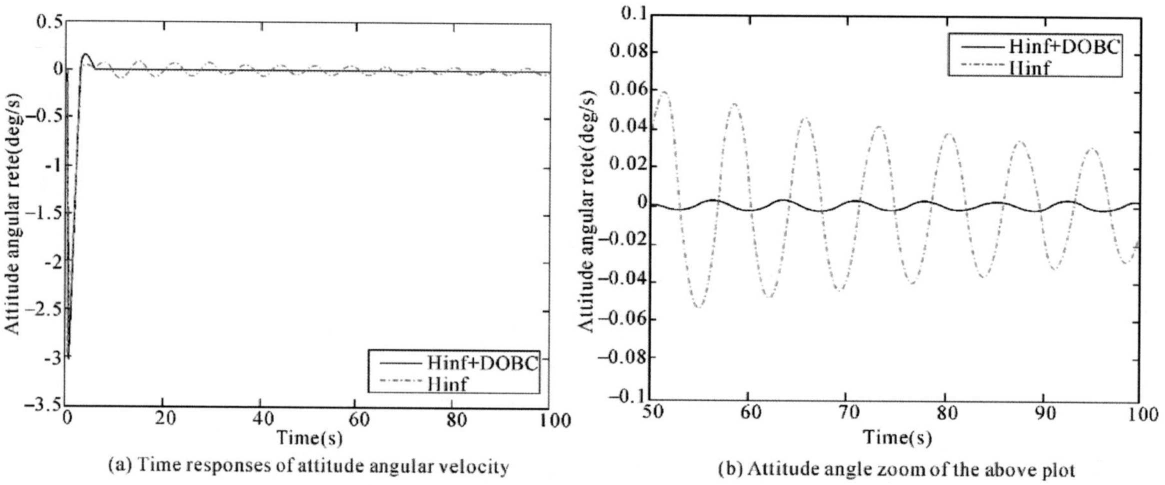

Fig. 5 Time responses of attitude angular velocity

5. Conclusions

In this paper, a composite attitude control scheme for flexible spacecraft is presented in the presence of model uncertainty, elastic vibration, and space environmental disturbances. The proposed composite controller possesses a hierarchical architecture, which consists of DOBC in the inner loop and H_∞ control in the outer loop. DOBC can reject the effect of the elastic vibration from the flexible appendages, and H_∞ control can attenuate the effect of the model uncertainty, noise from sensors, and external disturbances. Numerical simulations have shown that the composite controller can enhance the pointing accuracy and stabilization of the flexible spacecraft. This provides a promising and simple attitude control technique for flexible spacecraft.

It is noted that only H_∞ state feedback control is considered in this paper. A next step will attempt to use H_∞ output feedback control for more general cases. For the output feedback controllers, it is expected that the observer for the state could be designed together with the disturbance observer, similarly to the so-called full-order observer design discussed in reference [13]. Another point is that this paper has only considered two types of disturbance and has supposed that $d_1(t)$ and the derivative of $d_0(t)$ are bounded respectively. Although numerical simulations have shown that enhanced robustness can be achieved by using the proposed method, more general theoretical research and experimental simulations need to be carried out in the future.

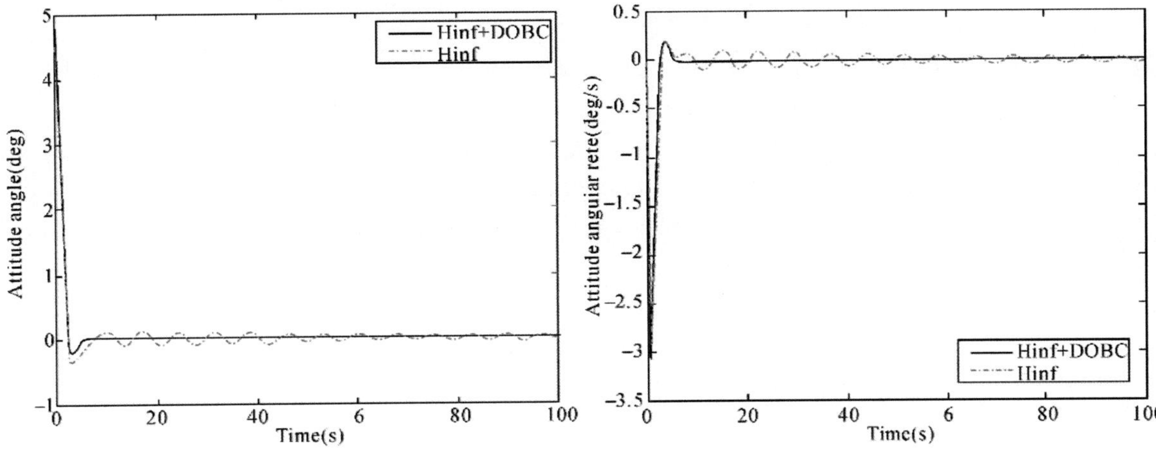

(a) Time responses attitude angle

(b) Time responses of attitude angular velocity

Fig. 6 Time responses of spacecraft attitude under inertia perturbation

Acknowledgements

The authors gratefully acknowledge the reviewers' comments.

Funding

This work was supported by the National Natural Science Foundation of China (NSFC) (grant no. 60925012), the National 973 programme (grant no. 2009CB724001), and the National 863 programme (grant no. 2008AA12A216).

References

[1] Chobotov, V. A. Spacecraft attitude dynamics and control, 1991 (Krieger, Malabar, Florida).

[2] Miller, A. J. , Gray, G. L. , and Mazzoleni, A. P. Nonlinear spacecraft dynamics with flexible appendage, damping, and moving internal submasses. J. Guid. Contr. Dynam. , 2001, 42(3), 605-615.

[3] Ben-Asher, J. , Burns, J. A. , and Clio, E. M. Time optimal slewing of flexible spacecraft. J. Guid. Contr. Dynam. , 1992,15, 360-367.

[4] Bolonkin, A. A. and Khot, N. S. Optimal bounded control design for vibration suppression. Acta Astronaut. , 1996, 38(10), 803-813.

[5] Hu, Q. L. and Ma, G. F. Variable structure control and active vibration suppression of flexible spacecraft during attitude maneuver. Aerosp. Sci. Technol. , 2005, 9, 307-317.

[6] Hu, Q. L. and Friswell, M. I. Robust variable structure attitude control with L2-gain performance for a flexible spacecraft including input saturation. Proc. IMechE, Part I: J. Systems and Control Engineering, 2010, 224(2), 2041-3041.

[7] Song, G. and Kotejoshyer, B. Vibration reduction of flexible structures during slew operations. Int. J. Acoust. Vibr. , 2002, 7(2), 105-109.

[8] Di Gennaro, S. Output stabilization of flexible spacecraft with active vibration suppression. IEEE Trans. Aerosp. Electron. Syst. , 2003, 39(3), 747-759.

[9] Kida, T. , Yamaguchi, I. , and Chida, Y. On-orbit robust control experiment of flexible spacecraft ETS-VI. J. Guid. Contr. Dynam. , 1997, 20(5), 865-872.

[10] Nagashio, T. , Kida, T. , and Ohtani, T. , et al. Design and implementation of robust symmetric attitude controller for ETS-VIII spacecraft. Contr. Engng Pract. , 2010, 18(12), 1440-1451.

[11] Yang, C. D. and Sun, Y. P. Mixed H_2/H_∞ state-feedback design for microsatellite attitude control. Contr. Engng Pract. , 2002, 10, 951-970.

[12] Charbonnel, C. HN and LMI attitude control design: towards performances and robustness enhancement. Acta Astronaut. , 2004, 54, 307-314.

[13] Guo, L. and Chen, W. H. Disturbance attenuation and rejection for systems with nonlinearity via DOBC approach. Int. J. Robust Nonlin. Contr. , 2005, 15, 109-125.

[14] Ishikawa,J. and Tomizuka, M. Pivot friction compensation using an accelerometer and a disturbance observer for hard disk. IEEE/ASME Trans. Mechatron. , 1998, 3, 194-201.

[15] Guo,L. , Feng, C. B. , and Chen, W. H. A survey of disturbance-observer-based control for dynamic nonlinear system. Dynam. Contin. Discr. Impuls. Syst. B Applic. Algor. , 2006, 13, 79-84.

[16] Chen, W. H. Nonlinear disturbance observer-enhanced dynamic inversion control of missiles. J. Quid. Contr. Dynam. , 2003, 26(1), 161-166.

[17] Chen, W. H. Disturbance observer based control for nonlinear systems. IEEE/ASME Trans. Mechatron, 2004, 9(4), 706-710.

[18] Back,J. and Shimb, H. Adding robustness to nominal output-feedback controllers for uncertain

nonlinear systems: A nonlinear version of disturbance observer. Automatica, 2008, 44, 2528-2537.

[19] Chen, X. K., Su, C. Y., and Fukuda, T. A nonlinear disturbance observer for multivariable systems and its application to magnetic bearing systems. IEEE Trans. Contr. Syst. Technol., 2004, 12(4), 569-577.

[20] Yang, Z. J. and Tsubakihara, H. A novel robust nonlinear motion controller with disturbance observer. IEEE Trans. Contr. Syst. Technol., 2008, 16(1), 137-147.

[21] Guo, L. and Wen, X. Y. Hierarchical anti-distance adaptive control for non-linear systems with composite disturbances and applications to missile systems. Trans. Inst. Meas. Contr., 2010, April. DOI: 10.1177/0142331210361555.

[22] Wei, X. and Guo, L. Composite disturbance-observer-based control and control for complex continuous models. Int. J. Robust Nonlin. Contr., 2010, 20(1), 106-118.

Geometry and Accuracy of Specular Points in GPS-Reflection Altimetry*

GUO Jia, YANG Dongkai, LING Keck Voon, SHAO Dingrong and ZHANG Yanzhong

School of Electronic and Information Engineering, Beihang University, Beijing, China

Abstract: The sea surface height recovery methods and error analysis of altimetry using GPS reflected signal are explored in this paper. For the case when the receiver is at low altitude positions, the analytical equation of the sea surface height error is obtained; and for the case when the receiver is at high altitude positions, a novel method which has a concise mathematical formulation and efficient solution is proposed. Moreover, the proposed method can obtain more accurate solution than methods currently available in the literature. Analysis and simulation show that the sea surface height error and all the measurement errors are of the same order of magnitude when the elevation angle is high.

Keywords: sea surface height; bistatic altimetry; global positioning system (GPS); reflected signal

1. Introduction

The concept of altimetry using GPS (Global positioning system) reflected signal was first proposed in 1993 by the European Space Agency as passive multistatic radar to monitor mesoscale ocean altimetry[1]. The scenario was that a receiver observed the signals emitted by GPS satellites, both direct rays and forward scattered rays after they were reflected from the sea surface. The difference in the time of arrival between direct and reflected signals provided information about altimetry. From then on, specific receivers have been designed[2,3] and experiments have been performed to validate the concept, from ground[4,5], aircraft[6,7] or space[8,9]. This technology is attractive because high temporal and spatial sampling would be achieved since many reflections can be simultaneously received, users can take advantage of the GPS infrastructure maintained for navigation purposes, the receiver technology is now in expensive since huge development costs have already been borne, and the signals, designed for high-accuracy navigation, have many of the characteristics desired for remote sensing.

A key challenge in GPS-Reflection altimetry is to locate the specular points. In the literature, there are three different methods:

* Published in Chinese Journal of Electronics, Vol. 21, No. 1, Jan. 2012: 91-96.

Y. Zhang, *Footprints in Cambridge and Aviation Industries of China*,
https://doi.org/10.1007/978-981-16-3176-4_30

In Ref. [10], the authors devised a two-step algorithm. In the first step an approximate nominal sea surface is assumed and in the second step the reflection point is refined with a more precise surface;

In Ref. [11], the authors modeled the difference of the reference lapse which is built over a reference surface near the local geoid and the measured one as a function of the height over the reference surface;

In Refs. [12, 13], the authors proposed an iterative approach which assume that the sea surface height is known to obtain a better approximation firstly, then repeat the process again and again.

In this paper, we also focus on locating the specular points and the associated error analysis. There are two main contributions in this paper:

For the case when the receiver is at low altitude positions, we obtained the analytical equation of the sea surface height error (Eq. (5)), and based on it, simulation showed that the sea surface height error and all the measurement errors are of the same order of magnitude;

For the case when the receiver is at high altitude positions, we formulated the problem of determining the position of the specular point as a constrained optimization problem (Eq. (7)) by exploiting a property of prolate spheroid. Our approach has two advantages: (1) we can use a more accurate model to describe the sea surface instead of the usual spherical model used in the literature and (2) we can use efficient methods to solve the problem.

The remaining parts of the paper are as follows. In Section 2, we review the method of altimetry from low altitude positions, and derive an analytical equation to calculate the error variance of the sea surface height. In Section 3, we show how the problem of determining the position of the specular point can be formulated as a constrained optimization problem. In addition, a special case when the sea surface is assumed to be spherical is considered in detail. The simulation results about error performance of our proposed method are given in Section 3. 3. In Section 4, we discuss the appropriateness of the flat sea surface assumption followed by conclusions in Section 5.

2. Altimetry from Low Altitude Positions

2. 1 Sea Surface Height Recovery

If the altitude of the receiver is much smaller than the radius of the earth, we could assume the sea surface to be flat. Fig. 1 shows the applicable geometry. The sea surface heightand the specular point position can be calculated as follows.

Firstly, the receiver computes its own position and the GPS satellite's position using the direct signal, then obtains the elevation angle θ. Secondly, the receiver estimates the path difference Δp between the direct and the reflected signal from the relative delay between them. Since $\Delta p = 2h \sin\theta$, the height h can be estimated. Finally, the sea surface height h_S is obtained by subtracting h from the receiver height h_R. The above analysis can be expressed as the following formula:

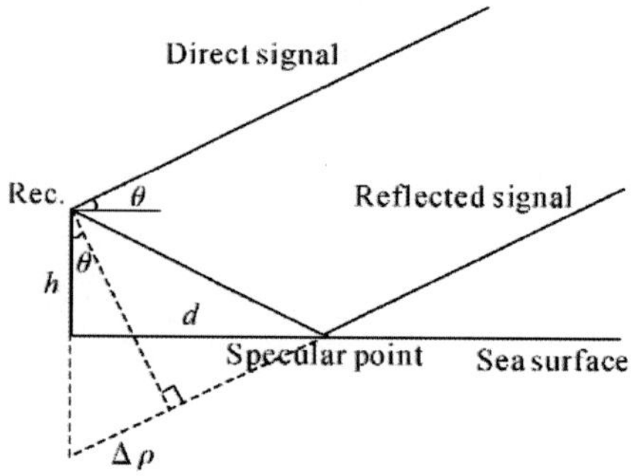

Fig. 1 Geometry of altimetry from low altitude positions

$$\begin{aligned} h_S &= h_R - h \\ &= h_R - \Delta\rho/2\sin\theta \\ &= h_R - \frac{\Delta\rho}{2\frac{r^{\mathrm{T}}(g-r)}{\sqrt{r^{\mathrm{T}}r}\sqrt{(g-r)^{\mathrm{T}}(g-r)}}} \end{aligned} \tag{1}$$

where $r=(x_R \quad y_R \quad z_R)^{\mathrm{T}}$ and $g=(x_G \quad y_G \quad z_G)^{\mathrm{T}}$ are position vectors of the receiver and the GPS satellite in the ECEF (Earth-Centered Earth-Fixed) system. Note that h_R can be obtained from r using the following transformation in the WGS 84 system:

$$\begin{cases} x=(N+h)\cos\psi\cos\lambda \\ y=(N+h)\cos\psi\sin\lambda \\ z=[N(1-e^2)+h]\sin\psi \end{cases} \tag{2}$$

$\Phi=(\psi \quad \lambda \quad h)^{\mathrm{T}}$ is the geographical latitude, longitude and height coordinates of any point, $(x \quad y \quad z)^{\mathrm{T}}$ is the Cartesian coordinates of the same point, $N=a(1-e^2\sin^2\psi)^{-1/2}$, and a and e are the major radius and eccentricity of the WGS 84 datum ellipsoid.

The length d can be estimated as:

$$d=h\cdot\cot\theta=\frac{\Delta\rho}{2\sin\theta}\cdot\cot\theta$$

and with the information of GPS satellite's azimuth, the specular point position is obtained approximately.

2.2 Error Analysis and Simulation

We see from Eqs. (1) and (2) that h_S is a function of $\Delta\rho$, g and r. Suppose that $\delta\Delta\rho$, δg and δr are the measurement errors, and δh_S and δh_R are the errors of height of the sea surface and receiver respectively. To find the dependency between these errors, we differentiate Eq. (1) to obtain:

$$\begin{aligned} \delta h_S = &-\frac{\sqrt{r^{\mathrm{T}}r}\sqrt{(g-r)^{\mathrm{T}}(g-r)}}{2r^{\mathrm{T}}(g-r)}\delta\Delta\rho+\delta h_R \\ &-\frac{\Delta\rho\sqrt{r^{\mathrm{T}}r}}{2}\cdot\frac{r^{\mathrm{T}}(g-r)(g-r)^{\mathrm{T}}-(g-r)^{\mathrm{T}}(g-r)r^{\mathrm{T}}}{r^{\mathrm{T}}(g-r)(g-r)^{\mathrm{T}}r\sqrt{(g-r)^{\mathrm{T}}(g-r)}}\delta g \\ &-\frac{\Delta\rho}{2}\cdot\{r^{\mathrm{T}}(g-r)[(g-r)^{\mathrm{T}}(g-r)r^{\mathrm{T}}-r^{\mathrm{T}}r(g-r)^{\mathrm{T}}] \end{aligned}$$

$$-r^{\mathrm{T}}r(g-r)^{\mathrm{T}}(g-r)(g-2r)^{\mathrm{T}}\}$$
$$\cdot\{r^{\mathrm{T}}(g-r)(g-r)^{\mathrm{T}}r\sqrt{r^{\mathrm{T}}r}\sqrt{(g-r)^{\mathrm{T}}(g-r)}\}^{-1}\delta r \tag{3}$$

Further, we can get the dependency between δh_R and δr by differentiating Eq. (2):

$$(\delta x \quad \delta y \quad \delta z)^{\mathrm{T}}=C(\delta\psi \quad \delta\lambda \quad \delta h)^{\mathrm{T}} \tag{4}$$

where C is a 3×3 matrix:

$$C=\begin{pmatrix} \frac{ae^2\sin\psi\cos\psi}{(1-e^2\sin^2\psi)^{\frac{3}{2}}}\cos\psi\cos\lambda-(N+h)\sin\psi\sin\lambda & -(N-h)\cos\psi\sin\lambda & \cos\psi\cos\lambda \\ \frac{ae^2\sin\psi\cos\psi}{(1-e^2\sin^2\psi)^{\frac{3}{2}}}\cos\psi\sin\lambda-(N+h)\sin\psi\cos\lambda & -(N-h)\cos\psi\cos\lambda & \cos\psi\sin\lambda \\ (1-e^2)\frac{ae^2\sin\psi\cos\psi}{(1-e^2\sin^2\psi)^{\frac{3}{2}}}\sin\psi+[N(1-e^2)+h]\cos\psi & 0 & \sin\psi \end{pmatrix}$$

Combining Eqs. (3) and (4), we get

$$\delta h_S=k_{\Delta\rho}\delta\Delta\rho+G^{\mathrm{T}}\delta g+R^{\mathrm{T}}\delta r$$

where

$$k_{\Delta\rho}=-\frac{\sqrt{r^{\mathrm{T}}r}\sqrt{(g-r)^{\mathrm{T}}(g-r)}}{2r^{\mathrm{T}}(g-r)}$$

$$G=-\frac{\Delta\rho\sqrt{r^{\mathrm{T}}r}}{2}\cdot\frac{(g-r)(g-r)^{\mathrm{T}}r-r(g-r)^{\mathrm{T}}(g-r)}{r^{\mathrm{T}}(g-r)(g-r)^{\mathrm{T}}r\sqrt{(g-r)^{\mathrm{T}}(g-r)}}$$

and R is a 3×1 vector:

$$R=C^{-\mathrm{T}}\cdot\begin{pmatrix}0\\0\\1\end{pmatrix}-\frac{\Delta\rho}{2}$$
$$\cdot\{r^{\mathrm{T}}(g-r)[(g-r)^{\mathrm{T}}(g-r)r-r^{\mathrm{T}}r(g-r)]-r^{\mathrm{T}}r(g-r)^{\mathrm{T}}(g-r)(g-2r)\}$$
$$\cdot\{r^{\mathrm{T}}(g-r)(g-r)^{\mathrm{T}}r\sqrt{r^{\mathrm{T}}r}\sqrt{(g-r)^{\mathrm{T}}(g-r)}\}^{-1}$$

Suppose that all of the measurement errors $\delta\Delta\rho$, δg and δr are independent, zero mean and their variance (or covariance matrixes) are $\sigma^2_{\Delta\rho}$, $\sum_g$ and $\sum_r$, then the mean of the sea surface height error is 0 and its variance is:

$$\begin{aligned}\sigma^2_{h_S} &= E(\delta h_S^2)\\ &= E[(k_{\Delta\rho}\delta\Delta\rho+G^{\mathrm{T}}\delta g+R^{\mathrm{T}}\delta r)^2]\\ &= k^2_{\Delta\rho}(\delta\Delta\rho^2)+G^{\mathrm{T}}E(\delta g\delta g^{\mathrm{T}})G+R^{\mathrm{T}}E(\delta r\delta r^{\mathrm{T}})R\\ &= k^2_{\Delta\rho}\sigma^2_{\Delta\rho}+G^{\mathrm{T}}\sum\nolimits_g G+R^{\mathrm{T}}\sum\nolimits_r R\end{aligned} \tag{5}$$

As assuming the measurement errors are independent, Eq. (5) shows that the sea surface height error is the superposition of these errors. Numerical simulation examples, based on Eq. (5), are shown in Fig. 2. For the three error sources discussed above, we give the results for the following cases:

Case 1: $\sigma_{\Delta\rho}=1$, $\sum_g=0$, $\sum_r=0$;

Case 2: $\sigma_{\Delta\rho}=0$, $\sum_g=I$, $\sum_r=0$;

Case 3: $\sigma_{\Delta\rho}=0$, $\sum_g=0$, $\sum_r=I$.

In the simulation, the true value of h is supposed to be 100. Fig. 2 shows that:

At low elevation angle the error of the sea surface height is dominated by the error of the path difference length (Case 1);

The error caused by the GPS satellites position can be ignored, since the GPS satellites are far away and the method is differential essentially (Case 2);

The error caused by the receiver position does not change with the elevation angle (Case 3).

Thus it can be concluded that when the elevation angle is high, the variance of the sea surface height error is of the same order of magnitude of the variances of the measurement errors.

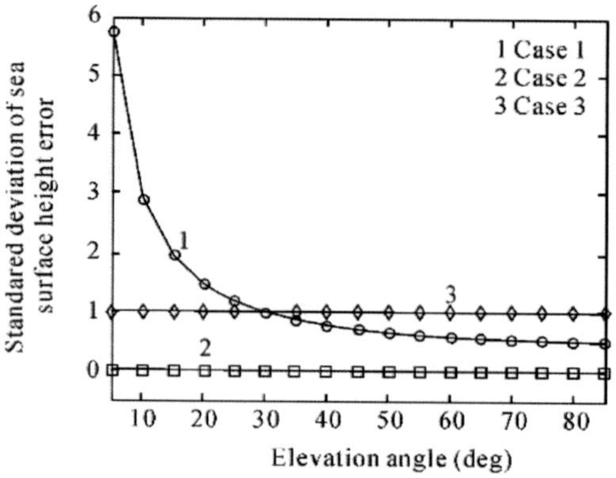

Fig. 2 Sea surface height error from measurement errors for the cases when the receiver is at low altitude positions

3. Altimetry from High Altitude Positions

3.1 Sea Surface Height Recovery

If the altitude of the receiver cannot be ignored compared with the radius of the earth, we could no longer model the sea surface to be flat. In this case, it is appropriate to model the sea surface as an ellipsoid using the WGS 84 datum, since the WGS 84 datum is a reasonable approximation to the Earth and the GPS receiver uses this model to determine its position.

Given the position vectors of the GPS satellite $g=(x_G \quad y_G \quad z_G)^T$ and the receiver $r=(x_R \quad y_R \quad z_R)^T$ in the ECEF system and the total path length of the reflected signal $\rho=\rho_1+\rho_2$, we seek the position vectors of the specular point $x=(x \quad y \quad z)^T$. The geometry of this problem is depicted in Fig. 3 and the coordinate of specular point x can be determined as follows.

Firstly, we note that the specular point is on a prolate spheroid as shown in Fig. 3. The major axe of the prolate spheroid passes through the receiver and the GPS satellite and the major radius is $\rho/2$. Secondly, note also that the incident signal, the reflected signal and the normal to the sea surface model at the specular point lie in the same plane and the angle which the incident signal makes with the normal is equal to the angle which the reflected signal makes to the same normal. But, prolate spheroid has the property that the normal at any point on the prolate spheroid bisects

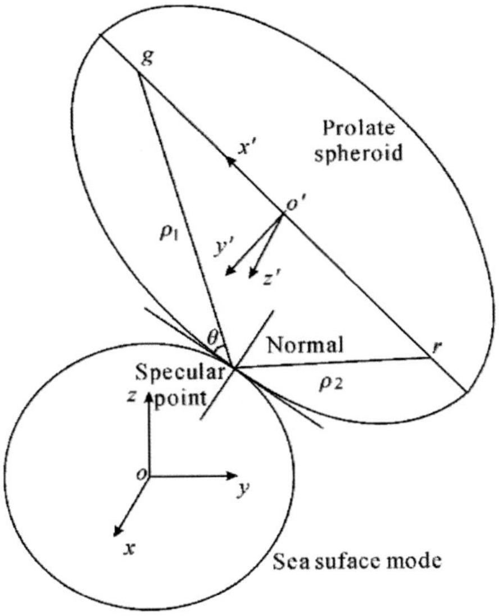

Fig. 3 Geometry of altimetry from high altitude positions

the angle formed by the lines joining the point and the two focuses ①. So the normal to the prolate spheroid and the normal to the sea surface model at the specular point overlap, i. e. the prolate spheroid and the sea surface model tangent at the specular point. The above two aspects yields:

$$\begin{cases} \sqrt{(x-x_G)^2+(y-y_G)^2+(z-z_G)^2}+\sqrt{(x-x_R)^2+(y-y_R)^2+(z-z_R)^2}=\rho \\ \dfrac{x^2+y^2}{a^2}+\dfrac{z^2}{b^2}=k^2 \end{cases} \tag{6}$$

where a and b are the major and minor radius of the WGS 84 datum ellipsoid, and $k>0$ parameterizes the sea surface model ellipsoid. From geometry, we know that k has two solutions and what we need is the smaller k corresponding to the smaller ellipsoid.

Eq. (6) cannot be solved using the method of elementary algebra, but it is equivalent to an optimization problem. Suppose that x is a movable point on the surface described by the first equation of Eq. (6) and the surface described by the second equation of Eq. (6) passes through this point as shown in Fig. 3, k has the smallest value when the two surfaces tangent. So Eq. (6) is equivalent to the following optimization problem:

$$\min_{(x,y,z)} \frac{x^2}{a^2}+\frac{y^2}{a^2}+\frac{z^2}{a^2}$$

$$\text{s. t. } \sqrt{(x-x_G)^2+(y-y_G)^2+(z-z_G)^2}+\sqrt{(x-x_R)^2+(y-y_R)^2+(z-z_R)^2}=\rho \tag{7}$$

This constrained optimization problem can be solved efficiently, for example, using the method of sequential quadratic programming.

Remark: For detecting a region, we can use a regional ellipsoid instead of the datum ellipsoid in WGS 84 which is a geocentric ellipsoid. As the regional ellipsoid fits the geoid best over the region

① This is the same property which derives the well-known conclusion that lights, which emit from the source at one of the prolatespheroid's foci, converge at the other foci through the reflection of the prolate spheroid.

of interest, the outcome will be more convenient and accurate. For using the regional ellipsoid, we only need to change the equation (the second equation of Eq. (6) and the objective function of Eq. (7)) describing the sea surface model.

3.2 A Special Case

The optimization problem Eq. (7) can in principle be solved using the method of Lagrange multipliers, but this is not practical as the constraint is too complicated. Although we can simplify the constraint by coordinate transformation, this would make the objective function complex unless the sea surface model is a sphere (i. e. $a=b$). We shall now derive the solution for this special case:

First step: Coordinate transformation.

Set the origin of the new coordinate system at the middle point of r and g, x-axis passing through r and g, y-axis in the plane through o, r and g, and z-axis with x-axis and y-axis forming a right-handed coordinate system (Fig. 3). Then problem (7) turns into problem (8) in the new coordinate system:

$$\begin{aligned} &\min_{(x',y',z')} (x'-x_o)^2+(y'-y_o)^2+(z'-z_o)^2 \\ &\text{s. t. } \frac{x'^2}{A^2}+\frac{y'^2}{B^2}+\frac{z'^2}{B^2}=1 \end{aligned} \tag{8}$$

where $A=\rho/2$, $B^2=(\rho/2)2-(\|r-g\|/2)^2$ and $(x_o\ y_o\ z_o)^{\mathrm{T}}$ is the coordinates of point o in the new coordinate system. The transformation between the new and old coordinate systems is as follows:

$$\begin{pmatrix} x' \\ y' \\ z' \end{pmatrix} = \left(\frac{r-g}{\|r-g\|} \quad \frac{r\times g}{\|r\times g\|}\times\frac{r-g}{\|r-g\|} \quad \frac{r\times g}{\|r\times g\|} \right)\cdot \begin{pmatrix} x \\ y \\ z \end{pmatrix} + \frac{r+g}{2} \tag{9}$$

Second step: Solve problem (8) using the method of Lagrange multipliers.

Formally, set $\Lambda(x',y',z',\lambda)=(x'-x_o)^2+(y'-y_o)^2+(z'-z_o)^2+\lambda(\frac{x'^2}{A^2}+\frac{y'^2}{B^2}+\frac{z'^2}{B^2}-1)$. Set the derivatives to zero, which yields the system of equations:

$$\begin{cases} \dfrac{\partial \Lambda}{\partial x'}=2(x'-x_o)+\dfrac{2\lambda}{A^2}x'=0 \\ \dfrac{\partial \Lambda}{\partial y'}=2(y'-y_o)+\dfrac{2\lambda}{B^2}y'=0 \\ \dfrac{\partial \Lambda}{\partial z'}=2(z'-z_o)+\dfrac{2\lambda}{B^2}z'=0 \\ \dfrac{x'^2}{A^2}+\dfrac{y'^2}{B^2}+\dfrac{z'^2}{B^2}-1=0 \end{cases} \tag{10}$$

The first three equations of Eq. (10) yield:

$$\begin{cases} x'=\dfrac{A^2}{A^2+\lambda}x_o \\ y'=\dfrac{B^2}{B^2+\lambda}y_o \\ z'=\dfrac{B^2}{B^2+\lambda}z_o \end{cases} \tag{11}$$

Substituting into the fourth equation of Eq. (10) yields:

$$\frac{A^2 x_o^2}{(A^2+\lambda)^2}+\frac{B^2 y_o^2}{(B^2+\lambda)^2}+\frac{B^2 z_o^2}{(B^2+\lambda)^2}-1=0$$

$$\Rightarrow A^2 x_o^2 (B^2+\lambda)^2+B^2 y_o^2 (A^2+\lambda)^2+B^2 z_o^2 (A^2+\lambda)^2-(A^2+\lambda)^2 (B^2+\lambda)^2=0 \qquad (12)$$

Eq. (12) is polynomial equation of degree four and we can obtain 4 solutions of it using explicit algebraic operations. Finally substituting the 4 solutions into Eq. (11) yields 4 coordinates, and the one which makes the objective function of Eq. (8) smallest is the final answer.

Third step: Substituting the final answer into Eq. (9) yields its coordinate in the old coordinate system.

Remark: If we apply other ellipsoid sea surface model, the degree of the polynomial equation in Eq. (12) will be greater than four and there is no general algebraic solution for it according to Abel Theorem. So the general problem cannot be solved analytically.

3.3 Error Analysis and Simulation

We see from Eqs. (6) and (7) that x is a function of p, g and r, but we cannot obtain an explicit form of it from these equations. As the error analysis for the general case is difficult, we use the Monte Carlo method here, i. e. suppose that the measurement errors $\delta\rho$, δg and δr are random variables, then the error of the specular point Cartesian coordinate δx and its mean and covariance matrix can be found numerically by solving Eq. (7). Further we will get the mean and covariance matrix of the latitude, longitude and height coordinates error $\delta\phi$ using Eq. (4):

$$E(\delta\phi)=E(C^{-1}\delta x)=C^{-1}E(\delta x)$$

$$E(\delta\phi\delta\phi^{\mathrm{T}})=E(C^{-1}\delta x\,\delta x^{\mathrm{T}}C^{-\mathrm{T}})=C^{-1}E(\delta x\delta x^{\mathrm{T}})C^{-\mathrm{T}}$$

Numerical simulation examples are shown in Fig. 4. For the three error sources discussed above, we give the results for the following cases:

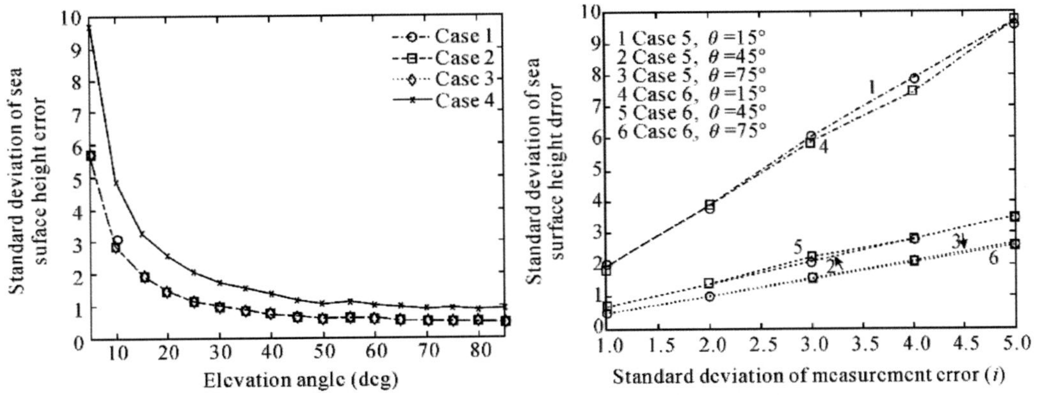

Fig. 4 Sea surface height error from measurement errors for the cases when the receiver is at high altitude positions

Case 1: $\sigma_p = 1, \sum_g = 0, \sum_r = 0$;

Case 2: $\sigma_p = 0, \sum_g = I, \sum_r = 0$;

Case 3: $\sigma_p = 0, \sum_g = 0, \sum_r = I$;

Case 4: $\sigma_p = 1, \sum_g = I, \sum_r = I$;

Case 5: $\sigma_p = i, \sum_g = 0, \sum_r = 0 \ (i = 1,2,3,4,5)$;

Case 6: $\sigma_p = 0, \sum_g = i^2 \times I, \sum_r = 0 \ (i = 1,2,3,4,5)$.

Other parameters are as follow: the position of the specular point is $(0° \quad 45° \quad 0)^T$ and the receiver altitude is 450km which is the same as in the CHAMP mission. The mean of the sea surface height error is 0. Fig. 4 shows that:

The sea surface height error caused by the three error sources individually is a monotonically decreasing function of the elevation angle (Case 1, 2 and 3);

The error caused by the receiver position and the GPS position are the same, which is caused by the geometric symmetry of the problem (Case 2 and 3)②;

The error caused by the three error sources together is not the superposition of these errors, but it is also a monotonically decreasing function of the elevation angle (Case 4);

The sea surface height error is proportional to the three error sources individually, and the ratios are decreasing function of the elevation angle (Case 5 and 6).

Thus it can be also concluded that when the elevation angle is high, the variance of the sea surface height error is of the same order of magnitude of the variances of the measurement errors.

4. Scope Analysis of the Method in Section 2

In Section 2, we derive a sea surface height recovery method which assumes a flat sea surface model when the altitude of the receiver is much smaller than the radius of the earth. In this section, we analyze the impact of this assumption. Fig. 5 shows the applicable geometry. Given the position of the GPS satellite and the receiver, we can obtain two specular points x' and x, and the associated sea surface heights h'_S and h_S using the methods in Section 2 and 3, where hs is more accurate than h'_S. Fig. 6 shows the sea surface height difference, i. e. $h'_S - h_S$, for different angular displacement α. For simplicity, we apply the spherical sea surface model when solving for h_S.

② From the simulation, it can be seen that the error caused by the path length is also nearly the same as the others. But this has no intuitive explanation.

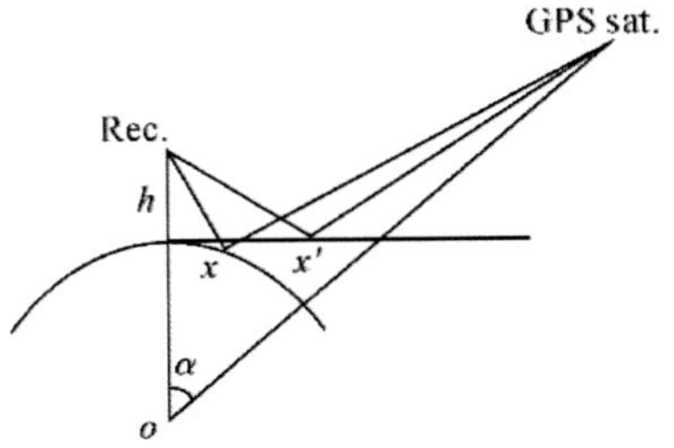

Fig. 5 Geometry of the scope analysis of the method in Section 2

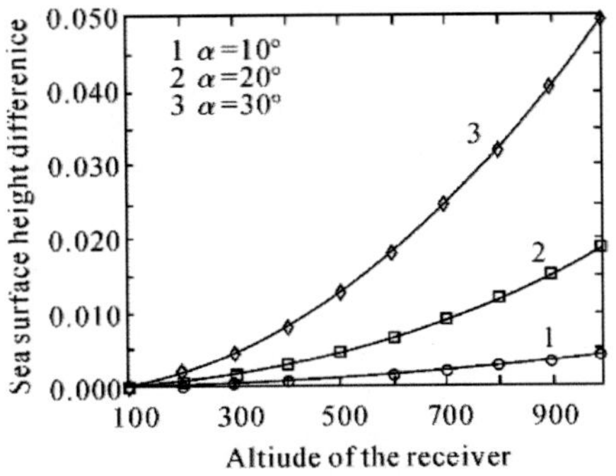

Fig. 6 Sea surface height difference from the altitude of the receiver

From Fig. 6, it can be seen that the sea surface height difference is a monotonically increasing function of the altitude of the receiver h, and the rate of increase is larger when the elevation angle is low (i. e. big α). Given an acceptable level of accuracy, Fig. 6 provides a guide to determine when a flat sea surface approximation is appropriate. For example, when $a=30°$ and the acceptable level of accuracy is 5cm, then flat sea surface approximation is appropriate when the altitude of the receiver is less than one kilometer.

5. Conclusion

The sea surface height recovery methods and error analysis of altimetry using GPS reflected signals are explored in this paper. We consider the case of altimetry from low and high altitude positions separately. The analytical equation of the sea surface height error is obtained when the receiver is at low altitude positions. A new method for the case of altimetry from high altitude positions which exploits a property of prolate spheroid is proposed. The new method has concise

mathematics formulation and efficient solution ways. Moreover, accurate sea surface models can be used in this method easily so that the solution is more accurate than other methods. Analysis and simulation show that the variance of the sea surface height error is of the same order of magnitude of the variances of the measurement errors when the elevation angle is high. So if the measurement errors are of the magnitude of centimeter, so the sea surface height error is.

References

[1] M. Martin-Neira, "A Passive reflectometry and interferometry system (PARIS)—Application to ocean altimetry", ESA Journal, Vol. 17, No. 4, pp. 331-355, 1993.

[2] S. T. Lowe, P. Kroger, G. Franklin, J. L. LaBrecque, "A delay/Doppler-mapping receiver system for GPS-reflection remote sensing", IEEE Transactions on Geoscience and Remote Sensing, Vol. 40, No. 5, pp. 1150-1163, 2002.

[3] O. Nogues-Correig, E. Cardellach Gali, J. Sanz Campderros, A. Rius, "A GPS-reflections

receiver that computes Doppler/delay maps in real time", IEEE Transactions on Geoscience and Remote Sensing, Vol. 45, No. 1, pp. 156-174, 2007.

[4] R. N. Treuhaft, S. T. Lowe, C. Zuffada, Y. Chao, "2-cm GPS altimetry over Crater Lake", Geophys. Res. Lett, Vol. 28, No. 23, pp. 4343-4346, 2001.

[5] M. B. Rivas, M. Martin-Neira, "Coherent GPS reflections from the sea surface", IEEE Geoscience and Remote Sensing Letters, Vol. 3, No. 1, pp. 28-31, 2006.

[6] S. T. Lowe, C. Zuffada, Y. Chao, P. Kroger, "5-cm precision aircraft ocean altimetry using GPS reflections", Geophys. Res. Lett., Vol. 29, No. 10, pp. 4359-4362, 2002.

[7] G. Ruffini, F. Soulat, M. Caparrini, O. Germain, "The eddy experiment: Accurate GNSS-R ocean altimetry from low altitude aircraft", Geophys. Res. Lett., Vol. 31, No. 1, pp. 123-126, 2004.

[8] S. T. Lowe, J. L. LaBrecque, C. Zuffada, L. J. Romans, "First spaceborne observation of an Earth-reflected GPS signal", Radio Science, Vol. 37, No. 1, pp. 701-728, 2002.

[9] S. Gleason, S. Hodgart, Y. Sun, C. Gommenginger, "Detection and processing of bistatically reflected GPS signals from low Earth orbit for the purpose of ocean remote sensing", IEEE Transactions on Geoscience and Remote Sensing, Vol. 43, No. 6, pp. 1229-1241, 2005.

[10] S. Wu, T. Meehan, C. Dunn, L. Young, "The potential use of GPS signals as ocean altimetry observables", Proceedings of the National Technical Meeting—Institute of Navigation, Santa Monica, CA, pp. 543-550, 1997.

[11] G. Ruffini, F. Soulat, M. Caparrini, O. Germain, "The GNSS-R eddy experiment I: Altimetry from low altitude aircraft", 2003 Workshop on Oceanography with GNSS Reflections, California, USA, pp. 1-8, 2003.

[12] C. Wagner, J. Klokovnik, "The value of ocean reflections of GPS signals to enhance satellite altimetry: data distribution and error analysis", Journal of Geodesy, Vol. 77, No. 3, pp. 128-138, 2003.

[13] J. Kostelecky, J. Klokocnik, C. A. Wagner, "Geometry and accuracy of reflecting points in bistatic satellite altimetry", Journal of Geodesy, Vol. 79, No. 8, pp. 421-430, 2005.

Accuracy Assessment of Single and Double Difference Models for the Single Epoch GPS Compass*

Wantong Chen, Honglei Qin, Yanzhong Zhang, Tian Jin
School of Electronic and Information Engineering, Beihang University, Beijing, China

Abstract: The single epoch GPS compass is an important field of study, since it is a valuable technique for the orientation estimation of vehicles and it can guarantee a total independence from carrier phase slips in practical applications. To achieve highly accurate angular estimates, the unknown integer ambiguities of the carrier phase observables need to be resolved. Past researches focus on the ambiguity resolution for single epoch; however, accuracy is another significant problem for many challenging applications. In this contribution, the accuracy is evaluated for the non-common clock scheme of the receivers and the common clock scheme of the receivers, respectively. We focus on three scenarios for either scheme: single difference model vs. double difference model, single frequency model vs. multiple frequency model and optimal linear combinations vs. traditional triple-frequency least squares. We deduce the short baseline precision for a number of different available models and analyze the difference in accuracy for those models. Compared with the single or double difference model of the non-common clock scheme, the single difference model of the common clock scheme can greatly reduce the vertical component error of baseline vector, which results in higher elevation accuracy. The least squares estimator can also reduce the error of fixed baseline vector with the aid of the multi-frequency observation, thereby improving the attitude accuracy. In essence, the "accuracy improvement" is attributed to the difference in accuracy for different models, not a real improvement for any specific model. If all noise levels of GPS triple frequency carrier phase are assumed the same in unit of cycles, it can be proved that the optimal linear combination approach is equivalent to the traditional triple-frequency least squares, no matter which scheme is utilized. Both simulations and actual experiments have been performed to verify the correctness of theoretical analysis.

Keywords: GPS; attitude determination; error analysis; single differences

1. Introduction

The heading and elevation indicator of a vehicle is one of the most important devices in

---

* Published in Advances in Space Research, 49 (2012): 725-738.

Y. Zhang, *Footprints in Cambridge and Aviation Industries of China*,
https://doi.org/10.1007/978-981-16-3176-4_31

navigation systems, which is indispensable and important for many challenging applications such as terrestrial, sea, air and space. The heading and elevation determination is the problem of estimating the orientation of an object with respect to a frame of reference. In the past decades, there are two widely used and radically different types of compass: the magnetic compass and the gyrocompass. The magnetic compass contains a magnet that interacts with the Earth's magnetic field and aligns itself to point to the magnetic poles. The gyrocompass contains a motorized gyroscope whose angular momentum interacts with the force produced by the Earth's rotation to maintain a north-south orientation of the gyroscopic spin axis, thereby providing a stable directional reference. However, the accuracy of the magnetic compass is affected by the magnetic field intensity nearby the equipment, and the gyroscopes suffer from the error drift. Compared with the above two devices, the Global Positioning System (GPS) compass can point to any desired direction without the above-mentioned shortcomings (Tu et al., 1997). For this technique, two antennas are attached to a vehicle, and then a baseline vector defined as a vector from reference antenna to another antenna can be determined using GPS relative positioning technique. The heading and elevation of a vehicle can be easily determined from this baseline vector and such an attitude determination system with single baseline vector is called GPS compass (Park et al., 1997). GPS compass system presents several obvious advantages: low-cost, less drift, maintenance-free and no ground reference.

In order to obtain accurate measurements of a platform attitude, the GPS carrier phase observations must be employed. However, the carrier phase observations are inherently affected by an integer ambiguity, and once this has been done successfully, the carrier phase data will act as very precise pseudorange data, thus making very precise attitude determination possible. Recent ambiguity resolution methods make use of the LAMBDA method (Lin et al., 2004; Monikes et al., 2005; Hide et al., 2007; Wang et al., 2009), as this method is known to be efficient and optimal (Teunissen, 1995,1999; Verhagen,2004). Since continuous carrier phases at enough epochs should be observed to obtain an accurate float solution for the LAMBDA method and cycle slips often occur in practical applications, the reparation of cycle slips is required to be achieved effectively. Any incorrect reparation of cycle slips will affect all subsequent observations. Ambiguity resolution in single epoch can guarantee a total independence from carrier phase slips and losses of lock, as this technique uses only the fractional value of carrier phase measurement. Thus, the challenge is to perform successful and efficient integer ambiguity resolution for single epoch. In general, the success rate of standard LAMBDA using single epoch data might be not as high as required by the specific application, since the standard LAMBDA method has been developed for unconstrained and/or linearly constrained GNSS models; it is therefore not necessarily optimal for the GNSS compass problem, for which often the baseline length is known as well (Teunissen et al., 2010a). In order to integrate the nonlinear baseline constraint into the ambiguity objective function, the constrained (C-) LAMBDA method has been presented for any GNSS compass model (Teunissen, 2007). The C-LAMBDA method is first applied for single epoch, single frequency attitude determination applications in Buist (2007) and the success rate of ambiguity resolution increases dramatically with this method. The implementation aspect of C-LAMBDA has been studied in Giorgi et al., 2008. The constrained integer least-squares theory for GNSS compass has been described in detail in Teunissen (2010b). The performance improvement of GPS compass is also studied in chen and Qin (in press), aiming to achieve a higher success rate of ambiguity resolution in land vehicle attitude determination. A multivariate constrained version of the LAMBDA method and its testing during an airborne remote sensing campaign are also studied, aiming to obtain a high success rate of ambiguity resolution by fully exploiting the known body geometry of the multi-antennae configuration (Giorgi et al., 2010).

It is very important to improve the success rates of ambiguity resolution for the single epoch GPS compass; however, the accuracy improvement should also be quite significant for many practical applications. In other words, once the ambiguity resolution is achieved, another important problem is 'how good is the measured attitude only using one epoch observation' and 'how to improve the accuracy further'. The error analysis of GPS compass is given in Park et al. (1997), which focuses on the baseline configuration and nominal attitude of vehicle, based on the single-frequency double-differenced GPS model. In this contribution, we focus on the error assessment of the single epoch GPS compass for the multi-frequency observation. It is obvious that the multi-frequency scheme can improve the accuracy, since more observations will result in a stronger model. However, the mathematical expression of the accuracy is required for the theoretical proof. We derived the expression of the multi-frequency observation accuracy and made the theoretical analysis for the extent of accuracy improvement. The experiments also demonstrate that the multi-frequency observation can effectively improve the accuracy of GPS compass. In particular, if all noise levels of triple frequency carrier phase are assumed the same in unit of cycles, the standard variation of each baseline component and attitude angle can be reduced to 0.68 of the single frequency case.

In many single-board dual-antenna GPS compass systems, both receivers feed a common clock, thus the accuracy of this scheme is also studied. Compared with the non-common clock scheme, this scheme can strongly improve the accuracy of the vertical component of baseline vector, resulting in the tremendous improvement of the elevation accuracy. For the non-common clock scheme of the receivers, it can be demonstrated that the double difference model is equivalent to the single difference model and the accuracies of east and north components are generally much better than that of the vertical component. However, if the single difference model with the common clock scheme is employed, the accuracy of the vertical component can be close to those of the east and north components. Moreover, the accuracy of GPS compass can be further improved by combining the multi-frequency observation and the single difference model with the common clock scheme.

Since the linear combination technique can also improve the short baseline precision by reducing the effects of thermal noise. Thus, the comparison of this approach and the traditional triple-frequency least squares method is performed. We also prove that both approaches are equivalent with the assumption that all noise levels of triple frequency carrier phase are assumed the same in unit of cycles, no matter the single or double difference scheme is utilized.

Note that the "accuracy improvement" is essentially attributed to the difference in accuracy for a number of different available models, not a real improvement for any specific model.

2. GNSS Compass Model and Integer Ambiguity Resolution for Single Epoch

2.1 Heading and Elevation Determination Using Single Baseline

For the GNSS compass system, two antennas are utilized to provide the observability of heading (or yaw) and elevation (or pitch). If the baseline vector from reference antenna to another antenna is parameterized with respect to the local East-North-Up frame, the heading ψ and the elevation θ can be computed from the baseline coordinates b_E, b_N and b_U as

$$\psi=\arctan(\frac{b_E}{b_N}),\theta=\arctan(\frac{b_U}{\sqrt{b_E^2+b_N^2}}) \tag{1}$$

2.2 GNSS Compass Model

For single baseline, the GNSS compass model reads as

$$E(Y)=Aa+Bb, D(Y)=Q_Y, a\in Z^n, b\in R^3, \|b\|=l \tag{2}$$

where Y is the given GNSS data vector, and a and b are the ambiguity vector and the baseline vector of order n and 3 respectively, and l denotes the known baseline length, which is assumed to be constant. $E(\cdot)$ and $D(\cdot)$ denote the expectation and dispersion operators, respectively, and A and B are the given design matrices that link the data vector to the unknown parameters. The variance matrix of Y is given by the positive definite matrix Q_Y. When solving the GNSS model (2), one usually applies the least-squares principle. This amounts to solving the following minimization problem:

$$\min_{a,b}\|Y-Aa-Bb\|_{Q_Y}^2, a\in Z^n, b\in R^3, \|b\|=l \tag{3}$$

where $\|(\cdot)\|_{Q_Y}^2=(\cdot)^T Q_Y^{-1}(\cdot)$. This problem was first introduced in Park and Teunissen (2003) and is referred to as a quadratically constrained (mixed) integer least-squares problem. Assuming that a is known, the conditional least-squares solution for b, can be written as (Teunissen, 2007)

$$\hat{b}(a)=(B^T Q_Y^{-1} B)^{-1} B^T Q^{-1}(Y-Aa) \tag{4}$$

The variance-covariance matrix is given by

$$Q_{\hat{b}(a)}=(B^T Q_Y^{-1} B)^{-1} \tag{5}$$

By the theory of integer least-squares principle with quadratic equality constraints, the most likely correct ambiguity vector a of model (2) can be given as (Park and Teunissen, 2009)

$$\check{a}=\arg\min_{a\in Z^n}(\|\hat{a}-a\|_{Q_{\hat{a}}}^2+\min_{b\in R^3, \|b\|=l}\|\hat{b}(a)-b\|_{Q_{\hat{b}(a)}}^2) \tag{6}$$

where the float solution $\hat{a}$ and its variance-covariance matrix $Q_{\hat{a}}$ can be obtained by the traditional least-squares method. Eq. (6) can be solved by the C-LAMBDA method with high efficiency and high success rate (Giorgi et al., 2008). Once the fixed solution $\check{a}$ has been computed, the baseline solution $\check{b}$ follows from substituting $\check{a}$ for a in (4):

$$\check{b}=\hat{b}(\check{a}) \tag{7}$$

3. Error Propagation of GNSS Compass

The error propagation of GPS compass for single epoch contains two parts: the error of baseline vector and the error of attitude transformation. The estimated variance-covariance matrix of the baseline vector represents the error of local level frame, which is affected by the geometry of satellites and the measurement precision of observables, see also (5). The error of attitude transformation is affected by the baseline length and the configuration of baseline. The first analysis of this problem is discussed in Park et al. (1997), and here the full steps are re-derived in detail for the following analysis.

The baseline vector in the local East-North-Up frame can be expressed as follows:

$$b=\begin{bmatrix} b_E \\ b_N \\ b_U \end{bmatrix}=\begin{bmatrix} l\sin\psi\cos\theta \\ l\cos\psi\cos\theta \\ l\sin\theta \end{bmatrix}=\Omega(\Lambda) \tag{8}$$

where Ω is the nonlinear operator and $\Lambda=(\psi\ \theta\ l)^T$. Linearization of these nonlinear observation equations can be given as

$$\begin{bmatrix} \delta b_E \\ \delta b_N \\ \delta b_U \end{bmatrix}=\begin{bmatrix} l_0\cos\psi_0\cos\theta_0 & -l_0\sin\psi_0\sin\theta_0 & \sin\psi_0\cos\theta_0 \\ -l_0\sin\psi_0\cos\theta_0 & -l_0\cos\psi_0\sin\theta_0 & \cos\psi_0\cos\theta_0 \\ 0 & l_0\cos\theta_0 & \sin\theta_0 \end{bmatrix}\times\begin{bmatrix} \delta\psi \\ \delta\theta \\ \delta l \end{bmatrix} \tag{9}$$

where the given Taylor point of expansion is $\Lambda_0=(\psi_0\ \theta_0\ l_0)^T$ and terms of second order and higher order have been neglected. The inverse of (9) becomes

$$\begin{bmatrix} \delta\psi \\ \delta\theta \\ \delta l \end{bmatrix}=\begin{bmatrix} \dfrac{\cos\psi_0}{l_0\cos\theta_0} & \dfrac{-\sin\psi_0}{l_0\cos\theta_0} & 0 \\ \dfrac{-\sin\psi_0\sin\theta_0}{l_0} & \dfrac{-\cos\psi_0\sin\theta_0}{l_0} & \dfrac{\cos\theta_0}{l_0} \\ \sin\psi_0\cos\theta_0 & \cos\psi_0\cos\theta_0 & \sin\theta_0 \end{bmatrix}\begin{bmatrix} \delta b_E \\ \delta b_N \\ \delta b_U \end{bmatrix} \tag{10}$$

With the law of variance and covariance (v-c) propagation, the v-c matrix of Λ can be calculated by

$$Q_\Lambda=J^TQ_{\hat{b}(a)}J \tag{11}$$

where J is the matrix that link the baseline error vector to the attitude error parameters in (10). Note that $Q_{\hat{b}(a)}$ is nothing to do with the baseline placement and it is determined by the satellite geometry and the noise levels of observables, see also (30) and (37). If the geometry of satellites is approximately symmetrical, the off-diagonal terms of $Q_{\hat{b}(a)}$ are generally very small, see also (71)—(80). In this case, they affect little in analysis of the baseline. Thus, we have the following expressions:

$$\sigma_{\delta\psi}^2=\frac{(\cos\psi_0)^2\sigma_E^2+(\sin\psi_0)^2\sigma_N^2}{l_0^2(\cos\theta_0)^2} \tag{12}$$

$$\sigma_{\delta\theta}^2=\frac{(\sin\psi_0\sin\theta_0)^2\sigma_E^2+(\cos\psi_0\sin\theta_0)^2\sigma_N^2+(\cos\theta_0)^2\sigma_U^2}{l_0^2} \tag{13}$$

$$\sigma_{\delta l}^2=(\sin\psi_0\sin\theta_0)^2\sigma_E^2+(\cos\psi_0\sin\theta_0)^2\sigma_N^2+(\sin\theta_0)^2\sigma_U^2 \tag{14}$$

Eqs. (12) and (13) imply that the accuracies of heading and elevation increases as the baseline length increases, and heading error is nothing to do with the altitude error σ_U. The baseline placement can also affect the accuracy of GPS compass, since the errors of heading and elevation are related to the direction of baseline vector. In general, the orientation of baseline depends on the motion of the vehicle, which may not be constant most of the time. In order to improve the precision of attitude determination in any case, the diagonal elements of the conditional variance-covariance matrix $Q_{\hat{b}(a)}$ should be as small as possible. A canonical theory for short GPS baselines has been studied in Teunissen (1997b). In the following section, we will further deduce the short baseline precision of various single epoch GPS compass models and then we demonstrate the difference in accuracy for those models.

4. The Short Baseline Precision of the Single Epoch GPS Compass Model

For the configuration of GPS compass system, one can use different clocks for the receivers, and also one can use a common clock for them. However, the accuracies of the two schemes may be different. In this contribution, we focus three scenarios for either scheme: single difference model vs. double difference model, single frequency model vs. multiple frequency model and optimal linear combinations vs. traditional triple-frequency least squares.

4.1 The Accuracy Assessment of the System with Different Clock Offsets of the Receivers

4.1.1 Single difference model vs. double difference model

For two nearby antennas A and B, the single-differenced carrier phase observation equation can be modeled as Misra and Enge (2006)

$$\lambda_i(\phi_{i,AB}^k + N_{i,AB}^k) = r_{AB}^k + c(\delta t_A - \delta t_B) + v_{i,AB}^k \tag{15}$$

where λ_i is the wavelength of L_i carrier; r_{AB}^k the single-differenced geometric range of two receivers for satellite k; $\phi_{i,AB}^k$ and $N_{i,AB}^k$ denote the single-differenced fractional phase and integer ambiguity respectively; $v_{i,AB}^k$ denotes the single-differenced phase observable noise; δt_A and δt_B are clock biases of receiver A and B; c is the velocity of light. In the same way, the single-differenced code observation equation can be expressed as

$$\rho_{i,AB}^k = r_{AB}^k + c(\delta t_A - \delta t_B) + \mu_{i,AB}^k \tag{16}$$

where $\rho_{i,AB}^k$ denotes the single-differenced code observable and $\mu_{i,AB}^k$ is the single-differenced code observable noise. Note that the unknown integer ambiguity is included on the left side of (15) for the reason that it might help some readers obtain an intuitive comparison of the carrier phase and code observation. Regardless of the noise level, the left side of Eq. (15) corresponds to that of (16). Since the altitude of a satellite is about 20,200 km above sea level and the baseline length for attitude determination is only several meters, the lines of sight (LOS) of two antennas to the same satellite are approximately parallel. Thus r_{AB}^k can be treated as the projection of the baseline vector b in the direction of LOS and expressed as $s^k \cdot b$ where s^k is the unit vector heading to the GPS satellite k (Misra and Enge, 2006). For m satellites in view, there are m independent single-differenced phase measurement equations. It is reasonable assumption that each channel of the receiver is independent and has the same characteristics for the same band. With $\sigma_{\phi,i}^2$ being the variance of the observed phase on GPS L_i band, all of the equations can be expressed in compact vector and matrix notation as

$$y_{i,S}^{\phi}=\frac{1}{\lambda_i}E\cdot b-N_{i,S}+e_m\beta+v_{i,S}^{\phi},v_{i,S}^{\phi}\sim N(0,2\sigma_{\phi,i}^2 I_m) \tag{17}$$

where $e_m=(1,1\cdots,1)^T$ of order m and the clock bias $\beta=c\cdot(\delta t_A-\delta t_B)/\lambda_1$ and

$$y_{i,S}^{\phi}=\begin{bmatrix}\phi_{i,AB}^1\\ \phi_{i,AB}^2\\ \vdots\\ \phi_{i,AB}^m\end{bmatrix},N_{i,S}=\begin{bmatrix}N_{i,AB}^1\\ N_{i,AB}^2\\ \vdots\\ N_{i,AB}^m\end{bmatrix},v_{i,S}^{\phi}=\begin{bmatrix}v_{i,AB}^1\\ v_{i,AB}^2\\ \vdots\\ v_{i,AB}^{m1}\end{bmatrix},E=\begin{bmatrix}(s^1)^T\\ (s^2)^T\\ \vdots\\ (s^m)^T\end{bmatrix} \tag{18}$$

Similarly, with $\sigma_{p,i}^2$ being the variance of code on GPS L_i band, all the code equations can be expressed as

$$y_{i,S}^{\rho}=\frac{1}{\lambda_i}E\cdot b+e_m\beta+v_{i,S}^{\rho},v_{i,S}^{\rho}\sim N(0,2\sigma_{p,i}^2 I_m) \tag{19}$$

where $y_{i,S}^{\rho}=[\rho_{i,AB}^1\quad \rho_{i,AB}^2\quad\cdots\quad \rho_{i,AB}^m]^T$ and $v_{i,S}^{\rho}=[\mu_{i,AB}^1\quad \mu_{i,AB}^2\quad\cdots\quad \mu_{i,AB}^m]^T$. To be consistent with carrier phase, there code is also expressed in units of cycles. The single difference model can be obtained by combining (17) and (19):

$$\begin{bmatrix}y_{i,S}^{\phi}\\ y_{i,S}^{\rho}\end{bmatrix}=\left[\begin{array}{cc|c}\frac{1}{\lambda_i}E & e_m & -I_m\\ \frac{1}{\lambda_i}E & e_m & 0\end{array}\right]\begin{bmatrix}b\\ \frac{\beta}{N_{i,S}}\end{bmatrix}+\begin{bmatrix}v_{i,S}^{\phi}\\ v_{i,S}^{\rho}\end{bmatrix} \tag{20}$$

Note that the entries of the baseline vector will consist of the baseline components (coordinates) and the clock bias and we use x as $(b^T\beta)^T$. This expression can be considered as the GNSS model $E(Y)=Aa+Bx$, if each term is expressed as follows:

$$Y=Y_{i,S}^{\phi\rho}=\begin{bmatrix}y_{i,S}^{\phi}\\ y_{i,S}^{\rho}\end{bmatrix},A=A_0=\begin{bmatrix}-I_m\\ 0\end{bmatrix},$$

$$B=B_i=\begin{bmatrix}\frac{1}{\lambda_i}E & e_m\\ \frac{1}{\lambda_i}E & e_m\end{bmatrix},Q_{Y\,i}=\begin{bmatrix}2\sigma_{\phi,i}^2 I_m & \\ & 2\sigma_{p,i}^2 I_m\end{bmatrix} \tag{21}$$

The subscript i denotes that the item depends on the given band L_i and the subscript 0 denotes that its structure does nothing with the given band. The variance-covariance matrix $Q_{\hat{x}(a)}$ can be obtained using (5) as follows:

$$Q_{\hat{x}(a)}=\left\{\left(\frac{1}{2\sigma_{\phi,i}^2}+\frac{1}{2\sigma_{p,i}^2}\right)\begin{Bmatrix}\frac{1}{\lambda_i^2}E^TE & \frac{1}{\lambda_i}E^Te_m\\ \frac{1}{\lambda_i}e_m^TE & e_m^Te_m\end{Bmatrix}\right\}^{-1}=\begin{Bmatrix}Q_{\hat{b}(a)} & Q_{\hat{b}(a)\beta(a)}\\ Q_{\beta(a)\hat{b}(a)} & Q_{\beta(a)\beta(a)}\end{Bmatrix} \tag{22}$$

Applying the well-known matrix inversion Lemma in block form, we can obtain the baseline precision of the single difference model:

$$Q_{\hat{b}(a)}=\left(\frac{1}{2\sigma_{\phi,i}^2}+\frac{1}{2\sigma_{p,i}^2}\right)^{-1}\left(\frac{1}{\lambda_i^2}E^TE-\left(\frac{1}{\lambda_i}E^Te_m\right)(e_m^Te_m)^{-1}\left(\frac{1}{\lambda_i}e_m^TE\right)\right)^{-1}=\lambda_i^2\omega_i^{-1}\left(E^T\left(I_m-\frac{1}{m}e_me_m^T\right)E\right)^{-1} \tag{23}$$

where $\omega_i=(\frac{1}{2\sigma_{\phi,i}^2}+\frac{1}{2\sigma_{p,i}^2})$.

Next, the baseline precision of the more common double difference model is deduced. In order to eliminate the clock biases of all the single-differenced observation equations, the double-differenced (DD) matrix operator will be used and it is defined as $D=(d_1\ d_2\cdots\ d_m)\in R^{(m-1)\times m}$ where each vector d_i is of order $m-1$. If satellite k is chosen as the "reference" satellite, thus the vector $d_k=-e_{m-1}$ and all the other vectors of D can be treated as I_{m-1}. In particularly, if we have chosen satellite 1 as the "reference" satellite, thus we have $D=(-e_{m-1}, I_{m-1})$. No matter which satellite is chosen as the "reference" satellite, the following property can be deduced, see also (Teunissen, 1997a):

$$D^T(DD^T)^{-1}D=I_m-\frac{1}{m}e_m e_m^T \tag{24}$$

After the multiplications on the left of (17) and (19) by the DD matrix operator, the following equations can be obtained:

$$y_{i,D}^{\phi}=H_i\cdot b-N_{i,D}+v_{i,D}^{\phi}, v_{i,D}^{\phi}\sim N(0, 2\sigma_{\phi,i}^2 Q) \tag{25}$$

$$y_{i,D}^{p}=H_i\cdot b+v_{i,D}^{p}, v_{i,D}^{p}\sim N(0, 2\sigma_{p,i}^2 Q) \tag{26}$$

where $y_{i,D}^{\phi}=Dy_{i,S}^{\phi}$, $y_{i,D}^{p}=Dy_{i,S}^{p}$, $v_{i,D}^{\phi}=Dv_{i,S}^{\phi}$, $v_{i,D}^{p}=Dv_{i,S}^{p}$, $N_{i,D}=DN_{i,S}$, $H_i=1/\lambda_i DE$ and $Q=DD^T$. In this model, the matrix Q would have 2s on the main diagonal and 1s in all off-diagonal positions due to the double differencing (Misra and Enge, 2006). Then the double difference model can be obtained by combining (25) and (26):

$$\begin{bmatrix} y_{i,D}^{\phi} \\ y_{i,D}^{p} \end{bmatrix}=\begin{bmatrix} H_i & -I \\ H_i & 0 \end{bmatrix}\begin{bmatrix} b \\ N_{i,D} \end{bmatrix}+\begin{bmatrix} v_{i,D}^{\phi} \\ v_{i,D}^{p} \end{bmatrix} \tag{27}$$

This expression can also be considered as the GNSS model if each term is expressed as follows:

$$Y=Y_{i,D}^{\phi p}=\begin{bmatrix} y_{i,D}^{\phi} \\ y_{i,D}^{p} \end{bmatrix}, A=A_0=\begin{bmatrix} -I \\ 0 \end{bmatrix},$$

$$B=B_i=\begin{bmatrix} H_i \\ H_i \end{bmatrix}, Q_{Y_i}=\begin{bmatrix} 2\sigma_{\phi,i}^2 Q & \\ & 2\sigma_{p,i}^2 Q \end{bmatrix} \tag{28}$$

With the expression (28), the ambiguity vector can be estimated by the C-LAMBDA method in single epoch. Once the correct ambiguity vector a is solved, the baseline vector can be calculated using (4) and the variance-covariance matrix can be obtained using (5). Thus, the baseline precision of the double difference model reads as

$$Q_{\hat{x}(a)}=\left((\frac{1}{2\sigma_{\phi,i}^2}+\frac{1}{2\sigma_{p,i}^2})(H_i^T Q^{-1}H_i)\right)^{-1}=\lambda_i^2\omega_i^{-1}(E^T D^T(DD^T)^{-1}DE)^{-1} \tag{29}$$

With (24), Eq. (29) is equal to (23). That indicates that if there are different clock offsets for the receivers, the single difference and double difference models are equivalent when the same original data is used in both approaches. In that case, the double difference operation is only performed to eliminate the different clock offsets of the receivers. The undifferenced or single differenced approach however, sometimes enables to use data that in the double difference approach has to be discarded, see also De Jonge (1998).

4.1.2 Single frequency model vs. multiple frequency model

With (24), the resulting variance-covariance matrix is independent of the choice of reference satellite. Here we define $T \equiv D^T(DD^T)^{-1}D$, the baseline precision of the single frequency model above can also be written as

$$Q_{\hat{x}(a)} = \lambda_i^2 \omega_i^{-1}(E^T TE)^{-1} \tag{30}$$

After a simple derivation, the following equation can be obtained:

$$\lambda_i^2 \omega_i^{-1} = 2\lambda_i^2 \frac{\sigma_{\phi,i}^2}{1+\frac{\sigma_{\phi,i}^2}{\sigma_{p,i}^2}} \tag{31}$$

Since the $\sigma_{\phi,i}^2 \ll \sigma_{p,i}^2$ the coefficient of $Q_{\hat{x}(a)}$ is approximately equal to $2\lambda_i^2\sigma_{\phi,i}^2$. This denotes that the code observation contributes too little to improve the precision of baseline vector. Thus, once the ambiguity vector is solved, we can use only the carrier phase to estimate the baseline vector. In this case, the conditional variance-covariance matrix can be written as follows:

$$Q_{\hat{x}(a)} = 2\lambda_i^2\sigma_{\phi,i}^2(E^T TE)^{-1} \tag{32}$$

Eq. (32) implies that the receivers with a smaller noise level can obtain a higher accuracy of baseline vector. The least squares solution is given by

$$\hat{b}(a) = \lambda_i (E^T TE)^{-1} E^T D(DD^T)^{-1}(y_{i,D}^{\phi} + N_{i,D}^{\phi}) \tag{33}$$

Assuming that n GPS frequencies are used to estimate the baseline vector, the standard GNSS model can also be obtained if the following expressions are considered:

$$Y = \begin{bmatrix} Y_1 \\ \vdots \\ Y_n \end{bmatrix}, B = \begin{bmatrix} B_1 \\ \vdots \\ B_n \end{bmatrix}, A = \begin{bmatrix} A_0 & & \\ & \ddots & \\ & & A_0 \end{bmatrix}, Q_Y = \begin{bmatrix} Q_{Y_1} & & \\ & \ddots & \\ & & Q_{Y_n} \end{bmatrix} \tag{34}$$

Once the ambiguity resolution is achieved, the conditional variance-covariance matrix is given by

$$Q_{\hat{b}(a)} = (\sum_{i=1}^{n} \omega_i (H_i^T Q^{-1} H_i)^{-1} = (\sum_{i=1}^{n} \lambda_i^{-2} \omega_i)^{-1} (E^T TE)^{-1} \tag{35}$$

If only the carrier phase observables of the n GPS bands are considered, the conditional variance-covariance matrix is obtained as follows:

$$Q_{\hat{b}(a)} = (\sum_{i=1}^{n} \frac{1}{2\sigma_{\phi,i}^2} (H_i^T Q^{-1} H_i))^{-1} = (\sum_{i=1}^{n} \frac{1}{\lambda_i^2} \frac{1}{2\sigma_{\phi,i}^2})^{-1} (E^T TE)^{-1} \tag{37}$$

This implies that more frequencies can reduce the conditional variance-covariance matrix, since the coefficient is smaller for more frequencies. After a simple derivation, we also obtain the following least squares solution and its equivalent expressions:

$$\hat{b}(a) = Q_{\hat{b}(a)} \sum_{i=1}^{n} \frac{1}{2\sigma_{\phi,i}^2} (H_i^T Q^{-1} (y_{i,D}^{\phi} + N_{i,D}^{\phi})) = (E^T TE)^{-1} E^T T (\sum_{i=1}^{n} \frac{1}{\lambda_i^2} \frac{1}{2\sigma_{\phi,i}^2})^{-1} \times \sum_{i=1}^{n} \frac{1}{\lambda_i^2} \frac{1}{2\sigma_{\phi,i}^2} (y_{i,S}^{\phi} + N_{i,S}^{\phi}) \tag{38}$$

Note that in our description of the second equivalent expansion, we have explicitly shown the presence of the single-differenced ambiguity vector. The reason for doing so lies in the fact that it will aid us in constructing the canonical forms involving the operator T.

Assuming that the noise level of carrier phase measurement is equal for each GPS band, thus we have the following equation for all three GPS bands (L_1, L_2 and L_5 bands)

$$\sigma_{\phi,1}^2 = \sigma_{\phi,2}^2 = \sigma_{\phi,3}^2 \tag{39}$$

With (37)

$$Q_{\hat{b}(a)} = 2\sigma_{\phi,i}^2 (\sum_{i=1}^{n} \frac{1}{\lambda_i^2})(E^T TE)^{-1} \tag{40}$$

With (38), the equivalent expression of least squares solution is given by

$$\hat{b}(a) = (E^T TE)^{-1} E^T D (DD^T)^{-1} (\sum_{i=1}^{n} \frac{1}{\lambda_i^2})^{-1} \times \sum_{i=1}^{n} \frac{1}{\lambda_i} (y_{i,D}^{\phi} + N_{i,D}^{\phi}) \tag{41}$$

4.1.3 Optimal linear combinations vs. traditional triple frequency least squares

There are many kinds of linear combination expressions for triple frequency GPS signals, some of which can reduce the error of the baseline vector. Assuming that k_i denotes the coefficient of frequency f_i, thus the following equation can be obtained:

$$k_i y_{i,D}^{\phi} + k_i N_{i,D}^{\phi} = k_i H_i \cdot b + k_i v_{i,D}^{\phi} \tag{42}$$

All the matrix and vector of linear combination observation can be given by

$$\begin{aligned} y_{*,D}^{\phi} &= \sum_{i=1}^{n} k_i y_{i,D}^{\phi}, N_{*,D} = \sum_{i=1}^{n} k_i N_{i,D}, v_{*,D}^{\phi} \\ &= \sum_{i=1}^{n} k_i v_{i,D}^{\phi}, H_* = \sum_{i=1}^{n} k_i H_i = \sum_{i=1}^{n} (\frac{k_i}{\lambda_i} DE) \\ &= \sum_{i=1}^{n} (\frac{k_i}{\lambda_i}) DE = \frac{1}{\lambda_*} DE \end{aligned} \tag{43}$$

where the wavelength of this combined signal λ^* is

$$\lambda_* = \frac{1}{\frac{k_1}{\lambda_1} + \frac{k_2}{\lambda_2} + \frac{k_3}{\lambda_3}} = \frac{c}{\frac{k_1 c}{\lambda_1} + \frac{k_2 c}{\lambda_2} + \frac{k_3 c}{\lambda_3}} = \frac{c}{k_1 f_1 + k_2 f_2 + k_3 f_3} = \frac{c}{f_*} \tag{44}$$

Thus, with (32), the least squares variance-covariance matrix can be written as

$$\sigma_{\phi,*}^2 = (k_1^2 + k_2^2 + k_3^2) \sigma_{\phi,1}^2 \tag{45}$$

$$Q_{\hat{b}(a)} = 2\lambda_*^2 \sigma_{\phi,*}^2 (E^T TE)^{-1} \tag{46}$$

where $\lambda_*^2 \sigma_{\phi,*}^2 = (k_1^2 + k_2^2 + k_3^2) \sigma_{\phi,1}^2 / (k_1 \lambda_1 + k_2 \lambda_2 + k_3 \lambda_3)^2$.

With (33), the least squares solution is given by

$$\hat{b}(a) = \lambda_* (E^T TE)^{-1} E^T D (DD^T)^{-1} (y_{*,D}^{\phi} + N_{*,D}^{\phi}) \tag{47}$$

Next the error analysis lf the specific linear combination approach will be performed. If $\sigma_{n,1}^2$ is assumed as the variance due to thermal noise on the L_1 signal in meterssquared, thus $\sigma_{\phi,1} = \sigma_{n,1}/\lambda_1$ can be deduced. With the assumption (39), the measurment precision of linear combination observables can be expressed in units of cycles as follows:

$$\sigma_*^{\phi} = \sqrt{k_1^2 + k_2^2 + k_3^2} (\frac{\sigma_{n,1}}{\lambda_1}) \tag{48}$$

Some linear combinations can greatly reduce the effect of noise when parameterized in length units. The following equation shows the amplification factor in length units for the noise component of a combined signal:

$$\sigma_*^2 [m^2] = (\lambda_* \sqrt{k_1^2 + k_2^2 + k_3^2} (\frac{\sigma_{n,1}}{\lambda_1}))^2 = \{\frac{\lambda_*^2}{\lambda_1^2} (k_1^2 + k_2^2 + k_3^2)\} \sigma_{n,1}^2 [m^2] \tag{49}$$

This term in the curly brackets is the amplification factor to be minimized. However, it has been proven in Richert and El-Sheimy (2007) that there is an absolute minimum value for this amplification factor. This minimum is 0.462 for triple frequency GPS. The combination coefficients that correspond to these minimum values satisfy the following equation:

$$\begin{bmatrix} k_1 \\ k_2 \\ k_3 \end{bmatrix} = \begin{bmatrix} f_1 \\ f_2 \\ f_3 \end{bmatrix} t = \begin{bmatrix} 154 \\ 120 \\ 115 \end{bmatrix} t', t \neq 0, t' \neq 0 \tag{50}$$

With (44), the least squares variance-covariance matrix can be obtained for the optimal linear combination. In fact, the baseline precision of this optimal combined signal is equal to the one using triple GPS bands. In other words, if the combination coefficients are chosen as (50), the result of (46) will be consistent with (40) under the assumption (39). Thus, the following equation must be satisfied:

$$\frac{1}{\left(\frac{k_1}{\lambda_1}+\frac{k_2}{\lambda_2}+\frac{k_3}{\lambda_3}\right)^2}(k_1^2+k_2^2+k_3^2) = \frac{1}{\frac{1}{\lambda_1^2}+\frac{1}{\lambda_2^2}+\frac{1}{\lambda_3^2}} \tag{51}$$

The equivalent expression can be written as

$$(f_1^2+f_2^2+f_3^2)\left(\frac{1}{\lambda_1^2}+\frac{1}{\lambda_2^2}+\frac{1}{\lambda_3^2}\right) = \left(\frac{f_1}{\lambda_1^2}+\frac{f_2}{\lambda_2^2}+\frac{f_3}{\lambda_3^2}\right)^2 \tag{52}$$

The detail proof is given as follows:

$$\begin{aligned}(f_1^2+f_2^2+f_3^2)\left(\frac{1}{\lambda_1^2}+\frac{1}{\lambda_2^2}+\frac{1}{\lambda_3^2}\right) &= (f_1^2+f_2^2+f_3^2)\left(\frac{f_1^2}{\lambda_1^2 f_1^2}+\frac{f_2^2}{\lambda_2^2 f_2^2}+\frac{f_3^2}{\lambda_3^2 f_3^2}\right) \\ &= \frac{1}{c^2}(f_1^2+f_2^2+f_3^2)^2 = \left(\frac{f_1 f_1}{\lambda_1 f_1}+\frac{f_2 f_2}{\lambda_2 f_1}+\frac{f_3 f_3}{\lambda_3 f_1}\right)^2 \\ &= \left(\frac{f_1}{\lambda_1}+\frac{f_2}{\lambda_2}+\frac{f_3}{\lambda_3}\right)^2\end{aligned} \tag{53}$$

With the assumption (39), the conditional least squares baseline vector from (47) is also equal to (41). This indicates that the following equations are satisfied:

$$\begin{aligned}\lambda_* y_{*,D}^{\phi} &= \left(\sum_{i=1}^{n}\frac{1}{\lambda_i^2}\right)^{-1}\sum_{i=1}^{n}\frac{1}{\lambda_i}y_{i,D}^{\phi} \\ \lambda_* N_{*,D}^{\phi} &= \left(\sum_{i=1}^{n}\frac{1}{\lambda_i^2}\right)^{-1}\sum_{i=1}^{n}\frac{1}{\lambda_i}N_{*,D}^{\phi}\end{aligned} \tag{54}$$

The detail proof of the first equation is given as follows:

$$\begin{aligned}\lambda_* y_{*,D}^{\phi} &= \frac{1}{\left(\frac{k_1}{\lambda_1}+\frac{k_2}{\lambda_2}+\frac{k_3}{\lambda_3}\right)^2}(k_1 y_{1,D}^{\phi}+k_2 y_{2,D}^{\phi}+k_3 y_{3,D}^{\phi}) \\ &= \left(\sum_{n=1}^{3}\frac{f_i}{\lambda_i}\right)^{-1}\sum_{n=1}^{3} f_i y_{i,D}^{\phi} \\ &= \left(\sum_{n=1}^{3}\frac{c}{\lambda_i \lambda_i}\right)^{-1}\sum_{n=1}^{3}\frac{c}{\lambda_i}y_{i,D}^{\phi} \\ &= \left(\sum_{n=1}^{3}\frac{1}{\lambda_i^2}\right)^{-1}\sum_{n=1}^{3}\frac{1}{\lambda_i}y_{i,D}^{\phi}\end{aligned} \tag{55}$$

In the same way, we can also prove the other equation. Therefore, the optimal linear combination gives the same result as the traditional least-squares result. It is inferred that the explicit usage of this specific combination is not needed as it does not bring anything extra.

4.2 The Accuracy Assessment of the System with a Common Clock (and without any Significant Linebias)

4.2.1 Single difference model vs. double difference model

If the two receivers of GPS compass feed a common clock and the cables have the same delay between two receivers, the clock biases between two receivers will be zero and can be omitted (Wang et al., 2007). Thus, the clock bias term can be eliminated from the single-differenced carrier phase and the single-differenced code observation equations as follows: (56)

$$y_{i,S}^{\phi}=\frac{1}{\lambda_i}E\cdot b-N_{i,S}+v_{i,S}^{\phi}, v_{i,S}^{\phi}\sim N(0,2\sigma_{\phi,i}^2 I_m) \tag{56}$$

$$y_{i,S}^{\rho}=\frac{1}{\lambda_i}E\cdot b+v_{i,S}^{\rho}, v_{i,S}^{\rho}\sim N(0,2\sigma_{\rho,i}^2 I_m) \tag{57}$$

Combining (56) and (57), the single difference model of this scheme can be expressed as

$$\begin{bmatrix} y_{i,S}^{\phi} \\ y_{i,S}^{\rho} \end{bmatrix}=\begin{bmatrix} H_i & -I \\ H_i & 0 \end{bmatrix}\begin{bmatrix} b \\ N_{i,S} \end{bmatrix}+\begin{bmatrix} v_{i,S}^{\phi} \\ v_{i,S}^{\rho} \end{bmatrix} \tag{58}$$

where $H_i=\frac{1}{\lambda_1}E$. Note that in this case the double difference model is the same as that of the non-common clock scheme.

The important point is that the single and double difference models are not equivalent, i.e. the redundancy is lower for the double difference model. In fact, since the clock offsets can be eliminated by the single difference operation, the double difference operation is meaningless and unnecessary in this case. However, compared to the double difference model of the non-common clock scheme, an improved performance is as expected, due to the higher redundancy in this single difference model. Hence, the accuracy improvement can be performed by choosing the common clock scheme, see also Section 5. In addition, in order to employ the single difference model accurately for this scheme, no significant linebias must be guaranteed for the system. Since linebias is caused by the uneven time delays when the electric signals travel from the antenna to the receiver. By making the antenna cables equal in length this disturbance is minimized.

4.2.2 Single frequency model vs. multiple frequency model

Since there is no double difference operation, the conditional variance-covariance matrix of L_i band is given by

$$Q_{\hat{b}(a)}=(\frac{1}{\lambda_i^2}\frac{1}{2\sigma_{\phi,i}^2})^{-1}(E^TE)^{-1} \tag{59}$$

The conditional variance-covariance matrix of n GPS frequencies can be written as

$$Q_{\hat{b}(a)}=(\sum_{i=1}^{N}\frac{1}{\lambda_i^2}\frac{1}{2\sigma_{\phi,i}^2})^{-1}(E^TE)^{-1} \tag{60}$$

Note that the coefficient of (60) is equal to that of (37), but the DD matrix operator is removed in the matrix due to the unrelated noise term of single-differenced model. In the same way, there is no DD matrix operator in the expression of the least squares solution:

$$\hat{b}(a) = (E^T E)^{-1} E^T (\sum_{i=1}^{n} \frac{1}{\lambda_i^2})^{-1} \sum_{i=1}^{n} \frac{1}{\lambda_i} (y_{i,S}^{\phi} + N_{i,S}^{\phi}) \tag{61}$$

4.2.3 Optimal linear combinations vs. traditional triple-frequency least squares

In the same way as (46) and (47), the least squares variance-covariance matrix and baseline vector can be written as

$$Q_{\hat{b}(a)} = 2\lambda_*^2 \sigma_{\phi,*}^2 (E^T E)^{-1} \tag{62}$$

$$\hat{b}(a) = \lambda_* (E^T E)^{-1} E^T (y_{i,S}^{\phi} + N_{i,S}^{\phi}) \tag{63}$$

Eq. (62) is equal to (60) due to the same proof (51)—Eq. (53). In the same way as (55), we can also prove the following equations:

$$\lambda_* y_{*,S}^{\phi} = (\sum_{i=1}^{n} 1 \frac{}{\lambda_i^2})^{-1} \frac{1}{\lambda_i} y_{i,S}^{\phi} \text{ and}$$

$$\lambda_* N_{*,S}^{\phi} = (\sum_{i=1}^{n} 1 \frac{}{\lambda_i^2})^{-1} \frac{1}{\lambda_i} N_{i,S}^{\phi} \tag{64}$$

Thus, the equivalence of both approaches still holds with the assumption (39).

5. The Accuracy Improvement for Single Epoch Attitude Determination

5.1 Accuracy Improvement with Triple Frequency

GPS modernization will expand the area of positioning service and will increase its convenience. Particularly, the third civil frequency namely L5 is expected to improve the performance of precise positioning using carrier phase observation. With triple frequency signals, not only the success rates of ambiguity resolution can be improved greatly, but also the accuracy of attitude determination for single epoch will increase. Using (32) and (37), the attitude-error ratio of single frequency L_1 and triple frequency can be defined as follows:

$$\eta = \frac{2\lambda_1^2 \sigma^2{}_{\phi,1}}{(\sum_{i=1}^{3} \frac{1}{\lambda_1^2} \frac{1}{2\sigma_{\phi,i}^2})^{-1}} = 1 + \frac{\lambda_1^2 \sigma_{\phi,1}^2}{\lambda_2^2 \sigma_{\phi,2}^2} + \frac{\lambda_1^2 \sigma_{\phi,1}^2}{\lambda_3^2 \sigma_{\phi,3}^2} \tag{65}$$

Since $\lambda_1^2/\lambda_2^2 = f_1^2/f_2^2$ and $\lambda_1^2/\lambda_5^2 = f_5^2/f_1^2$, thus the attitude-error ratio is given by

$$\eta = 1 + (\frac{120}{154})^2 \frac{\sigma_{\phi,1}^2}{\sigma_{\phi,2}^2} + (\frac{115}{154})^2 \frac{\sigma_{\phi,1}^2}{\sigma_{\phi,3}^2} \tag{66}$$

With the assumption (39), it is easy to obtain the attitude-error ratio as 2.1648 and the error is reduced to 0.462 of single frequency case. In other words, if triple-frequency observation is used to estimate the baseline vector, the standard variation of each baseline component and attitude angle can be reduced to 0.68 of single frequency case. This theoretical value is only determined by the number of frequencies and the wavelength of each frequency. For the dual-frequency case, the standard variation of each baseline component and attitude angle can be reduced to 0.7888 of single frequency case.

5.2 Accuracy Improvement with System Configuration

Note that the measurement precision of baseline vector may be different for different system configuration, see also (37) and (60). With (24), regardless of the coefficient, the matrix part of (37) can be expressed as

$$(E^T T E)^{-1} = (E^T E - (\frac{1}{\sqrt{m}} E^T e_m)(\frac{1}{\sqrt{m}} E^T e_m)^T)^{-1} \tag{67}$$

In order to expand this inverse matrix, we can apply the well-known matrix inversion Lemma (Abadir and Magnus, 2005):

$$(M - vv^T)^{-1} = M^{-1} + \frac{1}{1 - v^T M^{-1} v} M^{-1} v v^T M^{-1} \tag{68}$$

where $v^T M^{-1} v \neq 1$ and M is nonsingular matrix of order m and v is a vector of order m. For the symmetric matrix M, this Lemma has the following expression:

$$(M - vv^T)^{-1} = M^{-1} + \frac{1}{1 - v^T M^{-1} v} (M^{-1} v)(M^{-1} v)^T \tag{69}$$

Eq. (67) has the same structure as $(M - vv^T)^{-1}$ if each term is expressed as follows:

$$M = E^T E, v = \frac{1}{\sqrt{m}} E^T e_m \tag{70}$$

Since $E^T E$ is symmetric, we can obtain the following expanded expression by Lemma (69):

$$(E^T T E)^{-1} = (E^T E)^{-1} + \frac{1}{m - t} P P^T \tag{71}$$

where $p = (E^T E)^{-1} E^T e_m$ and $t = (e_m)^T E (E^T E)^{-1} E^T e_m$. Note that $E(E^T E)^{-1} E^T$ can be seen as the orthogonal projector P_E, which projects orthogonally onto the range of E:

$$P_E = E(E^T E)^{-1} E^T \tag{72}$$

It can be verified that

$$t = (P_E e_m)^T (P_E e_m) \tag{73}$$

Let us consider the problem from a geometric point of view. Since the length projected by the vector e_m is smaller than that of e_m, thus we can obtain the following inequality for $m > 3$:

$$0 < t = \| P_E e_m \|^2 < \| e_m \|^2 = m \tag{74}$$

Define $p \equiv (p_E \quad p_N \quad p_U)^T$ and we can obtain

$$pp^T = \begin{pmatrix} p_E^2 & p_E p_N & p_E p_U \\ p_N p_E & p_N^2 & p_N p_U \\ p_U p_E & p_U p_N & p_U^2 \end{pmatrix} \tag{75}$$

With (74) and (75), each diagonal element of $(E^T T E)^{-1}$ is larger than that of $(E^T E)^{-1}$

$$(E^T T E)_{ii}^{-1} > (E^T E)^{-1} \tag{76}$$

Thus, the baseline vector error of the non-common clock scheme is larger than that of the single difference model with the common clock scheme. The second term on the right side of (71) implies the gap between both schemes. In this contribution, the product of the diagonal elements and the coefficient are defined as the accuracy improvement indicator, which are written as

$$\tau_E = \frac{1}{m - t} p_E^2, \tau_N = \frac{1}{m - t} p_N^2, \tau_U = \frac{1}{m - t} p_U^2 \tag{77}$$

Next we will evaluate all the elements of pp^T. The vector p can be seen as the product of $(E^TE)^{-1}$ and E^Te_m, and both matrices are determined by the satellite geometry. If the unit vector heading to the GPS satellite i is expressed as $s^i=(s_E^i \quad s_N^i \quad s_U^i)^T$, the transpose of satellite geometry matrix E can be written as $E^T=[s^1 \quad s^2 \quad \cdots \quad s^m]$, and then we can obtain

$$E^TE = \sum_{i=1}^{m}\begin{pmatrix} s_E^i \\ s_N^i \\ s_U^i \end{pmatrix}(s_E^i \quad s_N^i \quad s_U^i)$$

$$= \begin{pmatrix} \sum_i^m (s_E^i)^2 & \sum_i^m s_E^i s_N^i & \sum_i^m s_E^i s_U^i \\ \sum_i^m s_N^i s_E^i & \sum_i^m (s_N^i)^2 & \sum_i^m s_N^i s_U^i \\ \sum_i^m s_U^i s_E^i & \sum_i^m s_U^i s_N^i & \sum_i^m (s_U^i)^2 \end{pmatrix} \tag{78}$$

However, due to the geometry of GPS satellites in the sky, s_E^i; s_N^i and s_U^i are roughly unrelated for different satellites, thus $\sum_i^m s_E^i s_N^i$, $\sum_i^m s_E^i s_U^i$ and $\sum_i^m s_N^i s_U^i$ are all close to zero for m visible satellites, especially m is very large. This implies that EE^T is approximately diagonal and the nondiagonal terms of its inverse matrix are very small. The other part of the product can be written as

$$E^Te_m = \left(\sum_i^m s_E^i \quad \sum_i^m s_N^i \quad \sum_i^m s_U^i\right)^T \tag{79}$$

Since a satellite with negative elevation is normally not received, the up component s_U^i is larger than zero for each visible satellite. However, the east and north components may be positive or negative, depending on the distribution of the visible satellites. Hence, $\sum_i^m s_E^i$ and $\sum_i^m s_N^i$ are generally very small and the following inequalities are almost correct unless most satellites get together in a certain direction:

$$\sum_i^m s_U^i \gg \sum_i^m s_E^i \quad \sum_i^m s_U^i \gg \sum_i^m s_N^i \tag{80}$$

Because the non-diagonal elements of $(E^TE)^{-1}$ is very small and the east and north components of E^Te_m are much smaller than the up component, we can obtain

$$p_U^2 > p_E^2 > 0 \quad p_U^2 > p_N^2 > 0 \tag{81}$$

This indicates that the gap of up-error between both schemes is much larger than that of the east-error and north-error. It can be inferred that if the geometry of satellites is approximately symmetrical, p_E and p_N are close to zero, thereby resulting in less improvement for the east-error and north-error.

5.3 Accuracy Improvement by Combing the Multi-Frequency Observation and the Single Difference Model with the Common Clock Sscheme

The multi-frequency scheme and the single difference model with the common scheme can both improve the accuracy of GPS compass, although the improvements of both schemes have different characteristics. The multifrequency scheme can diminish the error of the baseline vector by reducing the coefficient of (37) and the single difference model with the common scheme tries to improve the accuracy of the baseline coordinates by reducing the diagonal elements of the matrix part of (37). As their influences on accuracy are independent of each other, the accuracy of GPS compass should be higher by combining both approaches. Eq. (60) gives the accuracy of the combined scheme.

6. Experiments

The accuracy of GPS compass has been evaluated for the single-frequency and multiple-frequency cases, using the double difference model for the non-common clock scheme and the single difference model for the common clock scheme, respectively. The integer ambiguity estimation has been achieved using the C-LAMBDA method. After the ambiguity is fixed, the average and standard deviation of attitude angle and each component of the baseline vector are analyzed to evaluate the accuracy of different scenarios. Three simulated or actual experiments are given for verifying the accuracy improvement approaches presented.

6.1 Accuracy Assessment of Multi-frequency GPS Compass

In this section, we will introduce a simulation method proposed by Teunissen et al. (2010a) for generating multi-frequency GNSS data to evaluate the accuracy of GPS compass. In this method, the GNSS design matrices and observables needed for the simulations were constructed by means of the VISUAL software (Verhagen, 2006) using the assumed receiver locations and the actual GPS constellation as input. The different clock offsets of the receivers are considered in this simulation experiment, in order to simulate the non-common clock scheme. The baseline length given by the two assumed receiver locations is 1 m, and the baseline vector is $(1\ 0\ 0)^T$ with respect to the local East-North-Up frame. This denotes that the true heading is 90° and the elevation is 0°. We considered three scenarios: L1, L1/L2 combination and L1\L2\L5 combination. For the stochastic model, we assume the noise levels of all frequencies being 0.02 cycles for the undifferenced phase data, and this value is in the general case. For each simulation scenario, a set of 10,000 Gaussian distributed data vectors was generated with nine satellites. The double difference model is utilized in this experiment.

The average and standard deviation of attitude angle measurements are demonstrated in Table 1 for L1, L1\L2, L1\L2\L5, respectively. The average and standard deviation of east, north and up measurements of the baseline vector are demonstrated in Table 2. Comparing with the L1 case, L1\L2 and L1\L2\L5 combination can both improve the accuracy of heading and elevation of GPS compass. The L1\L2\L5 combination provides a higher accuracy than L1\L2 case. The reason is that more frequencies can reduce the conditional variance-covariance matrix of baseline vector, see also the coefficient of (37). The improvements of angular measurements are approximately the same, as well as each component of the baseline vector. Because the coefficient of (37) is common for all the baseline components, once it is reduced, all the elements of this matrix are reduced to the same extent.

Table 1 Mean values and standard deviations of the single epoch derived, ambiguity resolved, (constant) attitude angles using simulated data with the non-common clock scheme (1 m baseline and nine tracked satellites)

Methods	Average(degree)		Standard deviation (degree)	
	Heading	Elevation	Heading	Elevation
L1	89.9990	−0.0055	0.1956	0.4076
L1\L2	89.9994	0.0013	0.1559	0.3234
L1\L2\L5	90.0023	−0.0028	0.1353	0.2781

Table 2 Mean values and standard deviations of the single epoch derived, ambiguity resolved, (constant) baseline components using simulated data with the non-common clock scheme (1 m baseline and nine tracked satellites)

Methods	Average(m)			Standard deviation (mm)		
	b_E	b_N	b_U	b_E	b_N	b_U
L1	0.99996	1.8977e−5	−8.9861e−5	2.9941	3.4135	7.1150
L1\L2	1.00000	1.1410e−5	2.6764e−5	2.3453	2.7208	5.6446
L1\L2\L5	0.99999	−3.9786e−5	−4.5961e−5	2.0292	2.3608	4.8543

As is shown in Table 1, the accuracy of heading is better than that of elevation. This is generally true for the baseline that approximately lies in the plane of the local geodetic horizon. Because in that case the sine value of elevation in (13) is close to zero and the elevation error is mainly affected by σ_U, which is very large in the double difference model. However, the heading error is nothing to do with the altitude error σ_U, see also (12).

With triple-frequency GPS signals, the standard deviation of heading is reduce to 0.6917 of L1 case and the standard deviation of elevation is reduce to 0.6823 of L1 case, which are both very close to the theoretical value 0.68, see also the analysis below (66). With the assumption (39), the theoretical value is only determined by the number of frequencies and the wavelength of each frequency. We can also see that the standard deviation of heading and elevation is reduced to 0.7970 and 0.7934 of L1 case, respectively, which are both close to the theoretical value 0.7888. This demonstrates that more frequencies can obtain the higher accuracy.

6.2 Accuracy Improvement Using the Single Difference Model with the Common Clock Scheme

In order to verify the accuracy improvement of GPS compass attitude determination using the common clock scheme of the receivers, the actual single-frequency GPS compass experiment was achieved. In order to compare with the accuracy of simulation experiment, the 1 m baseline was also chosen for evaluating the practical accuracy. In the actual experiment, the baseline was placed in the plane of local geodetic horizon approximately, pointing to the northeast. In order to avoid the affect of multipath, data was collected on the roof. The GPS compass system was constructed with a set of low cost components and GPS software-defined receivers (SDR) developed by Beihang University.

The GPS software receivers have been developed for either post-process stored signals or operated in real-time. The detail SDR design in PC platform is given in Jin et al. (2007) and the design over stand-alone DSP platform is demonstrated in Jin et al. (2009a). For comparison purposes, the non-common clock scheme and the common clock scheme are achieved at the same time, with the configuration schematically shown in Fig. 1. For the noncommon clock scheme, two single-antenna GPS front-end modules are used for collecting data, indicating that the receivers have different clock offsets. For the common clock scheme, the hardware platform uses two front ends that feeding to a common clock to eliminate the clock offsets and makes the antenna cables equal in length to minimize the linebias. Similar design is also developed for some GPS attitude determination systems, without the theoretical analysis of the accuracy of attitude angle (Wang et al., 2007; Jin et al., 2009b). Note that the power divider is used to split the signal from the antenna into two branches equally. Thus, the same short baseline and the same satellite geometry can be guaranteed for both schemes. The standard deviation of L1 carrier phase observation is assumed as 0.018 cycles, which is the statistical result of zero baseline observation for about two hours, and the assessment of the observation noise with zero baseline tests is discussed in Bona and Tiberius (2000).

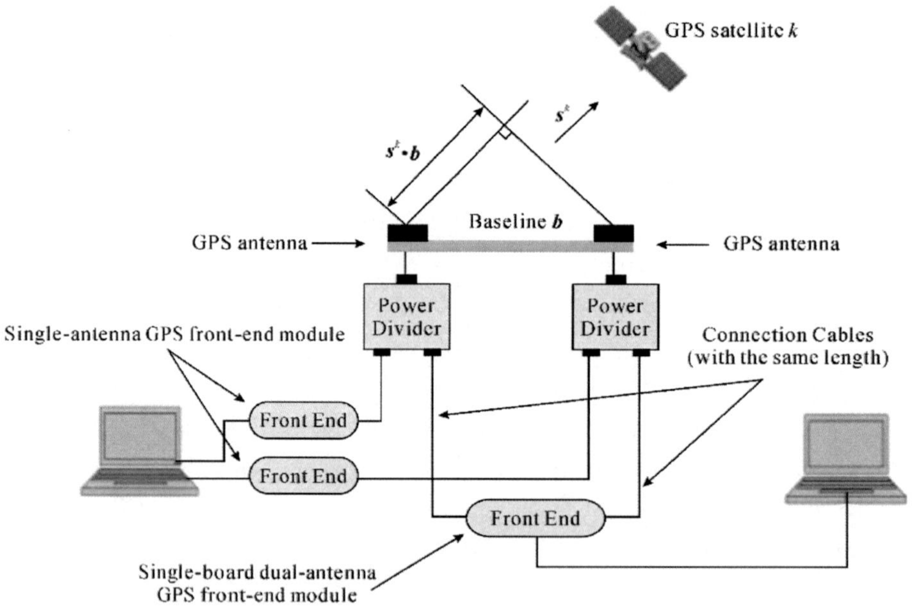

Fig. 1 GPS compass configuration for the non-common clock scheme and the common clock scheme

During about half an hour observation, the number of available satellites equals eight most of the time with a few drops to seven or six satellites. After removing the epochs that failed in the ambiguity resolution, we collected 1700 s data for the accuracy assessment. The heading and elevation using the double difference model with the noncommon clock scheme are given in Fig. 2, and the heading and elevation using the single difference model with the common clock scheme are given in Fig. 3. The average and standard deviation of attitude angle measurements and each component of the baseline vector are given in Table 3 and Table 4, respectively.

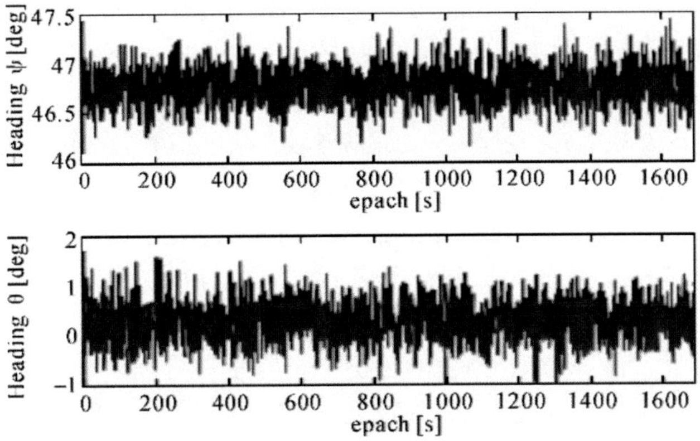

Fig. 2 Heading and elevation using the double difference model with the non-common clock scheme of the receivers

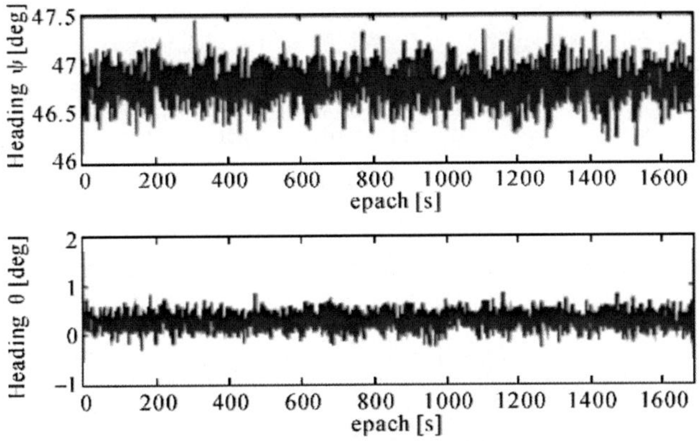

Fig. 3 Heading and elevation using the single difference model with the common clock scheme of the receivers

Table 3 Mean values and standard deviations of the single epoch derived, ambiguity resolved, (constant) attitude angles using actual L1 GPS compass data (1 m baseline)

Methods	Average(degree)		Standard deviation (degree)	
	Heading	Elevation	Heading	Elevation
DD/Non-common clock	46.7643	0.2772	0.19479	0.41902
SD/Common clock	46.7750	0.2769	0.18757	0.18364

Table 4 Mean values and standard deviations of the single epoch derived, ambiguity resolved, (constant) baseline components using actual L1 GPS data (1 m baseline)

Methods	Average(m)			Standard deviation (mm)		
	b_E	b_N	b_U	b_E	b_N	b_U
DD/Non-common clock	0.72846	0.68492	0.00485	3.7315	3.4956	7.3181
SD/Common clock	0.72863	0.68482	0.00483	3.4043	3.4913	3.2029

With (13), if the baseline approximately lies in the plane of the local geodetic horizon, the elevation error mainly depends on the altitude error b_U, which is much larger in the non-common clock scheme, no matter the single difference model or the double difference model is used, see also (23) and (29) and (71)—(81). Due to the enormous reduction of the altitude error, the accuracy of elevation is improved strongly by the single difference method with the common clock scheme of the receivers. However, for the accuracy of heading, the improvement is not obvious since the east and north components of baseline vector are improved too less. This is determined by the geometry of satellites, of which the east and north distribution is close to symmetry and the up components of almost all the satellites are positive with respect to the user position. The constellation of satellites in this experiment is shown in Fig. 4, and each satellite is discernible by its PRN number. The number of visible satellites is given in Fig. 5. Since the satellites in the north sky (PRN 12/20/4/23) correspond to the satellites in the south sky (PRN 28/10/13/17), it can be inferred that the sum of north components of all the visible satellites is close to zero. However, the sum of east components of all the visible satellites is larger than that of north components. The reason is that three pairs of satellites (PRN 4 and 17; PRN 10 and 28; PRN 12 and 20) are symmetrical and one pair is non-symmetrical in the east direction (PRN 23 and 13). This results in the less improvement of north-error than the east-error with the single difference model. Since the elevation of all the satellites are positive and not very low, the sum of all the up components is much larger than zero, thereby resulting in the tremendous improvement of elevation accuracy. The accuracy improvement indicator of each component of the baseline vector is given in the Fig. 6, which demonstrates that the improvement of up component is much larger than the north and east component. The sudden changes in curvature indicate that the number satellites change during that time.

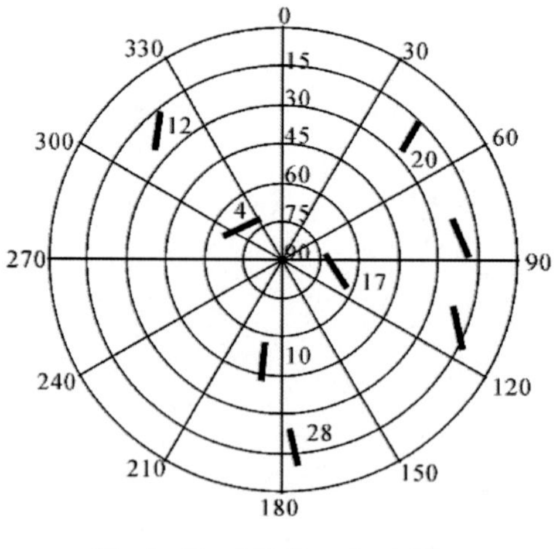

Fig. 4 Constellation of satellites

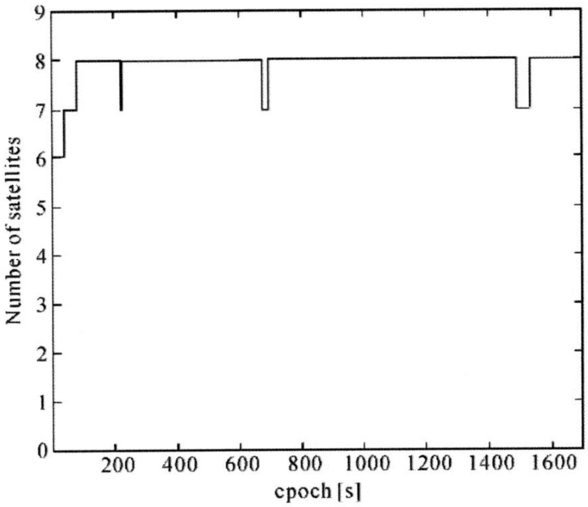

Fig. 5 Number of satellites

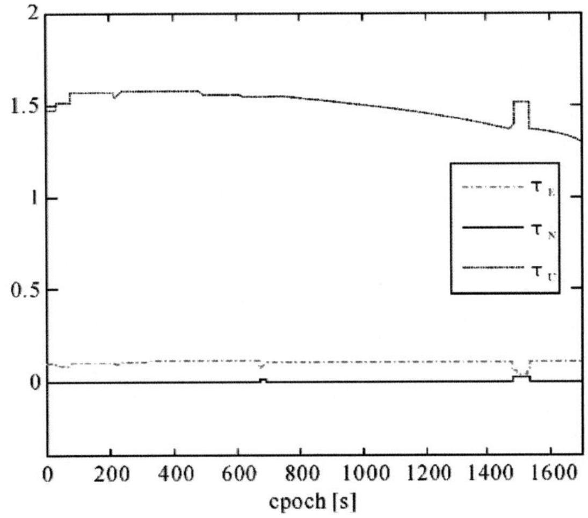

Fig. 6 Comparison of the accuracy improvement indicator

6.3 Accuracy Assessment of the Scheme that Combing the Multi-Frequency Observation and the Single Difference Model with a Common Clock

As is described in Section 5.3, the accuracy of GPS compass can be further improved by combining the multi-frequency scheme and the single difference model with the common clock scheme, see also the analysis of Section 5.3. Based on the experiment in Section 6.1, we generate new data that with the same clock offsets of the receivers, in order to simulate the common-clock scheme. As shown in Table 5, the average and standard deviation of attitude angle measurements of various methods are given. This indicates the L1\L2\L5 single difference model with a common clock can obtain the optimal accuracy compared to the other methods listed in the table. The accuracy of elevation is strongly improved compared to that of L1 DD method.

All the experiments above are achieved with the one-meter baseline, the accuracy of heading and elevation can be further improved for the longer baseline, since the baseline length is inversely proportional to the error of heading and elevation, see also (12) and (13). Based on the analysis above, on the accuracy improvement of GPS compass, we can consider four approaches as follows: (a) use a longer baseline; (b) use the receivers with a smaller noise level; (c) use more frequency observables if possible; (d) use the single difference model if the receivers feed a common clock.

Table 5 Mean values and standard deviations of the single epoch derived, ambiguity resolved, (constant) attitude angles using simulated GPS compass data with the non-common clock and the common clock schemes (1 m baseline)

Methods	Average(degree)		Standard deviation (degree)	
	Heading	Elevation	Heading	Elevation
L1 DD/ Non-common clock	89.9990	−0.0055	0.1956	0.4076
L1 SD/ Common clock	89.9991	−0.0008	0.1925	0.1759
L1\L2 DD/ Non-common clock	89.9994	0.0013	0.1559	0.3234
L1\L2 SD/ Common clock	89.9995	−0.0011	0.1532	0.1400
L1\L2\L5 DD/ Non-common clock	90.0023	−0.0028	0.1353	0.2781
L1\L2\L5 SD/ Common clock	90.0018	−0.0016	0.1331	0.1204

7. Conclusions

GPS compass is an important technology for providing accurate heading and elevation estimations of vehicles equipped with two antennas. Past researches make more efforts to achieve high success rate ambiguity resolution for the single epoch GPS compass model. This contribution focuses on the accuracy assessment of GPS compass and the difference in accuracy for a number of different available models. We analyze the precision of attitude estimation using the multi-frequency observation and using single or double difference model with the common clock scheme or the non-common clock scheme. If all noise levels of triple frequency carrier phase are assumed the same in unit of cycles, the resulting standard variation of each baseline component and each attitude angle can be approximately reduced to 0.68 of single frequency case. We also prove that the optimal linear combination method is equivalent to the traditional triple-frequency least squares method in this case. Compared with the single or double difference model of the non-common clock scheme, the single difference model of the common clock scheme can greatly reduce the elevation error of the single epoch GPS compass. In order to employ this scheme, a common clock is required and the same length should be guaranteed for cables.

References

[1] Abadir, K. M. , Magnus, J. R. Matrix Algebra. Cambridge Univ. Press, 2005.

[2] Bona, P. , Tiberius, C. An experimental comparison of noise characteristics of seven high-end dual frequency GPS receiver-sets, in: Proceedings of the IEEE Position Location and Navigation Symposium, San Diego, California, March 13-16, pp. 237-244, 2000.

[3] Buist, P. J. The baseline constrained LAMBDA method for single epoch, single frequency attitude determination applications, in: Proceedings of ION GPS, Fort Worth, TX, USA, pp. 2962-2973, 2007.

[4] Chen, W. , Qin, H. New method for single epoch, single frequency land vehicle attitude determination using low-end GPS receiver. In press, GPS Solutions, doi: 10. 1007/s10291-011-0234-8.

[5] De Jonge, P. J. A processing strategy for the application of the GPS in networks. NCG, Nederlandse Commissie voor Geodesie Netherlands Geodetic Commission, Delft, The Netherlands, pp. 3-5, 1998.

[6] Giorgi, G. , Teunissen, P. J. G. , Buist, P. J. A search and shrink approach for the baseline constrained LAMBDA: experimental results, in: Yasuda, A. (Ed.), Proceedings of the International Symposium on GPS/GNSS 2008, Tokyo University of Marine Science and Technology, pp. 797806, 2008.

[7] Giorgi, G. et al. Testing a new multivariate GNSS carrier phase attitude determination method for remote sensing platforms. J. Adv. Space Res. , doi:10. 1016/j. asr. 2010. 02. 023, 2010.

[8] Jin, T. , Qin, H. L. , Zhu, J. J. , Liu, Y. Real-time GPS software receiver correlator design, in: Proceedings of communications and Networking in China, pp. 549-553, 2007.

[9] Jin, T. , Shi, L. Single DSP hardware design in real-time multi-constellation satellite navigation receiver, in: Proceedings of Ninth IEEE International Conference on Computer and Information Technology, vol. 2, pp. 369-373, 2009a.

[10] Jin, W. R. , Zhai, C. R. , Wang, L. D. Hybrid analytical resolution approach based on ambiguity function for attitude determination. J. Zhejiang Univ. Sci. A 10 (7), 1038-1048, 2009b.

[11] Hide, C. , Pinchin, J. , Park, D. Development of a low cost multiple GPS antenna attitude system, in: Proceedings of ION GNSS, pp. 88-95, 2007.

[12] Lin, D. , Voon, L. K. , Nagarajan, N. Real-time attitude determination for microsatellite by LAMBDA method combined with Kalman filtering, in: 22nd AIAA international communication satellite systems conference, Monterey, US, pp. 8-9, 2004.

[13] P. Misra, P. Enge, Global positioning system, signals, measurements and performance, second ed. Ganga-Jamuna Press, Lincoln, pp. 238-239, 2006.

[14] Monikes, R. , Wendel, J. , Trommer G. F. A modified LAMBDA method for ambiguity resolution in the presence of position domain constraints, in: Proceedings of ION GPS, pp. 81-87, 2005.

[15] Park, C. , Kim, I. , Jee G. , Lee, J. G. An Error Analysis of GPS Compass, in: Proceedings

of the 36th SICE Annual Conference, pp. 1037-1042, 1997.

[16] Park, C., Teunissen, P. J. G. A new carrier phase ambiguity estimation for GNSS attitude determination systems, in: Proceedings of international GPS/GNSS symposium. Tokyo, pp. 8, 2003.

[17] Park, C., Teunissen, P. J. G. Integer least squares with quadratic equality constraints and its application to GNSS attitude determination systems. Int. J. Control Autom. 7(4), 566-576, 2009.

[18] Richert, T., El-Sheimy, N. Optimal linear combinations of triple frequency carrier phase data from future global navigation satellite systems. GPS Solutions 11(1), 11-19, 2007.

[19] Teunissen, P. J. G. The least-squares ambiguity decorrelation adjustment: a method for fast GPS integer ambiguity estimation. J. Geodesy 70(1), 65-82, 1995.

[20] Teunissen, P. J. G. GPS double difference statistics: with and without using satellite geometry. J. Geodesy 71(3), 138-148, 1997a.

[21] Teunissen, P. J. G. A canonical theory for short GPS baselines Part I: The baseline precision. J. Geodesy 71(8), 32-336, 1997b.

[22] Teunissen, P. J. G. An optimality property of the integer least-squares estimator. J. Geodesy 73(11), 587-593, 1999.

[23] Teunissen, P. J. G. The LAMBDA method for the GNSS compass. Artif. Satellites 41(3), 89-103, 2007.

[24] Teunissen, P. J. G., Giorgi, G., Buist, P. J. Testing of a new single-frequency GNSS carrier-phase compass method: land, ship and aircraft experiments. GPS Solutions 15(1), 15-28, 2010a.

[25] Teunissen, P. J. G. Integer least squares theory for the GNSS compass. J. Geodesy 84(7), 433-447, 2010b.

[26] Verhagen, S. Integer ambiguity validation: an open problem? GPS Solutions 8(1), 36-43, 2004.

[27] Verhagen, S. Visualization of GNSS-related design parameters: manual for the matlab user interface VISUAL. Delft University of Technology, available at http://www. lrtudelftnl/mgp, 2006.

[28] Tu, C. H., Tu, K. Y., Chang, F. R., Wang, L. S. GPS compass: novel navigation equipment. IEEE Trans. on Aero. Eelectron. Sys. 33(3), 1063-1068, 1997.

[29] Wang, Y. Q., Zhan, X. Q., Zhang, Y. H. Improved ambiguity function method based on analytical resolution for GPS attitude determination Meas. Sci. Technol. 18, 2985-2990, 2007.

[30] Wang, B., Miao, L., Wang, S., Shen, J. A constrained LAMBDA method for GPS attitude determination. GPS Solutions 13(2), 97-107, 2009.

Assessment of GNSS Orthogonal Transformation Model*

Wantong Chen and Yanzhong Zhang
School of Electronic and Information Engineering, Beihang University, Beijing, China

Abstract: GNSS relative positioning technique is an important field of study, in which the standard 'GNSS Baseline Model' is often used. Differencing between observation equations is used to construct the mathematical model, since this method can eliminate some common errors in the GNSS signal measurements. The 'Orthogonal Transformation' method can also construct the GNSS Baseline Model. However, as is described by some scholars, this model may avoid some drawbacks of Double Differencing (DD) while maintaining all the advantages. For comparison purposes, this model is evaluated and the theoretical equivalence of both approaches is proved for the short baseline from two aspects: the Integer Ambiguity Resolution and the conditional least-squares baseline vector.

Keywords: GNSS baseline model; integer ambiguity resolution; orthogonal transformation

1. Introduction

The 'GNSS Baseline Model' is widely used in GNSS computations involving the relative positioning technique (Tiberius, 1998; Wang, 2002; Leick, 2003; Misra and Enge, 2006; Hofmann-Wellenhof et al., 2008). The Double Differenced (DD) method is a very common approach for constructing the GNSS Baseline Model (Strang and Borre, 1997; Teunissen and Kleusberg, 1998); however, some scholars assume that there are some drawbacks in the DD method (Chang and Paige, 2003a). For example it is numerically slightly dubious, it makes the DD measurements correlated and it gives unnecessary prominence to the reference satellite. In (Chang and Paige, 2003b; Chang et al., 2004), an 'Orthogonal Transformation' approach is presented, which can avoid the first two drawbacks, while maintaining all the advantages. Both the DD and Orthogonal Transformation can be utilized to construct the GNSS Baseline Model. In order to evaluate both models, we focus on the 'Integer Ambiguity Resolution' and the conditional least-squares solution. The theoretical equivalence of both models is proved for the short baseline based on those two aspects.

* Published in Journal of Navigation, Vol. 65. No. 3, July 2012: 561-570.

Y. Zhang, *Footprints in Cambridge and Aviation Industries of China*,
https://doi.org/10.1007/978-981-16-3176-4_32

2. GNSS Baseline Model

2.1 The Standard GNSS Baseline Model

The standard GNSS Baseline Model is the mised integer model (Teunissen, 2003, 2011a). It is defined as:

$$E(Y)=Aa+Bb,\ D(Y)=Q_Y,\ a\in Z^n,\ b\in R^p \tag{1}$$

where

Y is the given GNSS data vector.

a and b are the unknown parameter vectors of order n and p respectively.

$E(\cdot)$ and $D(\cdot)$ denote the expectation and dispersion operators, respectively.

A and B are the given design matrices that link the data vector to the unknown parameters.

The variance matrix of Y is given by the positive definite matrix Q_Y. The n-vector a contains the integer DD ambiguities and the real-valued-vector b contains the remaining unknown parameters, such as for instance baseline components (coordinates) and possibly atmospheric delay parameters (troposphere, ionosphere). Many relative positioning methods make use of the Least-Squares Ambiguity Decorrelation Adjustment (LAMBDA) method to estimate the integer ambiguity, as this method is known to be efficient and to maximize the ambiguity success rate (Teunissen, 1995, 1999; Verhagan, 2004). In order to utilize the LAMBDA method, first regardless of the integer constraint of a, the float solution $\hat{a}$ and its variance-covariance (v-c) matrix $Q_{\hat{a}}$ should be obtained by the (weighted) least-squares method. The least-squares solution can be written as follows (Teunissen, 2006):

$$\hat{a}=(\overline{A}^TQ_Y^{-1}\overline{A})^{-1}\overline{A}^TQ_Y^{-1}Y \tag{2}$$

$$Q_{\hat{a}}=(\overline{A}^TQ_Y^{-1}\overline{A})^{-1} \tag{3}$$

where $\overline{A}=P_B^{\perp}A$ with the orthogonal projector $P_B^{\perp}=I-P_B, P_B=B(B^TQ_Y^{-1}B)^{-1}B^TQ_Y^{-1}$.

For the GNSS compass problem (Teunissen, 2006, 2010), the Constrained LAMBDA method is used since the baseline length is provided as well (Giorgi et al., 2010; Teunissen et al., 2011b). In this case, the conditional least-squares solution for b, which assumes that a is known, is also required for the estimator. The conditional least-squares solution and its v-c matrix read:

$$\hat{b}(a)=(B^TQ_Y^{-1}B)^{-1}B^TQ_Y^{-1}(Y-Aa) \tag{4}$$

$$Q_{\hat{b}(a)}=(B^TQ_Y^{-1}B)^{-1} \tag{5}$$

The notation $\hat{b}(a)$ emphasizes that we consider $\hat{b}(\cdot)$ a function. With the fixed integer ambiguity $\check{a}$, the fixed baseline vector can be obtained by:

$$\check{b}=\hat{b}(\check{a}) \tag{6}$$

In order to make the proof clearer, we firstly focus on the single frequency, single epoch and short baseline scenario for constructing the GNSS Baseline Model.

2.2 The Double-Differenced (DD) Model

For the short baseline and m visible satellites, with $\sigma_{\phi,1}^2\rho$, and $\sigma_{p,1}^2$ being the variance of carrier phase and code on band L_i, the Single-Differenced (SD) carrier phase and code equations can be expressed in compact vector and matrix notation as (Chang et al., 2004; Chen and Qin, 2011)

$$y_{i,S}^{\phi}=\frac{1}{\lambda_i}E\cdot b-N_{i,S}+e_m\beta+v_{i,S}^{\phi},v_{i,S}^{\phi}\sim N(0,2\sigma_{\phi,1}^2 I_m) \tag{7}$$

$$y_{i,S}^{p}=\frac{1}{\lambda_i}E\cdot b+e_m\beta+v_{i,S}^{p},v_{i,S}^{p}\sim N(0,2\sigma_{p,1}^2 I_m) \tag{8}$$

where

$y_{i,S}^{\phi}$ and $y_{i,S}^{p}$ are SD phase and code observations in units of cycles.

E is the $(m-1)\times 3$ matrix of normalized SD line-of-sight vectors.

$v_{i,S}^{\phi}$ and $v_{i,S}^{p}$ are SD phase and code noise.

$N_{i,S}$ is the SD integer ambiguity vector.

λ_i is the wave length.

e_m is a vector of order m for which each entry is 1 and β is the clock bias.

In order to eliminate the clock biases of all the SD equations, the $(m-1)\times m$ DD matrix operator will be defined as $D=(d_1\ d_2\cdots d_m)$ (Teunissen, 1997). This matrix can be set up in several ways, depending on the choice of reference satellite (Odijk, 2003). If satellite k is chosen as the reference satellite, thus the vector $d_k=-e_{m-1}$ where e_{m-1} is an $m-1$ vector for which each entry is 1, and all the other vectors of D can be treated as I_{m-1}. When the first satellite is selected as the reference satellite, it reads $D=(-e_{m-1}, I_{m-1})$. Pre-multiplications of Equations (7) and (8) by this operator give:

$$y_{i,D}^{\phi}=H_i\cdot b-N_{i,D}+v_{i,D}^{\phi},v_{i,D}^{\phi}\sim N(0,2\sigma_{\phi,1}^2 Q) \tag{9}$$

$$y_{i,D}^{p}=H_i\cdot b+v_{i,D}^{p},v_{i,D}^{p}\sim N(0,2\sigma_{p,1}^2 Q) \tag{10}$$

where

$y_{i,D}^{\phi}=Dy_{i,S}^{\phi}$

$y_{i,D}^{p}=Dy_{i,S}^{p}$

$v_{i,D}^{\phi}=Dv_{i,S}^{\phi}$

$v_{i,D}^{p}=Dv_{i,S}^{p}$

$N_{i,D}=DN_{i,S}$

$H_i=1/\lambda_i DE$

$Q=DD^T$

The DD model can be obtained by combining Equations (9) and (10):

$$\begin{bmatrix}y_{i,D}^{\phi}\\ y_{i,D}^{p}\end{bmatrix}=\begin{bmatrix}H_i\\ H_i\end{bmatrix}b+\begin{bmatrix}-I\\ 0\end{bmatrix}N_{i,D}+\begin{bmatrix}v_{i,D}^{\phi}\\ v_{i,D}^{p}\end{bmatrix} \tag{11}$$

Thus the standard GNSS model can be obtained if each term is expressed as follows:

$$Y=Y_i=\begin{bmatrix}y_{i,D}^{\phi}\\ y_{i,D}^{p}\end{bmatrix},A=A_0=\begin{bmatrix}-I\\ 0\end{bmatrix},B=B_i=\begin{bmatrix}H_i\\ H_i\end{bmatrix},Q_{Y_i}=\begin{bmatrix}2\sigma_{\phi,1}^2 Q & \\ & 2\sigma_{p,1}^2 Q\end{bmatrix} \tag{12}$$

The subscript i denotes that the item is depending on the given band and the subscript 0 denotes that its structure does nothing with the given band.

2.3 The Orthogonal Transformation Model

This model is based on an Orthogonal Transformation of single differences and the vector of DD integer ambiguities is still available (Chang and Paige, 2003b). The main steps are re-derived as follows (Chen and Qin, 2011). Let $P \in R^{m \times m}$ be an Orthogonal Transformation such that $Pe_m = \sqrt{m}u_1$ and the Householder transformation is used to form P as follows:

$$P = I_m - \frac{2uu^T}{u^T u}, u \equiv u_1 - \frac{1}{\sqrt{m}} e_m \tag{13}$$

where $u_1 = (1, 0, \cdots, 0)^T = (1\ 0)^T$.

By simple algebraic operations, we obtain for this matrix

$$P = \begin{bmatrix} \frac{1}{\sqrt{m}} & \frac{e_{m-1}^T}{\sqrt{m}} \\ \frac{e_{m-1}}{\sqrt{m}} & I_{m-1} + \frac{e_{m-1} \cdot e_{m-1}^T}{m - \sqrt{m}} \end{bmatrix} \equiv \begin{bmatrix} p_1 \\ \overline{P} \end{bmatrix} \tag{14}$$

where $\overline{P} = \begin{bmatrix} \frac{e_{m-1}}{\sqrt{m}} & I_{m-1} - \frac{e_{m-1} \cdot e_{m-1}^T}{m - \sqrt{m}} \end{bmatrix}$.

In order to eliminate the clock bias, applying P to Equatin (7), we obtain the initial Orthogonal Transformation of Equation(7):

$$\begin{bmatrix} p_1 y_{i,S}^{\phi} \\ \overline{P} y_{i,S}^{\phi} \end{bmatrix} = \begin{bmatrix} p_1 E \\ \overline{P} E \end{bmatrix} b - \begin{bmatrix} p_1 N_{i,S} \\ \overline{P} N_{i,S} \end{bmatrix} + \begin{bmatrix} 1 \\ 0 \end{bmatrix} \sqrt{m} \beta + \begin{bmatrix} p_1 v_{i,S}^{\phi} \\ \overline{P} v_{i,S}^{\phi} \end{bmatrix} \tag{15}$$

Note that only the first equation involves the clock bias term, the remaining part can be written as:

$$\overline{P} y_{i,S}^{\phi} = \overline{P} E b - \overline{P} N_{i,S} + \overline{P} v_{i,S}^{\phi} \tag{16}$$

Although double differencing is not used, the DD integer ambiguity vector can also be obtained by the following algebraic operations. Define the matrix F as follows:

$$F \equiv I_{m-1} - \frac{e_{m-1} \cdot e_{m-1}^T}{m - \sqrt{m}} \tag{17}$$

where F is nonsingular.

With Equation (14), it is easy to verify that:

$$\overline{P} = FD \tag{18}$$

where $D = [-e_{M-1}\ I_{m-1}]$ is the DD operator.

Thus the following formulation can be deduced:

$$\overline{P} N_{i,S} = FDN_{i,S} = FN_{i,D} \tag{19}$$

Replacing $\overline{P} N_{i,S}$ Equations (16) by (19), we obtain

$$\overline{P} y_{i,S}^{\phi} = \overline{P} E b - \overline{P} N_{i,D} + \overline{P} v_{i,S}^{\phi}, \overline{P} v_{i,S}^{\phi} \sim N(0, 2\sigma_{i,\phi}^2 I_{m-1}) \tag{20}$$

where the DD integer ambiguity vector exists.

The transformed noise vector still follows the same distribution because Orthogonal Transformation will not change the statistical properties of white noise (Chang and Paige, 2003b). Similarly, applying P to Equation (8), we obtain the following Orthogonal Transformation of SD code observation equation (Chang et al. , 2004):

$$\overline{P}y_{i,S}^{\rho}=\overline{P}Eb-\overline{P}N_{i,D},\overline{P}v_{i,S}^{\phi}\sim N(0,2\sigma_{\rho,i}^{2}I_{m-1}) \tag{21}$$

The combination expression of Equations (20) and (21) reads

$$\begin{bmatrix} \overline{y}_{i,D}^{\phi} \\ \overline{y}_{i,D}^{\rho} \end{bmatrix}=\begin{bmatrix} \overline{H}_{i} \\ \overline{H}_{i} \end{bmatrix}b+\begin{bmatrix} -F \\ 0 \end{bmatrix}N_{i,D}+\begin{bmatrix} \overline{v}_{i,D}^{\phi} \\ \overline{v}_{i,D}^{\rho} \end{bmatrix} \tag{22}$$

where

$\overline{y}_{i,D}^{\phi}=\overline{P}y_{i,S}^{\phi}$

$\overline{y}_{i,D}^{\rho}=\overline{P}y_{i,S}^{\rho}$

$\overline{v}_{i,D}^{\phi}=\overline{P}v_{i,S}^{\phi}$

$\overline{v}_{i,D}^{\rho}=\overline{P}v_{i,S}^{\rho}$

$N_{i,D}=FN_{i,S}$

$\overline{H}_{i}=1/\lambda_{i}\overline{P}E$

Note that both phase and code are expressed in units of cycles. This can be considered as the standard GNSS model if each term is expressed as follows:

$$Y=Y_{i}=\begin{bmatrix} \overline{y}_{i,D}^{\phi} \\ \overline{y}_{i,D}^{\rho} \end{bmatrix},A=A_{0}=\begin{bmatrix} -F \\ 0 \end{bmatrix},B=B_{i}=\begin{bmatrix} \overline{H}_{i} \\ \overline{H}_{i} \end{bmatrix},Q_{Y_{i}}=\begin{bmatrix} 2\sigma_{\phi,i}^{2}I_{m-1} & \\ & 2\sigma_{\rho,i}^{2}I_{m-1} \end{bmatrix} \tag{23}$$

3. Conditional Least-Square Solution Equivalence of Both Models

3.1 The Conditional Least-Squares Baseline Vector of the Double Difference Model

With Equations (4) and (5), the conditional least-squares solution reads

$$Q_{\hat{b}(a)}=((\frac{1}{2\sigma_{\phi,i}^{2}}+\frac{1}{2\sigma_{\rho,i}^{2}})(H^{T}Q^{-1}H_{i}))^{-1} \tag{24}$$

With $H_{i}=1/\lambda_{i}$DE and $Q=DD^{T}$, we obtain (Chen et al. , 2012)

$$Q_{\hat{b}(a)}=\lambda_{i}^{2}\omega_{i}^{-1}(E^{T}\Gamma E)^{-1} \tag{25}$$

where $\omega_{i}=(\frac{1}{2\sigma_{\phi,1}^{2}}+\frac{1}{2\sigma_{\rho,1}^{2}})$ and $\Gamma=D^{T}(DD^{T})^{-1}D$.

Since Equation (5) is part of Equation (4), the conditional least-squares solution reads:

$$\hat{b}(a)=Q_{\hat{b}(a)}\frac{1}{\lambda_{i}}(\frac{1}{2\sigma_{\phi,i}^{2}}E\Gamma(y_{i,S}^{\phi}+N_{i,S})+\frac{1}{2\sigma_{\rho,i}^{2}}E\Gamma y_{i,S}^{\rho}) \tag{26}$$

3.2 The Conditional Least-Squares Solution of the Orthogonal Transformation Model

In the same way, the conditional variance-covariance matrix of the Orthogonal Transformation model reads:

$$Q_{\hat{b}(a)}=((\frac{1}{2\sigma_{\phi,i}^{2}}+\frac{1}{2\sigma_{\rho,i}^{2}})(\overline{H}_{i}^{T}\overline{H}_{i}))^{-1}=\lambda_{i}^{2}\omega_{i}^{-1}(E^{T}KE)^{-1} \tag{27}$$

where the matrix $K=\overline{P}^{T}\overline{P}$.

The conditional least squares solution is expressed as the same structure of Equation (27):

$$\hat{b}(a)=Q_{\hat{b}(a)}\frac{1}{\lambda_i}\left(\frac{1}{2\sigma_{\phi,i}^2}EK(y_{i,S}^{\phi}+N_{i,S})+\frac{1}{2\sigma_{\rho,i}^2}EKy_{i,S}^{\rho}\right) \tag{28}$$

3.3 Equivalence of the Conditional Least-Squares Solution

In order to prove the equivalence of baseline vector, we should prove the equivalence of Equations (25) and (27) and also the equivalence of Equations (26) and (28). It is worth noting that they have similar expressions and only the matrix Γ and K are different for the DD model and the Orthogonal Transformation model. In order to prove $\Gamma=K$, we should prove the following Equation (29):

$$\overline{P}^T\overline{P}=D^T(DD^T)^{-1}D \tag{29}$$

With Equation (29), we can obtain the following expression:

$$\overline{P}^T\overline{P}=D^TF^TFD \tag{30}$$

With Equation (17), the following equation can be verified:

$$F^TF=I_{m-1}-\frac{1}{m}e_{m-1}e_{m-1}^T \tag{31}$$

For the DD operator, we have the same structure (Teunissen, 1997) as follows:

$$(DD^T)^{-1}=I_{m-1}-\frac{1}{m}e_{m-1}e_{m-1}^T \tag{32}$$

With Equations (31) and (32), we have:

$$F^TF=(DD^T)^{-1} \tag{33}$$

With Equations (30) and (33), Equation (29) can be verified, and so we prove the conditional least-square solution equivalence of both models.

4. Integer Ambiguity Resolution Equivalence of Both Models

In order to prove the equivalence of integer ambiguity resolution, we should focus on the float ambiguity vector $\hat{a}$ and its v-c matrix $Q_{\hat{a}}$, see also Equations (2) and (3). For the orthogonal projector $P_B^{\perp}$, the following properties can be verified:

$$Q^{-1}YP_B^{\perp}=(P_B^{\perp})^TQ_Y^{-1}=(P_B^{\perp})^TQ_Y^{-1}P_B^{\perp} \tag{34}$$

With Equation (34), the equivalent expression of Equation (3) is written as:

$$\begin{aligned}Q_{\hat{a}}&=(A^TQ_Y^{-1}(P_B^{\perp})A)^{-1}\\&=(A^TQ_Y^{-1}A-A^TQ_Y^{-1}B(B^TQ_Y^{-1}B)^{-1}B^TQ_Y^{-1}A)^{-1}\\&=(A^TQ_Y^{-1}A-A^TQ_Y^{-1}B\cdot Q_{\hat{b}(a)}\cdot(A^TQ_Y^{-1}B)^T)^{-1}\end{aligned} \tag{35}$$

We may also therefore write Equation (2) as:

$$\begin{aligned}\hat{a}&=Q_{\hat{a}}\cdot[(I-B(B^TQ_Y^{-1}B)^{-1}B^TQ_Y^{-1})A]^TQ_Y^{-1}Y\\&=Q_{\hat{a}}\cdot(A^TQ_Y^{-1}Y-A^TQ_Y^{-1}B(BQ_Y^{-1}B)^{-1}B^TQ_Y^{-1}Y\\&=Q_{\hat{a}}\cdot(A^TQ_Y^{-1}Y-A^TQ_Y^{-1}B\cdot Q_{\hat{b}(a)}\cdot B^TQ_Y^{-1}Y)\end{aligned} \tag{36}$$

4.1 Integer Ambiguity Resolution of the Double-Differenced (DD) Model

With the notations of Equation (12), the unknown matrices of Equation (35) can be calculated as follows:

$$A^T Q_Y^{-1} A = \frac{1}{2\sigma_{\phi,i}^2}(DD^T)^{-1} \text{ and } A^T Q_Y^{-1} B = -\frac{1}{2\sigma_{\phi,i}^2 \lambda_i}(DD^T)^{-1} DE \tag{37}$$

The unknown matrices of (36) read as

$$A^T Q_Y^{-1} Y = -(DD^T)^{-1} D \frac{1}{2\sigma_{\phi,i}^2} y_{i,s}^{\phi} \text{ and } B^T Q_Y^{-1} Y = \frac{1}{\lambda_i} E^T \Gamma \left(\frac{1}{2\sigma_{\phi,i}^2} y_{i,s}^{\phi} + \frac{1}{2\sigma_{p,i}^2} y_{i,s}^{p}\right) \tag{38}$$

4.2 Integer Ambiguity Resolution of the Orthogonal Transformation Model

With the notations of Equation (23), the matrices of Equation (35) can be calculated as follows:

$$A^T Q_Y^{-1} A = \frac{1}{2\sigma_{\phi,i}^2} F^T F \quad A^T Q_Y^{-1} B = -\frac{1}{2\sigma_{\phi,i}^2 \lambda_i} F^T (\overline{P} E) \tag{39}$$

The matrices of Equation (36) read as:

$$A^T Q_Y^{-1} Y = -F^T \overline{P} \frac{1}{2\sigma_{\phi,i}^2} y_{i,s}^{\phi} \quad B^T Q_Y^{-1} Y = \frac{1}{\lambda_i} T^T K \left(\frac{1}{2\sigma_{\phi,i}^2} y_{i,s}^{\phi} + \frac{1}{2\sigma_{p,i}^2} y_{i,s}^{p}\right) \tag{40}$$

4.3 Equivalence of Integer Ambiguity Resolution

With the derivation given in Equations (29) to (33), the equivalence of both operators is proved:

$$\Gamma = K \tag{41}$$

With Equation (33), the following equations can also be verified:

$$F^T \overline{P} = F^T F D = (DD^T)^{-1} D \tag{42}$$

With Equations (33) (41) and (42), the equivalence of both models can be verified for $Q_{\hat{a}}$ and $\hat{a}$, can all be proved. Thus the same float solution $\hat{a}$ and its variance-covariance matrix $Q_{\hat{a}}$ can be obtained for both models.

5. Assessment of General Cases

Assume that n GNSS bands are used to construct the GNSS Baseline Model, the single epoch GNSS model can be written as

$$Y = \begin{bmatrix} Y_1 \\ Y_2 \\ \vdots \\ Y_n \end{bmatrix}, B = \begin{bmatrix} B_1 \\ B_2 \\ \vdots \\ B_n \end{bmatrix}$$

$$A=\begin{bmatrix} A_0 & & & \\ & A_0 & & \\ & & \ddots & \\ & & & A_0 \end{bmatrix}, Q_Y=\begin{bmatrix} Q_{Y_1} & & & \\ & Q_{Y_2} & & \\ & & \ddots & \\ & & & Q_{Y_n} \end{bmatrix} \tag{43}$$

The multi-frequency observation of the same GNSS constellation will bring the summation operator 2 into the computation, for example:

$$Q_{\hat{b}(a)} = \left(\sum_{i=1}^{n}(B_i^T Q_{Y_i}^{-1} B_i)\right)^{-1} \text{ and } B^T Q_Y^{-1} Y = \sum_{i=1}^{n}(B_i^T Q_{Y_i}^{-1} Y_i) \tag{44}$$

For the computation of $Q_{\hat{a}}$ and $\hat{a}$, we obtain the following matrices:

$$A^T Q_Y^{-1} A=\begin{bmatrix} A_0^T Q_{Y_1} A_0 & & & \\ & A_0^T Q_{Y_2} A_0 & & \\ & & \ddots & \\ & & & A_0^T Q_{Y_n} A_0 \end{bmatrix}$$

$$A^T Q_Y^{-1} Y=\begin{bmatrix} A_0^T Q_{Y_1}^{-1} Y_1 \\ A_0^T Q_{Y_2}^{-1} Y_2 \\ \vdots \\ A_0^T Q_{Y_n}^{-1} Y_n \end{bmatrix} \quad A^T Q_Y^{-1} B=\begin{bmatrix} A_0^T Q_{Y_1}^{-1} B_1 \\ A_0^T Q_{Y_2}^{-1} B_2 \\ \vdots \\ A_0^T Q_{Y_n}^{-1} B_n \end{bmatrix} \tag{45}$$

Thus, in this case, the equivalence of both models does not alter. For k epochs, the standard GNSS model reads as follows (Giorgi et al., 2010)

$$Y=\begin{bmatrix} Y(1) \\ Y(2) \\ \vdots \\ Y(k) \end{bmatrix}, A=\begin{bmatrix} A(0) \\ A(0) \\ \vdots \\ A(0) \end{bmatrix}$$

$$B\begin{bmatrix} B(1) & & & \\ & B(2) & & \\ & & \ddots & \\ & & & B(k) \end{bmatrix}, Q_Y=\begin{bmatrix} Q_{Y(1)} & & & \\ & Q_{Y(2)} & & \\ & & \ddots & \\ & & & Q_{Y(k)} \end{bmatrix} \tag{46}$$

where

$A(0)$ denotes the constant A matrix for each epoch.

(k) denotes the corresponding matrix of the k epoch.

The multi-epoch observation will result in the Kronecker product operator (Giorgi et al., 2010), and in this case the equivalence of both models is also valid, since the basic matrix of Equations (41) and (42) still remain in the derivations.

6. Assessment of the Drawbacks And Advantages

Note that the cofactor matrix Q in Equations (9) and (10) has 2 s on the main diagonal and 1 s in all off-diagonal positions. Some scholars assume that the double differencing makes the DD measurements correlated, which is a drawback. However, this is not appropriate. Although the noise of Orthogonal Transformation model is white, the noise of DD model can still be white if an inverse-Cholesky factorization is used. The Cholesky decomposition of Q reads:

$$Q=LL^T \tag{47}$$

where

L is a lower triangular matrix with strictly positive diagonal entries.

L^T denotes the conjugate transpose of L.

Pre-multiplication of Equations (9) and (10) by inverse of L gives:

$$L^{-1}y_{i,D}^{\phi}=(L^{-1}H_i)\cdot b-L^{-1}\cdot N_{i,D}+L^{-1}v_{i,D}^{\phi},L^{-1}v_{i,D}^{\phi}\sim N(0,2\sigma_{\phi,i}^2 I_{m-1}) \tag{48}$$

$$L^{-1}y_{i,D}^{\rho}=(L^{-1}H_i)\cdot b+L^{-1}v_{i,D}^{\rho},L^{-1}v_{i,D}^{\rho}\sim N(0,2\sigma_{\rho,i}^2 I_{m-1}) \tag{49}$$

The structure of Equations (48) and (49) are the same as Equations (20) and (21). Although the noise levels of Equations (20) and (21) seem smaller than those of Equations (9) and (10), with the use of the Orthogonal Transformation, one does not avoid the higher noise level of double differencing, due to the equivalence of both models. Finally, although Orthogonal Transformation is a numerically stable approach, double differencing is also not numerically dubious if properly implemented.

7. Conclusions

The Double Difference (DD) method and the Orthogonal Transformation approach can both construct the standard GNSS Baseline Model. Some scholars assume that the Orthogonal Transformation approach is better than the DD method. In this paper, we prove the theoretical equivalence of both approaches for the short baseline. Assessment of some views about the model drawbacks and advantages is also given and discussed.

References

[1] Chang, X. W. and Paige, C. C. (2003a). An algorithm for combined code and carrier phase based GPS positioning. BIT Numerical Mathematics, 43, 915-927.

[2] Chang, X. W. and Paige, C. C. (2003b). An Orthogonal Transformation Algorithm for GPS Positioning. SIAM Journal of Scientific Computing, 24(5), 1710-1732.

[3] Chang, X. W., Paige, C. C. and Yin, L. (2004). Code and Carrier Phase Based Short Baseline GPS Positioning, Computational Aspects. GPS Solutions, 7, 230-240.

[4] Chen, W. and Qin, H. (2011). New Method for Single Epoch, Single Frequency Land Vehicle Attitude Determination Using Low-End GPS Receiver. GPS Solutions. In press, doi: 10.1007/s10291-011-0234-8.

[5] Chen, W., Qin, H., Zhang, Y. and Jin, T. (2012). Accuracy Assessment of Single and Double Difference Models for the Single Epoch GPS Compass. Advances in Space Research, 49(4), 725-738.

[6] Giorgi, G., Teunissen, P. J. G., Verhagen, S. (2010). Reducing the Time-To-Fix for Stand-Alone Single Frequency GNSS Attitude Determination. Proceedings of ION-ITM, 2010,

526-533.

[7] Hofmann-Wellenhof, B. , Lichtenegger, H. and Wasle, E. (2008). GNSS global navigation satellite systems. GPS, GLONASS. Galileo and More. Springer, Berlin.

[8] Leick, A. (2003). GPS Satellite Surveying, 3rd edition. Wiley, New York.

[9] Misra, P. and Enge, P. (2006). Global Positioning System Signals, Measurements and Performance. Ganga-Jamuna Press, Lincoln.

[10] Odijk, D. (2003). Ionosphere-Free Phase Combinations for Modernized GPS. Journal of Surveying Engineering, 129(4), 165-173.

[11] Strang, G. and Borre, K. (1997). Linear Algebra, Geodesy, and GPS. Wellesley-Cambridge Press, Wellesley.

[12] Teunissen, P. J. G. (1995). The Least-Squares Ambiguity Decorrelation Adjustment: a Method for Fast GPS Integer Ambiguity Estimation. Journal of Geodesy, 70, 65-82.

[13] Teunissen, P. J. G. (1997). GPS Double Difference Statistics: With and Without Using Satellite Geometry Journal of Geodesy, 71(3), 138-148.

[14] Teunissen, P. J. G. and Kleusberg, A. (1998). GPS for Geodesy. Springer, Berlin, Heidelberg, New York.

[15] Teunissen, P. J. G. (1999). An Optimality Property of the Integer Least-Squares Estimator. Journal of Geodesy, 73, 587-593.

[16] Teunissen, P. J. G. (2003). Towards a Unified Theory of GNSS Ambiguity Resolution. Journal of Global Positioning Systems, 2(1), 1-12.

[17] Teunissen, P. J. G. (2006). The LAMBDA Method for the GNSS Compass. Artificial Satellites, 41(3), 89-103.

[18] Teunissen, P. J. G. (2010). Integer Least Squares Theory for the GNSS Compass. Journal of Geodesy, 84(7), 433-447.

[19] Teunissen, P. J. G. (2011a). A New Method for DGPS Ambiguity Resolution? Journal of Navigation, 64, 375-379.

[20] Teunissen, P. J. G. , Giorgi, G. and Buist, P. J. (2011b). Testing of a New Single-Frequency GNSS Carrier-Phase Compass Method. Land, Ship and Aircraft Experiments. GPS Solutions, 15(1), 15-28.

[21] Verhagen, S. and Teunissen, P. J. G. (2006). New Global Navigation Satellite System Ambiguity Resolution Method Compared to Existing Approaches. Journal of Guidance, Control, and Dynamics, 29(4), 981-991.

[22] Tiberius, C. C. J. M. (1998). Recursive Data Processing for Kinematic GPS Surveying. PhD thesis, Department of Mathematical Geodesy and Positioning, Delft University of Technology.

[23] Wang, J. , Satirapod, C. and Rizos, C. (2002). Stochastic Assessment of GPS Carrier Phase Measurements for Precise Static Relative Positioning. Journal of Geodesy, 76(2), 95-104.

The Development of Large Aircraft C919 and C929*

ZHANG Yanzhong

Chinese Academy of Engineering, Beijing, China

Abstract: In the present article, the history of the development of large aircraft (LA) is summarized and the reasons for developing homemade large aircraft (HLA) in China are also analysed. Strategic targets of the Chinese large aircraft are proposed. The development situations of the airplane C919 and development strategies are also introduced in present paper. In addition, key technologies aircraft are proposed for large wide-bodied.

1. The History of the Development of HLA in the Whole World

The large passenger plane can be divided into the following class by the number of seats: 150 seats, 250 seats and over 300 seats, as shown Fig. 1. The airliners of 150 seats class (150SA) refer to the number of seats are 130～200, which primarily includes Boeing 737 series and Airbus A320 series. The airliners of 250 seats class (250SA) refer to the number of seats is 200～300, which

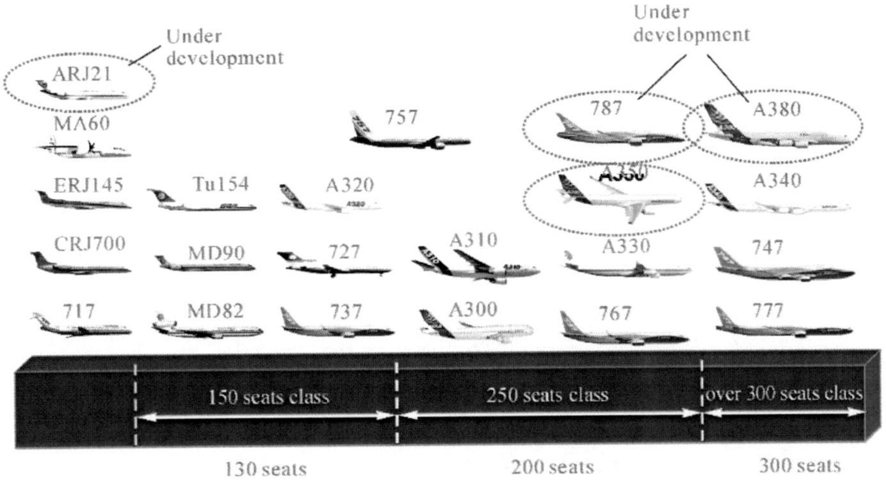

Fig. 1 The classification of seats class of airliners

includes Boeing 767 series, Airbus A300, A330 as well as Boeing 787 and Airbus A350 (These

* Published on the Seminar of Chinese Engineering & Technology, June 2015.

Y. Zhang, *Footprints in Cambridge and Aviation Industries of China*,
https://doi.org/10.1007/978-981-16-3176-4_33

airplanes are still in the development stage). The airliners of over 300 seats class include Boeing 777, 747, Airbus A340 and A380.

2. Why China Should Develop HLA?

2.1 Large Aircrafts Play a Significant Leading Role in Economy and Science, Which are Strategic Choices That Can Improve the Comprehensive National Power and International Competitiveness for Our Country

At present, our country is just in a historical period that economic and social developments are promoted by scientific and technological progress and innovation, thus some critical area for improving national competitiveness should be selected to realize leap-forward development. As for the aircraft industry, it has the characteristic of long industry chain, wide radiant surface and strong related effect. In aviation developed countries, large aircraft industry plays a significant role in the development of domestic economy and progress of science and technology and has become an important pillar industry.

According to international estimation, GDP can increase by 0.744%, as the saleroom of the civil aircraft industry as the primary part of the large aircraft industry increases 1%. In 2000, the United States produced 140 billion dollars gross domestic that can promote 900 billion dollars in production accounting for 9 percent of GDP and creating 11 billion jobs. According to the study of Japan Miti, as for the product value per unit weight, it is 1 for the ship industry, 9 for automotive industry, 50 for colour TV, 300 for electronic computers, 800 for jet airplane and 1400 for aircraft engines, respectively. Civil aircraft industry can provide 12 times of employment people for relevant industries and the 100 million dollars expenditure can promote related industry to output 8 billion dollars after 10 years according to the statistics of Rand Corporation of United States. The special project for LA in our country can effectively promote the growth of national economy as well as the adjustment of the industrial structure and the upgrade of technology.

LA has synthesized many high-tech achievements, which have the leading role in the technology development. Based on the comparison of the order of magnitude of the components and technical parameters, it is 104 for automotive industry, 105 for the rocket and 107 for the civil aircraft, respectively. The number of component parts in LA of Boeing 747-400 reaches 6 million. According to the investigation from the more than 500 technical cases by Japanese relevant institutions, the technology diffusion rate of automobile industry is 9.8%, while the one of aviation industry technology can reach 60%. Therefore, the development of large aircraft has a strong tractate effect on technological progress, leading to a breakthrough in the group of key technologies in new materials, advanced manufacture, electronic communication, automatic control and computers and promotion of many high-tech industries development.

The development of LA can significantly improve the progress of hydromechanics, solid mechanics, computational mathematics, thermal physics, chemistry, information sciences and environmental sciences or other related basic sciences, leading to the development of scientific and technological level of our country comprehensively. It is a strategic choice that the development of LA can promote the comprehensive strength and international competitiveness of our country.

2.2 The Demand of Domestic Civil Aviation Is Big and Unusual Historic Opportunity Is Met

The strong demand growth of Chinese civil aviation market will provide powerful traction and support for the development of large aircraft industry. It is our unique advantage for developing LA, as there is a great potential in civil aviation of our country, especially as compared to some country with great economic power such as Japan.

The civil aviation transportation industry of our country has a rapid growth with the quick increase of national economy. In 2004, the total turnover of Chinese civil aviation transportation and the total number of passenger transportation has jumped to the second place in the world. In our country, we have 863 civil aviation aircraft including 785 LA by the end of the year. The number of civil aviation aircraft in China is increased by 336 during the tenth-five-year plan period and 840 new planes have been purchased by civil aviation corporations during the 11th-five-year plan period. At present, the average annual growth has reached 14.6%.

The demand for LA is still growing in the next 20 years in China, and this period of opportunity will not last long. According to the 1997 International Monetary Fund (IMF) statistics as shown in Fig. 2, there is a period of high growth of air transport demand when GDP is between 1000 and 5000 dollars per capita, while the air transport growth will be slowed down after the GDP is more than 10000 dollars per capita and the number of taking plane reaching 1 per capita. The per capita number of taking plane in our country was 0.02 in 1997, while it was 0.1 in 2004. Accordingly, at present, our country is still in a period that the civil aviation market is increasing rapidly.

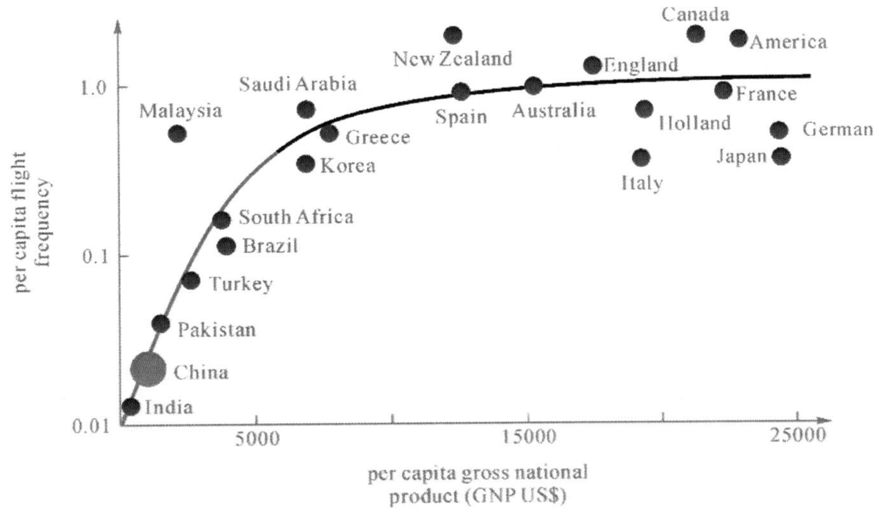

Fig. 2 The comparison of international air passenger demand (IMF report, 1997)

The development research institute of CAAC speculate that 4752 LA is increasing during 2007-2026 in our country. Boeing Co. predicts that 2319 passenger and cargo transport planes need to be added in China during 2004—2023, while Airbus predicts 1790 passenger and cargo transport planes need to be added over the next 20 years. Base on the mean value of these prediction, the civil aviation of our country needs to buy 2953 new aircrafts over the next 20 years, and their total worth is more than 100 million dollars, which will be enough to buy both Boeing and Airbus (Boeing and Airbus had net assets of 60 billion dollars and 20.8 billion dollars respectively at the end of 2005). If we do not catch the rare historical opportunity to develop LA, the great economic benefits of the rapid development of Chinese civil aviation market will fall into the hands of Boeing and Airbus.

3. The Strategic Target of LA

The strategic target of LA can be divided into two steps:

(1) First target step (2008—2017): develop 150SA (C919);

(2) Second target step (2017—2025): develop large wide-body airliner (LWBA) named as C929 with 280 seats class (280SA).

4. The Development of Aircraft C919

4.1 Market Requirement of 150 Seats Class Airliner

150SA is the main aircraft series for the domestic civil aviation market at present. In 2005, the total number of civil aviation fleet in China was 863, in which the number of 150SA is 600 accounting for 69.5% of the total number. The proportion of passenger plane of 150SA is 72.8% for the total volume of Chinese civil aviation transportation industry in 2005.

150SA is the principal demand of Chinese civil aviation market in the future. The Development Institute of CAAC predicts that the quantity of demand for 150SA will be up to 3712 and 250 seats class airliners is 829 in the domestic civil aviation market in the next 20 years. Especially, 2593 150SA and 570 250SA should be required in 2017—2026, as shown in Table 1. As the profit and loss equilibrium point number for LA is usually around 400, the equilibrium point can be exceeded as long as the domestic large airliners capture one-sixth of the 2593 expected domestic market.

Table 1 Civil aviation aircraft delivery forecast of China from 2007 to 2026

Year \ Type	150 seats class	250 seats class	300 seats class
2007—2016	1119	259	55
2017—2026	2593	570	156
Add up	3712	829	211

150SA is also the principal demand for the international civil aviation market. Boeing Co. predicts that 25700 airliners are required in the world over the next 20 years, in which 15300 single-aisle airliners (mainly 150SA and a handful of large regional airliners) are accounted for 59.5%. In addition, Airbus also predicts that 17000 airliners are required in the world over the next 20 years including 10902 single-aisle airliners accounted for 64.1% of the total.

4.2 150SA Has Wide Range of Routes and Airport Adaptability

The domestic routes are mainly on 800~2000 km at present. As shown in Fig. 3, among all main routes in 2005, there were 552 trunk liners in China with cruise distance of 800~2000 km, 136 trunk liners with cruise distance over 2000 km, and 336 feeder liners with cruise distance below 800 km. The routes between the major cities of Beijing, Shanghai, Guangzhou, Chongqing and Lanzhou are within 2000 km.

150SA is one of the most frequently used airplanes due to the route adaptability. As shown in Fig. 4, according to the domestic routes of 800-2000 km, operating routes with 150 seats class airliners, the number of flights and passengers accounts for over 50% of total number of routes, over 70% of the total flights, and over 70% of the total passenger transportation, respectively.

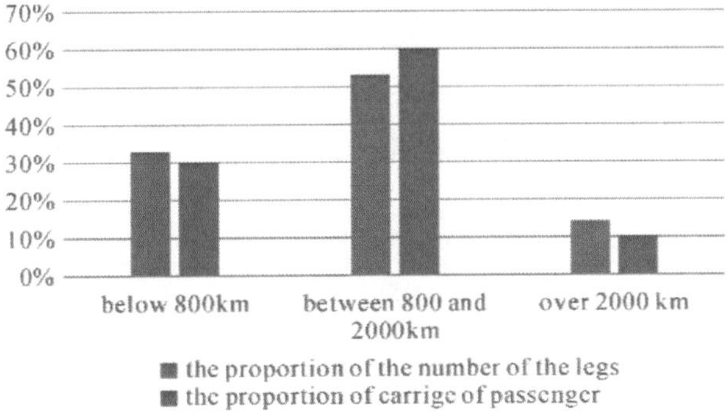

Fig. 3 The proportion of domestic airline

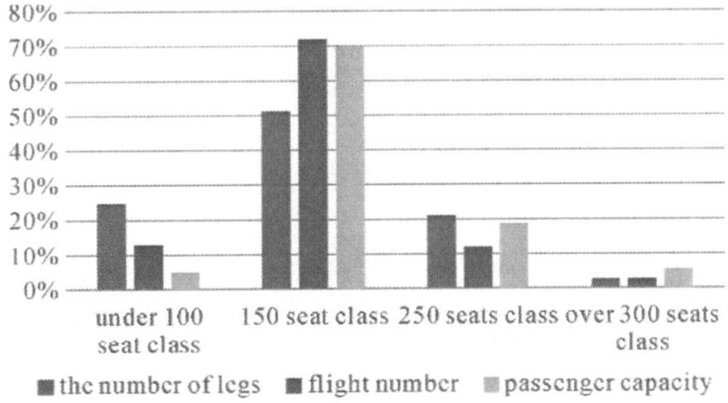

Fig. 4 The transportation of the main routes on 800~2000 km in China

150SA has extensive airport adaptability and coverage. Now, 82.6% airports in our country can meet the requirement of 150SA taking off and landing, while only 44.7% can meet the requirement of 250SA. Until the end of 2005, there were 132 domestic civil aviation airports in our country (excluding Hong Kong, Macao and Taiwan), as following:

(1) 25 4E class airports can permit 300 seats class airliners to take off and land;

(2) 59 4D class airports can permit 250SA to take off and land;

(3) 109 4C class airports can handle 150SA to take off and land;

(4) and 23 3C regional airports can handle the airliners below the Boeing 737.

As shown in Table 2, based on the development plan of civil aviation, over the next 15 years, the number of 4D class airports in China will be increased to 66 accounting for 29.5% of the total number of airports, and the number of over 4C class airports will be increased to 176.

Table 2 Domestic airport status and development plan for the next 15 years

The grade index in flight area	4F	4E	4D	4C	3C	add up
The biggest takeoff and landing aircraft	A380	747 A340	767 A300	737 A320	regional lines	
At the end of 2005	0	25	34	50	23	132
Huu7rpo	3	30	33	110	48	224

4.3 Preferable Market Opportunities for 150SA

Boeing and Airbus are scrambling to capture the strategic commanding heights of the aviation manufacturing industry. Now they severely compete with the new 250SA. Boeing is developing a 250-seat 787 expected to be launched in 2008, and an over 300 seats class 747-8 airliner with voyage of 15000 km. At the same time, Airbus is developing A380 with over 300 seats class and A350 with 250 seats class which already obtain orders from advanced users whose 150SA had been relatively aging at present.

The basic type of the Boeing 737 is 737-100 which was started to develop in 1964 and put in service in 1968. It is 39 years ago. In addition, next generation series 737 (600-900) were developed in the 1990s, but the fuselage configuration were not greatly improved. On the other hand, the basic of the Airbus A320 are A320-100 which were started to develop in 1979 and put in service in 1988. It is also 19 years ago. The successor of these airplanes is still in imagination stage and have not been started to develop. It is estimated that the successor of these airplanes will not be put into the market until 2017 if they are began to develop at present. Therefore, based on the need and comprehensive possibility analysis, it will be the best opportunity for us to independently develop large airliners entering the market. However, this opportunity will not last for a long period. So the domestic large airliners should be started as soon as possible and strived to enter the market before 2017. We should not miss this rare historic opportunity to independently develop our HLA.

Furthermore, as the economy of Brazil, Russia and other countries in the world is developing powerfully, they all have the urgent intention to develop large airliners due to their rapid expansion of the domestic civil aviation market. If we cannot catch this golden opportunity to manufacture homemade 150SA before 2017, We will face the fact that not only do we have to continue to buy the airliners from Boeing and Airbus, but also from Russia even countries such as Brazil.

4.4 150SA is Relatively Less Difficult to Develop Than 280SA

150SA has relatively small size and its size is more than 50% less than 250SA. This makes 150SA easy to meet the structural design requirement. This means that the requirement for CNC machine tools, large forging press equipment and heat pressure tanks are much higher in terms of manufacture. In addition, in the terms of engines, each engine for 150SA is 13000~15000 kg thrust while the one of the 280SA are 20000~25000 kg thrust which are more difficult to develop.

Generally, as compared to the 280SA, 150SA has the following advantage: (1) larger market capacity; (2) more opportunities in market; (3) stronger adaptability for cruises and airports; (4) less investment, so it is very suitable for our existing research ability and have a better overall condition. As a result, developing LA, we should choose 150SA as a priority to develop and lunch on the market before 2017; even gradually realize the serialization development.

As shown in Fig. 5, the basic technical characteristics of HLA are: 130~200 seats, 2000~5000 km cruise, 0.8 Ma, 11000 m cruise height and 60~100 ton of maximum takeoff weight. Furthermore, according to the market research, an aircraft series between 130 and 200 seats should be developed gradually including lengthen type, shorten type, cargo type and range extender types based on the basic type.

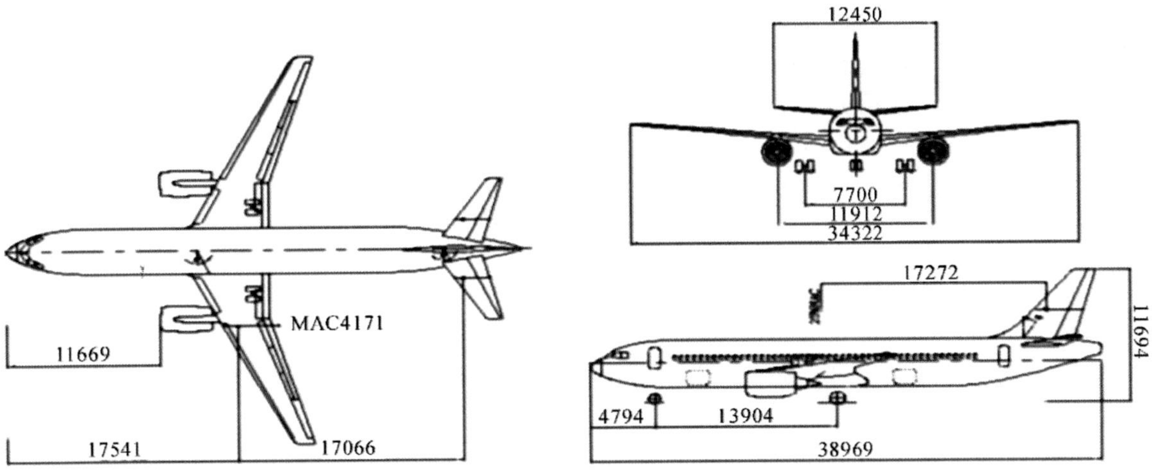

Fig. 5 A preliminary plans for 150 seats class airliners

Based on the forecast analysis of the next generation of the airliners of Airbus A320 and Boeing B737, it can be determined that in the light of the thinking of overall comparability and highlighted merits, the characteristic technical indicators of the domestic large airliners should ensure the safety, highlight the economical efficiency, improve comfort and pay more attention on environmental protection. The detailed explanations are as following:

(1) Ensure the safety: The aircraft design and manufacturing should meet the requirement of the airworthiness standards of CAAC, FAA or EASA. And the design, manufacturing, operation and maintenance of aircrafts should be strictly managed according to the airworthiness standards. In the service period of the aircraft put on the market, catastrophic incidences of millions hours should less than 0. 1%. This can provide a strong support for our air transport system to realize that major accidents rate of every millions flight hours should be lower than 0. 3 times, the low failure rate reaches the advanced level of the world and the rate of signing is over 99. 9%.

(2) Highlight the economical efficiency: as compared with the current Airbus A320, the cost of tonnage, single-airliner costs, direct use of cost (DOC) and the whole service period are reduced by 10%. This can increases the maintainability,

(3) Improve passenger comfort: The cabin space and seat width are increased and the cabin noise is reduced by 3 dB as well as the cabin layout is flexible for accommodating with multiple layouts as shown in Table 3.

(4) Pay more attention on environmental protection: Emission pollutants of engines should meet the requirement of the emission standards of International Civil Aviation Organization and the outer field noise meet the requirement of CCAR-36 (or FAR-36).

Table 3 the comparison of the width of seat

Airliner type	Width of the back of seat	The width of armrest	Walkway clearance
Proposed projects	18. 5″	2″	19. 5″
A320	18″	2″	19″
B737	17″	2″	18″

Security and economy is the core of the four characteristics as mentioned above. Ensuring safety is a fundamental requirement for aircrafts while the economic performance is the basis of commercial competitiveness. This requires that during the whole period of the design, manufacture and service, we should ensure aircraft security and economy and strictly implement airworthiness management, leading to make homemade aircraft price/performance ratio has better competitive advantage with the same type of aircrafts of Boeing and Airbus.

5. Development Strategy of the Iarge Wide-Bodied Airliner C929

5. 1 Necessity to Develop Large Wide-Bodied Airliners (LWBA)

Research shows that in the next 20 years, the total demand for civil aircrafts is 33500 with a value of 4. 06 trillion dollars, in which the demand for LWBA is 8150 with the worth of 2. 04 trillion dollars. It can be seen that the number of LWBA only accounts for only 24. 3% of the total number, but it accounts for 50. 2% of the total value. LWBA has become the product with the most marketable potential in civil airliner market and become the priority development after the single-aisle mainline airliners.

If we only have 150 seat single channel aircraft, we can not compete with Boeing and Airbus, as they have complete product types. We will face the fact that one log cannot prop up a shaky structure. Therefore, LWBA with international competitiveness must be developed to improve our competition in the civil aircraft market in the world. Thus, we can provide our customers with different levels of products based on market demand and improve the overall competitiveness in the civil aircraft industry by improving the type pedigree of civil aircrafts and enriching the series products of civil aircraft.

The main argument about LWBA is: How many the seats and voyage are perfect. Some suggests 220 seats and 8000 voyage, while our careful analysis can be seen as below.

5.2 Analysis of the Size of LWBA

As shown in Fig. 6, it can be observed that a total of 9256 LWBA will be delivered in 2023—2042, in which 7552 airliners are 250～350 seats class airliners accounting for 81% of the total. In 2013～2032, the 7547 LWBA will be delivered in the world, in which 250～350 seats class airliners accounting for approximately 77% of the total, while there will be 1268 LWBA in China, in which 250～350 seats class airliners accounts for approximately 85% of the total. Therefore, the development of 250～350 seats class airliners is more in line with market demand.

5.3 Analysis of the Voyage of LWBA

The voyage of 10000～12000 km of LWBA can meet more than 99% of the airlines requirements. The basic type of LWBA with 280SA and the 12000 km voyage are pretty good, as illustrated in Table 4.

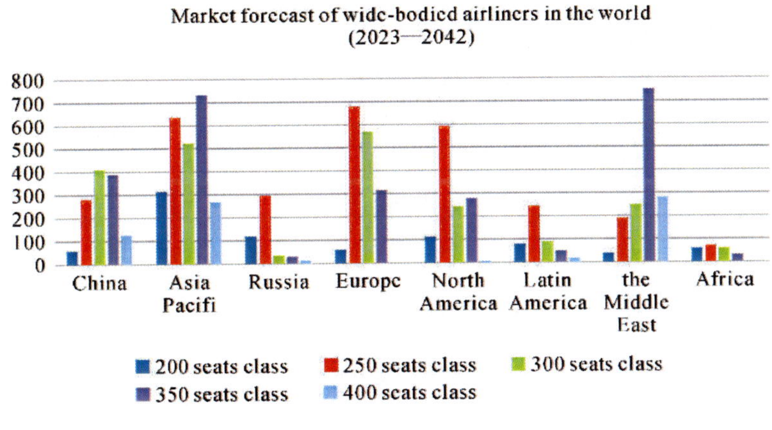

Fig. 6 The demand for global wide-bodied airliners market

Table 4 Analysis of voyage of wide-bodied airliners

Region of operators	<9000 km air lines	<10000 km air lines	<12000 km air lines
China	87%	95%	99%
Asia Pacific	91%	95%	99%
Russia	95%	99%	100%
Europe	86%	96%	100%
Latin America and Africa	92%	97%	99%
the Middle East	96%	97%	99%
North America	91%	93%	98%

5.4 The Overall Configuration of LWBA

It is recommended that the homemade LWBA can choose the configuration with double engines and low wing as shown in Fig. 7, according to the study about the international development.

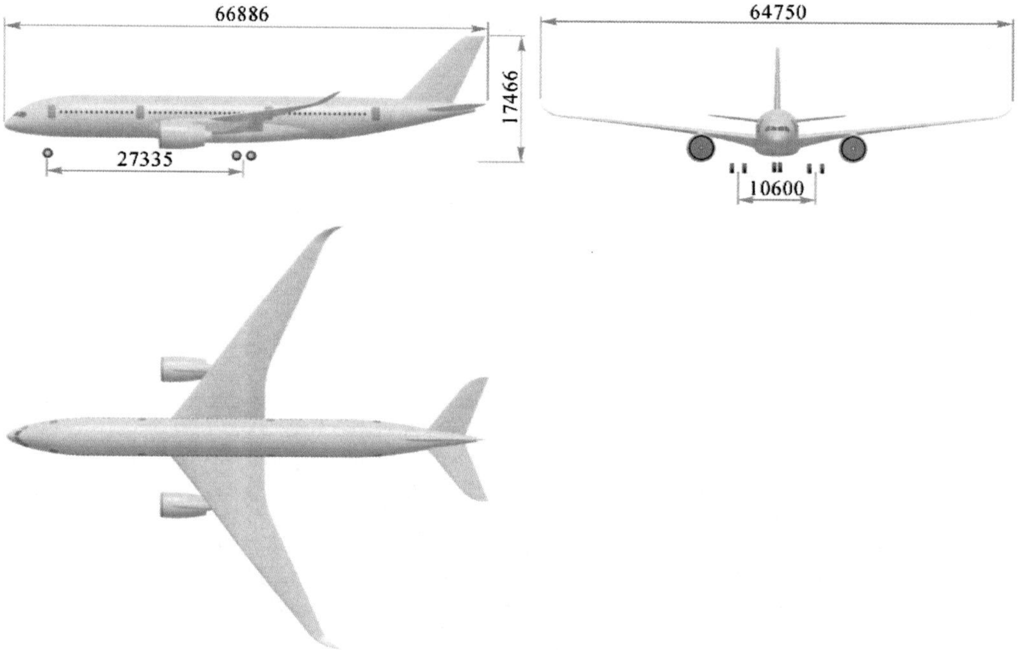

Fig. 7 A preliminary programmes for large wide-bodied airliner

5.5 Key Technologies for LWBA

The key technologies for LWBA can be summarized as following:

(1) The tradeoff design and validation of multiple projects for the overall aerodynamic

configuration of LWBA.

(2) Design, manufacture and verification technology of composite materials for the main bearing capacity of LWBA.

(3) System integrated design and verification technology based on the model.

(4) Integrated modular avionics and multi-electrical system application technology.

(5) Active control technology and fly-engine integrated design technology.

6. Conclusions

This paper explains why Chinese should manufacture HLA. The strategic targets of our LA are proposed that C919 with 150 seats should be constructed at first and then C929 LWBA with 280 seats should be further developed. The development of C919 and the development strategy of C929 LWBA are also introduced in present paper. The key technologies are proposed for LWBA.

Integration of GPS and Dead Reckoning Navigation System Using Moving Horizon Estimation*

Halima Mansour Omar, Zhang Yanzhong, Zhang Bo, Haris Ubaid Gul
Beihang University, Beijing, China

Abstract: In this paper, we propose an integration of GPS and dead reckoning navigation system based on moving horizon estimation (MHE) in order to improve its performance. The basic idea of MHE is to minimize an estimation cost function defined on a sliding window composed of a finite number of time stages. From simulation results we have done an objective comparison which leads us to confirm that MHE method is more accuracy than Unscented Kalman Filter (UKF) and Extended Kalman filter (EKF).

Keywords: dead reckoning; global positioning system; moving horizon estimation; Kalman filtering; data fusion

1. Introduction

Precise GPS positioning can be achieved only by receiving signals from three to four GPS satellites at the same time. When relying on GPS only positioning, there may be cases where the position accuracy is degraded or lost. For example, when the vehicle moves in areas where the GPS signals cannot be received (tunnels or underground passages) or where very strong multipath propagation occurs (areas surrounded by tall glass covered buildings). To overcome such limitations of GPS positioning, a Dead Reckoning (DR) solution is useful. It enables to keep high accuracy positioning by using information from various sensors (gyro sensor, accelerometer, odometer, etc.) to calculate the current position, even when GPS only positioning is difficult or impossible. Dead Reckoning solution is widely utilized in automotive navigation systems.

The main problem of realization GPS/DR integration system is definitely how to design information fusion scheme which is also how to obtain the optimal position information using fusion data. Two common integration methods can be set in here which are switching combination and data fusion[6]. In fact Kalman filter is widely used in fusion information. In addition, there are various filters originally based on Kalman filter such as Extended Kalman filter (EKF), Unscented Kalman Filter (UKF), and Particle filter (PF). Whereas big error can be produced due to linearization process as a fact of using EKF. Furthermore the distribution function is tough to be selected by using PF[6].

* Published in Proc. of Intl. Conf. IEEE. 2016: 553-556.

Y. Zhang, *Footprints in Cambridge and Aviation Industries of China*,
https://doi.org/10.1007/978-981-16-3176-4_34

Recently, the study of non-linear filtering algorithm has been steadily improved. Moving horizon estimation (MHE) became a powerful technique for facing the problem of state estimation of dynamic system in presence of nonlinearities and

disturbances. MHE is based on the idea of minimizing an estimation cost function defined on a sliding window composed of a finite number of time stages[2]. The cost function is generally made up of two contributions which are prediction error computed on a recent batch of inputs and outputs; and an arrival cost that serves the point of summarizing the past data[2].

In this paper, we are to use moving horizon estimation in the aim of improving the performance of GPS/DR navigation system. This document is organized as follow. Section 2 describes a clear approach in solving general moving horizon estimation problem through successive linearization. The model of GPS/DR integrated navigation system is elaborated in Section 3. Simulation results are discussed in Section 4. Finally this paper ends up with some conclusions.

2. Moving Horizon Estimation Design

This section describes a clear approach in solving general moving horizon estimation problem through successive linearization. The MHE problem is defined as follows[1]:

$$\underbrace{\min}_{x_k} \frac{1}{2}\sum_{k=2}^{N}\|y_k - h(x_k)\|^2_{W_{m,k}} + \frac{1}{2}\sum_{k=1}^{N-1}\|x_{k+1} - f(x_k, u_k)\|^2_{W_{p,k}} + \frac{1}{2}\|x_1^e - x_1\|^2_{W_a} \tag{1}$$

Subject to:

$$x_{k+1} = f(x_k, u_k) + w_k \text{ for } k=1,\cdots,N-1 \tag{2}$$

$$y_k = h(x_k) + v_k \text{ for } k=2,\cdots,N \tag{3}$$

$$l_k(x_k) = 0 \text{ for } k=1,\cdots,N \tag{4}$$

The state propagation discrete time has been described in equation (2), where x_k are the states to be estimated. In the same equation, u_k represents the known inputs, while w_k is the process noise with covariance matrix Q_k. Equation (3) represent the measurement equation, while y_k is the measurement, and v_i, R_k are the measurement noise and the measurement noise covariance respectively. According to equation (1) we can easily notice that x_1^e is the estimated value at the initial point in the horizon with covariance P_1. The last term in the cost function describes the arrival cost. We mention that the measurement at time point $k=l$ is not included in the cost function. As the fact that all the information gained from this measurement is already contained in the prior estimate xf. The equality constraints are given by equation (4). As result the output of MHE is x_N which is actually the estimate state vector at the final point in the horizon. This is thus a weighted least squares problem, in where the weights are the inverse of the covariance matrices[1]:

$$W_{m,k} = R_k^{-1} \tag{5}$$

$$W_{p,k} = Q_k^{-1} \tag{6}$$

$$W_a = (P_1)^{-1} \tag{7}$$

We admit the following notation:

$$[a]^T W[a] = \|a\|^2_W \tag{8}$$

The Gauss-Newton solution method is recommended to solve this nonlinear least-squares problem.

The measurement and state propagation functions are linearized about the previous solution $\bar{x}_k$ (or the best estimate, if no previous solution is available) in each iteration, with) $\Delta x_k = x_k - \bar{x}_k$ the deviation from this nominal state[1]:

$$\underbrace{\min}_{\Delta x_k} \frac{1}{2} \sum_{k=2}^{N} \left\| y_k - h(\bar{x}_k) - \frac{\partial h(\bar{x}_k)}{\partial x_k} \Delta x_k \right\|_{W_{m,k}}^2$$

$$+ \frac{1}{2} \sum_{k=1}^{N-1} \left\| \bar{x}_{k+1} - f(\bar{x}_k, u_k) + \Delta x_{k+1} - \frac{\partial f(\bar{x}_k, u_k)}{\partial x_k} \Delta x_k \right\|_{W_{p,k}}^2$$

$$+ \frac{1}{2} \| x_1^{(-)} - x_1 \|_{W_a}^2 \tag{9}$$

Subject to:

$$l_k(\bar{x}_k) + \frac{\partial l(\bar{x}_k)}{\partial x_k} \Delta x_k = 0 \text{ for } k = 1, \cdots, N \tag{10}$$

The optimization problem can be written as equation (11), with $\Delta X = [\Delta x_1^T, \Delta x_2^T, \cdots, \Delta x_N^T]$:

$$\underbrace{\min}_{\Delta x} \frac{1}{2} \| A_m \Delta X - B_m \|_{W_m}^2 + \frac{1}{2} \| A_p \Delta X - B_p \|_{W_p}^2 + \frac{1}{2} \| A_a \Delta X - B_a \|_{W_a}^2 \tag{11}$$

Subject to:

$$A_{eq} \Delta X - B_{eq} = 0 \tag{12}$$

where the matrices and vectors associated to the first term of the cost function are:

$$A_m = \text{diag}\left(-\frac{\partial h(\bar{x}_2)}{\partial x_2}, -\frac{\partial h(\bar{x}_3)}{\partial x_3}, \cdots, -\frac{\partial h(\bar{x}_N)}{\partial x_N}\right) \tag{13}$$

$$B_m = [(y_2 - h(\bar{x}_2))^T, (y_3 - h(\bar{x}_3))^T, \cdots, (y_N - h(\bar{x}_N))^T]^T \tag{14}$$

$$W_m = \text{diag}(R_2^{-1}, R_3^{-1}, \cdots, R_N^{-1}) \tag{15}$$

The matrices and vectors associated to the second term of the cost function are:

$$A_p = \begin{pmatrix} -\frac{\partial f(\bar{x}_1, u_1)}{\partial x_1} & 1 & 0 & \cdots & 0 \\ 0 & -\frac{\partial f(\bar{x}_2, u_2)}{\partial x_2} & 1 & \cdots & 0 \\ \vdots & \vdots & \ddots & \ddots & \vdots \\ 0 & 0 & \cdots & -\frac{\partial f(\bar{x}_{N-1}, u_{N-1})}{\partial x_{N-1}} & 1 \end{pmatrix} \tag{16}$$

$$B_p = [(f(\bar{x}_1, u_1) - \bar{x}_2)^T, (f(\bar{x}_2, u_2) - \bar{x}_3)^T, \cdots, (f(\bar{x}_N, u_N) - \bar{x}_N)^T]^T \tag{17}$$

$$W_p = \text{diag}(Q_2^{-1}, Q_3^{-1}, \cdots, Q_N^{-1}) \tag{18}$$

The matrices and vectors associated to the third term of the cost function are:

$$A_a = [-1, 0, \cdots, 0] \tag{19}$$

$$B_a = [\bar{x}_1 - x_1^e] \tag{20}$$

$$W_a = (P_1)^{-1} \tag{21}$$

Finally, the matrices and vectors associated to the equality constraints are:

$$A_{eq} = \text{diag}\left(\frac{\partial l_1(\bar{x}_1)}{\partial x_1}, \frac{\partial l_2(\bar{x}_2)}{\partial x_2}, \cdots, \frac{\partial l_N(\bar{x}_N)}{\partial x_N}\right) \tag{22}$$

$$B_{eq} = [-l(\bar{x}_1)^T, -l(\bar{x}_2)^T, \cdots, -l(\bar{x}_N)^T] \tag{23}$$

The linearised cost function in equation (11) can be combined into:

$$J(\Delta X = \frac{1}{2}\Delta X^T(A_m^T W_m A_m + A_p^T W_p A_p + A_a^T W_a A_a)\Delta X$$

$$- T(A_m^T W_m B_m + A_p^T W_p B_p + A_a^T W_a B_a)\Delta X + \frac{1}{2}C \quad (24)$$

$$= \frac{1}{2}[\Delta X H \Delta X^T - 2G\Delta X + C] \quad (25)$$

with C being a constant matrix, and H and G defined by the corresponding terms in equations (24) and (25).

This constrained linear least squares problem is solved using the approach of sequential quadratic programming (SQL) at each sampling instant[7].

3. GPS/DR System Model

3.1 State Vector Definition

We adopt a two dimensional plane motion model with fixed altitude direction which can greatly represent our GPS/DR integrated positioning system[4]. The states are selected as equation (26).

$$x = [e, v_e, a_e, n, v_n, a_n, \varepsilon_\omega, \psi] \quad (26)$$

Then we can note that e, v_e, a_e are the position, the velocity and the acceleration of the east respectively. And in the same way n, v_n, a_n are the position, the velocity and the acceleration of the north respectively. ε_ω is the random drift of Gyroscope and ψ is the calibration factor of odometer. The discrete state model is presented as:

$$x_{k+1} = f(x_k) + W_k \quad (27)$$

where f is the state transition function. For example, for the CA (motion model with constant acceleration), $f(x_k)$ is presented as:

$$f(x_k) = \begin{bmatrix} 1 & T & \frac{T^2}{2} & 0 & 0 & 0 & 0 & 0 \\ 0 & 1 & T & 0 & 0 & 0 & 0 & 0 \\ 0 & 0 & 1 & 0 & 0 & 0 & 0 & 0 \\ 0 & 0 & 0 & 1 & T & \frac{T^2}{2} & 0 & 0 \\ 0 & 0 & 0 & 0 & 1 & T & 0 & 0 \\ 0 & 0 & 0 & 0 & 0 & 1 & 0 & 0 \\ 0 & 0 & 0 & 0 & 0 & 0 & 1 & 0 \\ 0 & 0 & 0 & 0 & 0 & 0 & 0 & 1 \end{bmatrix}$$

W_k is the process noise with covariance matrix Q.

3.2 Observation Equation

GPS outputs provide the east and north positions e and n respectively, while Gyroscope offers the angular rate ω, S is the distance measured by the odometer. Then the equation (28) is the discrete observation equation.

$$Y_k = h(x_k) + V_k \tag{28}$$

where

$$h(x_k) = \begin{bmatrix} e_k \\ n_k \\ \omega_k = \dfrac{v_{n,k}a_{e,k} - v_{e,k}a_{n,k}}{v_{e,k}^2 + v_{n,k}^2} \\ S_k = \psi_k T\sqrt{v_{e,k}^2 + v_{n,k}^2} \end{bmatrix} + \varepsilon_{\omega,k}$$

Y_k is the measurement vector at time k, V_k is observation noise vector with covariance matrix $R_k = \text{diag}(\sigma_e^2, \sigma_n^2, \sigma_\omega^2, \sigma_s^2)$, $\sigma_e^2, \sigma_n^2, \sigma_\omega^2, \sigma_s^2$ are the covariance for each measurement.

Here the measurement equation is nonlinear. The Taylor series of $h(x_k)$ is given by:

$$h(x_k) \approx h(\bar{x}_k) + \frac{\partial h(\bar{x}_k)}{\partial x_k}\Delta x_k \tag{29}$$

Order

$$H_{(k)} = \frac{\partial h(\bar{x}_k)}{\partial x_k}$$

Hence

$$H_{(k)} = \begin{bmatrix} 1 & 0 & 0 & 0 & 0 & 0 & 0 & 0 \\ 0 & 0 & 0 & 1 & 0 & 0 & 0 & 0 \\ 0 & H_{32} & H_{33} & 0 & H_{35} & H_{36} & 1 & 0 \\ 0 & H_{42} & 0 & 0 & H_{45} & 0 & 0 & H_{48} \end{bmatrix} \tag{30}$$

where

$$H_{32} = \frac{a_{n,k}v_{e,k}^2 - 2v_{e,k}v_{n,k}a_{e,k} - a_{n,k}v_{n,k}^2}{(v_{e,k}^2 + v_{n,k}^2)}$$

$$H_{33} = \frac{a_{e,k}v_{e,k}^2 + 2v_{e,k}v_{n,k}a_{e,k} - a_{e,k}v_{n,k}^2}{(v_{e,k}^2 + v_{n,k}^2)}$$

$$H_{35} = \frac{v_{n,k}}{v_{e,k}^2 + v_{n,k}^2} \qquad H_{36} = \frac{-v_{e,k}}{v_{e,k}^2 + v_{n,k}^2}$$

$$H_{42} = \frac{\psi_k T v_{e,k}}{\sqrt{v_{e,k}^2 + v_{n,k}^2}} \qquad H_{45} = \frac{\psi_k T v_{n,k}}{\sqrt{v_{e,k}^2 + v_{n,k}^2}}$$

$$H_{48} = T\sqrt{v_{e,k}^2 + v_{n,k}^2}$$

4. Simulation and Discussion

In this section an example with constant acceleration motion model is presented with the aim to demonstrate the accuracy of the proposed algorithm. The filtering duration is about 1000 seconds with sampling time $T=1$ s. We suppose that the vehicle moves on a straight line road with constant acceleration $\sqrt{2}m/x^2$ and the heading is 45°. The initials values are as follow: the intial value of the east and north position is 0, the east velocity and north velocity is 1 m/s, and the calibration factor of the odometer is 1. The parameters related to the measurement noise are: $\sigma_e^2 = \sigma_n^2 = 10^2$, $\sigma_\omega^2 = (\frac{0.1\pi}{180})^2$,

$\sigma_s^2=1^2$, and the parameters related to the process noise are: $\sigma_e^2=\sigma_n^2=0.001^2$, $\sigma_{v_e}^2=\sigma_{v_n}^2=0.002^2$, $\sigma_{ae}^2=\sigma_{a_n}^2=0.1^2$, $\sigma_\omega^2=\sigma_\psi^2=0.01$.

Root mean square error (RMSE) of east direction is given by Fig. 1 which in fact shows simulation results after using EKF, UKF and MHE with length of horizon $N=5$ under the presence of Gaussian noise, while Fig. 2 shows the results using same filters with different length of horizon $N=10$. According to the two previous figures we can clearly observe that MHE performs much better than the two other filters.

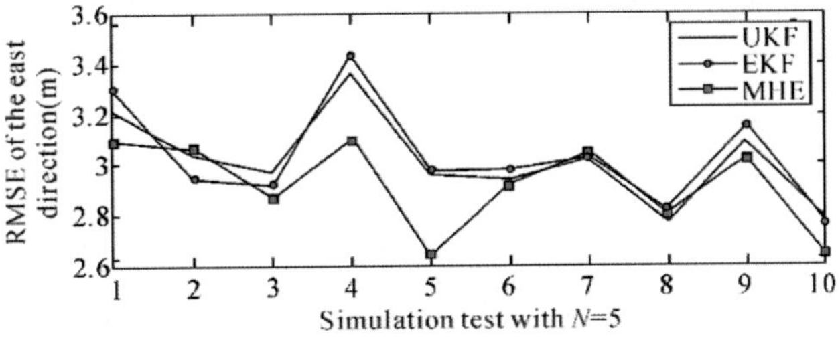

Fig. 1 RMSE of east direction position estimation with Gaussian noise and length of horizon $N=5$

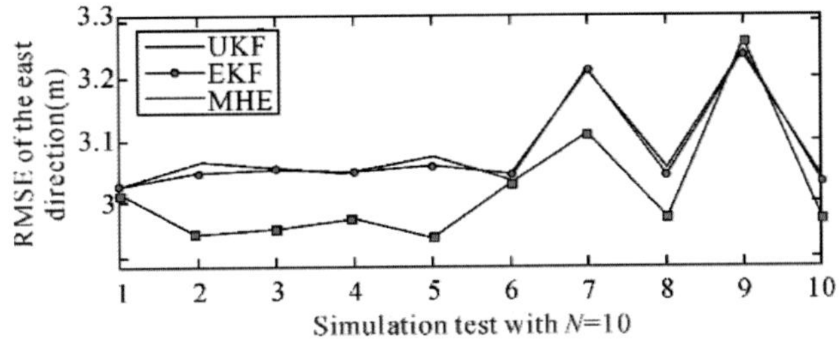

Fig. 2 RMSE of east direction position estimation with Gaussian noise and length of horizon $N=10$

Root mean square error (RMSE) of east direction is given by fig. 3 which in fact shows simulation results after using EKF, UKF and MHE with length of horizon $N=5$ under the presence of non-Gaussian noise. This figure also shows that UKF is more efficient than EKF while MHE provides much better results from both of them which means that MHE is a robust algorithm with non-Gaussian noise.

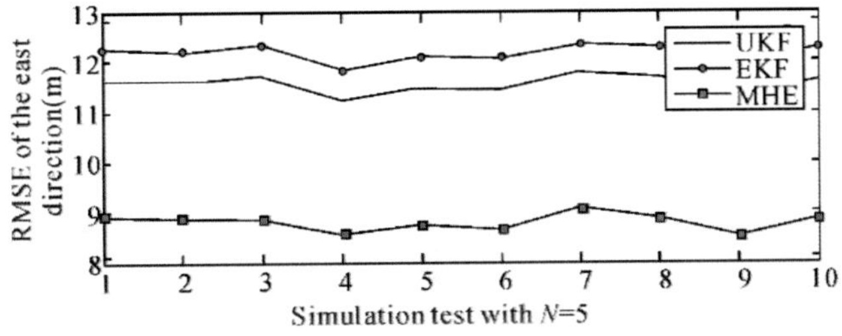

Fig. 3 RMSE of east direction position estimation with non-Gaussian noise and length horizon $N=5$

5. Conclusion

In this study, we propose an efficient approach named MHE in order to improve GPS/DR integrated positioning system performance. In MHE, the state of the moving vehicle can be estimated by solving an optimization problem using quadratic programming. We get a clear simulation results that can always prove and confirm the effectiveness of the mentioned approach in comparison to unscented Kalman filter and extended Kalman filter. The simulation results also guarantee the robustness of the proposed algorithm under non-Gaussian noise.

References

[1] J. Vandersteen, M. Diehl, C. Aerts, and J. Swevers, "Spacecraft Attitude Estimation and Sensor Calibration Using Moving Horizon Estimation," Journal of Guidance, Control, And Dynamics, Vol. 36, No. 3, 2013.

[2] A. Alessandri, M. Baglietto, G. Battistelli, and V. Zavala, "Advances in Moving Horizon Estimation for Nonlinear Systems," 49th IEEE Conference on Decision and Control, Altana, GA, USA, December 15-17, 2010.

[3] K. V. Ling, and K. W. Lim, "Receding Horizon State Estimation,", Automatic Control, IEEE Transactions on, Vol. 44, No. 9, pp. 1750-1753, 1999.

[4] D. K. Yang, X. L. Zhou, X. Liu, and Q. S. Zhang, "U-GPF Information Fusion Algorithm for GPS/DR Integrated Positioning System," Proceedings of the Sixth International Conference on Machine Learning and Cybernetics, Hong Kong, 19-22 August, 2007.

[5] W. Qiuping, G. Zhongyu, and W. Dejun, "An Adaptive Information Fusion Method to Vehicle Integrated Navigation,", Position Location and Navigation Symposium, IEEE, 2002. pp. 248-253.

[6] L. Zhao, and Z. Yuan, "GPS/DR Vehicle Integrated Navigation System Based on Central Difference Kalman Filter,", Journal of Information and Computational Science, Vol. 9, No. 10, 2012.

[7] P. Kuhl, M. Diehl, T. Kraus, J. P. Schloder, and H. G. Bock," A Real-Time Algorithm for Moving Horizon State and Parameter Estimation,", Journal of Computer and Chemical Engineering, Vol. 35, No. 1, pp. 71-83, 2011.

STRATEGY ISSUE OF DEVELOPMENT

Globalization of R&D: New Trend of Economy Globalization*

Zhang Yanzhong
Chinese Academy of Engineering, Beijing, China
UN, Geneva

The rapid development of economy globalization has exerted significant influence on the progress of the world. Globalization of R&D has become a new trend in economy globalization, which is thus led into a new stage. However, there exist some issues need to be studied and discussed jointly by all countries.

1. Profound Influence of Economy Globalization on the Progress of World

Last century, economy globalization experienced manufacturing globalization and market globalization. Transnational companies (TNCs) set up their manufacturing factories in developing countries for globalized production, which later extended into service industry gradually. Economy globalization benefits developed countries in terms of market expansion, taking advantages of cheap labor force and resources in developing countries, thus to reduce cost and gain rich profit on one hand; it accelerates the flow of those key elements, such as capital, product, technology, service, etc., to developing countries on the other hand. However, there arises an issue of protection of IPRs.

Economy globalization stimulates developing countries to introduce capital, increase employment opportunities, train talents and learn advanced manufacturing technology, management experience and modern marketing knowledge, which further drives the structure readjustment of economy and industry. Meanwhile, it also influenced the development of national industries in developing countries. However, it also engendered issues in environment, resources and human rights.

Since the reform and opening up policy was adopted in China, the utilization of foreign direct investment went through great changes from none to many, from small scale to large scale, from single mode to diversified mode, which has formed an all-directional, multi-layered and wide-ranging pattern. Currently, the countries and regions investing in China are more than 180. Over 480 of world top 500 TNCs have invested in China, and more than 40 of them has set up their regional headquarters in China. By 2006, China has approved the establishment of 613, 110 foreign

* Published in expert meeting, on impact of FDI on development, UNCTAD, Geneva, 24-26 Jan. 2005.

Y. Zhang, *Footprints in Cambridge and Aviation Industries of China*,
https://doi.org/10.1007/978-981-16-3176-4_35

investment enterprises accumulatively, with an amount of US $1078.168 billion of contractual foreign investment, of which US$723.8 billion executed.

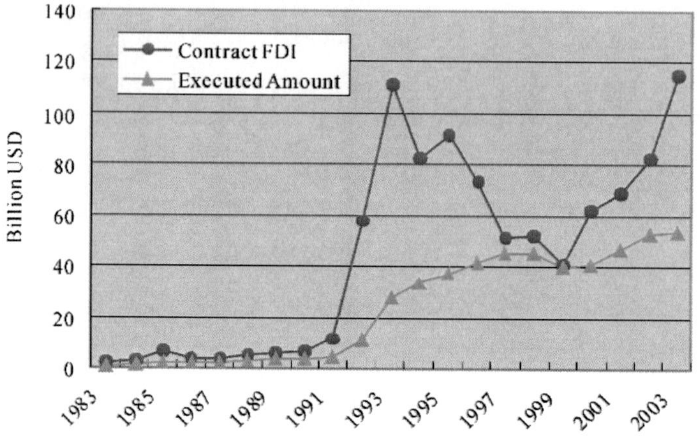

Fig. 1 FDI in China

In the first half of 2007, total export and import by foreign invested enterprises amounts to US $564.452 billion, increased by 21.31% in comparison with the corresponding period of 2006, and shares 57.54% of China's total import and export.

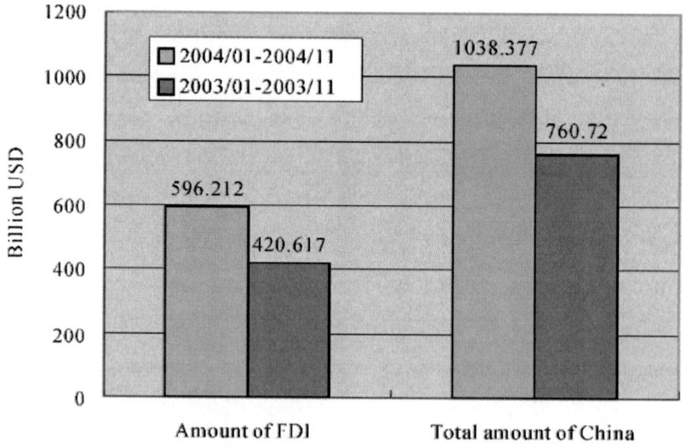

Fig. 2 Amount of export & import in China

2. Globalization of R&D—A New Stage of Economy Globalization

2.1 Driver from Developed Countries

With the rapid change of world S&T and market economy, the period of developing product is gradually shortened; the speed of upgrading product is getting faster and faster whilst the cost of

R&D raising constantly. It will not meet the needs of momentarily changing market and the development of global economy only by means of setting up manufacturing enterprises in developing countries. From the end of last century, discontented with the mode of investing only for setting up factories, many TNCs have shifted their attention to setting up R&D organizations in developing countries. They changed their gradual transition of investment strategy from labor intensive manufacture to capital and technology intensive R&D production, from simple technology transfer to R&D localization and from passive reluctant technology transfer to self-initiated technology investment. In the beginning of the 21st century, a growing number of TNCs has implemented the strategy of localization in developing countries to strengthen, in an all-directional and in-depth manner, the process of localized operation covering talents, materials, suppliers, manufacture, R&D and enterprises culture etc.

The development of world S&T and economy has promoted the improvement of technology and education level in developing countries, which consequently provides abundant human resources and a broad technology application market for developed countries when setting up R&D organizations in developing countries. Besides, the leap of IT industry, and especially the network technology in developing countries drives further the globalization of R&D.

The above-mentioned factors are the two drives of the speedy and fierce progress of R&D globalization. Economy globalization now enters into a new stage of R&D globalization.

2.2 Environment in Developing Countries

In developing countries, on one hand, the improvement in science education has provided low cost intelligence resources, and on the other hand, technology conditions, such as communication and internet, etc, changed rapidly, which also provides favorable conditions for the development of globalization of R&D.

FDI R&D can be categorized into 4 different types:

Table 1 Type of FDI R&D

	Developed countries (US, Europe, and Japan etc.)	Developing countries (China, India etc.)
Developed countries (US, Europe, and Japan etc.)	Collaboration (Type A)	Expansion (Type B)
Developing countries (China, India etc.)	Following (Type C)	Localization (Type D)

(1) Type A: Collaboration

This is the type in which developed countries (such as the US, European countries, Japan, etc.) collaborate with each other by making FDI in R&D among each other, thus to best utilize each other's specialties, for instance, Airbus aircraft is co-developed and produced under certain work share (France for final assembly, the UK for wing, Spain for tail, Germany for Airframe) among partner countries. In addition, Germany, Netherlands, France, England, Italy Sweden and Spain collaborate on basic R&D. Another example is ITER (International Thermo-nuclear Experimental Reactor), in which, Europe, the US, Russia, Japan, China, India and Korea join R & D. ITER device established in France.

(2) Type B: Expansion

This is the type in which developed countries set up R&D institutes in developing countries (such as China, India, etc.) to reduce R&D cost and expand market by localization. Take NOKIA as an example. NOKIA has set up 44 R&D centers in 22 countries, and hired 13,000 people which are about 30% of its overall overseas employees. Besides, companies such as Microsoft, IBM, GE, Siemens, Unilever, SONY and TOYOTA etc., have set up R&D institutes all over the world.

(3) Type C: Following

This is the type in which developing countries set up branches in developed countries to grasp information, to be aware of trend and to utilize local talents. For example, Haier, a Chinese company had set up design branch departments in the Silicon Valley of the US, Japan, France, the Netherlands and Canada etc., and set up information centers in South Korea, Austria, Australia, the US, Japan and the Netherlands etc.

(4) Type D: Localization

This is the type in which developing countries set up regional headquarters and R&D centres in developing countries aiming at developing products suitable for the local and realizing localization. Huawei, a Chinese company features this type. It set up 8 overseas regional headquarters, and established R&D centers in the US, Sweden, Russia and India to realize localization. Huawei has applied 2,635 PCT international and foreign patents (14,252 domestic patents)

Among all FDI R&D, type B is the most popular form, sharing 64% over total, whilst type D is the least applied form, only shares 3%, and type A and C share 25% and 8% respectively.

Table 2 Type of FDI R&D

	Developed countries (US, Europe, and Japan etc.)	Developing countries (China, India etc.)
Developed countries (US, Europe, and Japan etc.)	Type A 25%	Type B 64%
Developing countries (China, India etc.)	Type C 8%	Type D 3%

2.3 Rapid Development of TNCs' R&D in China

In China, foreign invested R&D organizations boom rapidly. Up to June 2006, R&D organizations established in every form have been over 750, with an accumulated R&D capital of US $4 billion. Nearly 40% are set up by the US, about 25% by Europe and Japan. Among those numerous TNCs, for instance:

IBM opened its IBM China R&D Center in September 1995, which is one of the eight research centers that IBM established worldwide. It mainly engages in researching and developing products and technologies including phonetic identification technology and text analysis, by using software and hardware with Chinese characteristics. In 1999, IBM set up the IBM China Software Development Lab in China with an employment of 800, mainly engaging in the development of software in areas such as E-business integration, distributed numeration, and LINUX, etc.

Microsoft R&D Center originally established in 1993 as the former Microsoft Beijing Testing Center, and it was formally opened in 1995 in the same name. Currently it has more than 150 employees, being the biggest software product R&D organization set up in China by a foreign company. The center consists of nine departments, including Windows Platform Department, Desktop Application Department, Chinese Technology Department, Windows System Server Department, MSTV Product Department, Program Developer Tool Product Department, Home Product & Retail Department, MSN International Product Department, and Mobile Product Department. Since 1995, the Microsoft R&D Center has successfully provided Chinese market with more than 230 Microsoft softwares in Chinese version, such as Windows operation systems, Office software, Venus software, and Microsoft Pinyin software.

GE (China) Research and Development Center is one of the three global research centers of GE. It has employed 700 people, which is planned to increase to 1200 in 2005, mainly engaging in research in areas of electric power, electronics, medical imaging technology, material science, advanced manufacturing technology, chemistry, etc. Its technology and product of R&D support all the organizations and businesses of GE over the world.

Nokia has set up 5 R&D organizations in China engaging in wireless technology research, such as Chinese character mobile design, CDMA software, and 3G/WCDMA net solution software platform, etc. Nokia has invested over RMB 10 billion to found a world-class high-tech industry zone, namely, Star Net International Industry Zone, to engage in R&D, manufacture, sale and service for mobile communication products as well as parts and components.

Airbus recently established in China an engineering center of around 200 engineers for the purpose of civil passenger aircraft development.

Please refer to appendixes for details of some other TNCs' R&D organizations.

TNCs' R&D organizations in China feature the following:

R&D organizations established at the initial stage, focusing on the real situation in China, mainly adapt the corresponding products in conformity to China's market rules and regulations, standards and language. For example: Microsoft, IBM, Nokia and other IT companies developed Chinese characterized computer, mobile phone, software and print equipment to sell in China; Sony, Panasonic and other household appliance manufactories converted the voltage of products sold in Japan and America from 110 volt to 220 volt and to cater for Chinese market's demand for PAL video and audio products; and many overseas car makers refit right-driving cars promoted in their own countries to left-driving ones to conform to China's driving rules; etc.

R&D organizations mainly centered in technology intensive industries, such as: computer and telecommunication equipment production, medical production, new materials and chemical production and transportation tools production, etc., and over 60% of R&D organizations are distributed in the IT industry, and about 10% or less than 10% are distributed in other industries, such as fine chemical, biotech, auto, and others.

The potential areas for Canada to cooperate with China in R&D can cover energy industry (especially reproducible energy), agriculture, environment protection, aviation industry and high-technology.

Human resources of R&D organizations gradually realized localization. Local R&D personnel employed work in TNCs' R&D organizations in China accounts for 95.1%.

R&D organizations located mainly in economy and technology better-developed areas, with comparatively good basic condition in terms of human resources, communication and transportation, etc. For example, R&D organizations set up in Beijing, Shanghai, Guangzhou and Shenzhen by overseas enterprises account for 80% of that in China. In addition, many are close to local universities and research institutions, or centered in development zone to enjoy relevant favorable policies. To take an example, in Beijing, R&D organizations of TNCs located in Zhongguancun area and in Beijing Economic and Technological Development Zone have reached 50. Zhongguancun is an area of dense intelligence on Mainland China where lies more than 200 research institutions and 69 universities.

In recent years, more and more TNCs have gradually leveraged their R&D units in China from China market oriented to global market oriented, thus to lift the position of those R&D units in their global strategy. For instance, the nuclear magnetic resonance, plastic and advanced manufacturing technology developed by GE's R&D center in Shanghai, which will support not only GE's products in China but also those in the world. Other examples also include Microsoft, Nokia, Sony Ericsson, and Panasonic, etc. Their R&D centers in China are also the global R&D centers of the group.

3. Opportunities and Challenges of R&D Globalization

3.1 Opportunities and Challenges to Developed Countries

(1) Favorable for occupying and expanding market

By establishing R&D organizations in developing countries, TNCs benefit from strengthened competitiveness in the world, R&D cost reduction and higher profit obtained.

(2) Favorable for industrial localization

Setting up R&D organizations is beneficial to develop products on the basis of existing ones and in conformity with local standard, criterion, and language products to meet local market's requirement, thus to realize R&D localization; beneficial to utilize local resources to realize localization of material supply and production; and beneficial to the employment of local talents to realize talent location.

(3) Favorable for cutting cost

In developed countries, the continuous increase of technology contents of product and the rise of R&D labor cost lead to growing R&D cost to develop products. For example, the R&D cost over the total product cost for developed countries is respectively: 60%~70% for software industry, 20%~30% for aviation products, 5%~20% for automobile products. While developing countries have the advantages of abundant skill and human resources for R&D, lower labor cost (In developing countries it is 1/10 ~ 1/3 of that in developed countries), lower service charge, convenient transportation, short market access distance and cheap transportation costs, etc., which provides favorable conditions for TNCs to set up R&D organizations and to cut R&D cost.

(4) Favorable for utilizing local intelligence resources

Some developing countries, such as China and India, have seen fast development of domestic higher education since the eighties last century. Graduates and postgraduates have been booming in both qualities and quantities, with obviously improved foreign language commanding capability. Besides, there are many oversea returned students. For example, in China, there are 3.5 million graduates, and over 0.3 million postgraduates and tens of thousands of oversea returned students every year. They are not only versed in Chinese but also familiar with English and own certain specialized knowledge, who provide R&D organizations set up by developed countries in developing countries with rich and low cost intellectual resources, with no immigrant problems.

(5) Acquiring market information to develop new technology and product

Help TNCs to understand consuming custom and culture tradition of developing countries and grasp local market information, allowing developing products right for the market and easily accepted by consumer. For example, Chinese people prefer multi-functional and long life household electrical appliances and economical sedans of low price, high quality and more seats.

(6) Problems existed

There are some problems about the protection of intellectual property right.

3.2 Opportunities and Challenges Facing Developing Countries

(1) Favorable for improving the utilization efficiency of foreign direct investment and stimulating structure optimization and industry upgrading.

(2) Favorable for meeting ever growing diversified demand of domestic consumers, stimulating consuming and promoting economy development.

(3) Favorable for attracting foreign investment, introducing equipment, undertaking all-directional joint investment cooperation in a broader field, strengthening localization and magnifying the technology overflowing effect to speed up development.

(4) Favorable for training local staff and attracting overseas scholars back to start businesses in home countries.

(5) Favorable for acquiring information, learning advanced technologies and improving.

(6) Problems existed: competition among high-tech talents, the ability for independent innovation and the formation of proprietary intellectual property rights.

3.3 Boost and Effect on Economy Globalization

As for the effect of R&D globalization, its advantages for developed countries include acquiring market information to develop new technology and product, using human resources of developing countries, cutting cost, realizing localization of product, occupying and expanding market and gaining excessive profit. Whilst, developing countries are able to attract foreign direct investment, introduce equipment, undertake all-directional joint investment cooperation in a broader field, train local staff and attract overseas scholars back to create businesses in home countries, acquire information, learn advanced technology and raise levels. Viewing from overall world trade, it would be difficult for economy globalization to develop further to a higher stage without the globalization of R&D. Therefore, globalization of R&D is a new trend to drive economy globalization and to stimulate the development of world economy, which is really good more than harm to both developed countries and developing countries, resulting in a "win-win" situation. All countries in the world should make joint efforts and with full energy to push it forward.

4. Policies Encouraging Setting up R&D Institutions in Developing Countries

Nowadays, the globalization of R&D progresses rapidly in the whole world, but there still exists some imbalance among countries as well as regions, and lack efficient communication scheme bridging developed countries with developed counties. Therefore, it is necessary to study policies and measures to support and encourage setting up R&D organizations in developing countries.

4.1 Measures Suggested Taken by Developed Countries

(1) The restrictions on high-and new-tech industries for establishing R&D institutions in developing countries should be loosened.

(2) Formulate policies to support TNCs to set up R&D organizations in developing countries, and to encourage the flow of capital, products, technology and R&D to developing countries.

(3) Help developing countries to speed up talents training and improve scientific education level.

4.2 Measures Suggested Taken by Developing Countries

(1) Infrastructure, such as: Internet, communication, transportation, finances, and for daily life, should be enhanced to create necessary basic conditions for setting up R&D organizations.

(2) "Soft" environment attracting developed countries to set up R&D units should be created. More favorable policies than supporting simply investing for factories should be formulated, such as: to support domestic scientific research institutes to jointly set up R&D organizations with TNCs; exempt tax of non-profit making specialized equipment that TNC invested to R&D organizations; to grant national treatment to R&D organizations set up by TNCs, extend protection scope of intellectual property and encourage the mobility of talents.

(3) Laws and regulations in respect of how to develop independent innovation abilities and proprietary intellectual property rights when jointly setting up R&D organizations with TNCs should be studied and promulgated.

(4) Talents training and reinforcing research condition construction as well as scientific

education development should be emphasized.

4.3 Measures Taken by Chinese Government

(1) Equipment and supporting technologies, spare parts imported within total investment for self-use and is unproductive equipment not forming production scale are exempted from import tariff and import linkage tax.

(2) Income from transferring technology independently developed by R&D organization is exempt from turnover tax.

(3) If the expenditure on technology development increased by 10% or above than that of the previous year, the taxable income of that year can be deducted by 50% of the actual expenditure on technology development.

(4) A small amount of high-tech products produced by the parent company are permitted to be imported and sold by R&D unit for the purpose of performing market survey.

4.4 Measures Suggested Taken by UN

(1) To support and encourage developed countries gradually remove the restrictions on high technology transfer and advocate TNCs to found or jointly found R&D organizations in developing countries should be embraced into the agenda of UN trade development as an important content to promoting the development of world trade.

(2) Bolster developing countries to accelerate the development of S&T and education. Help developing countries to jointly found R&D organizations with TNCs to be win-win. Promote the globalization of R&D and narrow the gap between developing countries and developed countries.

(3) To establish communication and negotiation mechanism on R&D organizations between developed and developing countries for timely studying and coordinating problems in the process of R&D globalization. To work out measures for a clear delimitation of intellectual property rights. To set game rules and determine protection policies for R&D globalization and to promote the development of both economy and R&D globalization.

5. FDI R&D is One Important Part of Host Country's Innovation System

FDI R&D is one important part of host country's innovation system for the following reasons:

(1) Localization of FDI R&D promotes international cultural communications, renewal of knowledge.

(2) Technology Spillover Effect of FDI R&D contributes to increasing technology and management level in host countries.

(3) Innovation orientated FDI R&D contributes to host countries' knowledge innovation.

(4) FDI R&D activities contribute more to host countries' economic and S&T development than joint ventures.

To support trans-national innovation orientated R&D activities, countries could take measures, such as:

(1) Preferential policy support, such as in taxation, to innovation orientated R&D activities;

(2) Perfect polices on Venture Capital Investment;

(3) Reinforce IPR protection;

(4) Encourage hastening the transfer and application of R&D achievements;

(5) Encourage medium- and small-TNCs' R&D.

6. Conclusions

The rapid development of economy globalization has exerted significant influence on the progress of the world. Globalization of R&D has become the new trend of the economy globalization. Opportunities and challenges co-exist in GRD. Special attention will be paid to the development trends and drivers of the GRD, its impacts on innovation and technology transfer, and its benefits and costs for both home and host economies.

As the largest developing country, China applauds measures taken by UN for boosting the globalization of R&D. More TNCs are welcomed to set up high-tech R&D organizations in China which would benefit both sides, thus to make joint contributions to globalization of R&D.

Appendix I Part (82) of R&D organizations set up in China

(Up to August 2002)

No.	R&D Organizations	Industry	Country/ Region	Location
01	Novonordisk Biology R&D Center	Biological medicine	Denmark	Beijing
02	Novozymes China R&D Center	Biological medicine	Denmark	Beijing
03	SAP Beijing Software Co. R&D Center	IT	Germany	Beijing
04	Siemens Mobile Communication Pioneer Technologies Center	IT	Germany	Beijing
05	Schlumberger (China) Technologies Co. Ltd.	Optical, mechanical and electronic integration	France	Beijing
06	Servier (Beijing) Medical R&D Co. Ltd.	Biological medicine	France	Beijing
07	Nokia (China) R&D Center	IT	Finland	Beijing
08	LG Electronic Department Research	IT	Korea	Beijing
09	Samsung SDS Beijing R&D Center	IT	Korea	Beijing
10	Beijing Samsung Telecom Technologies Research Co. Ltd.	IT	Korea	Beijing
11	Nortel Networks R&D Center	IT	Canada	Beijing
12	IBM China R&D Center	IT	US	Beijing
13	Peopletech (China) R&D center	IT	US	Beijing

(To be continued)

Appendix I

No.	R&D Organizations	Industry	Country/ Region	Location
14	Sun China Engineering Research Center	IT	US	Beijing
15	Agilent Technologies Software Co. Ltd.	IT	US	Beijing
16	Agilent Technologies Co. Ltd.	IT	US	Beijing
17	GE Hangwei Medical Systems Co. Ltd.	Biological medicine	US	Beijing
18	Beijing P&G Technologies Co. Ltd.	Fine chemistry	US	Beijing
19	Beijing Petrolor Equipment Co. Ltd.	Oil exploration	US	Beijing
20	Rainbow China R&D Center	IT	US	Beijing
21	Delphi Technologies R&D Center	Automobile	US	Beijing
22	Qualcomm Beijing CDMA Center	IT	US	Beijing
23	HP China Lab	IT	US	Beijing
24	Honeywell Beijing Software R&D Center	IT	US	Beijing
25	Lucent technologies (China) Co. Ltd. Bell Labs Innovation	IT	US	Beijing
26	Fourth Shift China R&D Center	IT	US	Beijing
27	Microsoft R&D Center	IT	US	Beijing
28	Microsoft Research	IT	US	Beijing
29	Intel China R&D center	IT	US	Beijing
30	JVC (Beijing) Technology R&D Center	IT	Japan	Beijing
31	SMC-Qinghua University Pneumatics Center	Optical, mechanical and electronic integration	Japan	Beijing
32	Beijing Shimadzu Analysis Center	Biological Medicine	Japan	Beijing
33	Mitsubishi Telecom Co. Ltd. R&D Center	IT	Japan	Beijing
34	Beijing SUOHONG Electronics Co. Ltd.	IT	Japan	Beijing
35	NEC-CAS Software Laboratories Co. Ltd.	IT	Japan	Beijing
36	Fujitsu R&D Center Co. Ltd.	IT	Japan	Beijing
37	Panasonic China Co. Ltd. R&D Department	IT	Japan	Beijing
38	Panasonic R&D Co. Ltd.	IT	Japan	Beijing
39	Sony Ericsson Mobile Communications (China) Co. Ltd.	IT	Japan	Beijing
40	Shiseido Research Center (Shiseido Liyuan Cosmetic Company)	Fine chemistry	Japan	Beijing
41	Toshiba China R&D Center	IT	Japan	Beijing
42	Ericsson Mobile Multi-Media Communication Open Lab	IT	Sweden	Beijing

(To be continued)

Appendix I

No.	R&D Organizations	Industry	Country/ Region	Location
43	Amersham Biosciences	Biological Medicine	Sweden	Beijing
44	VIA Technologies, Inc. , Beijing	IT	Taiwan, China	Beijing
45	Kinpo Electronics (Beijing) Co. , Ltd.	IT	Taiwan, China	Beijing
46	Softstar Technology (Beijing) Co. , Ltd.	IT	Taiwan, China	Beijing
47	Acer Computer Beijing R&D Center	IT	Taiwan, China	Beijing
48	Beijing Recbok Technology Development Co. Ltd.	IT	UK	Beijing
49	Motorola China R&D Institute	IT	US	Beijing
50	Guangdong Nortel R&D Centre	IT	Canada	Guangzhou
51	Guangzhou Honda Technology Centre	Automobile	Japan	Guangzhou
52	Nokia (China) R&D Centre	IT	Finland	Hang Zhou
53	BASF Colorants and Chemicals Co. Ltd.	Fine chemistry	Germany	Shanghai
54	Bayer Shanghai Polymer R&D Center	Fine chemistry	Germany	Shanghai
55	Alcatel Shanghai Bell Co. Ltd.	IT	France	Shanghai
56	RHODIA SILICONES SHANGHAI CO. , LTD/Rhone-Polenc (CHINA) CO. , LTD. SHANGHAI BRANCH	Fine chemistry	France	Shanghai
57	IBM China Software Development Lab (CDSL)	IT	US	Shanghai
58	Bell Labs Research China	IT	US	Shanghai
59	Delphi Corporation	Automobile	US	Shanghai
60	DuPont Shanghai Science Co. Ltd.	Fine chemistry	US	Shanghai
61	Toshiba Electronics (Shanghai) Co. Ltd.	IT	US	Shanghai
62	GM-Shanghai Jiao Tong University Automotive Powertrain Technology Institute	Automobile	US	Shanghai
63	Honda Motorcycle R&D China Co. Ltd.	Automobile	Japan	Shanghai
64	Shanghai Toshiba Co. Ltd.	IT	Japan	Shanghai
65	Ericsson Communication Software R&D. (Shanghai) Co. , Ltd.	IT	Sweden	Shanghai
66	Shanghai Givaudan Ltd.	Fine chemistry	Switzerland	Shanghai
67	Unilever R&D Shanghai	Fine chemistry	UK	Shanghai
68	Shenzhen Oracle China Development Centre	IT	US	Shenzhen
69	Compaq Shenzhen R&D Centre	IT	US	Shenzhen
70	Lucent Science & Technology Shenzhen R&D Centre	IT	US	Shenzhen
71	Fujitsu Shenzhen Co.	IT	Japan	Shenzhen

(To be continued)

Appendix I

No.	R&D Organizations	Industry	Country/ Region	Location
72	VTech Holdings	IT	Hong Kong, China	Shenzhen
73	Philips R&D Center	IT	Holland	Suzhou
74	BenQ (Suzhou) Ltd. Co.	IT	Taiwan	Suzhou
75	Ambow Software Suzhou Corp. Ltd.	IT	US	Suzhou
76	Philips Research East Asia Xi'an	IT	Holland	Xi'an
77	HP E-business R&D Centre	IT	US	Xi'an
78	Xi'an NEC Radio Communications Equipment	IT	Japan	Xi'an
79	Agilent Technologies Qingdao R&D Center	IT	France	Chengdu
80	Alcatel	IT	US	Qingdao
81	Qingdao DIC Finechemicals Co., Ltd. Research Institute	Fine Chemistry	Japan	Qingdao
82	United Nations Nantong Pesticied Formulation Development Center	Biological medicine	—	Nantong

Building a Conservation-Minded Society by Means of Science and Technology Innovation*

Dr. Zhang Yanzhong
The "Building of a Conservation-Minded Society" Project Group
Chinese Academy of Engineering

Abstract: To build a conserving-minded society is one of the basic policies of China, an important practice in the construction of socialism with Chinese characteristics and also an obligatory duty for the engineering and technology fields.

China has a large population of 1.3 billion, accounting for 20% of the total population of the world. However, our country is short of natural resources compared to the population. The natural resources per capita of China are much lower than the average figure of the world. The energy consumption of the major energy-consumed products in China is about 20% higher than that in the advanced countries, and the general efficiency of our energy system is about 10% to 20% lower than that of the advanced systems in the world. China has to import a large amount of raw materials, such as petroleum, iron and nonferrous ore etc. Furthermore, our country has not formed a recycling economy yet. For each year, more than 2 million vehicles and more than 15 million television sets, wash machines and refrigerators are discarded as useless. In the countryside, about 700 million tons of straw are waiting for treatment every year. The rapid development of economy and the utilization of resources are accompanied with environmental pollution, including air, water and land pollution. As the emission of sulfur dioxide is so great, about one third of the areas of China is under the threaten of acid rain.

Facing this critical circumstance and under the scientific concept of development, China must change the current economy growth mode to build a resource-conserving and environment-friendly society so as to avoid the development mode with high energy & resources consumption and heavy pollution.

The targets for the building of a conservation-minded society are as follows:

By the end of the 11th "Five-Year Plan", China with double the GDP per capita in 2010 compared with that in 2000, will decrease the consumption of energies per unit GDP by 20% compared with that of 2005. The emission for unit GDP will be 10% lower than that of 2005.

The GDP per capita will be RMB 15656 in 2010, twice as RMB 7828 in 2000. The unit consumption of energies will equal to 1.144 tons of standard coal equivalent (tce) for GDP of per

* Published in CAE Congress, Beijing, June 2006.

Y. Zhang, *Footprints in Cambridge and Aviation Industries of China*,
https://doi.org/10.1007/978-981-16-3176-4_36

RMB 10000 in 2010, decreasing from 1.43 tons of standard coal (tce) for GDP of per RMB 10000 in 2005.

By 2020, the GDP per capita will be four times as that of 2000, and the consumption of energies per unit GDP will decrease by 50% from that of 2005.

There is a gap between the supply and the demand of resources and energies in China in the realization of the targets for the year 2020.

The total energy resources demanded by China will be 2.8 to 3.2 billion tce, while the domestic supply of China could only provide 2.3 to 2.7 billion tce. In details, the domestic supply of China could satisfy the demand for coal and electricity; the domestic supply of petroleum is 180 to 200 million tons, about 40% of the total demand of 450 million tons; the domestic supply of natural gas is 120 billion m^3, 60% of the total demand of 200 billion m^3. China's energies will still mainly rely on coal in future, while the proportion accounted by hydroelectricity, reproducible energies and nuclear electricity will grow up. There is a great lack of petroleum and natural gas, which should be addressed by multi-way solution. The annual consumption of water should be controlled within 630 billion m^3, the consumption of water for agricultural irrigation should be endeavored to realize zero increase and the consumption of water for industries should keep an average annual increase within 1%. There is a gap about 50 million mu of plow land in China. The whole country will need 330 million tons of steel while the domestic market could only provide less than 50% row materials. The ratio of domestic supply of copper, aluminum, lead and zinc will be 35%, 50%, 60% and 85% respectively.

As there is some gap between the supply and the demand of energy and resources, the Chinese government must take measures and try its best to save resources.

The targets on saving resources by the end of the 11th "Five-Year Plan" are as follows:

The GDP produced by unit resource for 15 resources will increase by 25% compared with that of 2003; the average effective utilization coefficient for agricultural irrigation will rise to 0.5; the water used for industrial increased value per RMB 10 thousand will decrease to 120 m^3; the total recycling ratio and general utilization ratio for ores and minerals will rise 5% respectively; the general utilization ratio for industrial solid waste will rise to 60%; the regenerated copper, aluminum and lead will take 35%, 25% and 30% in their total output respectively; the ratio for the recycling and utilization of major regenerated resources will be over 65%; the storage of solid industrial waste and those under treatment will be restrained to 450 million tons; and the increase for urban household garbage will be controlled within around 5%.

To reach the aim of a quadrupled GDP per capita before 2020, China should not only rely on its resources and energies, but also follow out the scientific concept of development, building an innovation-oriented country, accelerate the readjustment of the economical structure, change the mode of economic growth and strengthen the development of tertiary industry and hi-tech industry. In a word, the development of China's economy should depend on enhancing science & technology and building a conservation-minded society.

As to the situation in China, before 2020, the population will continue growing at a rate of about 10 million per year, and about 10 million to 20 million people will become urban inhabitant because of the urbanization. Therefore, houses, automobiles and household appliances will remain as key growth points in the domestic demands for a good long time. Further more, as the development of Western China and the growth of Central China, which demand the construction of infrastructure firstly, the process manufacturing industries will remain to be the largest consumer for energies and resources in a certain period. Therefore, the gap between the supply and demand of energies and resources will be a long-term problem, which could be solved by science and technology innovation, saving resources and energies, and the implementation of conserving manufacture.

Saving resources (water, timbers and lands). The water per capita in China is less than one third of that of the world average figure, and 16 provinces are short of water resources. The effective utilization coefficient for agricultural irrigation is 45%. 20% of China's water areas are suffering heavy pollution. The ratio for the utilization of timbers is only 60%, lower than that of 85% for the developed countries. The ratios for the recycling on wasted timbers and wooden products are at low levels, for example, the ratio for the recycling on wasted paper is only 28.7%, far behind 70% of the developed countries. The plow land of China is reducing year by year and 40% of the plow land has degenerated. About 300 million mu of farmland has been polluted, of which about 100 million mu is polluted by three kinds of industrial wastes. The Chinese government must take measures and try its best to save water, timbers and land resources.

Saving energies (especially coal and petroleum). Power plants are big customers for coal in China, whose consumption takes 52% of the total. In the future, electricity generated by coal will continually account for a major proportion and, by 2020, it will account for 66.2% of the total electricity. Through improving coal burning efficiency and adding de-sulfur equipment to generators which use coal with more than 1% sulfur, the amount of coal used could be reduced by 5%. With the technologies on generating heat and electricity simultaneously and advanced electric power and electronic technologies, the consumption of coal for every kilowatt-hour will be reduced to 330 grams from 380 grams so as to save 100 million tons of coal in total each year.

The energy consumed in buildings (including heating and air conditioning) accounts for 27.6% of the total energies in China, which is twice to three times as that of countries at the same latitude. Reasons for such large energy consumption include not only the climate but also low efficiency and disperse of boilers, poor heat preservation ability of houses, and doors, windows and glass which are easy to transfer heat. Therefore, China should develop energy saving buildings, such as, by using the high efficient heat supply system applied in the North; the heating pump system suitable for the areas along the Yangtze River; a new central heating system with separate adjustment by end user; the solution to construct build-in solar water heater; the air conditioning system that can control temperature and humidity separately; the transportation and distribution systems in large public buildings with high efficient and energy saving abilities; and the energy saving control administration on machinery and electric systems within large public buildings. The application of those systems and technologies can reduce the energy consumption of buildings by 50%.

The petroleum consumed by automobiles accounts for 40% of the total consumption in China, which can be reduced by, on one hand, the use of small automobiles which should be strongly encouraged by the government; on the other hand, the development and application of new energy saving technologies. For example, we may develop automobile with hybrid energy, fuel system for automobile engines, develop and apply high efficient ternary emission transforming system, promote diesel automobile and research automobile using alternative fuels. The weight of vehicle body could be decreased by the appliance of aluminum, plastic and composite materials. Using new fuel saving technologies for automobile could save 5 million tons of petroleum and 5 million tons of alternative fuels by 2010 if the efficiency could rise by 15%.

Strengthening the development of reproducible energies, such as hydrodynamic, wind power and solar power. China has rich water resources of 676 million kilowatt-hours, but most of which scatter in the middle and west parts. China has enormous wind power resources, of which 1 billion kilowatt-hours are available. There are also a great potential for using solar power. About two thirds territory of China has a daily solar radiation of 4 kW · h/m^2 per year. At present, the technologies for water heater using solar power are mature and the technologies for generating electricity by solar power are developing.

Conserving manufacture. Manufacturing industry has consumed most of China's resources and energies, accounting for 54%, while its emission of sulfur dioxide accounts for 60% of the total in China. Steel, construction materials, petrochemicals, chemicals and nonferrous metal industries are all resource and energy consuming industries. The unit energy consumptions for thermal power, steel, ethylene, synthetic ammonia and cement are lower than the world advanced levels by 18%, 11%, 27%, 19% and 29% respectively. The general recycling rate for mineral resources is about 30%, 20% less than that of the world advanced rate.

The process manufacturing industries will remain as the biggest consumer in China for resources and energies during a certain period. This problem should be solved through science and technology innovation and conserving manufacture. First, develop resource-conserved products, such as energy saving houses, automobiles and equipments. Second, practice conserving production, for example, we should reduce the energy consumption in steel, nonferrous metal, petrochemical, chemical and construction material manufactures. Third, develop and use new materials to reduce the weight of automobile and new materials used in buildings. Fourth, extend the usage life of buildings and improve the quality of durable products, such as appliances, containers, automobiles, ships and aircraft. Fifth, develop recycling, re-utilization and re-production to improve general efficiency, such as the integrated utilization within manufacturing industry, the recycling and re-utilization in production area and consumption area, and the recycling and re-utilization of water, timbers and paper and mineral resources.

After the stipulation of a strategic decision on the building of a conservation-minded society by Chinese government, the Chinese Academy of Engineering has set up a project group comprising more than 100 CAE members and experts since May 2005 to carry out researches on resources and energies saving, the improvement of efficiency for the utilization of resources and energies, decreasing energy consumption by equipments and the recycling of resources. Through several rounds of discussions and reviews, the group has put forward the following 17 key projects:

(1) Project on water-saving and high efficient utilization in agriculture;

(2) Project on treating polluted water and turning polluted water into resources;

(3) Project on improving fields with low or medium yields;

(4) Project on recycling wasted timbers and paper;

(5) Project on high efficient exploitation and utilization of low or medium quality mines and

complex co-existing mines;

(6) Project on improving the proven and recovery efficiency on non-reproducible resources and energies;

(7) Project on large scale and high efficient electricity generation, transportation and distribution system;

(8) Project on automobile's environmental protection and petroleum saving and alternative fuels;

(9) Project on utilization of re-producible energies by large scale;

(10) Project on production of new generation multi-function steel;

(11) Project on re-production of nonferrous metals;

(12) Project on improving the utilization efficiency of petrochemical resources;

(13) Project on saving and all-around utilization of phosphorus resources;

(14) Project on energy saving production of cement and wall applied materials;

(15) Project on energy saving in buildings;

(16) Project on recycling urban waste;

(17) Project on recycling and re-production of machineries and electric products.

If the above mentioned 17 key projects and those key technologies proposed by the project group are successfully completed and implemented, by 2020, China will save a large quantity of resources and energies, which means to save energies equivalent to 550 million tce, provide re-producible energies equivalent to 450 million tce, improve 300 million mu fields with low or medium yields, save 60 billion m^3 of water, save 20 million tons of non-ferrous metal ores and 10 million tons of phosphorus ores, and keep the total building area in cities and towns within 30 billion m^3, saving a lot of energies & resources.

PhD Thesis

Digital Signal Processing Systems with Finite State Machine Realization

ZHANG Yan-Zhong
A dissertation submitted to the University of Cambridge
for the degree of Doctor of Philosophy

Contents

Y. Zhang, *Footprints in Cambridge and Aviation Industries of China*,
https://doi.org/10.1007/978-981-16-3176-4_37

Preface

The research contained in this dissertation is a result of my own work at the Cambridge University Engineering Department. The work is original except where reference is made to the work of others, and includes nothing which is the outcome of work done in collaboration: It has not been submitted, in part or whole, to any other university for any degree.

Y. Z. Zhang
Trinity College
Cambridge
June 1984

Acknowledgements

I would like to thank my supervisor, Dr. P. J. W. Rayner, for introducing me to this subject and for his guidance, advice and encouragement during the period of this research. I am indebted to Professor W. A. Mair for his help at the time of my application. I am deeply obliged to Professor P. S. Brandon for many illuminating discussions and suggestions.

I am most grateful to the Chinese Government for sending me to Cambridge to study and for its financial support at the beginning of this research.

Many thanks are due to Trinity College and the Engineering Department, Cambridge University, for their financial support. Thanks are also extended to the Committee of Vice-Chancellors and Principals of the Universities of the U. K. for providing an Overseas Research Student award for the partial remission of tuition fees.

Particular acknowledgement is given to the Board of Graduate Studies of Cambridge University for granting me one term on leave to study in China during the Easter Term 1983.

I would like to express my sincere appreciation to all the staff and research students in the Communications Laboratory of Cambridge University. Finally, I wish to thank Mrs. P. Lister and Mrs. H. Thomson for their excellent typing of the manuscript.

Y. Z. Zhang

Summary

Digital signal processing systems with finite state machine realization—a new and more general realization—are systematically studied.

Methods for representing digital filters and discrete Fourier transforms as a finite state machine are presented. The optimum coefficient and word length choice are discussed.

An analytical method for transforming a state table of machines into a Reed-Muller polynomial is presented by using a DFT. The cyclic subgroup properties of some polynomials are presented and can considerably reduce the size of memory and amount of computations. A method is developed for converting a multiple-variable polynomial into a single-variable one and vice versa.

A new fast DFT algorithm is developed by converting a DFT of length $(p^M - 1)$ into several subgroup cyclic convolutions. The properties of cyclic subgroups can considerably simplify the algorithm structures.

An efficient algorithm for minimization of Exclusive-OR functions is developed. This algorithm is $n/[l+(n-1)\,2^{-n}]$ times faster than the fast Reed-Muller transform. A logical method is proposed for determining the common terms among multiple-output functions. An efficient algorithm for minimization of Reed-Muller polynomials over $GF(2^M)$ is developed. The average number of multiplications for this algorithm is reduced to $M * 2^{M-2}$.

The direct and indirect synthesis methods of stable feedback shift registers are respectively proposed. The synthesis of maximum transient FSRs is developed. A large class of maximum transient FSRs is presented.

A linearization method of non-linear machines is introduced by means of a similarity transformation having DFT properties. A polynomial method for stabilization of an unstable machine is developed by breaking cycles, forming a root and connecting branches as a tree. A connection matrix method is presented for detecting and locating limit cycles. The limit cycles can be removed with minimal increase in system noise.

Keywords

Digital signal processing, effect of finite word length, IIR digital filters, limit cycle oscillations, discrete Fourier transformations, fast DFT algorithms, circular convolutions, subgroup convolutions, recursive algorithm, word length consideration.

Finite state machines, feedback shift registers, synthesis of stable FSRs, maximum transient FSRs, stability of FSMs, linearization, stabilization, suboptimum successors, connection matrix methods.

Digital systems, minimization of switching functions, exclusive-OR Logic, optimum polarity, Gray code, multiple-value logic, common terms, Galois fields, Reed-Muller polynomials, least common extension fields, computer-aided circuit design.

Abbreviations and Notation

Principal abbreviations and notation only are given below. Further notation is stated in the relevant context.

Abbreviations

CRT	Chinese Remainder Theorem
DFT	Discrete Fourier Transform
DSP	Digital Signal Processing
EOR	Exclusive-OR Logic

FFT	Fast Fourier Transform
FIR	Finite Impulse Response
FSM	Finite State Machine
FSR	Feedback Shift Register
IIR	Infinite Impulse Response
LCM	Least Common Multiple
LCO	Limit Cycle Oscillation
LFSM	Linear Finite State Machine
LFSR	Linear Feedback Shift Register
MOD	Modulo
PFA	Prime Factor Algorithm
PLA	Programmable Logic Array
ROM	Read Only Memory
S/N	Signal-to-Noise Ratio
VLSI	Very Large Scale Integration
WFTA	Winograd Fourier Transform Algorithm

Notation

a_i	Coefficients of next state polynomials
b_i	Coefficients of output polynomials
$C_A(\lambda_i)$	Characteristic polynomial of matrix $\underline{A}$
C_L	Connection matrix
D	Optimum coefficient factor
$E(f)$	Entropy of function f
$\underline{f}$	Next state mapping
$\underline{g}$	Output mapping
g, $g(n)$	Output of FSMs
$GF(q)$	Galois Fields, $q=p^M$
$h(n)$	Impulse response of digital filters
$H(z)$	Transfer function of digital filters
K_i, L_i	Coefficients of IIR digital filters
$\underline{N}$	Next state operator
p	Prime number
$p_M(x)$	Primitive polynomial of order M
Q	Quantization
R,R_o	Range factors
r	Order of Reed-Muller polynomials, $r=2^M-i$
s, $s(u)$	Present state of FSMs
$s',s(n+1)$	Next state of FSMs
$\underline{T}_n$	Reed-Muller transform matrix of n variables
$u,u(n)$	Input of FSMs

w	Weight of Reed-Muller polynomials
W	Complex coefficient of DFTs, $W = e^{-j2\pi/N}$
$x(n)$	Input signal of DFTs and digital filters
$X(k)$	Frequency component of DFTs
$Y(n)$	Output of digital filters
$Z(N)$	Integer ring modulo N
α	Primitive element of extension field $GF(2^M)$
β, r	Primitive elements of subfields $GF(2^N)$, $GF(2^k)$
ξ, η	Power ratio of primitive elements of subfields
$w(n)$	Mediate variable of IIR filters
$\phi(N)$	Euler's function
$\delta(i)$	Dirac function
Δ	Step of quantization
$\in$	Member of set
$\subset$	Subset
$\forall$	Universal quantifier
$\vee$	Logical OR operation
$\wedge$	Logical AND operation
$\bar{x}_i$	Logical NOT operation on binary variable x_i
$\oplus$	Exclusive-OR operation
$\otimes$	Kronecker products
$\oplus$ Σ	Modulo-2 additions
$\sum_{GF(q)}$	Sum for all the elements of $GF(q)$
$\sum_{GF'(q)}$	Sum for non-zero elements of $GF(q)$

Chapter 1
Introduction

1.1 General Introduction

During the past twenty years, Digital Signal Processing (DSP) has been an extremely active and dynamic field. The theory and methods of DSP have been widely used in communication, speech processing, image analysis, acoustic, sonar, radar, automatic control, vibration, geophysics, biomedical engineering and other fields of science and technology. Advances in computer architecture and integrated circuit technology have greatly motivated the development and application of digital signal processing.

Digital signal processing possesses many advantages of accuracy, high speed and programmability etc. A digital signal processing system may be realized in terms of either a computational program on general-purpose computers, or a special-purpose digital hardware. Because of the effects of finite word length, both the software and the hardware realization of a DSP system has to approximate the infinite precision arithmetic operations by means of finite word length arithmetic operations. This approximation may lead to noise, errors and instabilities[J7, J9].

A Finite State Machine (FSM) is an algebraic model of finite storage systems. It is an infinite accurate system[Q1] and has been successfully used in automatic control, code theory and digital computers etc. In general, any digital system with memory may be realized in terms of a finite state machine.

This dissertation is concerned with the realization of a digital signal processing system as a finite state machine which is the least error approximation to the ideal system and is without oscillations. This idea connects the two previously unrelated concepts: DSP and FSM, and presents a new way of solving the problems arising from finite word length in DSP systems

1.2 Effects of Finite Word Length on Digital Signal Processing Systems

Any digital signal processing system may be specified by a difference equation which is discrete in time and continuous in amplitude, or by a z-transform function. A linear time invariant digital filter is described by the linear constant coefficient difference equation[P1]

$$y(n) = \sum_{i=1}^{m} K_i y(n-i) + \sum_{i=1}^{m} L_i x(n-i) \tag{1.1}$$

or by the z-transfer function

$$H(z) = \frac{\sum_{i=1}^{m} L_i z^{-i}}{1 - \sum_{i=1}^{m} K_i z^{-i}} \tag{1.2}$$

where $x(n-i)$ and $y(n-i)$ are respectively the input and output at the $(n-i)$th clock instant, z is a complex variable, K_i and L_i are constant coefficients. When all the coefficients $\{K_i\}$ in eqn. (1.1) are zero, the filter is called a Finite Impulse Response (FIR) digital filter, as its impulse response is non-zero for only a finite duration less than m samples. On the other hand, if at least one of the coefficients $\{K_i\}$ in eqn. (1.1) is non-zero, then the filter has an infinite impulse response. It is called an Infinite Impulse Response (IIR) digital filter. All the input $x(n)$, output $y(n)$ and coefficients K_i and L_i in eqns. (1.1) and (1.2) are assumed to have an infinite accuracy. In practice, since the word length is finite in a DSP system, the inputs, outputs, coefficients and arithmetic operations in eqn. (1.1) have to be represented with only a finite number of bits. The effects of finite word length on a DSP system are as follows.

(a) The quantization noise of input signals.

(b) Roundoff or truncation noise of multiplication operations.

(c) The errors of coefficient quantizations.

(d) Limit cycle and overflow oscillations.

1.2.1 Input Signal Quantization Noise

The sampled input signal is first converted into a digital signal at the input end of a system. An analog-to-digital converter is the commonly used quantizer; this is a uniform quantizer.

The amplitude of quantizing errors is related to the quantization method (roundoff or truncation) and the number representation (fixed-point or floating-point, one's or two's complement etc.). The usual approach for treating the effect of input quantization is to regard the quantizing error as white random noise. It is assumed that:

(a) The error signal is a stationary random process.

(b) The error signal and the input signal are mutually uncorrelated.

(c) The error signals are uncorrelated by themselves.

(d) The probability density of error signals is uniform.

Theory and experiments have shown that this assumption gives a very good statistical model of input quantization noise[P2, P3]. The roundoff quantization noise has zero mean and variance $\Delta^2/12$: here Δ is a quantizing step. The signal-to-noise ratio of input signals depends on the word length of the quantizer and increases at 6 dB per bit[P2].

A non-uniform quantization on input signals may lead to better signal-to-noise ratio. An example shows that an optimum quantization filter of order 6 offers an improvement of 7 dB in signal-to-noise ratio for Gaussian random signals[H8].

Many authors[M1-M8] are concerned with the optimum quantization for minimum noise. However, a non-linear quantization filter cannot be realized in terms of the usual linear arithmetic structure. A non-linear finite state machine is ideal for the realization of optimum quantization digital filters[H8].

1.2.2 Roundoff or Truncation Noise of Multiplications

As eqn. (1.1) shows, any digital filter has some multiplication operations between its coefficients and variables. Since the register length is finite, roundoff or truncation errors of multiplications are introduced. e.g. a b-bit data sample multiplied by a b-bit coefficient results in a product which is $2b$-bit long. This product is to be stored in a b-bit register, so its least significant b bits must be rounded or truncated.

In order to model the effects of multiplication quantizations, certain assumptions concerning the statistical independence of multiplying quantization noise are made as follows.

(a) Any two different samples from the same noise are uncorrelated.

(b) Any two different noise sources are regarded as random and uncorrelated.

(c) All the noise sources are uncorrelated with the input sequences.

It is clear that the assumptions are not always valid. In particular, if the input is a constant (or zero), all the assumptions break down. In such cases the quantizing noise is no longer uncorrelated with the input sequence. This is the question of correlated errors or limit cycles and will be introduced in Section 1.2.4.

The assumption is reasonable for most varying signals. The accumulation error of multiplications is related to the quantization methods, the number representations and the structures (direct, cascade or parallel etc.) Studies[J5] show that for a fixed word length, the accuracy achievable with a direct realization of a high-order filter is considerably less than that with either cascade or parallel realization of the same filter.

In order to avoid the overflow, scaling factors are commonly used. Jackson[J4] studied the interaction between roundoff noise and dynamic range. He shows that the output noise variance is dependent on the sequential ordering of the sub-sections as well as the exact way in which numerator and denominator sections are paired together in cascade implementations. The optimum ordering and pairing can be obtained by minimizing the output noise. Rader[J18] proposed a Graph Theory method for obtaining the optimum ordering of sub-sections in cascade realizations.

1.2.3 Coefficient Quantization Errors

The frequency response of a digital filter is determined by its coefficients which are usually obtained by some theoretical design procedure and are assumed to have infinite precision. For practical realizations, the coefficients have to be quantized to a finite number of bits. As a result, the frequency response of the filter realization deviates from that which would have been obtained with infinite precision coefficients.

There are two general approaches to the analysis and synthesis of digital filters with finite precision coefficients. The first approach is to treat the coefficient quantization error as a random quantity: the mean difference of the frequency responses between the actual and the ideal filters can be easily evaluated.

The coefficient sensitivity of a digital filter is related to its structure and pole-zero patterns. Direct form has larger coefficient sensitivity than cascade and parallel form. High order filters are more sensitive to the coefficients than the lower order filters. High Q filters with poles near the unit circle in z-plane have very high coefficient sensitivity. Many filter structures with low coefficient sensitivity have been proposed[J5, J16]. Digital wave filters as proposed by Fettweis[L7] appear to have much less coefficient sensitivity.

The second approach to coefficient quantization is to study each individual filter with finite precision coefficients. The optimum filters with finite precision coefficients may be obtained by minimizing the difference between the ideal and the actual frequency response[J11-J12]. The advantage of this method is that the best finite precision representation of the desired frequency response may be obtained without general studies of coefficient quantization[J15].

1.2.4 Overflow and Limit Cycle Oscillations

The discussion of multiplication quantization in digital filters assumed that the samples of quantization noise were uncorrelated with each other and with the input sequences. It is clear that there are many cases where the assumption is no longer valid, e. g. when the input is constant (or zero), the non-linear roundoff error is not independent and is accumulated in the feedback loop, so that the filters may produce deadbands or limit cycle oscillations[K1-K3]. For example, the first order IIR filter

$$y(n) = -0.8y(n-1) + x(n)$$

If the register word length is 3 bits, and roundoff is used in output quantizations, then the output sequences for zero inputs and from the initial value $y(0) = \frac{7}{8}$ are

$$\frac{7}{8}, -\frac{6}{8}, \frac{5}{8}, -\frac{4}{8}, \frac{3}{8}, -\frac{2}{8}, \frac{2}{8}, -\frac{2}{8}, \frac{2}{8} \ldots\ldots$$

A limit cycle oscillation of amplitude $\pm 2/8$ at half sampling frequency appears in the filter.

There are two different oscillations in the filters. The first is due to overflow in the addition operations. The second is from the roundff or the truncation of signal-coefficient, products at the output of multipliers.

The existence of overflow is dependent on the coefficient values. Ebert et al. [K1] and Jackson[K3] show that the necessary and sufficient conditions for absence of overflow oscillations in a second-order IIR filter are $|K_1|+|K_2|<1$. They also show that the overflow oscillation may be eliminated by introducing saturation arithmetic in its adders.

The second type of oscillation consists of limit cycles resulting from rounding or truncating in the multiplication operations. This phenomenon is dependent on the filter configuration, the type of arithmetic, the quantization method, and the actual filter coefficient values, etc.

Many authors are concerned with estimating the bounds on amplitudes, and frequency of limit cycles. Jackson[K3] proposed that the limit cycles are caused by moving the effective pole positions of filters to the unit circle in z-plane, and he gave an effect model and determined the amplitude and the frequency of limit cycles. Hess[K4] introduced the successive-value phase plane and noted symmetries which exist in limit cycle sequences. From these he developed a new bound based on Lyapnov methods and matrix methods. Sandberg and Kaiser[K5] presented a new bound on the maximum root-mean-square value of limit cycle amplitudes. Long and Trick[K7] refined bounds on the amplitude for the second-order filter. A tighter bound is given by [K13, K17].

Claasen et al. [K10-K12] proposed the concept of accessibility or reachability of limit cycles, a more general approach to the existence of limit cycles is formulated in the frequency domain for nth order systems. Kaneko[K8] dealt with the limit cycles in floating-point digital filters. An interesting experiment about limit cycle phenomena in a practical tenth-order digital filter is given by Kieburtz[K9]. His results showed good agreement between the actual behaviour and the theoretical bound of limit cycles.

1.2.5 Suppression of Limit Cycle Oscillations

Overflow oscillations may be prevented by introducing saturation arithmetic in adders[K1].

The suppression of limit cycles which are caused by rounding or truncations of multiplications have been studied by many authors[L1-L16].

The limit cycles are dependent on the quantization methods. In the first-order IIR filter, limit cycle oscillations may be suppressed by using truncation quantization rather than the rounding. Claasen et al. [L1] show that some limit cycles in the second-order IIR filters may be avoided by using only one magnitude-truncation quantizer at the output end of the two multipliers. But this method cannot suppress some high Q critical frequency near zero or half sampling rate limit cycles[L1]. Butterweck[L2], Verkroof[L3], Lawrence and Mina[L4-L5] proposed controlled rounding arithmetics that eliminated all 0-input limit cycle oscillations in second-order digital filters. This method increases the output noise and the hardware which is needed for controlling the rounding.

Several new structures[L8, L12] which are free of limit cycles have been the focus of recent attention. Fettweis[L7] shows that limit cycle oscillations cannot occur in some particular classes of wave digital filters.

An error feedback technique[L9-L11] is proposed for reducing the rounding noise and suppressing the limit cycle oscillations. It is based on the principle of saving the truncated portion of products and feeding it back multiplied by appropriate coefficients and considering it when forming the output at sequent iterations. This technique increases the complexity of hardware.

Papers [L6, L14] show that the limit cycle oscillations may be broken up or suppressed by introducing a very low random dither. This method causes an increase in output noise about 2 dB[L14]. Wong and King[L15] show that the multirate digital filters with periodically varying coefficients can also cause limit cycle to be suppressed.

Most work is concerned with suppressing the 0-input limit cycle oscillations. The suppression of limit cycles for non-zero inputs has received relative limited studies. All the methods which have been proposed for suppression of limit cycles increase either the output noise or the complexity of hardware. The FSM realization digital filter may remove the limit cycles from the machine with minimum increase in output noise, and with hardly any increase in the complexity of machines, as it involves only a reallocation of machine states or a modification of polynomial coefficients (Chapter 8).

1.3 Finite State Machines

1.3.1 Mealy Machines

A finite state machine is a mathematical model. of sequential systems. It comprises a finite input set $U \in (U_0, U_1, \cdots U_{N_U-1})$, a finite output set $G \in (G_0, G_1, \cdots, G_{N_G-1})$, a finite state set $S \in (S_0, S_1, \cdots, S_{N_S-1})$ and two mappings $\underline{f}$ and $\underline{g}$ defined by[Q1]

$$\underline{s}(n+1) = \underline{f}[\underline{s}(n), \underline{u}(n)] \tag{1.3}$$

$$\underline{g}(n) = \underline{g}[\underline{s}(n), \underline{u}(n)] \tag{1.4}$$

where $\underline{u}(n) \in U$, $\underline{g}(n) \in G$ and $\underline{s}(n) \in S$ are respectively the input, output and present state of systems at the nth clock instant. $\underline{s}(n+1) \in S$ denotes the state at the $(n+1)$ th clock instant and is named the next state. The mapping $\underline{f}$ maps the present state $\underline{s}(n)$ and the input $\underline{u}(n)$ into the next state $\underline{s}(n+1)$, and is called the next state mapping, $\underline{g}$ is referred to as the output mapping which produces the present output $\underline{g}(n)$ from the present state $\underline{s}(n)$ and input $\underline{u}(n)$. This model is called the Mealy model[G1]. Another finite state machine model is the Moore model[G11] which produces the output $\underline{g}(n)$ from only the present state $\underline{s}(n)$ as follows:

$$\underline{g}(n) = \underline{g}[\underline{s}(n)] \tag{1.5}$$

It has been proved that any Mealy model machine is equivalent to a Moore machine[G2]. The Moore machine might need more states than the Mealy machine for performing the same computation[Q1]. Therefore Mealy machines are used in this dissertation. A Mealy finite state machine is shown in Fig. 1.1.

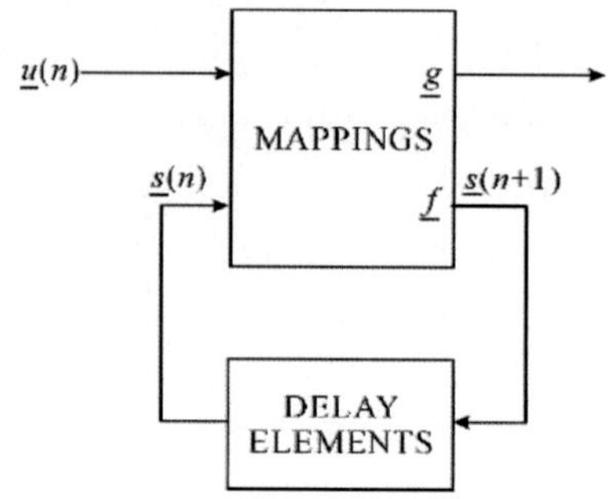

Fig. 1. 1 A mealy machine

where $\underline{s}(n)$ and $\underline{s}(n+1)$ are m-dimensional state vectors,

$\underline{u}(n)$ is an n-dimensional input vector,

$\underline{g}(n)$ is a k-dimensional output vector,

$\underline{f}$ denotes an m-dimensional next state mapping,

$\underline{g}$ denotes a k-dimensional output mapping.

Fig. 1. 1 may be viewed as a "black box". It may have different "internal" structures, but its "external" properties of machines are determined by the two mappings; the next state mapping and the output mapping.

If there are no inputs to a finite state machine, e. g. $u(n)=0$, then the machine is called an autonomous, and it can be represented as follows:

$$\underline{s}(n+1)=f[\underline{s}(n)] \tag{1.6}$$

$$\underline{g}(n)=\underline{g}[s(n)] \tag{1.7}$$

1. 3. 2 Description Methods for Finite State Machines: The State Table and the State Transition Graph

A finite state machine may be described by a state table or by a state transition graph. The state table has N_S rows and N_U columns. Each row corresponds to each member of the finite state set S, and each column corresponds to each member of the finite input set U. The entries of the table denote the next states $s(n+1)$ and the machine's output $g(n)$.

The state transition graph is a directed graph which has N_S vertices and $N_S \times N_U$ arcs. Each vertex represents a member of the finite state set S and the arcs with arrays represent the state transitions from the present states to the next states with the corresponding inputs U_i and outputs G_i indicated by the label U_i/G_i. An example of a Mealy model is shown in Fig. 1. 2.

next state / output	Input U_0	Input U_1
present state S_0	S_0/G_0	S_3/G_1
S_1	S_0/G_1	S_2/G_1
S_2	S_1/G_3	S_0/G_2
S_3	S_2/G_1	S_2/G_0

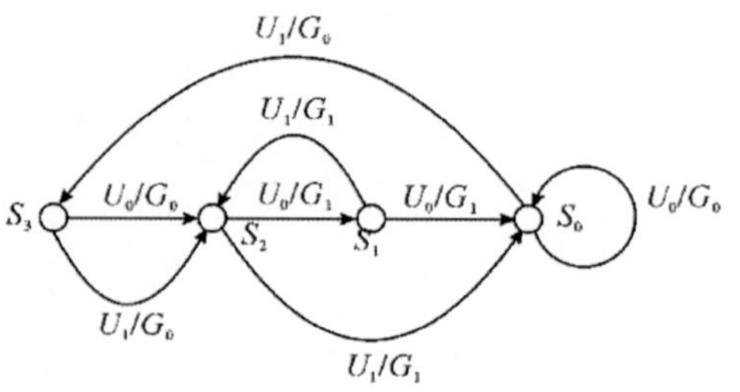

Fig. 1. 2 Example of state table and state transition graph

Sometimes, the above table may be separated into two sub-tables: a next state table and an output table. They can separately show the next state and the output mapping and are often used in the dissertation.

The state table may be obtained from some physical or mathematical models and it can be stored on computers. The disadvantage of state table representation is that it cannot show clearly the dynamic behaviour and stability of systems. The state transition graph can clearly show the dynamic behaviour and the stability of systems, but it is difficult to draw a state transition graph when the state set S and the input set U is large. An analytical representation of finite state machines is necessary for realizing the systems and studying the stability of machines (Chapters 3 and 7).

1.3.3 Previous Work

Many authors[Q4-Q15] have been concerned with finite state machines. Moore[Q11] and Mealy[G1] proposed different FSM models.

Bryant and Killick[F1] gave the transition matrices and characteristic functions of machines from state graphs. Booth[F3] presented the polynomial representation of finite state machines. Some authors[Q10-Q15] dealt with the properties and realization of FSMs. Gill[G3] proposed a system method for minimization of linear finite state machines. Herman[G7, G8] proved that every finite sequential machine was linearly realizable. A number of workers have been concerned with the linearization of non-linear FSMs[G15, G20].

Non-linear feedback shift registers are a special class of non-linear finite state machines. Feedback shift registers are easily implemented. Some workers[G9, F7, F8, F12, F13] showed that many finite state machines could be realised in terms of feedback shift registers.

Golomb[Q6] systematically studied non-linear feedback shift register sequences. Massay[F6] proposed the equivalence of different non-linear feedback shift registers. Green, Dimond and Kelsch[F16-F18] dealt with nonlinear product-feedback shift registers, their polynomial representations and compositions. Mayleby[F2] presented the synthesis of nonlinear feedback shift registers, but did not study the synthesis of stable FSRs which are very important for stable finite state machine realisation.

Finite state machines have been successfully used in automatic control, code theory[Q8] and computer science[Q16]. Lately, several authors proposed the finite state machine approximation to the dynamic processes[H1] and continuous plant behaviours[H3-H6]. Depeyrot et al.[H2] gave an automation-theoretical approach to the fast Fourier Transform.

Rayner[H7, H8, H10] studied the application of finite state machines to the design of digital processing systems. An optimum finite-state digital transmission system is developed by Gaarder[H9].

The advantages of using finite state machines are:

The states, inputs and outputs of FSMs are the elements of finite sets. The next state mapping and the output mapping may be defined over the finite sets. These mappings are infinite accuracy operations which are free of errors. Because the next state of an FSM depends on not only the present inputs but also the past states, an FSM may simulate any finite storage system in terms of an input map f_1 and an output map f_2[Q1] as shown in Fig. 1.3. One can design several "standard" machines which are simple and easily implemented. A complex machine may be realized by some

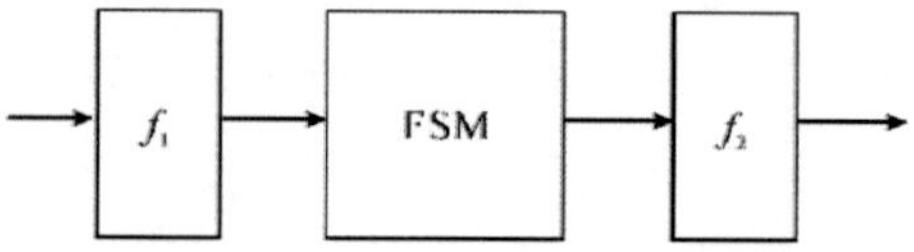

Fig. 1.3 Machine simulations

"standard" machines or their composition. e. g. a stable maximum length machine of length less than 2^M may simulate any stable autonomous machine of length less than 2^M in terms of an output map (Chapter 6).

A finite state machine comprises two mappings: the next state mapping and output mapping which are combinational circuits and easily implemented by Very Large Scale Integration (VLSI), Read Only Memory (ROM), or Programmable Logic Arrays (PLA), etc. They possess advantages of high speed and programmability etc.

1.4 Finite State Machine Realization of Digital Signal Processing Systems

1.4.1 FSM Representation of Digital Signal Processing Systems

A finite state machine may be used to represent any digital signal processing system in which the output at any clock instant is a function of past and present values of inputs and machine states. When the continuous algebraic equation description of systems is given, one method of approach is to evaluate the system output for all possible combinations of finite inputs and system states using the continuous equation. The output calculated in this manner will not, of course, be in the same finite set as the input and state so that some form of approximation is needed to achieve this. However, the system output values determined in this way can be within $\pm 1/2$ least significant bit over the representable range of outputs. For example, consider a first-order all pole digital filter described by the following difference equation

$$y(n)=K_1 y(n-1)+x(n) \tag{1.8}$$

where the present output $y(n)$ is a function of the past output $y(n-1)$ and the present input $x(n)$. Compare eqn. (1.3) and(1.8), and

(a) Choose the word length of $y(n)$ and $x(n)$.

(b) Give the quantization method (roundoff or truncation etc.).

(c) Assign to each output quantization level $\bar{y}(n)$ a member of the state set S, and to each input quantization level $\bar{x}(n)$ a member of the input set U of a machine, then the first-order filter may be represented as a finite state machine, and the next state may be obtained by quantizing the infinite accuracy computation result of eqn. (1.8)

$$\bar{y}(n)=Q[K_1 \bar{y}(n-1)+\bar{x}(n)] \tag{1.9}$$

where Q denotes the quantization operation.

If the coefficient $K_1=0.55$, input $x(n)=0$ in eqn. (1.8), let the output $\bar{y}(n)$ be represented in 3-bit word then the state table may be set up by rounding the following equation for every possible $\bar{y}(n-1)$ as follows:

$$y(n)=0.55\bar{y}(n-1)$$

Table 1.1 State table of the example

$\bar{y}(n-1)$	$y(n)$	$\bar{y}(n)$
0	0	0
1	0.55	1
2	1.1	1
3	1.65	2
−4	−2.2	−2
−3	−1.65	−2
−2	−1.1	−1
−1	−0.55	−1

This table may be realized by a one-dimensional FSM. The machine outputs for all combinations of inputs and past states have at most $\pm 1/2$ bit error from the filter outputs with infinite accuracy. In this simple example, the obtained outputs are no different from those which would be obtained from the arithmetic realization. However, in a system containing more than one multiplier the results obtained from the above method would more accurately match the results from an infinite precision filter. Chapter 2 will show that the nth order IIR filter may be represented as an n-dimensional FSM.

1.4.2 Advantages of FSM Realization

The main advantages for FSM realization of IIR digital filters are as follows.

(a) The effect of coefficient quantizations on pole-positions of digital filters is avoided, as the state table is calculated from infinite accuracy coefficients.

(b) The system errors are reduced to minimum, since the next state assignments in the state table have only 1/2 bit error from the infinite accuracy computation of eqn. (1.1), and the state table is realized by a finite state algorithm structure which is free of error. For certain error requirements, the FSM realization may give the shortest word length.

(c) Limit cycle oscillations may be removed by stabilizing the machine with minimum increase in noise and hardly any increase in the complexity of hardware implementations (Chapter 7 and 8).

(d) Non-linear quantization for optimum signal-to-noise ratio may be used for the input signal. The optimum quantization filter cannot be realised in terms of the normal linear arithmetic operations, but can be easily realised by a non-linear finite state machine where each quantization level is assigned to a member of the machine state set[H8].

1.4.3 Major Problems of System Realization

The finite state machine for realizing a digital signal processing system is rather complex. There are many aspects of the system realization problems which should be solved. The main problems may be separated into three groups as follows.

(a) System Representation

There are two representation problems: the FSM representation of digital signal processing systems, and the analytical representation of finite state machines. The first one is devoted to representing an IIR digital filter or a discrete Fourier transformation as an FSM in the form of state tables. The second one enables the state table to be represented as a polynomial which could be easily minimized for circuit implementations, and should be convenient for investigating the dynamic behaviour of systems.

(b) System Realization

In order to realize a system, the coefficients of its polynomial representation should be obtained. Efficient transform algorithms should be developed for mapping the system outputs into the polynomial coefficients.

The main problem for the practical realization of systems is complexity. For example, even a simple system such as a single pole filter with input and output variables of 10 bits length may have at most 2^{10} coefficients for the polynomial representation. Efficient methods for minimization of the polynomial representations should be developed and the methods should be suitable for computer-aided minimization.

Another approach for realizing FSMs is the feedback shift register (FSR) which is the simplest FSM. The study on synthesis of stable FSRs may lead to more simple realization of stable systems in terms of direct or indirect algorithms.

(c) System Stability

The FSM for realizing a digital signal processing system should be stable. There exist two stability problems in a system. One is the detection of stability, the other is the stabilization of an unstable FSM. For an IIR digital filter, limit cycles should be detected and removed with minimum increase in noise and complexity of the system.

1.5 Outline of the Dissertation

This dissertation is concerned with the finite state machine realization of digital signal processing systems. Attention will be focused on Infinite Impulse Response (IIR) digital filters. There are two reasons for this; firstly, IIR digital filters exhibit limit cycle oscillations when realized in the normal way; secondly, the complexity of IIR digital filters is reasonable, when measured in terms of the number of variables, compared with other signal processing systems such as finite impulse response filters and discrete transform computations.

There are nine chapters in this dissertation. Its structure is shown in Fig. 1.4. The main contents of the dissertation may be separated into three parts:

(a) Part Ⅰ includes Chapters 2 and 3, and is devoted to the representations of systems.

(b) Part Ⅱ includes Chapters 4, 5 and 6, and deals with the realizations of systems. This is the central part of the dissertation.

(c) Part Ⅲ includes Chapters 7 and 8, and is concerned with the stability of systems.

Chapter 1 is an introduction to the dissertation. The effect of finite word length on digital signal processing systems is first reviewed. The finite state machine is briefly introduced. The basic

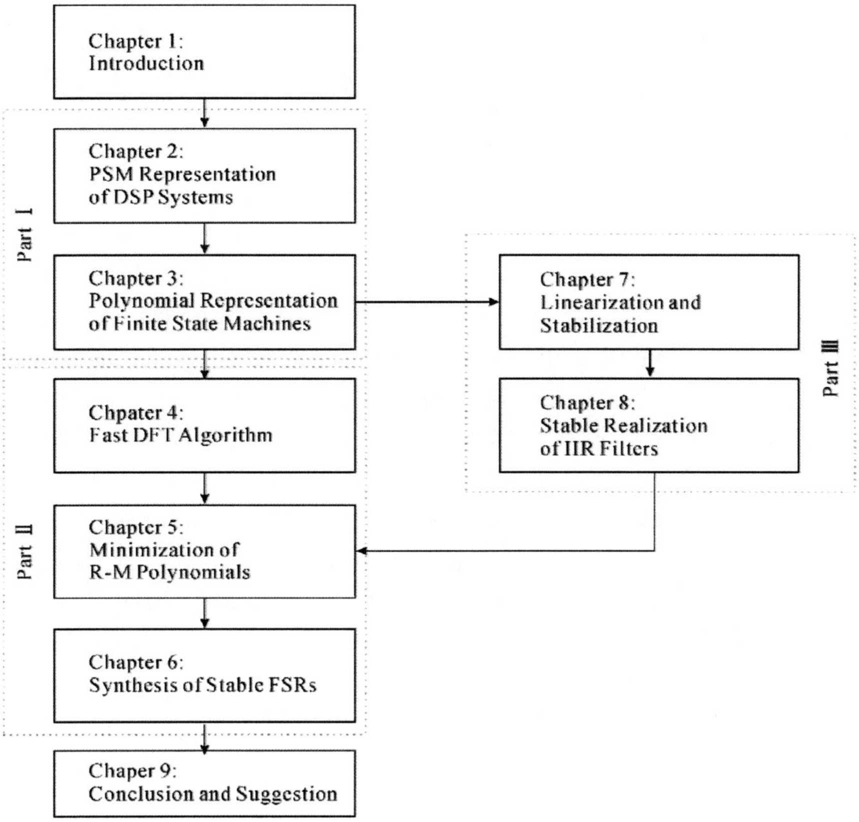

Fig. 1.4 Layout of the dissertation

concept, advantages and problems for FSM realizations of DSP systems are described.

Chapter 2 is devoted to the FSM representation of DSP systems. The n-dimensional Mealy Machine representation of n-order IIR filters is presented. The dynamic range and word length choice is discussed. The two-dimensional Moore Machine representation of a DFT is developed. The optimum coefficient for computing a prime length DFT is presented.

Chapter 3 is concerned with the polynomial representation of FSMs. An analytical method is presented for transforming a state table of FSMs into a polynomial over $GF(2^M)$ by using a DFT of length (2^M-1). The polynomial whose inputs and outputs are in different fields is particularly considered. The cyclic sub-group property of its coefficients is presented and can considerably reduce the size of memory and amount of computation. Much attention is paid to converting a multiple-variable machine into a single-variable one and vice versa. The single-variable polynomials are used to study the stability of systems in Chapter 7. The minimization of Reed-Muller polynomials is considered in Chapter 5.

A new fast DFT algorithm is developed in Chapter 4. The cyclic sub-groups of Ring $z(p^M-1)$ are investigated. A method for converting the DFT of length (P^M-1) into several sub-group cyclic convolutions is presented. The number and length of cyclic sub-groups are discussed. Most of these sub-group convolutions have the same length which can considerably simplify the program structures. This efficient algorithm can be employed in computing conventional DFTs over the complex field, or determining the coefficients of Reed-Muller polynomials over GF(p^M).

Chapter 5 deals with the minimization of Reed-Muller polynomials. An efficient algorithm is developed for minimization of a Ring-sum expansion with fixed polarity. The comparison of operational numbers for different algorithms is presented. With regard to common gates, the minimization of multiple-output functions is particularly considered. A method for determing the the common terms is developed by means of certain logical operations. An efficient algorithm for minimization of Reed-Muller polynomials over GF(2^M) is presented. The average number of multiplications is reduced to $M * 2^{M-2}$. The above methods can be employed in computer-aided logic circuit design.

Chapter 6 is devoted to synthesis of stable feedback shift registers. Two distinct methods for synthesis of stable FSRs—direct and indirect algorithm—are respectively presented. The synthesis of maximum transient FSRs is developed. A large class of maximum transient FSRs is presented. They can be used as stable generators of field elements in indirect algorithms.

Chapter 7 deals with the stability of FSMs. Firstly, a linearization method is introduced so that the stability of nonlinear systems can be investigated by using linear system theory. Secondly, the stabilization of unstable machines is investigated. The polynomial methods are developed for breaking a cycle, forming a root and connecting the branches as a tree in the state graph.

The stable FSM realization of IIR digital filters is studied in Chapter 8. The theory of stability in the above chapter can be used in the FSM realization of IIR filters. The connection matrix method is used to detect and locate limit cycle oscillations. The polynomial method is used to remove limit cycles with minimal increase in system noise.

Chapter 9 gives the conclusions and suggestions for further research.

Chapter 2
Finite State Machine Representation of Digital Signal Processing Systems

2.1 Introduction

Digital filters and Discrete Fourier Transforms (DFT) are two of the most important fields in digital signal processing. Both are required to process signals which are discrete in time and continuous in amplitude. The usual method of realising a digital signal processing system is to approximate the continuous amplitude arithmetic by means of finite word length binary arithmetic. Roundoff and truncation errors are introduced. The error behaviour is dependent on the structures and properties of systems. In some feedback systems, such as Infinite Impulse Response (IIR) filters, this approximation may lead to instability[K2, K3].

A Finite State Machine (FSM) is a finite arithmetic structure of sequential systems. It has no arithmetic errors and can be used to realize a finite storage systems. Rayner[H7, H10] first proposed the application of FSMs to design of IIR digital filters. This chapter is concerned with the FSM representation of digital signal processing systems. Section 2.2 describes the FSM representation of IIR digital filters, the dynamic range and word length consideration in fixed-point implementations are discussed in Section 2. 3. Section 2. 4 presents the FSM representation of discrete Fourier transformations.

2. 2 FSM Representation of Digital Filters

2. 2. 1 FSM Representation of IIR Digital Filters

From eqn. (1. 2), the transfer function of the mth order IIR digital filters may be rewritten as follows:

$$H(z) = \frac{\sum_{i=1}^{m} L_i z^{-i}}{1 - \sum_{i=1}^{m} K_i z^{-i}} = \frac{-L_m}{K_m} + \frac{\sum_{i=0}^{m-1} (L_i - K_i \frac{L_m}{K_m}) z^{-i}}{1 - \sum_{i=1}^{m} K_i z^{-i}} \tag{2. 1}$$

where $K_0 = -1, K_m \neq 0$.

The above equation may be considered as the parallel connection of two sections: a scaling factor $\frac{-L_m}{K_m}$ and a filter with m poles and $(m-1)$ zeros. The denominator of the second section represents an all-pole filter.

Introducing a variable $w(n)$, the output sequence $y(n)$ may be written as the following difference equations:

$$w(n) = x(n) + \sum_{i=1}^{m} K_i w(n-i) \tag{2. 2}$$

$$y(n) = \frac{-L_m}{K_m} x(n) + \sum_{i=0}^{m-1} (L_m - K_i \frac{L_m}{K_m}) w(n-i) \tag{2. 3}$$

where $K_0 = -1, K_m \neq 0$.

This is a modified canonic form of IIR filters. The modification is that its output is related to the input $x(n)$, but unrelated to the mth delayed variable $w(n-m)$. Its structure is shown in Fig. 2. 1.

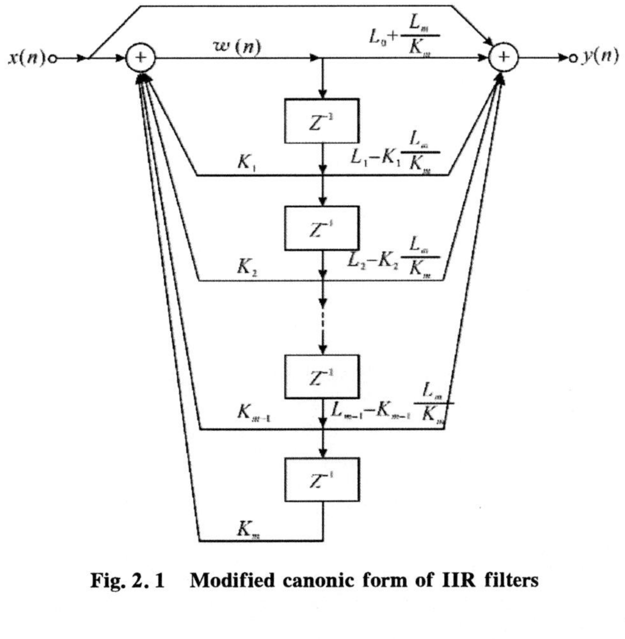

Fig. 2. 1 Modified canonic form of IIR filters

Let $\bar{x}(n)$, $\bar{y}(n)$ and $\bar{w}(n)$ denote the quantized discrete amplitude of $x(n)$, $y(n)$ and $w(n)$ respectively. From eqns. (2.2) and (2.3), $\bar{w}(n)$ and $\bar{y}(n)$ may be obtained by

$$\bar{w}(n) = Q\{\bar{x}(n) + \sum_{i=1}^{m} K_i \bar{w}(n-i)\} \tag{2.4}$$

$$\bar{y}(n) = Q\{\frac{-L_m}{K_m}\bar{x}(n) + \sum_{i=0}^{m-1}(L_i - K_i \frac{L_m}{K_m})\bar{w}(n-i)\} \tag{2.5}$$

where Q denotes the quantization of amplitudes.

The discrete output determined in this way can be within $\pm 1/2$ least significant bit over the representable range. An m-dimensional vector $\underline{s}(n)$ is defined by

$$\underline{s}(n) = \{s_0(n), s_1(n), \cdots, s_{m-1}(n)\}^t$$

where

$$\begin{aligned} s_0(n) &= \bar{w}(n-m+1) \\ s_1(n) &= \bar{w}(n-m+2) \\ &\vdots \\ s_{m-1}(n) &= \bar{w}(n) \end{aligned} \tag{2.6}$$

t denotes transpose

Let

$$\begin{aligned} u(n) &= \bar{x}(n) \in U \\ g(n) &= \bar{y}(n) \in G \\ \bar{w}(n) &\in S \end{aligned} \tag{2.7}$$

where U, G and S are finite sets.

If $\bar{x}(n)$ and $\bar{w}(n-i)$ in eqns. (2.4) and (2.5) are all the possible values of the finite sets U and S respectively, then a state table may be obtained. Eqns. (2.4) and (2.5) may be represented as the following two mappings:

$$\underline{s}(n) = \underline{f}[\underline{s}(n-1), u(n)] \tag{2.8}$$

$$g(n) = g[\underline{s}(n), u(n)] \tag{2.9}$$

where $\underline{f}$ is an m-dimensional next state mapping,

g is a one-dimensional output mapping.

This is a Mealy Machine. It is shown that any one-dimensional IIR digital filter of order m may be represented as a special class of Mealy Machines which possess m-dimensional states, a one-dimensional input and a one-dimensional output. This machine is shown in Fig. 2.2.

In Fig. 2.2, $u(n)$ denotes the one-dimensional input to the machine and is defined to N bits. $\{s_0(n-1), s_1(n-1), \cdots, s_{m-1}(n-1)\}^t$ and $\{s_0(n), s_1(n), \cdots, s_{m-1}(n)\}^t$ are respectively the state vectors at the $(n-1)$th and nth instant. The state vector has m dimensions. The word length of each dimension is M bits. $g(n)$ represents the one-dimensional output. It has K-bit word length.

The total number of machine states is 2^{mM}. For a second order IIR digital filter, if the word length is 8 bits, then the machine has 65,536 states.

The FSM realization of digital filters is different from their conventional realization. The FSM realization has no "internal structure" problems. (e.g. pairing and ordering etc.), and the machine may be considered as a "black box". The conventional multiplications and additions in IIR filters can be replaced by two mappings: the next state mapping and output mapping in. FSMs which can be

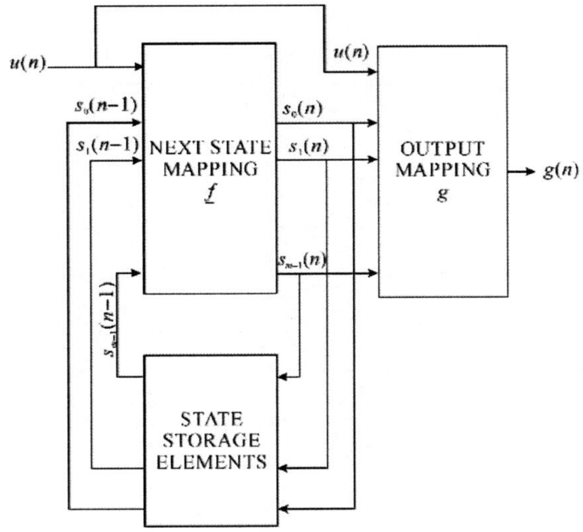

Fig. 2. 2 A mealy machine for representing IIR filters

implemented in terms of finite arithmetic structures and have no arithmetic errors. The only error is caused by calculating the state table from the continuous amplitude equations. This error is less $\pm 1/2$ bit and has no enlargements and accumulations within the FSMs.

Example 2. 1

The difference equation of a 1st order IIR filter is given by

$$y(n)=0.55y(n-1)+0.5x(n)-0.44x(n-1)$$

From eqns. (2. 2) and (2. 3), it may be written in the modified canonic form as follows:

$$w(n)=0.55w(n-1)+x(n)$$
$$y(n)=-0.3w(n)+0.8x(n)$$

Let $\bar{w}(n)\in(-4,3)$ and $\bar{x}(n)\in(-2,1)$. (The dynamic range and word length consideration of systems will be discussed in Section 2. 3). From the all-pole equation, the following table may be obtained.

Table 2. 1 Outputs of $w(n)$ in Example 2. 1

$w(n)$		$\bar{x}(n)$			
		0	1	−2	−1
$\bar{w}(n-1)$	0	0	1	−2	−1
	1	0. 55	1. 55	−1. 45	−0. 45
	2	1. 1	2. 1	−0. 9	0. 1
	3	1. 65	2. 65	−0. 35	0. 65
	−4	−2. 2	−1. 2	−4. 2	−3. 2
	−3	−1. 65	−. 65	−3. 65	−2. 65
	−2	−1. 1	−0. 1	−3. 1	−2. 1
	−1	−0. 55	0. 45	−2. 55	−1. 55

Quantizing the above table, the state table of an FSM may be obtained as follows:

Table 2.2 State table of Example 2.1

$s(n)$		$u(n)$			
		0	1	−2	−1
$s(n-1)$	0	0	1	−2	−1
	1	1	2	−1	0
	2	1	2	−1	0
	3	2	3	0	1
	−4	−2	−1	−4	−3
	−3	−2	−1	−4	−3
	−2	−1	0	−3	−2
	−1	−1	0	−3	−2

Similarly, the truth table of output functions may be obtained from the second equation as follows:

Table 2.3 Truth table of output functions in Example 2.1

$g(n)$		$u(n)$			
		0	1	−2	−1
$s(n)$	0	0	1	−2	−1
	1	0	1	−2	−1
	2	−1	1	−2	−1
	3	−1	0	−3	−2
	−4	1	0	0	0
	−3	1	2	−1	0
	−2	1	1	−1	0
	−1	0	1	−1	−1

These tables have only $\pm 1/2$ bit error from the difference equation. Since $u(n)$, $g(n)$ and $s(n)$ are finite, the above tables may be realized in terms of a Mealy Machine.

If $L_m = 0$ and $K_m \neq 0$ in eqn. (2.1), i.e. the number of the filter zeros is less than that the number of its poles, then eqns. (2.2) and (2.3) become:

$$w(n) = x(n) + \sum_{i=1}^{m} K_i w(n-i) \tag{2.10}$$

$$y(n) = \sum_{i=0}^{m-1} L_i w(n-i) \tag{2.11}$$

This form is very simple. It is a special case of the modified canonic form of IIR filters with $L_m = 0$. This is named a "degenerative" canonical form of IIR filters. Its output $y(n)$ is independent of both the input $x(n)$ and the mth delayed variable $w(n-m)$. Its structure is shown in Fig. 2. 3.

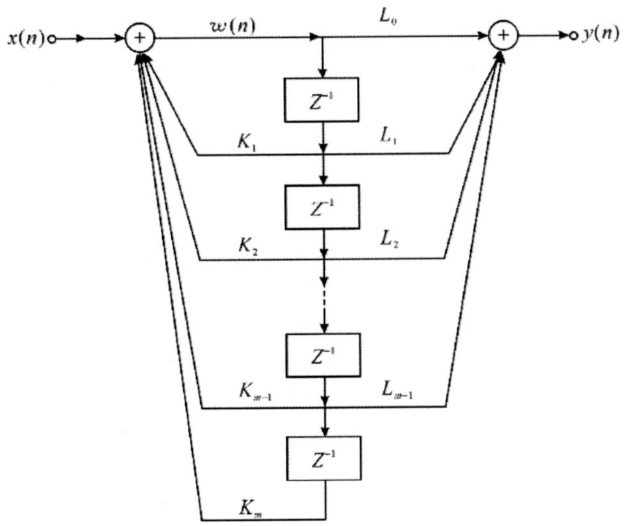

Fig. 2. 3 "Degenerative" canonical form of IIR filters with $L_m=0$

Using the definition of the m-dimensional state $\underline{s}(n)$ in eqn. (2. 5) and the discrete variables $u(n)$, $g(n)$ in eqn. (2. 7), two mappings may be obtained from eqns. (2. 10) and (2. 11) as follows:

$$\underline{s}(n)=\underline{f}[\underline{s}(n-1), u(n)]$$
$$g(n)=g[\underline{s}(n)] \tag{2. 12}$$

This is a Moore Machine. It can be successfully used to represent discrete Fourier Transform (Section 2. 4) and the IIR digital filter whose number of poles is greater than the number of zeros. But the above machine can not be used to represent the filters which have the same number of poles and zeros. The Mealy Machine is more general, and will be used in this dissertation.

2. 2. 2 Generalized Feedback Shift Register Realization of High Order IIR Filter

From the modified canonic form of IIR filters and the state vector in eqn. (2. 6), the next state mapping in eqn. (2. 8) may be written as follows:

$$s_0(n)=s_1(n-1)$$
$$s_1(n)=s_2(n-1)$$
$$\vdots$$
$$s_{m-2}(n)=s_{m-1}(n-1)$$
$$s_{m-1}(n)=f[s_0(n-1), s_1(n-1), \cdots, s_{m-1}(n-1), u(n)] \tag{2. 13}$$

The output mapping may be written as follows:

$$g(n)=g[s_0(n), s_1(n), \cdots, s_{m-1}(n), u(n)] \tag{2. 14}$$

The eqn. (2. 13) may be considered as a generalized feedback shift register (FSR), the next state components are shifted versions of present states except the last component $s_{m-1}(n)$. This structure is called a generalized FSR, as the state variable $s_i(n)$ is an M-bit word rather than a simple bit. The generalized FSR realisation, with output mapping, of a high order IIR digital filter is shown in Fig. 2. 4.

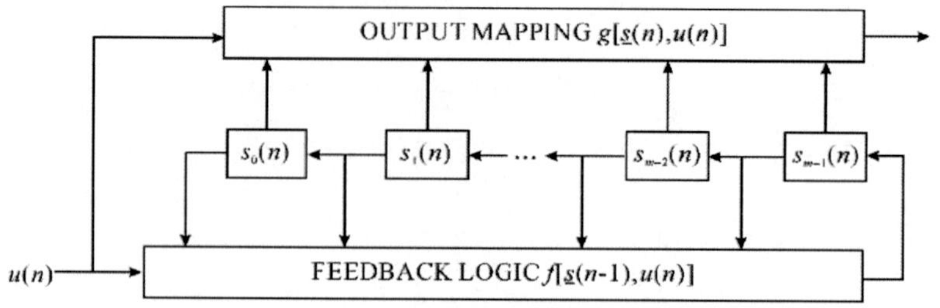

Fig. 2.4 Generalized FSR realization of IIR filters

It is shown that a one-dimensional IIR digital filter of order m may be represented as a generalized FSR having m stages. The truth tables of the feedback logic and output mapping of FSRs may be obtained from the difference equations (2.4) and (2.5), when the quantization methods and word lengths are determined.

Example 2.2

A second order all-pole IIR digital filter is given by

$$w(n)=0.7w(n-1)-0.5w(n-2)+x(n)$$

As shown in eqn. (2.6), let $s_1(n)=\overline{w}(n)$, $s_0(n)=\overline{w}(n-1)$, then $s_1(n-1)=s_0(n)=\overline{w}(n-1)$, $s_0(n-1)=\overline{w}(n-2)$. This is a generalized feedback shift register of length 2. Its state table with zero input is shown in Table 2.4.

Table 2.4 Stable table of Example 2.2

$s_1(n-1)$	$s_0(n-1)$	$w(n)$	$s_1(n)$	$s_0(n)$
0	0	0.0	0	0
0	1	−0.5	0	0
0	−2	1.0	1	0
0	−1	0.5	1	0
1	0	0.7	1	1
1	1	0.2	0	1
1	−2	1.7	1	1
1	−1	1.2	1	1
−2	0	−1.4	−1	−2
−2	1	−1.9	−2	−2
−2	−2	−0.4	0	−2
−2	−1	−0.9	−1	−2
−1	0	−0.7	−1	−1
−1	1	−1.2	−1	−1
−1	−2	0.3	0	−1
−1	−1	−0.2	0	−1

2.3 Dynamic Range and Word Length Consideration

In fixed-point implementation of IIR digital filters, overflow may be caused during addition operations. The signal level at certain point of digital filters must be scaled so that overflow is unlikely to occur under most normal inputs and operating conditions. Let $h_s(n)$ be the impulse response of the filter at point s then the output at point s may be written as

$$y_s(n) = \sum_{i=0}^{\infty} h_s(n-i)x(i) \tag{2.15}$$

If $x_{\max}$ denotes the maximum input of the filter, then its maximum output $y_{s\max}$ at point s is

$$y_{s\max} \leq x_{\max} \sum_{i=0}^{\infty} |h_s(i)| \tag{2.16}$$

In order to prevent overflow in digital filters a scaling factor

$$A_s = \frac{1}{\sum_{i=0}^{\infty} |h_s(i)|} \quad 0 < A_s < 1 \tag{2.17}$$

must be introduced in the input of filters.

The introduction of the scaling factor A_s prevents digital filters from overflowing. But it decreases the effective word length of input signals, and decreases the signal-to-noise ratio of systems. The interaction of round off noise and dynamic range in digital filters was studied by Jackson[J4].

In FSM realizations of IIR digital filters, the state table of machines may be computed over all the possible integer values of inputs $u(n)$ and states $\underline{s}(n)$. The use of scaling factor $A_s(0<A_s<1)$ in filter inputs is equivalent to a decrease in input amplitude ranges. Therefore, the shorter input word length in FSMs should be used rather than the state and output word length, such that overflows are unlikely to occur under most combinations of inputs and states. Let $x_{\max}$, $y_{\max}$ and $w_{\max}$ denote the maximum amplitudes of inputs $x(n)$, output $y(n)$ and variable $w(n)$ in IIR filters respectively, then

$$w_{\max} \leq x_{\max} \sum_{n=0}^{\infty} |h_0(n)| = x_{\max} R_0 \tag{2.18}$$

$$y_{\max} \leq x_{\max} \sum_{n=0}^{\infty} |h(n)| = x_{\max} R \tag{2.19}$$

where R_0 and R are named the range factors, $h_0(n)$ and $h(n)$ are impulse responses respectively determined by the following z-transform functions

$$H_0(z) = \frac{1}{1 - \sum_{i=1}^{m} k_i z^{-i}} \tag{2.20}$$

$$H(z) = \frac{\sum_{i=0}^{m} L_i z^{-i}}{1 - \sum_{i=0}^{m} k_i z^{-i}} \tag{2.21}$$

Let N, K and M denote the word length of Inputs $u(n)$, output $g(n)$ states $s(n)$ in FSMs respectively, then

$$M = N + \Delta b_0 \tag{2.22}$$

$$K = N + \Delta b \tag{2.23}$$

where Δb_0 and Δb are minimum integers such that

$$2^{\Delta b_0} \geq \sum_{n=0}^{\infty} |h_0(n)| > 2^{\Delta b_0 - 1} \tag{2.24}$$

$$2^{\Delta b} \geq \sum_{n=0}^{\infty} |h(n)| > 2^{\Delta b - 1} \tag{2.25}$$

The above equations present the methods for determining the dynamic ranges and word lengths of FSMs, so that overflows may be prevented. There are no overflows in the internal structures of FSMs, as the next state and output mapping can be realized in terms of finite arithmetic structures. This is an essential difference of FSM realizations from the conventional realization of filters.

The first and second order IIR filters are basic blocks of digital filters. They are discussed as follows.

2.3.1 First Order IIR Digital Filters

The 1st order all-pole filter may be written as the following difference equation

$$w(n) = k_1 w(n-1) + x(n) \tag{2.26}$$

Its impulse response $h_0(n)$ and range factor are

$$h_0(n) = k_1^n \quad \text{for} \quad w(-1) = 0$$

$$R_0(n) = \frac{w_{\max}}{x_{\max}} = \sum_{n=0}^{\infty} |h_0(n)| = \frac{1}{1 - |k_1|} \qquad |k_1| < 1 \tag{2.27}$$

The word length of state variables in FSMs should be Δb_0 bits longer than that of input variables such that

$$2^{\Delta b_0} \geq \frac{1}{1 - |k_1|} > 2^{\Delta b_0 - 1}$$

The relationship between Δb_0 and the coefficient k_1 is shown in Table 2.1.

Table 2.1 Relationship between k_1 and Δb_0

k_1	$w_{\max}/x_{\max}$	Δb_0 (bits)
0.1	1.11	1
0.2	1.25	1
0.3	1.43	1
0.4	1.67	1
0.5	2.0	1
0.6	2.5	2
0.7	3.33	2
0.75	4.0	2
0.8	5.0	3
0.875	8.0	3
0.9	10.0	4
0.9375	16.0	4
0.96875	32.0	5

The general form of the 1st order IIR filter is

$$y(n)=k_1 y(n-1)+L_0 x(n)+L_1 x(n-1) \tag{2.28}$$

Its impulse response $h(n)$ and range factor R are as follows:

$$h(n)=\begin{cases} L_0 & n=0 \\ L_0 k_1^n + L_1 k_1^{n-1} & n\geq 1 \end{cases}$$

$$R=\frac{y_{\max}}{x_{\max}}=\sum_{n=0}^{\infty} |h(n)| = \frac{|L_0|+|L_1|}{1-|k_1|} \tag{2.29}$$

It is shown that the maximum output amplitude should be $|L_0|+|L_1|$ times the maximum amplitude of state variables in FSMs. The factor $|L_0|+|L_1|$ is dependent on the zeros of filters, but independent of the pole. When $|L_0|+|L_1|<1$, the word length of output variable may be less than that of state variables in FSMs.

2.3.2 Second Order IIR Digital Filters

The all-pole and general form of and IIR filters may be written as the following difference equations

$$w(n)=k_1 w(n-1)+k_2 w(n-2)+x(n) \tag{2.30}$$

$$y(n)=k_1 y(n-1)+k_2 y(n-2)+L_0 x(n)+L_1 x(n-1)+L_2 x(n-2) \tag{2.31}$$

Their impulse responses depend on the properties of poles: complex poles or real poles.

2.3.2.1 Two complex poles

When $\frac{k_1^2}{4}+k_2<0(-1<k_2<0)$, the above two equations have two complex poles $re^{\pm j\theta}$ where $r=\sqrt{-k_2}(0<r<1)$, $\theta=\cos^{-1}(k_1/2r)$. The impulse response $h_0(n)$ and range factor R_0 of the 2nd order all-pole filters are as follows:

$$\begin{aligned} h_0(n) &= \frac{r^n \sin(n+1)\theta}{\sin\theta} \quad \text{for } y(-2)=y(-1)=0 \\ R_0 &= \frac{w_{\max}}{x_{\max}} = \sum_{n=0}^{\infty} |h_0(n)| < \frac{1}{(1-r)\sin\theta} \quad 0<r<1 \end{aligned} \tag{2.32}$$

The word length of state variables in FSMs should be Δb_0 bits longer than that of input variables such that

$$2^{\Delta b_0} \geq \frac{1}{(1-r)\sin\theta} > 2^{\Delta b_0 -1} \tag{2.33}$$

some example are shown in Table 2.2.

The impulse response $h(n)$ of general 2nd order IIR filters in Equation (2.31) is as follows:

$$h(n)=\begin{cases} L_0 & n=0 \quad \text{for } y(-2)=y(-1)=0 \\ \frac{1}{\sin\theta}[L_0 r^n \sin(n+1)\theta + L_1 r^{n-1}\sin(n\theta) + L_2 r^{n-2}\sin(n-1)\theta] & n\geq 1 \end{cases} \tag{2.34}$$

for $n\geq 1$, $h(n)$ may be rewritten as

$$h(n)=\frac{c r^{n-1}\sin(n\theta+\phi)}{\sin\theta} \quad n\geq 1 \tag{2.34}$$

Table 2.2 Range factor of 2nd order IIR filters

$\theta=\frac{\pi}{6}$			$r=0.75$		
r	w_{max}/x_{max}	Δb_0 (bits)	θ	w_{max}/x_{max}	Δb_0 (bits)
0.1	2.2222	1	0.05π	25.27	5
0.2	2.5	2	0.1π	12.94	4
0.3	2.8571	2	0.15π	8.811	4
0.4	3.3333	2	0.2π	6.805	3
0.5	4.0	2	0.25π	5.657	3
0.6	5.0	3	0.3π	4.944	3
0.7	6.6667	3	0.35π	4.489	3
0.75	8.0	3	0.4π	4.206	3
0.8	10.0	4	0.45π	4.050	3
0.9	20.0	5	0.5π	4.0	2

where

$$c=\{[(L_0 r+\frac{L_2}{r})\cos\theta+L_1]^2+[(L_0 r-\frac{L_2}{r})\sin\theta]^2\}^{\frac{1}{2}}$$

$$\phi=\cos^{-1}[(L_0 r+\frac{L_2}{r})\cos\theta+L_1]/c$$

From the above equation, the range factor R may be obtained as

$$R=\sum_{n=0}^{\infty}|h(n)|<|L_0|+\frac{c}{(1-r)\sin\theta} \quad (2.35)$$

It is shown that the output amplitude may be $[|L_0|(1-r)\sin\theta+c]$ times larger than that of state variables. Here the factor $[|L_0|(1-r)\sin\theta]$ is dependent on both poles and zeros of filters.

2.3.2.2 Real poles

When $\frac{k_1^2}{4}+k_2>0$, the 2nd order filter has two real poles: $r_{1,2}=\frac{k_1}{2}\pm\sqrt{\frac{k_1^2}{4}+k_2}$. When $\frac{k_1^2}{4}+k_2=0$, it has two real coinciding poles $r_1=r_2=\frac{k_1}{2}$. The impulse responses $h_0(n)$ of the 2nd order all-pole filter in Equation (2.30) are as follows:

$$\begin{cases} h_0(n)=\frac{1}{r_1-r_2}(r_1^{n+1}-r_2^{n+1}) & \text{for } r_1\neq r_2 \\ h_0(n)=(n+1)r^n & \text{for } r_1=r_2=r \end{cases} \quad (2.36)$$

The range factors may be obtained by

$$\begin{cases} R_0=\frac{w_{max}}{x_{max}}=\sum_{n=0}^{\infty}|h_0(n)|\leq\frac{1}{1-|k_1|-k_2} & \text{for } \frac{k_1^2}{4}+k_2>0 \\ R_0\leq\frac{1}{(1-|\frac{k_1}{2}|)^2} & \text{for } \frac{k_1^2}{4}+k_2=0 \end{cases} \quad (2.37)$$

Some examples are shown in Table 2. 3.

Table 2. 3 Range factor of 2nd order IIR filters

k_2	k_1	w_{max}/x_{max}	Δb_0 (bits)
0. 9	0. 06875	32. 0	5
0. 8	0. 1375	16. 0	4
0. 7	0. 175	8. 0	3
0. 6	0. 15	4. 0	2
0. 5	0. 25	4. 0	2
0. 4	0. 1	2. 0	1
0. 3	0. 2	2. 0	1
0. 2	0. 3	2. 0	1
−0. 1	0. 4	2. 0	1
−0. 1	0. 65	2. 0	1
−0. 2	0. 95	4. 0	2
−0. 3	1. 175	8. 0	3
−0. 4	1. 275	8. 0	3
−0. 5	1. 4375	16. 0	4
−0. 75	1. 734375	64. 0	6

The impulse responses $h(n)$ of general 2nd IIR filters in Equation (2. 31) are as follows:

$$\begin{cases} h(n)\begin{cases} L_0 & n=0 \\ \dfrac{1}{r_1-r_2}[L_0(r_1^{n+1}-r_2^{n+1})+L_1(r_1^n-r_2^n)+L_2(r^{n-1}-r_2^{n-1})] & n\geq 1 \end{cases} & \text{for } r_1\neq r_2 \\ h(n)=\begin{cases} L_0 & n=0 \\ L_0(n+1)r^n+L_1nr^{n-1}+L_2(n-1)r^{n-2} & n\geq 1 \end{cases} & \text{for } r_1=r_2=r \end{cases}$$

The range factors may be obtained from

$$\begin{cases} R=\dfrac{y_{max}}{x_{max}}=\sum_{n=0}^{\infty}|h(n)|\leq\dfrac{|L_0|+|L_1|+|L_2|}{1-|k_1|-k_2} & \dfrac{k_1^2}{4}+k_2>0 \\ R\leq\dfrac{|L_0|+|L_1|+|L_2|}{(1-\left|\dfrac{k_1}{2}\right|)^2} & \dfrac{k_1^2}{4}+k_2=0 \end{cases} \tag{2.39}$$

It is shown that the output range y_{max} may be $[|L_0|+|L_1|+|L_2|]$ times the state range. The factor $[|L_0|+|L_1|+|L_2|]$ is dependent on zeros of the filter, but independent of its poles.

2.4 FSM Representation of Discrete Fourier Transforms

2.4.1 FSM Representation of DFTs with Prime Length

The discrete Fourier transform of N points may be written as follows:

$$x(k) = \sum_{n=0}^{N-1} x(n)w_N^{nk}$$
$$= [\cdots[[W_N^k x(N-1) + x(N-2)]W_N^k + x(N-3)]W_N^k + \cdots + x(1)]W_N^k + x(0) \quad (2.40)$$

The above equation may be expressed as the following recursive form:

$$y_k(n)=W_N^k y_k(n-1)+u(n)$$
$$y_k(0)=u(0) \quad k=0,1,2,\cdots,(N-1). \quad (2.41)$$

where $u(n)=x(N-1-n)$, $n=0,1,2,\cdots,(N-1)$.

The kth frequency component $X(k)$ of DFTs is the $(N-1)$th output of the recursive sequences $\{y_k(n)\}$

$$X(k)=y_k(N-1) \quad k=0,1,2,\cdots,(N-1)$$

This is Goertzel algorithm[P3, Ch. 6]. The advantages of the Goertzel algorithm are that only one coefficient W_N^k is required to compute each frequency component $X(k)$, and the algorithm can be realized by using the simple recursive equation. However, the algorithm needs N coefficients for computing the whole N frequency components. Curts[A18] proposed that when the length of DFTs $N=p$ is prime, only one coefficient is required to compute $(p-1)$ frequency components by permuting the input sequence $\{x(n)\}$. This can be described by the following example.

Consider a five-point DFT

$$X(k) = \sum_{n=0}^{4} x(n)W^{nk}, \ k = 0,1,2,3,4$$

where $W=\exp(-j2\pi/5)$.

The non-zero frequency components may be written as the following matrix form:

$$\begin{bmatrix} X(1) \\ X(2) \\ X(3) \\ X(4) \end{bmatrix} = \begin{bmatrix} 1 & w^1 & w^2 & w^3 & w^4 \\ 1 & w^2 & w^4 & w^1 & w^3 \\ 1 & w^3 & w^1 & w^4 & w^2 \\ 1 & w^4 & w^3 & w^2 & w^1 \end{bmatrix} \begin{bmatrix} x(0) \\ x(1) \\ x(2) \\ x(3) \\ x(4) \end{bmatrix}$$

It is obvious that each row of the coefficient matrix is a permutation of the first row. Rearrange the ordering of the input sequences $\{x(n)\}$, the above matrix can be rewritten as follows:

$$\begin{bmatrix} X(1) \\ X(2) \\ X(3) \\ X(4) \end{bmatrix} = \begin{bmatrix} x(0) & x(1) & x(2) & x(3) & x(4) \\ x(0) & x(2) & x(1) & x(4) & x(3) \\ x(0) & x(3) & x(4) & x(1) & x(2) \\ x(0) & x(4) & x(3) & x(2) & x(1) \end{bmatrix} \begin{bmatrix} 1 \\ w^1 \\ w^2 \\ w^3 \\ w^4 \end{bmatrix}$$

One coefficient can be employed in recursively computing all the 4 non-zero frequency components by permuting the input data. For prime length DFTs, the recursive equation (2.41) may be changed to

$$y_k(n) = W_p^D y_k(n-1) + u_k(n)$$
$$y_k(0) = u_k(0) \quad k = 1, 2, \cdots, (p-1) \tag{2.42}$$

where $u_k(p-1-n) = X(\langle nk^{-1}D\rangle_p), n = 1, 2, \cdots, (p-1)$, (2.43)

k^{-1} is the inverse element of k in GF(p), $k^{-1}k = 1 \bmod p$,

D is an integer, $1 \leq D < p$.

The above recursive equation may be viewed as a first order IIR filter with a complex coefficient. Its z-transform function is as follows:

$$H(z) = \frac{1}{1 - W_p^D z^{-1}} \tag{2.44}$$

It may be expressed as a real coefficient function by

$$H(z) = \frac{1}{1 - W_p^D z^{-1}} \frac{1 - W_p^D z^{-1}}{1 - W_p^D z^{-1}}$$
$$= \frac{1 - \cos(2\pi D/p)z^{-1} - j\sin(2\pi D/p)z^{-1}}{1 - 2\cos(2\pi D/p)z^{-1} + z^{-2}} \tag{2.45}$$

Eqn. (2.45) may be considered as a 2nd order IIR filter as shown in Fig. 2.5.

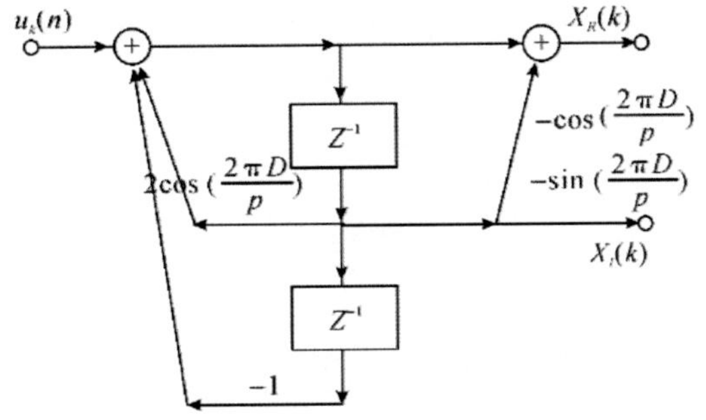

Fig. 2.5 The 2nd order recursive realizations of DFTs

In Fig. 2.5, $X_R(k)$ and $X_I(k)$ denote respectively the real and the imaginary components of DFT outputs. Obviously, this is a "degenerative" canonic form of 2nd order IIR filters. Having introduced a variable $w(n)$ the function (2.45) may be written as the following difference equations:

$$w(n) = 2\cos\left(\frac{2\pi D}{p}\right)w(n-1) - w(n-2) + u_k(n) \tag{2.46}$$

$$\begin{cases} X_R(k) = w(p-1) - \cos\left(\frac{2\pi D}{p}\right)w(p-2) \\ X_I(k) = -\sin\left(\frac{2\pi D}{p}\right)w(p-2) \end{cases} \tag{2.47}$$

Let $\underline{s}(n)$ and $\underline{g}(k)$ be respectively two-dimensional state and output vectors defined by

$$\underline{s}(n) = \begin{bmatrix} s_0(n) \\ s_1(n) \end{bmatrix} = \begin{bmatrix} \overline{w}(n-2) \\ \overline{w}(n-1) \end{bmatrix}$$

$$\underline{g}(k) = \begin{bmatrix} g_0(k) \\ g_1(k) \end{bmatrix} = \begin{bmatrix} \underline{X}_R(k) \\ \underline{X}_I(k) \end{bmatrix}$$

then a state table may be obtained by quantizing eqn. (2. 46). This state table can be represented as two mappings as follows:

$$\begin{cases} s_0(n+1)=s_1(n) \\ s_1(n+1)=f[s_0(n), s_1(n), u_k(n)] \end{cases} \tag{2.48}$$

$$\begin{cases} g_0(k)=g_0[s_0(p), s_1(p)] \\ g_1(k)=g_1[s_0(p)] \end{cases} \tag{2.49}$$

This is a two-dimensional Moore Machine for representing the DFT. The machine has three features:

(a) The input sequence $\{u_k(n)\}$ is finite, and is a permutation of the original sequence $\{u(n)\}$.

(b) The next state mapping is required to have $(p-1)$ recursive computations for each frequency componente.

(c) The output mapping needs only one calculation for each frequency component.

The machine is shown in Figure 2. 6.

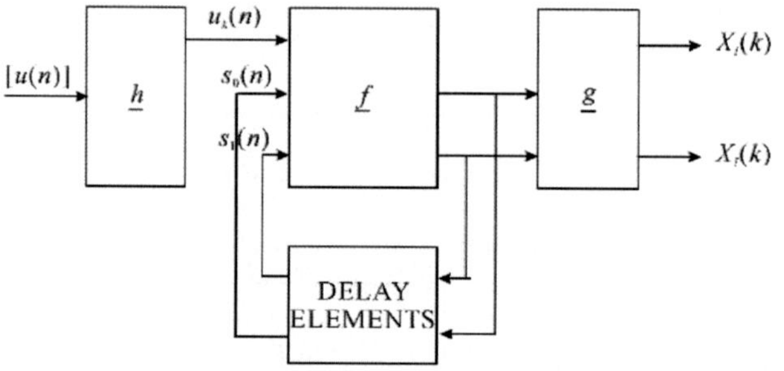

Fig. 2. 6 Moore model for representing DFTs

In the above figure, $\underline{h}$ is an input mapping which permutes the inputs $\{u(n)\}$ into the new ordering $\{u_k(n)\}$ based eqn. (2. 43) for computing each frequency component. It may be precomputed and stored as address maps or tables. $\underline{f}$ is a next state mapping. Since only one coefficient is required to compute all the $(p-1)$ non-zero frequency components in eqn. (2. 47), so the next state tables are unchanged for different frequency. The mapping $\underline{f}$ may be implemented in terms of high speed circuits which can be faster than the conventional multipliers used in IIR filters. $\underline{g}$ is a two-dimensional output mapping which can be implemented in terms of low speed devices.

2. 4. 2 Dynamic Range and Optimum Coefficient Choice

As shown in eqns. (2. 42) and (2. 44), the recursive calculation of DFTs may be considered as a process of finding the spectrum of input signals at a fixed point A on a unit cycle in z-plane (see Figure 2. 7). For a prime p, all the $(p-1)$ frequency components can be computed at the same point A $(=W_p)$ in the z-plane by permuting the input data. Since p is prime, any one of the $(p-1)$ pole points $B(=W_p^D, 1\leq D<p)$ on unit circle can be used as the fixed-point instead of the point A for recursively computing the DFT.

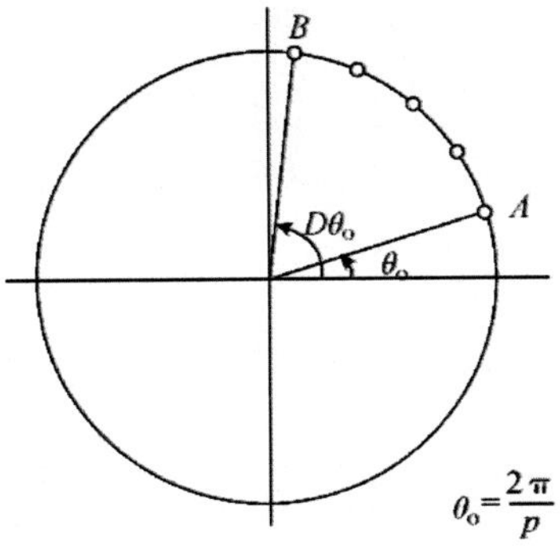

Fig. 2. 7 Choice of pole positions

For the floating-point implementation of the recursive eqn. (2. 46), any coefficients $W_p^D(1\leq D<p)$ can be used in the system, and they have no effect on the final results. However, in the fixed-point implementation, the coefficients W_p^D will affect the signal-to-noise ratio and dynamic range of systems.

In order to prevent overflows in fixed-point implementation, a scaling factor A_D should be introduced in the input of the second order all-pole filter (2. 46), such that

$$A_D<\frac{1}{\sum_{n=0}^{P-1}|h_D(n)|}\quad 0<A_D<1 \tag{2.50}$$

where $h_D(n)$ is the impulse response of the all-pole filter (2. 46).

The signal-to-noise ratio of systems is proportional to the scaling factor A_D. For the all-pole filter, its impulse response $h_D(n)$ and scaling factor A_D are as follows:

$$h_D(n)=\frac{\sin\frac{2\pi(n+1)D}{p}}{\sin\frac{2\pi D}{p}} \tag{2.51}$$

$$A_D=\frac{\left|\sin\frac{2\pi D}{p}\right|}{\sum_{n=0}^{p-1}\left|\sin\frac{2\pi(n+1)D}{p}\right|} \tag{2.52}$$

Since p is prime, $\sum_{n=0}^{p-1}\sin\frac{2\pi(n+1)D}{p}$ is independent on the factor D, namely

$$\sum_{n=0}^{p-1}\left|\sin\frac{2\pi(n+1)D}{p}\right|=C_p<p \tag{2.53}$$

where C_p is a constant and $C_p<p$.

Using the above equation, the scaling factor A_D can be written as

$$A_D=\left|\sin\frac{2\pi D}{p}\right|/C_P \tag{2.54}$$

It is obvious that when $\frac{2\pi D}{p}$ approaches $\pm\pi/2$, A_D has the maximum value, and the system possesses the optimum signal-to-noise ratio without overflows.

Table 2. 4 gives the optimum D and the recursive coefficient $\cos(\frac{2\pi D}{p})$ for all $p<100$. Some results for $p>100$ are shown in Table 2.5.

Table 2.4 Optimum coefficients for $p<100$

p	D	$\cos(\frac{2\pi D}{p})$
3	1	−0.5
5	1	0.309017
7	2	−0.222521
11	3	−0.142315
13	3	0.120537
17	4	0.092268
19	5	−0.825793
23	6	−0.068242
29	7	0.054139
31	8	−0.050649
37	9	0.042441
41	10	0.029303
43	11	−0.036522
47	12	−0.033415
53	13	0.029633
59	15	−0.026621
61	15	0.025748
67	17	−0.023443
71	18	−0.022122
73	18	0.021516
79	20	−0.019882
83	21	−0.018924
97	24	0.016193

Table 2.5 Optimum coefficients for $p>100$

p	D	$\cos(\frac{2\pi D}{p})$
127	32	-1.5831×2^{-7}
257	65	1.5647×2^{-8}
509	127	1.5801×2^{-9}
1021	255	1.5754×2^{-10}

(To be continued)

Table 2.5

p	D	$\cos(\frac{2\pi D}{p})$
2039	510	1.5777×2^{-11}
4093	1023	1.5720×2^{-12}
8191	2048	-1.5710×2^{-13}

Similarly, for the FSM implementation of DFTs, the word length of state variables should be Δb_0 bits longer than that of input variables such that

$$2^{\Delta b_0} \geq \frac{C_p}{\left|\sin\frac{2\pi D}{p}\right|} > 2^{\Delta b_0 - 1} \tag{2.55}$$

From the output equation (2.47), it is shown that the word length of the output variables should be 1 bit longer than that of the state variable.

2.5 Conclusion

IIR digital filters and DFTs may be represented as finite state machines. A modified canonic form of IIR filters has been presented, such that the IIR filter can be easily represented as a Mealy Machine in the form of state tables. This state table can be computed from the filter difference equation and only possesses $\pm 1/2$ least significant bit error over the representable range of variables. The dynamic range and word length consideration in fixed-point implementation have been discussed. The proper choice of word lengths can prevent the FSM from overflows under most combinations of variables. The mth order IIR filter may be represented as a generalized feedback shift register of stage m. Some FSMs may be unstable, the stability of machines will be studied in Chapter 7. The recursive algorithm of DFT with prime length has been discussed. Only one coefficient in the algorithm is needed for computing $(p-1)$ frequency components. This algorithm may be realized in terms of an FSM and an input mapping. The dynamic range and optimum coefficient choice have been discussed. The coefficients with maximum signal-to-noise ratio and without overflows for fixed-point implementation have been presented.

Chapter 3
Finite Field Representation of Finite State Machines①

3.1 Introduction

Digital signal processing systems may be represented as a Finite State Machine (FSM) in the

① The contents of this chapter have been published in the Research Report of Cambridge University Engineering Department, CUED/B-Elect/TR70, 1984, pp. 1-44[Z5].

form of state tables. However, the state tables do not clearly show the dynamic behaviour of systems. In order to implement finite state machines in logical circuits, it is desired that the state tables can be represented in analytical forms.

The analytical form of a system may be obtained from its state table, conversely, the state table should be easily found from its analytical representation. Boolean OR/AND/NOT functions may be used to represent the next state and the output function of a finite state machine from its state table. However, the FSM representation of digital signal processing systems requires a large number of Boolean variables, the Boolean function description of systems is very complex, and it is not easily tested and minimized for the number of variables $n>6$[Q19]. Moreover, Boolean functions are not well-suited to investigating the dynamic performance of systems. One algebraic structure that is well-suited to the analytical representation of FSMs is the Galois Field.

Galois Field GF(p^M) is a finite algebraic structure which has $q=p^M$ elements for any prime $p>1$ and for any integer $M>0$, and two operations, which are called addition and multiplication, defined on all p^M elements. Both operations are closed, commutative and associative, and multiplication is distributive over addition. Each field contains two special elements: 0 and 1, which act as null and identity elements for addition and multiplication operations respectively. For every element of GF (p^M), there exists an inverse element in the same field.

All finite fields of order p^M are isomorphic[Q1]. If $p_M(x)$ is a primitive polynomial of order M over GF(p), then its root α is a primitive element of GF(p^M). All the (p^M-1) non-zero elements of GF(p^M) may be represented as powers of α, or polynomials of order ($M-1$) as follows[Q1]:

$$\alpha^i = x_{M-1}\alpha^{M-1} + x_{M-2}\alpha^{M-2} + \cdots + x_1\alpha + x_0 \mod p_M(\alpha)$$
$$i=0,1,2,\cdots,(q-2) \quad \alpha^{q-1}=1 \tag{3.1}$$

From the above equation, an important property of the Galois Field can be obtained as follows:

$$x^{n(q-1)+1} = x \text{ for } \forall x \in GF(q), \ n\text{-integer} \tag{3.2}$$

This property is very useful for the representation of finite state machines. The arithmetic operations of Galois Fields may be implemented in terms of Exclusive-OR and AND logic circuits[C3].

This chapter is concerned with the Galois Field representation of finite state machines. The advantages of Galois Field representations are as follows.

(a) Galois Fields possess properties which may be efficiently employed in computing the polynomial coefficients, minimizing machines and investigating the dynamic behaviours of systems.

(b) The functions over Galois Fields may be implemented in terms of Exclusive-OR and AND logic circuits which are more easily tested in comparison with vertex networks[D10], since the change of any input to the Exclusive-OR function will always propagate a change through to the function output, unlike vertex gates which require specific input pattern change to sensitise a path to output.

(c) The Exclusive-OR implementation of digital circuits may be more economical than the conventional OR/AND/NOT implementation either in the number of gates or in the number of gate interconnections.

This advantage may be justified by using the Entropy definition of OR and Exclusive-OR logic. The entropy $E(y)$ of a binary function y may be defined as follows[D9]:

$$E(y) = -p_0\log_2 p_0 - p_1\log_2 p_1$$

where p_0 and p_1 are respectively the probability of being "0" and "1" for the function output. For a logic OR function $y_0 = x_0 \vee x_1$, if the independent variables x_0 and x_1 each have an equal probability of being "0" and "1", then the logic OR function has probability 0.25 of being "0" and 0.75 of being "1". The entropy of logical OR function y_0 is

$$E(y_0) = -0.25\log_2 0.25 - 0.75\log_2 0.75 = 0.8113$$

Similarly, the entropy of logical Exclusive-OR function $y_e = x_0 \oplus x_1$ is as follows:

$$E(y_e) = -0.5\log_2 0.5 - 0.5\log_2 0.5 = 1$$

The latter is larger than the former.

Booth[C2] first proposed an analytical representation of signals in linear sequential networks. Gill and Jacob[C6] presented a mapping polynomial for Galois Fields, but only single-variable polynomials were discussed. Green and Taylor[C9] developed the modular representation of multiple-valued logic systems. The mappings between polynomial and operational domains were presented, but they did not reveal the DFT property for computing polynomial coefficients. Benjanthrit and Reed[C13, C15] presented the fundamental structure of Galois Field functions and a systematic method for calculating the coefficients of multiple-variable polynomials. This method calculates the constant term a_0 and the highest coefficient a_r using a separating formula from the other coefficients. This will be improved in this chapter. A number of authors are concerned with the Galois Field functions[C12-C18]. They only studied the case that the functions and the independent variables are in the same field. A finite state machine is a complex system, and in general, the states, inputs and outputs of machines are in different fields. In order to study the performance of systems, it is required to convert a multi-variable machine into a single-variable one and vice versa.

Section 3.2 shows the representation of next state machines. It may be considered as a mapping from the present state s to the next state s' over the same field GF(2^M). The computation of polynomial coefficients may be converted into a DFT by rearranging the ordering of coefficients and field elements. Section 3.3 introduces the output functions of autonomous machines, this is a mapping from present state $s \in$ GF(2^M) to the output $g \in$ GF(2^K). The coefficients of the polynomial representation of the output function are in the common extension field GF(2^L) of GF(2^M) and GF(2^K) and possesses cyclic subgroup properties which can considerably reduce the memory size and amount of computation. The next state function of a general FSM is a mapping from present state s and input u to the next state s' and will be discussed in Section 3.4. The coefficients of the polynomial representation of the next state functions may be computed in terms of a two-dimensional DFT and 2 one-dimensional DFTs. The representation of output mappings of a general FSM is shown in Section 3.5. Any multiple-variable machine may be converted into a single-variable machine in terms of a variable transformation and vice versa. This is discussed in Section 3.6.

3.2 Representation of Autonomous Machines

Autonomous machines show the behaviour of systems under zero-input conditions. The next state of an autonomous machine is a function of only the present states. This is a simple, but important case, as the behaviour of IIR filters under zero-inputs can be described in terms of an autonomous machine. Zero-input limit cycle oscillations of IIR filters may be studied by means of the stabilities of autonomous machines. From eqn. (1.6), one-dimensional autonomous machines may be written as follows:

$$s' = f(s) \tag{3.3}$$

$$g = g(s) \tag{3.4}$$

where s', $s \in GF(2^M)$ are the next and present states,

$g \in GF(2^K)$ is the output of the machine.

Equation (3.3) is the next state function of an autonomous machine. It is a mapping from $GF(2^M)$ to $GF(2^M)$. Equation (3.4) is the output function of an autonomous machine. It is a mapping from $GF(2^M)$ to $GF(2^K)$, and will be studied in Section 3.3.

3.2.1 An Important Property of Finite Field GF(*q*)

We begin with the properties of Finite Fields. In 1978, Pradham[C14], Benjanthrit and Reed[C15] introduced the following lemma:

Lemma 3.1

The sum of the ith power of all the nonzero elements over the finite field $GF(q)$ is null for $0<i<q-1$, and -1 for $i=q-l$, mathematically,

$$\sum_{x \in GF'(q)} x^i = \begin{cases} 0 & \text{for } 0<i<q-1 \\ -1 & \text{for } i=q-1 \end{cases} \tag{3.5}$$

where $\sum_{x \in GF'(q)}$ denotes the sum for the nonzero elements of $GF(q)$.

We improve upon the above lemma as follows.

Lemma 3.2

The sum of the $(r-j)$th power of all the elements over the field $GF(q)$ is null for $0<j\leq q-1$ and -1 for $j=0$, mathematically,

$$\sum_{x \in GF(q)} x^{r-j} = -\delta(j) = \begin{cases} 0 & \text{for } 0<j<q-1 \\ -1 & \text{for } j=0 \end{cases} \tag{3.6}$$

where $r=q-1$,

$\sum_{x \in GF(q)}$ denotes the sum for whole field elements including null,

δ is a Dirac symbol, $\delta(j) = \begin{cases} 0 & \text{for } j\neq 0,\ 0<j<q-1 \\ -1 & \text{for } j=0 \end{cases}$

This lemma can be derived from Lemma 3.1. (Note: $0^0=1$, $0^r=0$.) Three improvements are introduced in Lemma 3.2:

(a) The sum $\sum_{x \in GF(q)}$ in Lemma 3.2 is over all the field elements including null, but the sum $\sum_{x \in GF'(q)}$ in Lemma 3.1 is for the nonzero elements of fields. This improvement leads to the next one.

(b) The power of variables in Lemma 3.2 covers all the integer $0<(r-j)\leq q-1$, but Lemma 3.1 does not include the power $i=0$. This improvement avoids separately treating the coefficient a_{q-1} as in previous work[15].

(c) The power $(r-j)$ is used in Lemma 3.2 instead of i in Lemma 3.1. This will reduce the complexity for further analysis, especially in multiple-variable machines.

In eqn. (3.6), $q=p^M$, here p is a prime. In this chapter, we only deal with the case of $q=2^M$. However, all the results can be extended to $GF(p^M)$.

3.2.2 The Mapping of $GF(2^M) \to GF(2^M)$

The next state function of autonomous machines is a mapping from $GF(2^M)$ to $GF(2^M)$. It may be represented as a polynomial over $GF(2^M)$ by means of the following theorem:

Theorem 3.1

The next state mapping of autonomous machines may be represented as a polynomial of order $r=2^M-1$ over $GF(2^M)$ as follows:

$$s' = f(s) = \sum_{i=0}^{2^M-1} a_i s^i \tag{3.7}$$

where $s',\ s \in GF(2^M)$,

$a_i \in GF(2^M)$, $i=0,1,2,\cdots,(2^M-1)$,

and the coefficient a_j is determined by

$$\begin{aligned} a_0 &= f(0) \\ a_j &= \sum_{s\in GF(2^M)} f(s) s^{r-j} \\ j &= 1,2,\cdots,(2^M-1),\ r = 2^M-1 \end{aligned} \tag{3.8}$$

Proof

Substituting $f(s)$ of eqn. (3.8) by eqn. (3.7), then the right-hand side of eqn. (3.8) is equal to

$$\sum_{s\in GF(2^M)} \left(\sum_{i=0}^{2^M-1} a_i s^i\right) s^{r-j} = \sum_{i=0}^{2^M-1} a_i \sum_{s\in GF(2^M)} s^{r-(j-i)}$$

Using eqn. (3.6) it follows that

$$= \sum_{i=0}^{2^M-1} a_i \delta(j-i) = a_j \quad j = 1,2,\cdots,(2^M-1)$$

This is the left-hand side of eqn. (3.8).

It is easy to prove from eqn. (3.7) that

$$a_0 = f(0)$$

Note:

(a) $a_0 \neq \sum_{s\in GF(2^M)} f(s) s^{r-0} = a_0 + a_r$.

(b) Because of $s^0=1$ for all the field elements including zero,

$$s^r = \begin{cases} 1 & \text{for } s\neq 0, s\in GF(2^M) \\ 0 & \text{for } s=0 \end{cases}$$

The terms a_0 and $a_r s^r$ play different roles in the polynomials.

Example 3.1

The state table of an autonomous machine is as follows.

Table 3.1 The state table of Example 3.1

s	$s'=f(s)$
0	0
1	0

(To be continued)

Table 3. 1

s	$s'=f(s)$
2	1
3	2
−4	−2
−3	−2
−2	−1
−1	0

If α is a primitive root of the polynomial $(\alpha^3+\alpha+1)$ in $GF(2^3)$, then, from eqn. (3. 1), the nonzero elements of $GF(2^3)$ may be represented as follows:

$$\alpha^i = s_2\alpha^2 + s_1\alpha + s_0 \quad \mathrm{mod}(\alpha^3+\alpha+1)$$

$$i=0,1,\cdots,6 \qquad s_2, s_1, s_0 \in GF(2)$$

Let $N(s)=s_2 2^2+s_1 2+s_0$ be a decimal number ($N(0)=0$), then each field element code (s_2, s_1, s_0) and the number $N(s)$ are in one-to-one correspondence. This is named the positive-polarity assignment of field elements, as (s_2, s_1, s_0) are all in positive (non-complementary) polarities.

If the numbers in Table 3. 1 are represented in two's complements and the positive-polarity assignment of field elements is used, then the above state table can be rewritten as follows.

Table 3. 2 The state table of Example 3. 1 over GF(2^3)

s				$s'=f(s)$			
GF(2^3)	s_2	s_1	s_0	s'_2	s'_1	s'_0	GF(2^3)
0	0	0	0	0	0	0	0
1	0	0	1	0	0	0	0
α	0	1	0	0	0	1	1
α^3	0	1	1	0	1	0	α
α^2	1	0	0	1	1	0	α^4
α^6	1	0	1	1	1	0	α^4
α^4	1	1	0	1	1	1	α^5
α^5	1	1	1	0	0	0	0

From eqns. (3. 7) and (3. 8), the polynomial coefficients may be computed and the state table may be represented as the following polynomial:

$$s'=f(s)=\alpha^3 s+s^2+\alpha^2 s^3+\alpha s^4+\alpha^4 s^5+\alpha^4 s^6+\alpha^2 s^7$$

$$s', s \in GF(2^3)$$

As a special case, any linear autonomous machine over $GF(2^M)$ may be represented as a linear function over $GF(2^M)$ as follows:

$$s'=a_1 s+a_0$$

where $s', s \in GF(2^M)$.

a_1 and a_0 are elements of $GF(2^M)$ and may be represented as powers of α which is a primitive element of $GF(2^M)$. The constant term a_0 in the above equation may be eliminated by introducing a linear variable transformation. Therefore, a linear autonomous machine may be written as follows:

$$s' = \alpha^n s \tag{3.9}$$

Note that if n and $(2^M - 1)$ are relatively prime, then the machine is a primitive linear machine having a state graph consisting of two cycles: a maximum cycle of length $(2^M - 1)$ and a cycle of length 1. There are $\phi(2^M - 1)$ primitive linear finite state machines over $GF(2^M)$. ϕ is Euler's function.

3.2.3 Discrete Fourier Transformation for Computing the Polynomial Coefficients

Theorem 3.1 presents the Galois Field representation of autonomous machines. The polynomial coefficients may be obtained by computing eqn. (3.8) over all the elements of $GF(2^M)$. If the field elements and the polynomial coefficients $\{a_i\}$ are rearranged according to the following new ordering

$$\begin{gathered} 0, (\alpha^r = 1), \alpha, \alpha^2, \cdots, \alpha^{r-1} \\ a_0, a_r, a_1, a_2, \cdots, a_{r-1} \end{gathered} \tag{3.10}$$

where α is a primitive element of $GF(2^M)$,

$$r = 2^M - 1$$

then the next state mapping in eqn. (3.7) may be written as the following matrix form:

$$\begin{bmatrix} f(0) \\ f(1) \\ f(\alpha) \\ f(\alpha^2) \\ \vdots \\ f(\alpha^{r-1}) \end{bmatrix} = \begin{bmatrix} 1 & 0 & 0 & 0 & \cdots & 0 \\ 1 & 1 & 1 & 1 & \cdots & 1 \\ 1 & 1 & \alpha & \alpha^2 & & \alpha^{r-1} \\ 1 & 1 & \alpha^2 & \alpha^4 & & \alpha^{2(r-1)} \\ \vdots & \vdots & \vdots & \vdots & \ddots & \vdots \\ 1 & 1 & \alpha^{r-1} & \alpha^{2(r-1)} & \cdots & \alpha^{(r-1)^2} \end{bmatrix} \begin{bmatrix} a_0 \\ a_r \\ a_1 \\ a_2 \\ \vdots \\ a_{r-1} \end{bmatrix} \tag{3.11}$$

Let

$$\begin{aligned} & x(k) = f(\alpha^k) - a_0 \quad && k = 0, 1, 2, \cdots, (2^M - 2) \\ & x(0) = a_r && \\ & x(i) = a_i && i = 1, 2, \cdots, (2^M - 2) \end{aligned} \tag{3.12}$$

then eqn. (3.7) may be represented as follows:

$$f(0) = a_0 \tag{3.13a}$$

$$x(k) = \sum_{i=0}^{2^M-2} x(i)\alpha^{ik} \quad k = 0, 1, 2, \cdots, (2^M - 2) \tag{3.13b}$$

Because α is a primitive unit root of order $(2^M - 1)$, so eqn. (3.13b) is a DFT of length $(2^M - 1)$.

Similarly, if the independent variable s and the polynomial coefficients a_i in eqn. (3.8) are reordered according to eqn. (3.10), then the computation of coefficients a_i in eqn. (3.8) may be written as the following matrix form:

$$
\begin{bmatrix} a_0 \\ a_r \\ a_1 \\ a_2 \\ \vdots \\ a_{r-1} \end{bmatrix} = \begin{bmatrix} 1 & 0 & 0 & 0 & \cdots & 0 \\ 1 & 1 & 1 & 1 & \cdots & 1 \\ 0 & 1 & \alpha^{-1} & \alpha^{-2} & & \alpha^{-(r-1)} \\ 0 & 1 & \alpha^{-2} & \alpha^{-4} & & \alpha^{-2(r-1)} \\ \vdots & \vdots & \vdots & \vdots & \ddots & \vdots \\ 0 & 1 & \alpha^{-(r-1)} & \alpha^{-2(r-1)} & \cdots & \alpha^{-(r-1)^2} \end{bmatrix} \begin{bmatrix} f(0) \\ f(1) \\ f(\alpha) \\ f(\alpha^2) \\ \vdots \\ f(\alpha^{r-1}) \end{bmatrix} \tag{3.14}
$$

Using the new variables in eqn. (3.12), the above matrix may be written as follows:

$$a_0 = f(0) \tag{3.15a}$$

$$x(i) = \sum_{k=0}^{2^M-2} x(k)\alpha^{-ik} \quad i = 0, 1, 2, \cdots, (2^M - 2) \tag{3.15b}$$

It is obvious that eqn. (3.15b) is an inverse DFT of length (2^M-1).

For Example 3.1,

$$a_0 = f(0) = 0$$

the other 7 coefficients may be computed in terms of an inverse DFT of length 7 as follows:

$$
\begin{bmatrix} a_7 \\ a_1 \\ a_2 \\ a_3 \\ a_4 \\ a_5 \\ a_6 \end{bmatrix} = \begin{bmatrix} 1 & 1 & 1 & 1 & 1 & 1 & 1 \\ 1 & \alpha^{-1} & \alpha^{-2} & \alpha^{-3} & \alpha^{-4} & \alpha^{-5} & \alpha^{-6} \\ 1 & \alpha^{-2} & \alpha^{-4} & \alpha^{-6} & \alpha^{-1} & \alpha^{-3} & \alpha^{-5} \\ 1 & \alpha^{-3} & \alpha^{-6} & \alpha^{-2} & \alpha^{-5} & \alpha^{-1} & \alpha^{-4} \\ 1 & \alpha^{-4} & \alpha^{-1} & \alpha^{-5} & \alpha^{-2} & \alpha^{-6} & \alpha^{-3} \\ 1 & \alpha^{-5} & \alpha^{-3} & \alpha^{-1} & \alpha^{-6} & \alpha^{-4} & \alpha^{-2} \\ 1 & \alpha^{-6} & \alpha^{-5} & \alpha^{-4} & \alpha^{-3} & \alpha^{-2} & \alpha^{-1} \end{bmatrix} \begin{bmatrix} 0 \\ 1 \\ \alpha^4 \\ \alpha \\ \alpha^5 \\ 0 \\ \alpha^4 \end{bmatrix} = \begin{bmatrix} \alpha^2 \\ \alpha^3 \\ 1 \\ \alpha^2 \\ \alpha \\ \alpha^4 \\ \alpha^4 \end{bmatrix}
$$

3.3 Output Functions of Autonomous Machines

3.3.1 The Output Mapping of GF(2^M) → GF(2^K)

The output of autonomous machines is a function of present states. In general, the state and output of a finite state machine are in different fields, e.g. the state s is in GF(2^M), the output g is in GF(2^K). Therefore, the output function is a mapping from GF(2^M) to GF(2^K).

In order to represent the output mapping of autonomous machines, an Extension Field GF(2^L) is introduced such that both GF(2^M) and GF(2^K) are the subfields of GF(2^L). Thus, the output function becomes a mapping from extension field GF(2^L) to GF(2^L) [exactly a mapping from a subfield to another subfield within the extension field GF(2^L)] as shown in Fig. 3.1.

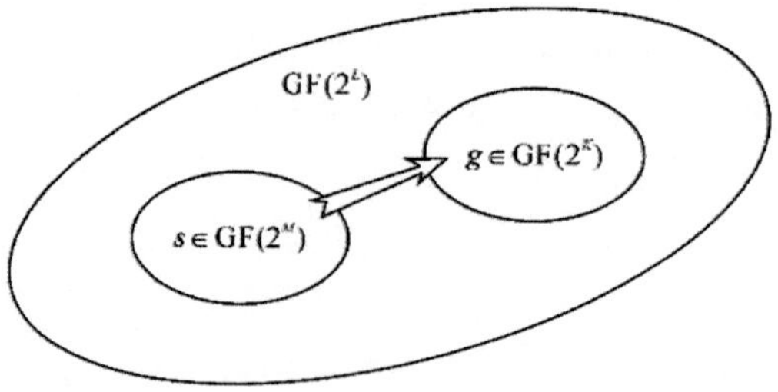

Fig. 3.1 The mapping between different subfields

Before giving the representation of output mappings, we introduce the following lemma[Q8]:

Lemma 3.3

$GF(2^K)$ is a subfield of $GF(2^L)$, if and only if K divides L. If α is a primitive element of $GF(2^L)$, then the primitive element β of subfield $GF(2^K)$ is determined by

$$\beta=\alpha^{\eta} \quad \eta=(2^L-1)/(2^K-1) \tag{3.16}$$

From the above lemma, if L is the Least Common Multiple (LCM) of K and M,

$$L=\mathrm{LCM}(K,M)$$

then $GF(2^K)$ and $GF(2^M)$ both are the subfields of $GF(2^L)$. $GF(2^L)$ is the least common extension field of $GF(2^K)$ and $GF(2^K)$.

Using the above properties, the output mapping of autonomous machines may be obtained by the following theorem.

Theorem 3.2

The output function of an autonomous machine may be represented as a polynomial with coefficients in the common extension field as follows:

$$g=g(s)=\sum_{i=0}^{2^M-1} b_i s^i \tag{3.17}$$

where $s\in GF(2^M)$, $g\in GF(2^K)$,

$b_i\in GF(2^L)$, $L=\mathrm{LCM}(K,M)$, $i=0,1,2,\cdots,(2^M-1)$.

The coefficients b_j are defined in least common extension field $GF(2^L)$ and determined by

$$\begin{aligned} b &= g(0) \\ b_j &= \sum_{s\in GF(2^M)} g(s)s^{r-j} \\ j &= 1,2,\cdots,(2^M-1) \quad r=2^M-1 \end{aligned} \tag{3.18}$$

Proof

See Appendix 3.2.

Theorem 3.2 is somewhat similar to Theorem 3.1, but the meaning of Theorem 3.2 is quite different from that of Theorem 3.1. The output g and state s in Theorem 3.2 are in different fields $GF(2^K)$ and $GF(2^M)$. The coefficients b_j of polynomial (3.17) may be neither in $GF(2^K)$ nor in $GF(2^M)$. In general, b_j is the element of extension field $GF(2^L)$ such that the polynomial (3.17) maps $GF(2^M)$ into $GF(2^K)$.

Example 3.2

The output function table of a FSM is shown in Table 3.3.

Table 3.3 Truth table of Example 3.2

s	$g=g(s)$
0	0
1	0
2	1
3	1
−4	−2

(To be continued)

Table 3.3

s	$g=g(s)$
-3	-1
-2	-2
-1	-1

Using the positive-polarity assignment of field elements and two's complement, the above table may be rewritten as follows.

Table 3.4 Truth table of Example 3.2 over GF(2^6)

s				$g=g(s)$		
GF(2^6)	s_2	s_1	s_0	g_1	g_0	GF(2^6)
$0=0$	0	0	0	0	0	$0=0$
$1=1$	0	0	1	0	0	$0=0$
$\beta=\alpha^9$	0	1	0	0	1	$1=1$
$\beta^3=\alpha^{27}$	0	1	1	0	1	$1=1$
$\beta^2=\alpha^{18}$	1	0	0	1	0	$\gamma=\alpha^{21}$
$\beta^6=\alpha^{54}$	1	0	1	1	1	$\gamma^2=\alpha^{42}$
$\beta^4=\alpha^{36}$	1	1	0	1	0	$\gamma=\alpha^{21}$
$\beta^5=\alpha^{45}$	1	1	1	1	1	$\gamma^2=\alpha^{42}$

In Table 3.4, γ, β and α are respectively the primitive elements of GF(2^2), GF(2^3) and GF(2^6). Because $6=\text{LCM}(3,2)$, from Lemma 3.3 GF(2^6) is the least common extension field of GF(2^2) and GF(2^3) and it holds that

$$\beta=\alpha^{(2^6-1)/(2^3-1)}=\alpha^9$$
$$\gamma=\alpha^{(2^6-1)/(2^2-1)}=\alpha^{21}$$

From Theorem 3.2, the above truth table may be represented as a polynomial over GF(2^6) as follows:

$$g=g\ (s)=\alpha^{36}s+\alpha^9 s^2+\alpha^{38}s^3+\alpha^{18}s^4+\alpha^{26}s^5+\alpha^{41}s^6$$

where $b_3=\alpha^{38}$, $b_5=\alpha^{26}$ and $b_6=\alpha^{41}$ are not the elements of either GF(2^2) or GF(2^3), but the polynomial maps the above truth table from GF(2^3) into GF(2^2).

Theorem 3.2 may also be represented as a DFT. Let the state s and the coefficients $\{b_j\}$ be rearranged in the following ordering:

$$0,1,\alpha^\eta,\alpha^{2\eta},\cdots,\alpha^{(r-1)\eta}$$
$$b_0,b_r,b_1,b_2,\cdots,b_{r-1} \tag{3.19}$$

where

$$\eta=(2^L-1)/(2^M-1)$$
$$r=2^M-1$$

Then eqn. (3.17) and eqn. (3.18) may be written in matrix form as follows:

$$\begin{bmatrix} g(0) \\ g(1) \\ g(\alpha^{\eta}) \\ g(\alpha^{2\eta}) \\ \vdots \\ g(\alpha^{(r-1)\eta}) \end{bmatrix} = \begin{bmatrix} 1 & 0 & 0 & 0 & \cdots & 0 \\ 1 & 1 & 1 & 1 & 1 & 1 \\ 1 & 1 & \alpha^{\eta} & \alpha^{2\eta} & \cdots & \alpha^{(r-1)\eta} \\ 1 & 1 & \alpha^{2\eta} & \alpha^{4\eta} & \cdots & \alpha^{2(r-1)\eta} \\ \vdots & \vdots & \vdots & \vdots & \ddots & \vdots \\ 1 & 1 & \alpha^{(r-1)\eta} & \alpha^{2(r-1)\eta} & \cdots & \alpha^{(r-1)^2\eta} \end{bmatrix} \begin{bmatrix} b_0 \\ b_r \\ b_1 \\ b_2 \\ \vdots \\ b_{r-1} \end{bmatrix} \quad (3.20)$$

$$\begin{bmatrix} b_0 \\ b_r \\ b_1 \\ b_2 \\ \vdots \\ b_{r-1} \end{bmatrix} = \begin{bmatrix} 1 & 0 & 0 & 0 & \cdots & 0 \\ 1 & 1 & 1 & 1 & 1 & 1 \\ 1 & 1 & \alpha^{-\eta} & \alpha^{-2\eta} & \cdots & \alpha^{-(r-1)\eta} \\ 1 & 1 & \alpha^{-2\eta} & \alpha^{-4\eta} & \cdots & \alpha^{-2(r-1)\eta} \\ \vdots & \vdots & \vdots & \vdots & \ddots & \vdots \\ 1 & 1 & \alpha^{-(r-1)\eta} & \alpha^{-2(r-1)\eta} & \cdots & \alpha^{-(r-1)^2\eta} \end{bmatrix} = \begin{bmatrix} g(0) \\ g(1) \\ g(\alpha^{\eta}) \\ g(\alpha^{2\eta}) \\ \vdots \\ g(\alpha^{(r-1)\eta}) \end{bmatrix} \quad (3.21)$$

Equations (3.20) and (3.21) contain a Discrete Fourier Transform of length (2^M-1).

3.3.2 Cyclic Subgroup Properties of the Coefficients $\{b_j\}$

Matrix (3.20) may be considered as a mapping from 2^M coefficients $\{b_j\}$ into the output vector $\{g(\alpha^{i\eta})\}$ of length 2^M. However, if $2^K < 2^M$ and g $(\alpha^{i\eta}) \in GF(2^K)$, then $g(\alpha^{i\eta})$ have at most 2^K different values, the other values are not independent. Therefore, the 2^M coefficients $\{b_j\}$ are also dependent. In general, the coefficients $\{b_j\}$ are elements of extension field $GF(2^L)$, but they are not arbitrary, some constrains exist such that the polynomial (3.17) is a mapping from $GF(2^M)$ to $GF(2^K)$.

Because $g=g(s)$ is an element of $GF(2^K)$, from eqn. (3.2) it holds

$$g^{2^k}=g \quad (3.22)$$

Applying the above constraint to eqn. (3.17), may give some cyclic properties of the coefficients $\{b_j\}$.

Before presenting these properties, a lemma is introduced[Q1].

Lemma 3.4

Let $g(x)$ be a polynomial over $GF(2^M)$ as follows:

$$g(x) = \sum_{i=0}^{2^M-1} b_i x^i$$

where g, x and b_i are elements of any extension field of $GF(2^M)$. Then it holds

$$[g(x)]^{2^n} = \sum_{i=0}^{2^M-1} b_i^{2^n} x^{i2^n} \quad (3.23)$$

where n is any integer.

Theorem 3.3

If the output function of an autonomous machine is

$$g = g(s) = \sum_{i=0}^{2^M-1} b_i s^i$$

where $g \in GF(2^K)$, $s \in GF(2^M)$, and $b_j \in GF(2^L)$, $L = \mathrm{LCM}(K, M)$, then the polynomial coefficients have the following cyclic properties:

$$b_j = b_i^{2^k} \text{ for } j = i * 2^k \bmod (2^M-1) \quad (3.24)$$

Proof

Because g is an element of $GF(2^L)$, from eqn. (3.22) it follows that:

$$[g(s)]^{2^k} = g(s)$$

and using Lemma 3.4, the above equation may be written as:

$$\sum_{i=0}^{2^M-1} b_i^{2^k} s^{i*2^k} = \sum_{i=0}^{2^M-1} b_j s^j$$

Compare the coefficients of the two polynomials in the above equation. If $s^j = s^{i*2^K} \bmod (2^M - 1)$, then their coefficients are equal, i. e.

$$b_j = (b_i)^{2^K} \text{ for } j = i * 2^K \bmod (2^M - 1)$$

There are two special cases for the above result, they are:

$$b_0 = b_0^{2^K} \text{ and } b_r = b_r^{2^M} \quad r = 2^M - 1 \tag{3.25}$$

as $0 = 0 * 2^K$ and $r = r * 2^K$, mod $(2^M - 1)$.

For Example 3.2, $2^K = 4$, the polynomial coefficients have the following relations:

$$(\alpha^{36} s)^4 = \alpha^{18} s^4, \quad (\alpha^{18} s^4)^4 = \alpha^9 s^2, \quad (\alpha^9 s^2)^4 = \alpha^{36} s$$

$$(\alpha^{38} s^3)^4 = \alpha^{26} s^5, \quad (\alpha^{26} s^5)^4 = \alpha^{41} s^6, \quad (\alpha^{41} s^6)^4 = \alpha^{38} s^3$$

The polynomial may be rewritten as the sum of two cyclic subgroups as follows:

$$g = g(s) = (\alpha^{36} s + \alpha^{18} s^4 + \alpha^9 s^2) + (\alpha^{38} s^3 + \alpha^{26} s^5 + \alpha^{41} s^6)$$

From the above example, it is shown that only the coefficients b_1 and b_3 are independent, they should be computed by means of Theorem 3.2. The other coefficients can be obtained by using their cyclic subgroup properties as shown in Theorem 3.3. The cyclic subgroup properties of coefficients $\{b_j\}$ considerably reduce the size of computer memory and the amount of calculation.

3.4 Representation of Next State Mappings

3.4.1 The Next State Mapping of $[GF(2^M), GF(2^N)] \rightarrow GF(2^M)$

The next state of a general FSM is a function of present states and inputs. In general, the state s and input u are in different fields. The next state mapping of one-dimensional FSMs may be represented as a two-variable function as follows:

$$s' = f(s, u) \tag{3.26}$$

where s', $s \in GF(2^M)$, $u \in GF(2^N)$. This is a mapping from $GF(2^M)$ and $GF(2^N)$ into $GF(2^M)$.

In order to represent next state mappings, a common extension field $GF(2^{L_1})$ of subfields $GF(2^M)$ and $GF(2^N)$ is introduced, so the function (3.26) may be considered as a mapping from two subfields to one subfield within the extension field.

The next state mapping of multiple-dimensional FSMs is a multiple-variable function. For the sake of brevity, only the two-variable function is dealt with in this section. The multiple-variable function will be discussed in Appendix 3.1.

Theorem 3.4

The next state mapping of one-dimensional FSMs may be represented as a two-variable polynomial with coefficients in a common extension field as follows:

$$s' = f(s,u) = \sum_{i=0}^{2^M-1} \sum_{j=0}^{2^N-1} a_{ij} s^i u^j \tag{3.27}$$

where

$$s's \in GF(2^M), \qquad u \in GF(2^N)$$
$$a_{ij} \in GF(2^{L_1}), \qquad L_1 = LCM(M,N)$$
$$i=0,1,2,\cdots,(2^M-1), j=0,1,2,\cdots,(2^N-1)$$

The polynomial coefficients are determined by:

$$a_{00} = f(0,0) \tag{3.28a}$$

$$a_{i0} = \sum_{s\in GF(2^M)} f(s,0) s^{r_1-i} \qquad r_1 = 2^M - 1 \tag{3.28b}$$

$$a_{0j} = \sum_{s\in GF(2^N)} f(0,u) u^{r_2-j} \qquad r_2 = 2^N - 1 \tag{3.28c}$$

$$a_{ij} = \sum_{s\in GF(2^M)} \sum_{s\in GF(2^N)} f(s,u) s^{r_1-i} u^{r_2-j} \tag{3.28d}$$

$$i=1,2,\cdots,(2^M-1), \qquad j=1,2,\cdots,(2^N-1)$$

where $\sum$ denotes the sum over all the field elements including zero.

Proof

See Appendix 3.2.

Example 3.3

The state table of a FSM is as follows.

Table 3.5 The state table of example 3.3

s	u			
	0	1	−2	1
0	0	1	−2	−1
1	1	2	−1	0
2	1	2	−1	0
3	2	3	0	1
−4	−2	−1	−4	−3
−3	−2	−1	−4	−3
−2	−1	0	−3	−2
−1	−1	0	−3	−2

Let the state s be in $GF(2^3)$, the input u in $GF(2^2)$, then the polynomial coefficients a_{ij} will be in the common extension field $GF(2^6)$, as $6 = LCM(2,3)$. If the primitive elements of $GF(2^6)$, $GF(2^3)$ and $GF(2^2)$ are α, β and γ respectively. From Lemma 3.3 it holds that $\beta=\alpha^9$, $\gamma=\alpha^{21}$. Using two's complement and positive-polarity assignment of field elements, then the above state table may be written in the following form:

Table 3.6 The state table of Example 3.3 over GF(2^6)

	0	1	α^{21}	α^{42}
0	0	1	α^{36}	α^{45}
1	1	α^{9}	α^{45}	0
α^{9}	1	α^{9}	α^{45}	0
α^{27}	α^{9}	α^{27}	0	1
α^{18}	α^{36}	α^{45}	α^{18}	α^{54}
α^{54}	α^{36}	α^{45}	α^{18}	α^{54}
α^{36}	α^{45}	0	α^{54}	α^{36}
α^{45}	α^{45}	0	α^{54}	α^{36}

From Theorem 3.4, the next state function of the above example may be obtained as follows:

$$\begin{aligned} s' = f(u,s) &= \sum_{i=0}^{7}\sum_{j=0}^{3} a_{ij} s^i u^j \\ &= (\alpha^{62}u + \alpha^{55}u^2 + \alpha^{27}u^3) + (\alpha^{27}u + \alpha^{27}u^2 + \alpha^{27}u^3)s \\ &\quad + (\alpha^9 + \alpha^{61}u + \alpha^{47}u^2 + \alpha^{36}u^3)s^2 + (\alpha^{18} + \alpha^{38}u + \alpha^{52}u^2 + \alpha^{45}u^3)s^3 \\ &\quad + (\alpha^{27} + \alpha^{33}u + \alpha^{12}u^2 + \alpha^{27}u^3)s^4 + (\alpha^{54} + \alpha^{41}u + \alpha^{13}u^2 + \alpha^{45}u^3)s^5 \\ &\quad + (\alpha^{54} + \alpha^{18}u + \alpha^{18}u^2)s^6 + (\alpha^9 + \alpha^{14}u + \alpha^{49}u^2 + \alpha^{45}u^3)s^7 \end{aligned}$$

where

$$s's \in GF(2^3), \qquad u \in GF(2^2)$$

3.4.2 Two-Dimensional DFT for Computing the Coefficients of Next State Polynomials

The polynomial coefficients of next state functions may be obtained by computing eqn. (3.28) over all the elements of GF(2^M) and GF(2^N). If α, β and γ are primitive elements of GF(2^{L_1}), GF(2^M) and GF(2^N) respectively, let the variables and coefficients of polynomials be rearranged in the following ordering;

$$\begin{aligned} &s = 0, 1, \beta, \beta^2, \cdots, \beta^{2^M-2} \qquad \beta = \alpha^{(2^{L_1}-1)/(2^M-1)} \\ &u = 0, 1, \gamma, \gamma^2, \cdots, \gamma^{2^N-2} \qquad \gamma = \alpha^{(2^{L_1}-1)/(2^N-1)} \end{aligned} \tag{3.29}$$

$$\begin{aligned} &i = 0, (2^M-1), 1, 2, \cdots, (2^M-2) \\ &j = 0, (2^N-1), 1, 2, \cdots, (2^N-2) \end{aligned} \tag{3.30}$$

Then eqn. (3.28) may be written as follows:

$$a_{00} = f(0,0) \tag{3.31a}$$

$$a_{i0} = f(0,0)\delta(r_1 - i)\sum_{k=0}^{2^M-2} f(\beta^i, 0)\beta^{-ik} \tag{3.28b}$$

$$a_{0j} = f(0,0)\delta(r_2 - j)\sum_{k=0}^{2^N-2} f(0, \gamma^j)\gamma^{-jk} \tag{3.28c}$$

$$a_{ij} = f(0,0)\delta(r_1 - i)\delta(r_2 - j)\sum_{k=0}^{2^M-2}\sum_{k=0}^{2^N-2} f(\beta^i, \gamma^j)\beta^{-ik}\gamma^{-jk} \tag{3.28d}$$

where δ is Dirac function,

$$\begin{aligned} &r_1 = 2^M - 1, r_2 = 2^N - 1 \\ &i = (2^M-1), 1, 2, \cdots, (2^M-2) \\ &j = (2^N-1), 1, 2, \cdots, (2^N-2) \end{aligned}$$

It is obvious that eqn. (3.31b) is a one-dimensional DFT of length (2^M-1), eqn. (3.31c) is a one-dimensional DFT of length (2^N-1), and eqn. (3.31d) is a two-dimensional DFT of size $(2^M-1)\times(2^N-1)$.

For Example 3.3, using the ordering (3.29) and $\beta=\alpha^9$, $\gamma=\alpha^{21}$, the computation of polynomial coefficients may be represented in the matrix form:

$$\underline{A}=\begin{bmatrix}1&0&0&0&0&0&0&0\\1&1&1&1&1&1&1&1\\0&1&\alpha^{-9}&\alpha^{-18}&\alpha^{-27}&\alpha^{-36}&\alpha^{-45}&\alpha^{-54}\\0&1&\alpha^{-18}&\alpha^{-36}&\alpha^{-54}&\alpha^{-9}&\alpha^{-27}&\alpha^{-45}\\0&1&\alpha^{-27}&\alpha^{-54}&\alpha^{-18}&\alpha^{-45}&\alpha^{-9}&\alpha^{-36}\\0&1&\alpha^{-36}&\alpha^{-9}&\alpha^{-45}&\alpha^{-18}&\alpha^{-54}&\alpha^{-27}\\0&1&\alpha^{-45}&\alpha^{-27}&\alpha^{-9}&\alpha^{-54}&\alpha^{-36}&\alpha^{-18}\\0&1&\alpha^{-54}&\alpha^{-45}&\alpha^{-36}&\alpha^{-27}&\alpha^{-18}&\alpha^{-9}\end{bmatrix}\underline{F}$$

where $\underline{F}$ is the state table matrix reordering according to eqn. (3.29).

$$\underline{F}=\begin{bmatrix}0&1&\alpha^{36}&\alpha^{45}\\1&\alpha^{9}&\alpha^{45}&0\\1&\alpha^{9}&\alpha^{45}&0\\\alpha^{36}&\alpha^{45}&\alpha^{18}&\alpha^{54}\\\alpha^{9}&\alpha^{27}&0&1\\\alpha^{45}&0&\alpha^{54}&\alpha^{36}\\\alpha^{45}&0&\alpha^{54}&\alpha^{36}\\\alpha^{36}&\alpha^{45}&\alpha^{18}&\alpha^{54}\end{bmatrix}$$

The above computation contains a two-dimensional DFT of size (7×3), a one-dimensional DFT of length 3 and a one-dimensional DFT of length 7. The coefficient matrix $\underline{A}$ of (8×4) can be obtained in the new ordering (3.30) as follows:

$$\underline{A}=\begin{bmatrix}a_{0,0}&a_{0,3}&a_{0,1}&a_{0,2}\\a_{7,0}&a_{7,3}&a_{7,1}&a_{7,2}\\a_{1,0}&a_{1,3}&a_{1,1}&a_{1,2}\\a_{2,0}&a_{2,3}&a_{2,1}&a_{2,2}\\a_{3,0}&a_{3,3}&a_{3,1}&a_{3,2}\\a_{4,0}&a_{4,3}&a_{4,1}&a_{4,2}\\a_{5,0}&a_{5,3}&a_{5,1}&a_{5,2}\\a_{6,0}&a_{6,3}&a_{6,1}&a_{6,2}\end{bmatrix}=\begin{bmatrix}0&\alpha^{27}&\alpha^{62}&\alpha^{55}\\\alpha^{9}&\alpha^{45}&\alpha^{14}&\alpha^{49}\\0&\alpha^{27}&\alpha^{27}&\alpha^{27}\\\alpha^{9}&\alpha^{36}&\alpha^{61}&\alpha^{47}\\\alpha^{18}&\alpha^{45}&\alpha^{38}&\alpha^{52}\\\alpha^{27}&\alpha^{27}&\alpha^{33}&\alpha^{12}\\\alpha^{54}&\alpha^{45}&\alpha^{41}&\alpha^{13}\\\alpha^{54}&0&\alpha^{18}&\alpha^{18}\end{bmatrix}$$

3.5 Representation of Output Mappings

The output of one-dimensional FSMs is a function of present states and inputs. In general, the output g, state s and input u are in different fields. The output mapping of one-dimensional machines may be represented as follows:

$$g=g(s,u) \tag{3.32}$$

where

$$g \in GF(2^K),\ s \in GF(2^M),\ u \in GF(2^N)$$

This is a mapping from $GF(2^M)$ and $GF(2^N)$ into $GF(2^K)$.

In order to represent the output mapping, a common extension field $GF(2^{L_2})$ of the three subfields $GF(2^K)$, $GF(2^M)$ and $GF(2^N)$ is introduced, such that eqn. (3.32) may be considered a mapping from two subfields to another subfield within the extension field. This mapping may be represented

$$g = g(s,u) = \sum_{i=0}^{2^M-1}\sum_{j=0}^{2^N-1} b_{ij}s^iu^j \tag{3.33}$$

where

$$g \in GF(2^K),\ s \in GF(2^M),\ u \in GF(2^N)$$
$$b_{ij} \in GF(2^{L_2}), L_2 = \mathrm{LCM}(K,M,N)$$
$$i=0,1,\cdots,(2^M-1), j=0,1,\cdots,(2^N-1)$$

The coefficients b_{ij} are in extension field $GF(2^{L_2})$ and are determined by:

$$\begin{aligned}
b_{00} &= g(0,0) \\
b_{i0} &= \sum_{s\in GF(2^M)} g(s,0)s^{r_1-i} \qquad r_1 = 2^M-1 \\
b_{0j} &= \sum_{s\in GF(2^N)} g(0,u)u^{r_2-j} \qquad r_2 = 2^N-1 \\
b_{ij} &= \sum_{s\in GF(2^M)}\ \sum_{s\in GF(2^N)} g(s,u)s^{r_1-i}u^{r_2-j} \\
i &= 1,2,\cdots,(2^M-1), \qquad j = 1,2,\cdots,(2^N-1)
\end{aligned} \tag{3.34}$$

This function is similar to the function in eqns. (3.27) and (3.28), but the coefficients b_{ij} here are over the extension field of three subfields. The proof is shown in Appendix A3.2.

If β_1, β_2, β_3 and α are respectively the primitive elements of $GF(2^M)$, $GF(2^N)$, $GF(2^K)$ and extension field $GF(2^{L_2})$, let the states, inputs and outputs of machines be

$$\begin{aligned}
s &= 0,1,\beta_1,\beta_1^2,\cdots,\beta_1^{2^M-2} \qquad \beta_1 = \alpha^{(2^{L_2}-1)/(2^M-1)} \\
u &= 0,1,\beta_2,\beta_2^2,\cdots,\beta_2^{2^N-2} \qquad \beta_2 = \alpha^{(2^{L_2}-1)/(2^N-1)} \\
S &= 0,1,\beta_3,\beta_3^2,\cdots,\beta_3^{2^K-2} \qquad \beta_3 = \alpha^{(2^{L_2}-1)/(2^K-1)}
\end{aligned} \tag{3.35}$$

and let the subscripts of polynomial coefficients and powers of variables be reordered as shown in eqn. (3.30), then eqns. (3.33) and (3.34) may be represented as two-dimensional DFT forms.

The output mapping of multiple-dimensional FSMs is shown in Appendix 3.1.

3.6 Conversion Between Single-Variable and Multiple-Variable Machines

3.6.1 Converting a Multiple-Variable Machine into a Single-Variable Machine

A high-order IIR digital filter may be represented as a multiple-variable FSM with polynomial coefficients which can be computed by means of multipl-dimensional DFTs. Single-variable machines are well studied. From the theoretical viewpoint, the coefficient computation, minimization, linearization and dynamic behaviour analysis of single-variable machines are much simpler than that of multiple-variable machines. The conversion between multiple-variable and single-variable machines is useful for further study.

We begin with two-variable machines. The next state of one-dimensional machines is a function of present states s and inputs u as shown in eqn. (3.27). In general, such a machine may be considered as a two-variable machine with states $s \in GF(2^M)$ and inputs $u \in GF(2^N)$. The polynomial coefficients are in extension field $GF(2^{L_1})$. All the elements of subfields $GF(2^M)$ and $GF(2^N)$ are members of extension field $GF(2^{L_1})$. If a mapping from the two subfields to the extension field is defined such that each pair of elements of subfields $GF(2^M)$ and $GF(2^N)$ corresponds with a member of extension field $GF(2^{L_1})$, then the next state function may be mapped into a single-variable polynomial over extension field $GF(2^{L_1})$. There are a number of these mappings, most of them are non-linear. One of the simplest mappings is the following linear mapping:

$$x = u\alpha + s \tag{3.36}$$

where α is a primitive element of extension field $GF(2^{L_1})$.

$x \in GF(2^{L_1})$ is an extension field variable, it is similar to the complex variable which consists of two real variables. In the complex field, the real and imaginary components can be obtained in terms of linear combinations of the complex variable variable and its conjugate. In a Galois Field, there is no such linear relationship. The subfield variables s and u may be represented as functions of the extension field variable x. In general, they are non-linear and can be obtained by using the field properties: $s^{2^M} = s$ and $u^{2^N} = u$. Let

$$R = \mathrm{LCM}[(2^M - 1), (2^N - 1)] + 1 \tag{3.37}$$

From Lemma 3.3, the Rth power of eqn. (3.36) is as follows:

$$x^R = u\alpha^R + s \tag{3.38}$$

The variables s and u may be obtained by solving the linear algebraic equations (3.36) and (3.38) as follows:

$$s = (\alpha^R x + \alpha x^R)/(\alpha^R + \alpha) \tag{3.39}$$
$$u = (x + x^R)/(\alpha^R + \alpha)$$

These are two mappings from extension field $GF(2^{L_1})$ to subfields $GF(2^M)$ and $GF(2^N)$. Substituting eqn. (3.39) into the next state functions, the machine with single variable x over extension field $GF(2^{L_1})$ is obtained.

This method can also be used in to case that the variables s and u are in the same fields, but the extension field is $GF(2^{2M})$ instead of $GF(2^{L_1})$.

Example 3.4

The state table of two-variable machines is given by:

Table 3.7 State table of Example 3.4

u	s	$s' = f(s, u)$
0	0	0
0	1	0
0	−2	1
0	−1	1

(To be continued)

Table 3.7

u	s	$s'=f(s,u)$
1	0	1
1	1	0
1	-2	1
1	-1	1
-2	0	-1
-2	1	-2
-2	-2	0
-2	-1	-1
-1	0	-1
-1	1	-1
-1	-2	0
-1	-1	0

Let β be a primitive element of GF (2^2). Using the positive-polarity assignment of field elements, the above table may be written as follows.

Table 3.8 State table of Example 3.4 over GF(2^2)

u	s	$s'=f(s,u)$
0	0	0
0	1	0
0	β	1
0	β^2	1
1	0	1
1	1	0
1	β	1
1	β^2	1
β	0	β^2
β	1	β
β	β	0
β	β^2	β^2
β^2	0	β^2
β^2	1	β^2
β^2	β	0
β^2	β^2	0

From Theorem 3. 3, the above state table may be represented as a polynomial with two variables over $GF(2^2)$ as follows:

$$s'=f(s,u)=(s+s^2)+(1+\beta s+\beta^2 s^3)u+(\beta s+\beta^2 s^2)u^2+(\beta^2 s^2+\beta s^3)u^3$$

where $s',s,u\in GF(2^2)$.

Let x be a variable of extension field $GF(2^4)$, defined by

$$x=u\alpha+s$$

where α is a primitive root of the primitive polynomial $(\alpha^4+\alpha+1)$ over extension field $GF(2^4)$. From eqn. (3. 39) the variables s and u may be represented as follows:

$$s=\alpha^4 x+\alpha x^4$$

$$u=x+x^4$$

Substituting the above two equations into the function $f(s,u)$, and using $\beta=\alpha^5$ (see Lemma 3. 3), the next state function is as follows:

$$s'=\alpha^3x^3+x^5+\alpha^4x^6+\alpha^3x^7+\alpha x^9+\alpha^5x^{10}+\alpha^5x^{11}+\alpha^{12}x^{12}+\alpha^{12}x^{13}+\alpha^5x^{14}+\alpha^5x^{15}$$

where $s'\in GF(2^2)$ and $x\in GF(2^4)$.

From Theorem 3. 3, the coefficients of the above polynomial have the following cyclic subgroup properties:

$$(\alpha^3x^3)^4=\alpha^{12}x^{12},\qquad (\alpha^4x^6)^4=\alpha x^9$$

$$(\alpha^3x^7)^4=\alpha^{12}x^{13},\qquad (\alpha^4x^{11})^4=\alpha^5x^{14}$$

$$(x^5)^4=x^5,\qquad (\alpha^5x^{10})^4=\alpha^5x^{10},\qquad (\alpha^5x^{15})^4=\alpha^5x^{15}$$

The above result for two-variable machines can be extended to multiple-variable cases. An m-variable autonomous machine is given by:

$$s'=f(s_0,s_1,\cdots,s_{m-1})=\sum_{i=0}^{r}\sum_{i=0}^{r}\cdots\sum_{i_{m-1}=0}^{r}a_{i_0i_1\cdots i_{m-1}}s_0^{i_0}s_1^{i_1}\cdots s_{m-1}^{i_{m-1}} \tag{3.40}$$

where $s_0,s_1,\cdots,s_{m-1}$, $a\in GF(2^M)$, $r=2^M-1$.

An extension field $GF(2^M)$ is introduced, α is its primitive element and a variable x over $GF(2^{mM})$ is defined by:

$$x=s_0+s_1\alpha+\cdots+s_{m-1}\alpha^{m-1}$$

Using the Galois Field property in eqn. (3. 2): $s_j^{n(q-1)+1}=s_j$ $(q=2^M,\ n\geq 0)$ and Lemma 3. 4, the m linear algebraic equations of subfields $(s_0,s_1,\cdots,s_{m-1})$ can be obtained as follows:

$$\begin{aligned} x^q&=s_0+s_1\alpha^q+\cdots+s_{m-1}\alpha^{q(m-1)}\\ x^{2q-1}&=s_0+s_1\alpha^{2q-1}+\cdots+s_{m-1}\alpha^{(2q-1)(m-1)}\\ &\vdots\\ x^{(q-1)(m-1)+1}&=s_0+s_1\alpha^{(q-1)(m-1)+1}+\cdots+s_{m-1}\alpha^{(q-1)(m-1)^2+(m-1)} \end{aligned} \tag{3.41}$$

Eqns. (3. 41) are m linear algebraic equations of variables $(s_0, s_1, \cdots, s_{m-1})$. Solving the linear equations and substituting their solutions into eqn. (3. 40), a single-variable machine over extension field $GF(2^{mM})$ may be obtained.

3. 6. 2 Converting a Single-Variable Machine into Multiple-Variable Machines

A single-variable machine over extension field $GF(q^m)$ may be converted into a number of multiple-variable machines over subfields $GF(q)$ by the following variable transformation:

$$s=s_{m-1}\alpha^{m-1}+\cdots+s_1\alpha+s_0 \tag{3.42}$$

where

$$s \in \mathrm{GF}(q^m) \qquad \alpha \text{ is its primitive element}$$

$$s_i \in \mathrm{GF}(q) \qquad i=0,1,\cdots,(m-1) \qquad q=2^M$$

Substituting eqn. (3.42) into the single-variable machine

$$s' = \sum_{i=0}^{q^m-1} a_i s^i \tag{3.43}$$

where $a_i \in GF(q^m)$ can be represented as power of α.

Let

$$s' = s'_{m-1}\alpha^{m-1} + \cdots + s'_1\alpha + s'_0 \tag{3.44}$$

Compare the coefficients of α^i, $i=0,1,\cdots,(m-1)$, m machines with m variables over subfield GF(q) are obtained.

For Example 3.1, let

$$s = s_2\alpha^2 + s_1\alpha + s_0$$

$$s' = s'_2\alpha^2 + s'_1\alpha + s'_0$$

Substituting the above new variables into the extension field polynomial, three polynomials over subfields may be obtained as follows:

$$s'_0 = s_1 + s_0 s_1$$

$$s'_1 = s_2 + s_0 s_1$$

$$s'_2 = s_2 + s_0 s_1 s_2$$

In general, a FSM with a single-variable over extension field GF(2^n) may be converted into a number of multiple-variable machines over GF(2), which can be represented as n-variable polynomials over GF(2). This polynomial is called the modulo-2 ring-sum expansion, it is of particular interest for practical implementation of machines, and will be discussed in Chapter 5.

3.7 Conclusion

A systematic study of the representation of finite state machines as Galois Field polynomials is presented in this chapter. The polynomial coefficients can be computed from the state table of machines. A number of authors[C12-C15] have been concerned with the structures of Galois Field functions. The contributions of this chapter are:

(a) An improvement in the computation of polynomial coefficients is introduced in that the sum in Lemma 3.2 is for all the field elements including null element, and s^{r-i} is used instead of s^{-i}. This improvement avoids separately treating the coefficient a_{q-1} and considerably simplifies the representation of coefficient computations.

(b) A DFT of length $(2^M - 1)$ for computing the polynomial coefficients is described by rearranging the ordering of coefficients and function output vectors.

(c) The mapping between different fields GF(2^M) and GP (2^K) is developed in terms of a polynomial with coefficients which are in the common extension field GF(2^L). These mapping polynomials are very useful for representation of the machines with states, inputs and outputs which are in different fields.

(d) The cyclic subgroup property of the common extension polynomials is revealed. This property reduces considerably the size of memory and amount of computations.

(e) It is proved that a general FSM may be represented as a two-variable polynomial and the coefficients may be computed in terms of a two-dimensional DFT of size $(2^M-1)\times(2^N-1)$, and 2 one-dimensional DFTs of length (2^M-1) and (2^N-1) respectively by rearranging the ordering of coefficient and function output matrices.

(f) A method for converting a multiple-variable machine into single-variable machines is given by means of a non-linear transformation which maps the extension field variable into several subfield variables. These non-linear transformations can be obtained by solving a set of linear equations.

Appendix 3.1: Representation of Multiple-Variable Finite State Machines

A finite state machine with n-input, k-output and m-state variables may be represented as a set of polynomials as follows:

$$s'_i = f_i(s_0, s_1, \cdots, s_{m-1}, u_0, u_1, \cdots, u_{n-1}) = \sum_{i_0=0}^{r_1}\cdots\sum_{i_{m-1}=0}^{r_1}\sum_{j_0=0}^{r_2}\cdots\sum_{j_{n-1}=0}^{r_2} a^{(i)}_{i_0\cdots i_{m-1}j_0\cdots j_{n-1}} s_0^{i_0}\cdots s_{m-1}^{i_{m-1}} u_0^{j_0}\cdots u_{n-1}^{j_{n-1}}$$

$i = 0, 1, \cdots, (m-1)$ (3.45)

$$g_i = g_j(s_0, s_1, \cdots, s_{m-1}, u_0, u_1, \cdots, u_{n-1}) = \sum_{i_0=0}^{r_1}\cdots\sum_{i_{m-1}=0}^{r_1}\sum_{j_0=0}^{r_2}\cdots\sum_{j_{n-1}=0}^{r_2} b^{(j)}_{i_0\cdots i_{m-1}j_0\cdots j_{n-1}} s_0^{i_0}\cdots s_{m-1}^{i_{m-1}} u_0^{j_0}\cdots u_{n-1}^{j_{n-1}}$$

$j = 0, 1, \cdots, (k-1)$ (3.46)

where

$$s_0, s_1, \cdots, s_{m-1} \in GF(2^M) \qquad r_1 = 2^M - 1$$

$$u_0, u_1, \cdots, u_{n-1} \in GF(2^N) \qquad r_2 = 2^N - 1$$

$$g_0, g_1, \cdots, g_{k-1} \in GF(2^K)$$

The coefficients

$$a^{(i)}_{i_0\cdots i_{m-1}j_0\cdots j_{n-1}} \in GF(2^{L_1}) \qquad L_1 = LCM(M, N)$$

$$b^{(j)}_{i_0\cdots i_{m-1}j_0\cdots j_{m-1}} \in GF(2^{L_2}) \qquad L_1 = LCM(K, M, N)$$

$$0 \leq i_0, i_1, \cdots, i_{m-1} \leq r_1$$

$$0 \leq j_0, j_1, \cdots, j_{n-1} \leq r_2$$

The coefficients of polynomials are determined by:

$$\left.\begin{aligned}
a_{00\cdots 0}^{(i)} &= f_i(0,0,\cdots,0)\\
a_{i_0 0\cdots 0}^{(i)} &= \sum_{s_0} f_i(s_0,0,\cdots,0)s_0^{r_1-i_0}\\
&\vdots\\
a_{0\cdots 0 j_{n-1}}^{(i)} &= \sum_{u_{n-1}} f_i(0,\cdots,0,u_{n-1})u_{n-1}^{r_2-j_{n-1}}\\
a_{i_0 i_1\cdots 0}^{(i)} &= \sum_{s_0}\sum_{s_1} f_i(s_0,s_1,\cdots,0)s_0^{r_1-i_0}s_1^{r_1-i_1}\\
&\vdots\\
a_{0\cdots 0 i_{m-1} j_0\cdots 0}^{(i)} &= \sum_{s_{m-1}}\sum_{u_0} f_i(0,\cdots,0,s_{m-1},u_0,0,\cdots,)s_{m-1}^{r_1-i_{m-1}}u_0^{r_2-j_0}\\
&\vdots\\
a_{0\cdots 0 j_{n-2} j_{n-1}}^{(i)} &= \sum_{u_{n-2}}\sum_{u_{n-1}} f_i(0,\cdots,0,u_{n-2},u_{n-1})u_{n-2}^{r_2-j_{n-2}}u_{n-1}^{r_2-j_{n-1}}\\
&\vdots\\
a_{i_0\cdots i_{m-1} j_0\cdots j_{n-1}}^{(i)} &= \sum_{s_0}\cdots\sum_{s_{m-1}}\sum_{u_0}\cdots\sum_{u_{n-1}} f_i(s_0,\cdots,s_{m-1},u_0,\cdots,u_{n-1})s_0^{r_1-i_0}\cdots s_{m-1}^{r_1-i_{m-1}}u_0^{r_2-j_0}\cdots u_{n-1}^{r_2-j_{n-1}}\\
i &= 0,1,\cdots,(m-1)
\end{aligned}\right\}\quad(3.47)$$

$$\left.\begin{aligned}
b_{00\cdots 0}^{(j)} &= g_j(0,0,\cdots,0)\\
b_{i_0 0\cdots 0}^{(j)} &= \sum_{s_0} g_j(s_0,0,\cdots,0)s_0^{r_1-i_0}\\
&\vdots\\
b_{0\cdots 0 j_{n-1}}^{(j)} &= \sum_{u_{n-1}} g_j(0,\cdots,0,u_{n-1})u_{n-1}^{r_2-j_{n-1}}\\
b_{i_0 i_1\cdots 0}^{(j)} &= \sum_{s_0}\sum_{s_1} g_j(s_0,s_1,\cdots,0)s_0^{r_1-i_0}s_1^{r_1-i_1}\\
&\vdots\\
b_{0\cdots 0 i_{m-1} j_0\cdots 0}^{(j)} &= \sum_{s_{m-1}}\sum_{u_0} g_j(0,\cdots,0,s_{m-1},u_0,0,\cdots,)s_{m-1}^{r_1-i_{m-1}}u_0^{r_2-j_0}\\
&\vdots\\
b_{0\cdots 0 j_{n-2} j_{n-1}}^{(j)} &= \sum_{u_{n-2}}\sum_{u_{n-1}} g_j(0,\cdots,0,u_{n-2},u_{n-1})u_{n-2}^{r_2-j_{n-2}}u_{n-1}^{r_2-j_{n-1}}\\
&\vdots\\
b_{i_0\cdots i_{m-1} j_0\cdots j_{n-1}}^{(j)} &= \sum_{s_0}\cdots\sum_{s_{m-1}}\sum_{u_0}\cdots\sum_{u_{n-1}} g_j(s_0,\cdots,s_{m-1},u_0,\cdots,u_{n-1})s_0^{r_1-i_0}\cdots s_{m-1}^{r_1-i_{m-1}}u_0^{r_2-j_0}\cdots u_{n-1}^{r_2-j_{n-1}}\\
j &= 0,1,\cdots,(n-1),0\le i_0,i_1,\cdots,i_{m-1}\le r_1,0\le j_0,j_1,\cdots,j_{n-1}\le r_2
\end{aligned}\right\}\quad(3.48)$$

where $\sum_x$ denotes the sum for all the elements of variable x.

Appendix 3.2 will show the proof for two-variable polynomials. It is easily extended to the general case.

If β_1, β_2 and β_3 are respectively the primitive elements of GF(2^M), GF(2^N) and GF(2^K), let the states, inputs and outputs of machines respectively be

$$\begin{aligned}
&0,1,\beta_1,\beta_1^2,\cdots,\beta_1^{2^M-2} && \text{for states } s_0,s_1,\cdots,s_{m-1}\\
&0,1,\beta_2,\beta_2^2,\cdots,\beta_2^{2^N-2} && \text{for input } u_0,u_1,\cdots,u_{n-1}\\
&0,1,\beta_3,\beta_3^2,\cdots,\beta_3^{2^K-2} && \text{for output } g_0,g_1,\cdots,g_{k-1}
\end{aligned}\quad(3.49)$$

and let the subscripts of polynomial coefficients and powers of variables in the above equations be reordered as follows:

$$0,(2^M-1),1,2,\cdots,(2^M-2) \quad \text{for } i_0, i_1, \cdots, i_{m-1} \qquad (3.50)$$
$$0,(2^N-1),1,2,\cdots,(2^N-2) \quad \text{for } j_0, j_1, \cdots, j_{n-1}$$

Then the above equations may be written in DFT forms.

Appendix 3.2: Mathematical Proof of Theorems

Proof of Theorem 3.2

Since L is Least Common Multiple of N and K, from Lemma 3.3, $GF(2^N)$ and $GF(2^K)$ are both the subfields of Extension Field $GF(2^L)$. Therefore, $s \in GF(2^M) \subset GF(2^L)$ and $f \in GF(2^K) \subset GF(2^L)$, eqn. (3.17). and eqn. (3.18) are arithmetic operations within Extension Field $GF(2^L)$.

Substituting $g(s)$ of eqn. (3.17) into eqn. (3.18), the right side of eqn. (3.18)

$$\sum_s \Big(\sum_{i=0}^{2^M-1} b_i s^i\Big) s^{r-j} = \sum_{i=0}^{2^M-1} b_i \Big(\sum_s s^{r-(i-j)}\Big)$$

Using Lemma 3.2, it follows

$$= \sum_{i=0}^{2^M-1} b_i \delta(i-j) = b_j \qquad j = 1,2,\cdots,2^M-1$$

and is equal to the left-hand side of eqn. (3.18).

It is easy to prove that $b_0 = g(0)$.

Proof of Theorem 3.4

Since both $GF(2^M)$ and $GF(2^N)$ are subfields of $GF(2^L)$, Theorem 3.4 may be considered as an arithmetic operation within extension field $GF(2^L)$.

Substituting $f(s,u)$ of eqn. (3.27) into the right-hand side of eqn. (3.28d), it follows.

$$\sum_s \sum_u f(s,u) s^{r_1-i} u^{r_2-j} = \sum_s \sum_u \Big(\sum_{i'=0}^{2^M-1} \sum_{j'=0}^{2^N-1} a_{i'j'} s^{i'} u^{j'}\Big) s^{r_1-i} u^{r_2-j}$$
$$= \sum_{i'=0}^{2^M-1} \sum_{j'=0}^{2^N-1} a_{i,j} \Big(\sum_s s^{r_1-(i-i')}\Big)\Big(\sum_u u^{r_2-(j-j')}\Big)$$

Using Lemma 3.3, it follows

$$= \sum_{i'=0}^{2^M-1} \sum_{j'=0}^{2^N-1} a_{i,j} \delta(i-i')\delta(j-j') = a_{ij}$$

It is equal to the left-hand side of eqn. (3.28d).

Similarly, using $0^0 = 1$, and substituting $f(0,u)$ of eqn. (3.27) into the right-hand side of eqn. (3.28c), it follows

$$\sum_u f(0,u) u^{r_2-j} = \sum_u \Big(\sum_{j'=0}^{2^N-1} a_{0j'} u^{j'}\Big) u^{r_2-j}$$
$$= \sum_{j'=0}^{2^N-1} a_{0j'} \Big(\sum_u u^{r_2-(j-j')}\Big) = \sum_{j'=0}^{2^N-1} a_{0j'} \delta(j-j') = a_{0j}$$

This is eqn. (3.28c).

Substituting $f(s,0)$ of eqn. (3.27) into the right-hand side of eqn. (3.28b), it follows

$$\sum_s f(s,0) s^{r_1-i} = \sum_s \Big(\sum_{i'=0}^{2^M-1} a_{i'0} s^{i'}\Big) s^{r_1-i}$$
$$= \sum_{i'=0}^{2^M-1} a_{i'0} \Big(\sum_s s^{r_1-(i-i')}\Big) = \sum_{i'=0}^{2^M-1} a_{0i'0} \delta(i-i') = a_{i0}$$

This is the left-hand side of eqn. (3.28b).

It is easy to prove that $a_{00} = f(0,0)$.

Chapter 4
New Fast Fourier Transform Algorithm Using Subgroup Convolutions①

4.1 Introduction

The Discrete Fourier Transform (DFT) of a sequence $\{x(n)\}$ of length N is defined by

$$X(k) = \sum_{n=0}^{N-1} x(n) w^{nk} \qquad k = 0, 1, \cdots, (N-1) \tag{4.1}$$

where w is a primitive Nth root of unity in the field or ring under consideration. In the case of the conventional DFT, $w=\exp(-j2\pi/N)$.

The discrete Fourier transform is a central operation in digital signal processing. It can be used to analyse and estimate the spectrum of signals, to design and implement finite impulse response digital filters and to compute convolution and correlation functions etc. The above chapter shows that the coefficients of Reed-Muller polynomials can be computed by using a DFT of length (p^M-1).

The direct evaluation of an N-point DFT requires $(N-1)^2$ complex multiplications and $N(N-1)$ complex additions[P1]. Thus for reasonably large values of N, direct evaluation of the DFT requires an inordinate amount of computations. The idea of fast algorithms is to find some special length N of sequences such that the DFT needs fewer computations. In 1965, Cooley and Tukey[A1] proposed that when $N=2^M$ is a highly composite number, the 2^M-point DFT can be efficiently computed in terms of $M \times 2^M$ complex multiplications. In 1968, Rader[A2] proposed that the DFT, when its length $N=p$ is a prime, can be changed into a circular convolution of length $(p-1)$ by rearranging the ordering of data. Later Winograd[A6], McClellan and Rader[P4], showed that DFTs of length $N=p^M$, where p is prime, can be converted into one convolution of length $p^{M-1}(p-1)$, two convolutions of length $p^{M-2}(p-1)$, four convolutions of length $p^{M-3}(p-1)$, $\cdots$, terminating with p^{M-1} convolutions of length $(p-1)$.

The computation of small N convolutions with minimum number of multiplications has been reported by Winograd[A12]. Using the finite field theory, he proved that the minimum number of multiplications for computing a convolution of length N is $(2N-k)$, where k is the number of distinct irreducible factors of the modulo polynomial of order N. For large N convolutions, Agarwal and Cooley[B5] showed that one dimensional circular convolution, whose length is the product of relatively prime integers, can be converted into a multi-dimensional circular convolution by means of the Chinese Remainder Theorem (CRT) reconstruction.

For long sequence DFTs, when its length N is the product of relatively prime integers, e. g. $N=N_1N_2$, Good[A4] proposed a Prime Factor Algorithm (PFA) which transforms a one-dimensional DFT of length N into a multi-dimensional DFT, using an index mapping[A10] based on the Chinese Remainder Theorem. For the case of two factors, it is a two-dimensional DFT of size $N_1 \times N_2$, and

① The contents of this chapter have been published in Signal Processing Ⅱ: Theory and Applications, edited by H. W. Schussler, Elsevier Science Publishers B. V. (North-Holland), 1983, pp. 719-724[Z1].

can be computed in terms of N_1 one-dimensional DFT of length N_2, and N_2 one-dimensional DFT of length N_1. This algorithm needs less multiplications than the FFT algorithm[A11]. Winograd[A8] proposed another algorithm called the Nested Algorithm, which is a reordering of operations in the prime factor algorithm to merge all the multiplications and thus reduce the total number of multiplications relative to the prime factor algorithm. This algorithm is named the Winograd Fourier Transform Algorithm (WFTA). The WFTA and PFA require the computation of different length circulation convolutions and the small-N algorithm used in WFTA cannot be performed in place. The different circular convolution lengths lead to a rather complex program structure.

This chapter is devoted to the DFT of length (p^M-1). Following the above description of the properties of DFTs with length 2^M, p and p^M, it will be shown in this chapter that the DFT of length (p^M-1) may be converted in several subgroup circular convolutions of length M or divisor of M. Most of these convolutions for the DFT of length (p^M-1) have the same length which can considerably simplify the program structure.

Another reason for consideration of the DFT of length (p^M-1) is that the Reed-Muller polynomials over GF(p^M) for representing a FSM can be computed by using a DFT with length (p^M-1). Here (p^M-1) is no longer a highly composite number, the Cooley and Tukey's FFT algorithm cannot be used, PFA and WFTA require the computation of different length convolution and lead to a rather complex program. A new efficient algorithm is required.

In this chapter, $W=\exp(-j2\pi/N)$ (a primitive unity root of order $N=p^M-1$ over the complex field) and the discussion is concerned with the computation of the conventional DFTs of length (p^M-1). However, all the results of this chapter can be employed in computing Reed-Muller polynomials, when w in eqn. (4.1) is replaced by a primitive element α of GF(p^M).

The cyclic subgroup property of ring $Z(p^M-1)$ is introduced in Section 4.2. All the non-zero elements of ring $Z(p^M-1)$, whether (p^M-1) is prime or not, can be separated into several cyclic subgroups of length M or divisor of M. The DFTs of length $N=p^M-1$, with N prime can be converted into subgroup convolutions of length M, the algorithm concerning this conversion is presented in Section 4.3. Section 4.4 shows that any DFT of length (p^M-1), with N not prime, can also be converted into subgroup convolutions. The number and length of subgroups is discussed in Section 4.5.

4.2 Cyclic Subgroup Properties of Ring $Z(p^M-1)$

It is well known that all the non-zero elements of GF(p^M) may be separated into a number of subsets of length M or divisor of M. For example, the non-zero elements of GF(2^4) may be separated into the following subsets:

$$\begin{array}{cccc} \alpha & \alpha^2 & \alpha^4 & \alpha^8 \\ \alpha^{11} & \alpha^7 & \alpha^{14} & \alpha^{13} \\ \alpha^6 & \alpha^{12} & \alpha^9 & \alpha^3 \\ \alpha^5 & \alpha^{10} & & \end{array}$$

$$\alpha^{15}=\alpha^0=1$$

where α is a primitive element of GF(2^4).

Two features in the above subsets may be noted:

(a) Each element of a subset is the square of its predecessor, and they are cyclic within the subsets. For the above example,

$$(\alpha^8)^2 = \alpha$$
$$(\alpha^{13})^2 = \alpha^{11}$$
$$(\alpha^3)^2 = \alpha^6$$
$$(\alpha^{10})^2 = \alpha^5$$
$$(\alpha^{15})^2 = \alpha^{15} = 1$$

This result can be obtained by computing their power modulo 15.

(b) The length of each subset is M or divisor of M, and most subsets have the length M. For the above example, there are three subsets of length $4(=M)$, one subset of length $2(=M/2)$ and $\alpha^{15} = \alpha^0 = 1$ may be considered as a unity subset of length 1.

Since all the powers in GF(p^M) form a ring $Z(p^M - 1)$, the above properties of Galois Fields imply that all the non-zero elements of ring $(p^M - 1)$ can be separated into a number of cyclic subgroups with respect to its modulo multiplication. The zero element 0 of ring $(p^M - 1)$ always forms a zero subgroup of length 1, it corresponds to $w^0 = 1$ in DFT. Since no multiplication operation is required to multiply by w^0, the zero subgroup in this section is not discussed. The non-zero cyclic subgroups of ring $Z(p^M - 1)$ are respectively discussed, for both the case $(p^M - 1)$ is prime and $(p^M - 1)$ is composite.

4.2.1 Subgroups of Ring $Z(p^M - 1)$, when $(p^M - 1)$ is Prime

All the non-zero elements of ring $Z(N)$, when $N = (p^M - 1)$ is a prime, may be separated into m subsets of length M as follows:

$$\begin{matrix} p^0 & p^1 & p^2 & \cdots & p^{M-1} \\ gp^0 & gp^1 & gp^2 & \cdots & gp^{M-1} \\ g^2p^0 & g^2p^1 & g^2p^2 & \cdots & g^2p^{M-1} \\ \vdots & \vdots & \vdots & \cdots & \vdots \\ g^{m-1}p^0 & g^{m-1}p^1 & g^{m-1}p^2 & \cdots & g^{m-1}p^{M-1} \end{matrix} \quad \text{mod } N \tag{4.2}$$

where $m = \phi(p^M - 1)/M = \dfrac{N-1}{M}$, ϕ is Euler function. g is a unity root of order m, where m is the least integer such that $g^m = 1 \text{ mod}(p^M - 1)$.

Since $(p^M - 1)$ is a prime, $Z(N)$ is a field, so there exists a primitive element β such that

$$\beta^{\phi(N)} = \beta^{N-1} = 1 \text{ mod } N$$

Let

$$g = \beta^{\phi(N)/m} = \beta^M \tag{4.3}$$

then g is a unity root of order m on ring $Z(p^M - 1)$. For example,

$$\text{Ring } Z(31) \quad N = 2^5 - 1 = 31$$

is a prime, the subgroup number $m = \phi\ (31)/M = 6$. The primitive element of GF(31) is $\beta = 3$ (it holds that $3^{30} = 1$ mod 31). The unity root of order 6 on ring $Z(31)$ is $g = 3^5 = 26$ mod 31. 26 is a generating element of subgroups, $26^0 = 1$, $26^1 = 26$, $26^2 = 25$, $26^3 = 30$, $26^4 = 5$, $26^5 = 6$, $16^6 = 1$ mod 31. From eqn. (4.2), the 30 non-zero elements of ring $Z(31)$ may be separated into 6 subsets of length 5 as follows:

$$\begin{matrix} 1 & 2 & 4 & 8 & 16 \\ 26 & 21 & 11 & 22 & 13 \\ 25 & 19 & 7 & 14 & 28 \\ 30 & 29 & 27 & 23 & 15 \\ 5 & 10 & 20 & 9 & 18 \\ 6 & 12 & 24 & 17 & 3 \end{matrix} \quad \bmod 31$$

4.2.2 Subgroups of Ring $Z(p^M-1)$, when (p^M-1) is Not Prime

There are $\phi(N)$ elements that are prime to N on ring $Z(N)$. These $\phi(N)$ elements may be represented as m_1 subsets of length M as follows:

$$\begin{matrix} p^0 & p^1 & p^2 & \cdots & p^{M-1} \\ gp^0 & gp^1 & gp^2 & \cdots & gp^{M-1} \\ g^2p^0 & g^2p^1 & g^2p^2 & \cdots & g^2p^{M-1} \\ \vdots & \vdots & \vdots & \cdots & \vdots \\ g^{m_1-1}p^0 & g^{m_1-1}p^1 & g^{m_1-1}p^2 & \cdots & g^{m_1-1}p^{M-1} \end{matrix} \quad \bmod N \tag{4.4}$$

where $m_1=\phi(N)/M$, g is a unity root of order m_1.

The remaining $(N-\phi(N)-1)$ non-zero elements that are not prime to N may be represented as m_2 subsets

$$\begin{matrix} g_0p^0 & g_0p^1 & g_0p^2 & \cdots & g_0p^{d_0-1} \\ g_1p^0 & g_1p^1 & g_1p^2 & \cdots & g_1p^{d_1-1} \\ g_2p^0 & g_2p^1 & g_2p^2 & \cdots & g_2p^{d_2-1} \\ \vdots & \vdots & \vdots & \cdots & \vdots \\ g_{m_2-1}p^0 & g_{m_2-1}p^1 & g_{m_2-1}p^2 & \cdots & g_{m_2-1}p^{d_{m_2-1}-1} \end{matrix} \quad \bmod N \tag{4.5}$$

where $d_i=M$ or divisor of M,

g_i is the factor of N, or multiples of the factors,

$i=0,1,2,\cdots,(m_2-1)$,

$m_2=m-m_1$,

m is the total number of subgroups on ring $Z(p^M-1)$.

It is equal to the total number of irreducible polynomials whose degree is M, or divisor of M over GF(p) and will be described in Section 4.5.

For example,

$$N=2^4-1=15$$

$$\phi(15)=\phi(3)\phi(5)=8, m_1=\phi(15)/4=2$$

$11^2=1 \bmod 15$ is a unity root of order 2, from eqn. (4.4), the 8 elements that are prime to 15 may be separated into 2 subsets of length 4 as follows:

$$\begin{matrix} 1 & 2 & 4 & 8 \\ 11 & 7 & 14 & 13 \end{matrix} \quad \bmod 15$$

From eqn. (4. 5) the remaining 6 elements may be divided into 2 subsets of length 4 and 2 respectively:

$$\begin{matrix} 3 & 6 & 12 & 9 \\ 5 & 10 & & \end{matrix} \quad \text{mod } 15$$

4. 3 Fast DFT Algorithm of Length $N = p^M - 1$, with N Prime, Using Subgroup Convolutions

From eqn. (4. 1), the DFT of length $N = p^M - 1$ may be written as follows:

$$\begin{aligned} X(0) &= \sum_{n=0}^{N-1} x(n) \\ X(k) &= x(0) + \overline{X}(k) \quad k = 1,2,\cdots,(N-1) \end{aligned} \tag{4.6}$$

where

$$\begin{aligned} X(k) &\overset{N-1}{\underset{n=0}{=}} x(n) w^{nk} \\ w &= e^{-j2\pi/N} \end{aligned} \tag{4.7}$$

It will be proved that eqn. (4. 7) may be converted into a number of subgroup circular convolutions, when $N = p^M - 1$.

4. 3. 1 Converting the DFT of Length $(p^M - 1)$ into Subgroup Convolutions, when $(p^M - 1)$ is a Prime Mersenne Number

When $N = p^M - 1$ is prime, then N is a Mersenne number. For example:

$N = 3,\ 7,\ 31,\ 127,\ 8191$, etc.

From the cyclic subgroup property of ring $Z(N)$ as shown in eqn. (4. 2), the variables n and k in eqn. (4. 7) may be written as follows:

$$\begin{aligned} n &= (n_1, n_2) = g^{n_1} p^{n_2} \quad n_1 = 0,1,\cdots,(m-1) \quad n_2 = 0,1,\cdots,(M-1) \\ k &= (k_1, k_2) = g^{k_1} p^{k_2} \quad k_1 = 0,1,\cdots,(m-1) \quad k_2 = 0,1,\cdots,(M-1) \end{aligned} \tag{4.8}$$

where $m = \phi(N)/M$, g is a unity root of order m.

Using eqn. (4. 8), eqn. (4. 7) may be expressed as follows:

$$\overline{x}(k_1, k_2) = \sum_{n_1=0}^{m-1} \sum_{n_2=0}^{M-1} x(n_1, n_2) w^{(g^{k_1+n_1} p^{k_2+n_2})} \tag{4.9}$$

where $k_1 = 0,1,\cdots,(m-1)$; $k_2 = 0,1,\cdots,(M-1)$.

Eqn. (4. 9) may be considered as a 2-dimensional circular convolution of size $M \times m$. For example,

$$N = 2^3 - 1 = 7 \text{ is prime}$$

$$m = \phi(7)/3 = 2 \quad M = 3$$

Since $6^2 = 1 \bmod 7$, 6 is a unity root of order 2. From eqn. (4. 3) the two subsets are as follows:

1 2 4 6 5 3

Rearranging the order in accordance with the above subsets eqn. (4. 7) may be written in matrix form as:

$$
\begin{bmatrix} \bar{x}(1) \\ \bar{x}(2) \\ \bar{x}(4) \\ \bar{x}(6) \\ \bar{x}(5) \\ \bar{x}(3) \end{bmatrix} = \begin{bmatrix} w^1 & w^2 & w^4 & w^6 & w^5 & w^3 \\ w^2 & w^4 & w^1 & w^5 & w^3 & w^6 \\ w^4 & w^1 & w^2 & w^3 & w^6 & w^5 \\ w^6 & w^5 & w^3 & w^1 & w^2 & w^4 \\ w^5 & w^3 & w^6 & w^2 & w^4 & w^1 \\ w^3 & w^6 & w^5 & w^4 & w^1 & w^2 \end{bmatrix} \begin{bmatrix} x(1) \\ x(2) \\ x(4) \\ x(6) \\ x(5) \\ x(3) \end{bmatrix}
$$

where $w = e^{-j2\pi/7}$.

This may be converted into 2 convolutions of length 3:

$$
\begin{bmatrix} \bar{x}(1) \\ \bar{x}(2) \\ \bar{x}(4) \\ \bar{x}(6) \\ \bar{x}(5) \\ \bar{x}(3) \end{bmatrix} = \begin{bmatrix} 1 & & & 1 & & \\ & 1 & & & 1 & \\ & & 1 & & & 1 \\ 1 & & & -1 & & \\ & 1 & & & -1 & \\ & & 1 & & & -1 \end{bmatrix}
$$

$$
\times \frac{1}{2} \begin{bmatrix} (w^1+w^{-1}) & (w^2+w^{-2}) & (w^4+w^{-4}) & & & \\ (w^2+w^{-2}) & (w^4+w^{-4}) & (w^1+w^{-1}) & & & \\ (w^4+w^{-4}) & (w^1+w^{-1}) & (w^2+w^{-2}) & & & \\ & & & (w^1-w^{-1}) & (w^2-w^{-2}) & (w^4-w^{-4}) \\ & & & (w^2-w^{-2}) & (w^4-w^{-4}) & (w^1-w^{-1}) \\ & & & (w^4-w^{-4}) & (w^1-w^{-1}) & (w^2-w^{-2}) \end{bmatrix}
$$

$$
\times \begin{bmatrix} 1 & & & 1 & & \\ & 1 & & & 1 & \\ & & 1 & & & 1 \\ 1 & & & -1 & & \\ & 1 & & & -1 & \\ & & 1 & & & -1 \end{bmatrix} \begin{bmatrix} x(1) \\ x(2) \\ x(4) \\ x(6) \\ x(5) \\ x(3) \end{bmatrix}
$$

$$
= \begin{bmatrix} 1 & & & 1 & & \\ & 1 & & & 1 & \\ & & 1 & & & 1 \\ 1 & & & -1 & & \\ & 1 & & & -1 & \\ & & 1 & & & -1 \end{bmatrix} \begin{bmatrix} \cos\theta & \cos 2\theta & \cos 4\theta & & & \\ \cos 2\theta & \cos 4\theta & \cos\theta & & & \\ \cos 4\theta & \cos\theta & \cos 2\theta & & & \\ & & & j\sin\theta & j\sin 2\theta & j\sin 4\theta \\ & & & j\sin 2\theta & j\sin 4\theta & j\sin\theta \\ & & & j\sin 4\theta & j\sin\theta & j\sin 2\theta \end{bmatrix}
$$

$$
\times \begin{bmatrix} 1 & & & 1 & & \\ & 1 & & & 1 & \\ & & 1 & & & 1 \\ 1 & & & -1 & & \\ & 1 & & & -1 & \\ & & 1 & & & -1 \end{bmatrix} \begin{bmatrix} x(1) \\ x(2) \\ x(4) \\ x(6) \\ x(5) \\ x(3) \end{bmatrix}
$$

where $\theta = 2\pi/7$.

This example shows, that the DFT of length $(2^3 - 1)$ can be converted into two subgroup circular convolutions of length 3. There are three important features:

(a) All the circular convolutions have length $3(=M)$.

(b) All elements in each circular convolution belong to the same subgroup, and each element is a square of its predecessor, say,

$$\theta_{i+1} = 2\theta_i \qquad i+1 \bmod M$$
$$i = 0, 1, \cdots, (M-1)$$

This feature holds even if $N = p^M - 1$ is not prime.

c) All the subgroup convolutions are real (or pure imaginary) arithmetic operations.

The regular structure of the decomposition can lead to appreciable advantages in terms of programming complexity and computation speed for large N.

4.3.2 Algorithm for Reordering the Rows and Columns

The ordering may be rearranged from the following equation:

$$g^i p^0, g^i p^1, g^i p^2, \cdots, g^i p^{M-1} \tag{4.10}$$

where $\{i\}$ is a permutation of $\{0, 1, 2, \cdots, (m-1)\}$.

If $m = \phi\ (p^M - 1)/M$ can be factored into s mutually prime factors, say,

$$m = p_1 \cdot p_2 \cdot \cdots \cdot p_s \tag{4.11}$$

then the permutation $\{i\}$ may be determined by the Chinese Remainder Theorem (CRT).

Step 1

Determine the permutation $\{i\}$. Let

$$j = 0, 1, 2, \cdots, (m-1)$$

and

$$(j_1, j_2, \cdots j_s) = (j \bmod p_1,\ j \bmod p_2, \cdots,\ j \bmod p_s) \tag{4.12}$$

then the permutation $\{i\}$ may be constructed from lexicographic ordering of $(j_1, j_2, \cdots j_s)$ by the Chinese Remainder Theorem[P4]

$$i = \sum_{k=1}^{s} j_k N_k M_k \bmod m \tag{4.13}$$

where

$$M_k = \frac{m}{p_k}, N_k M_k = 1 \bmod p_k,\ k = 1, 2, \cdots, s \tag{4.14}$$

Step 2

Find a unity root g of order m, as in eqn. (4.3), where m is the least integer such that

$$g^m = 1 \bmod N \tag{4.15}$$

The generating element of the subsets is as follows:

$$\{g^i\} \tag{4.16}$$

Here $\{i\}$ is a permutation of $\{0, 1, \cdots, (m-1)\}$ as shown in eqn. (4.13).

Step 3

Determine the new ordering of rows and columns in terms of eqn. (4.10).

The flow chart of the procedure is shown in Fig. 4.1.

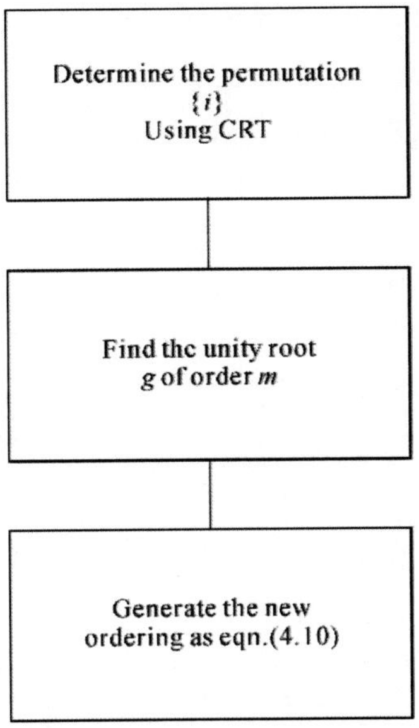

Fig. 4. 1 Flow chart for generating new ordering

4.3.3 Example

Compute the DFT of length $N=31$, $31=2^5-1$ is prime. The number of subgroups: $m=\phi(31)/M=6$, $p_1=3$, $p_2=2$.

Step 1

Determine permutation $\{i\}$. From eqn. (4. 12), it follows

$$j=0,1,2,3,4,5$$

$$(j_1,j_2)=(j \bmod 3, j \bmod 2)=(0,0),(1,1),(2,0),(0,1),(1,0),(2,1)$$

From eqn. (4. 14),

$$p_1=3,\quad M_1=2,\quad N_1=2$$

$$p_2=1,\quad M_2=3,\quad N_2=1$$

Using eqn. (4. 13), the permutation $\{i\}$ may be constructed from the lexicographic ordering of (j_1,j_2) by means of the Chinese Remainder Theorem as follows:

$$\{j_1,j_2\}=(0,0),(1,1),(2,0),(0,1),(1,0),(2,1)$$

$$\{i\}=\{0,3,4,1,2,5\}$$

Step 2

Find a unity root of order 6. Since 3 is a primitive root of GF (31), say $3^{30}=1 \bmod 31$. So $26=3^5$ is a unity root of order 6,

$$26^6=1 \bmod 31$$

From eqn. (4. 16), the generating elements of subsets are as follows:

$$\{g^i\}=\{26^0,26^3,26^4,26^1,26^2,26^5\} \bmod 31=\{1,30,5,26,25,6\}$$

Step 3

Compute the new ordering of rows and columns.

From eqn. (4. 10)f the new ordering of rows and columns is obtained as follows:

$$
\begin{aligned}
k,n= \ & 1, && 2, && 4, && 8, && 16,\\
& 30, && 29, && 27, && 23, && 15,\\
& 5, && 10, && 20, && 9, && 18,\\
& 26, && 21, && 11, && 22, && 13,\\
& 25, && 19, && 7, && 14, && 28,\\
& 6, && 12, && 24, && 17, && 3.
\end{aligned}
$$

By rearranging the input and output sequences of DFTs according the above ordering, the DFT of length 31 may be written in the following matrix form:

$$
\left[\begin{array}{ccccc|ccccc|ccccc|ccccc|ccccc|ccccc}
w^{1}&w^{2}&w^{4}&w^{8}&w^{16}&w^{30}&w^{29}&w^{27}&w^{23}&w^{15}&w^{5}&w^{10}&w^{20}&w^{9}&w^{18}&w^{26}&w^{21}&w^{11}&w^{22}&w^{13}&w^{25}&w^{19}&w^{7}&w^{14}&w^{28}&w^{6}&w^{12}&w^{24}&w^{17}&w^{3}\\
w^{2}&w^{4}&w^{8}&w^{16}&w^{1}&w^{29}&w^{27}&w^{23}&w^{15}&w^{30}&w^{10}&w^{20}&w^{9}&w^{18}&w^{5}&w^{21}&w^{11}&w^{22}&w^{13}&w^{26}&w^{19}&w^{7}&w^{14}&w^{28}&w^{25}&w^{12}&w^{24}&w^{17}&w^{3}&w^{6}\\
w^{4}&w^{8}&w^{16}&w^{1}&w^{2}&w^{27}&w^{23}&w^{15}&w^{30}&w^{29}&w^{20}&w^{9}&w^{18}&w^{5}&w^{10}&w^{11}&w^{22}&w^{13}&w^{26}&w^{21}&w^{7}&w^{14}&w^{28}&w^{25}&w^{19}&w^{24}&w^{17}&w^{3}&w^{6}&w^{12}\\
w^{8}&w^{16}&w^{1}&w^{2}&w^{4}&w^{23}&w^{15}&w^{30}&w^{29}&w^{27}&w^{9}&w^{18}&w^{5}&w^{10}&w^{20}&w^{22}&w^{13}&w^{26}&w^{21}&w^{11}&w^{14}&w^{28}&w^{25}&w^{19}&w^{7}&w^{17}&w^{3}&w^{6}&w^{12}&w^{24}\\
w^{16}&w^{1}&w^{2}&w^{4}&w^{8}&w^{15}&w^{30}&w^{29}&w^{27}&w^{23}&w^{18}&w^{5}&w^{10}&w^{20}&w^{9}&w^{13}&w^{26}&w^{21}&w^{11}&w^{22}&w^{28}&w^{25}&w^{19}&w^{7}&w^{14}&w^{3}&w^{6}&w^{12}&w^{24}&w^{17}\\
\hline
w^{30}&w^{29}&w^{27}&w^{23}&w^{15}&w^{1}&w^{2}&w^{4}&w^{8}&w^{16}&w^{26}&w^{21}&w^{11}&w^{22}&w^{13}&w^{5}&w^{10}&w^{20}&w^{9}&w^{18}&w^{6}&w^{12}&w^{24}&w^{17}&w^{3}&w^{25}&w^{19}&w^{7}&w^{14}&w^{28}\\
w^{29}&w^{27}&w^{23}&w^{15}&w^{30}&w^{2}&w^{4}&w^{8}&w^{16}&w^{1}&w^{21}&w^{11}&w^{22}&w^{13}&w^{26}&w^{10}&w^{20}&w^{9}&w^{18}&w^{5}&w^{12}&w^{24}&w^{17}&w^{3}&w^{6}&w^{19}&w^{7}&w^{14}&w^{28}&w^{25}\\
w^{27}&w^{23}&w^{15}&w^{30}&w^{29}&w^{4}&w^{8}&w^{16}&w^{1}&w^{2}&w^{11}&w^{22}&w^{13}&w^{26}&w^{21}&w^{20}&w^{9}&w^{18}&w^{5}&w^{10}&w^{24}&w^{17}&w^{3}&w^{6}&w^{12}&w^{7}&w^{14}&w^{28}&w^{25}&w^{19}\\
w^{23}&w^{15}&w^{30}&w^{29}&w^{27}&w^{8}&w^{16}&w^{1}&w^{2}&w^{4}&w^{22}&w^{13}&w^{26}&w^{21}&w^{11}&w^{9}&w^{18}&w^{5}&w^{10}&w^{20}&w^{17}&w^{3}&w^{6}&w^{12}&w^{24}&w^{14}&w^{28}&w^{25}&w^{19}&w^{7}\\
w^{15}&w^{30}&w^{29}&w^{27}&w^{23}&w^{16}&w^{1}&w^{2}&w^{4}&w^{8}&w^{13}&w^{26}&w^{21}&w^{11}&w^{22}&w^{18}&w^{5}&w^{10}&w^{20}&w^{9}&w^{3}&w^{6}&w^{12}&w^{24}&w^{17}&w^{28}&w^{25}&w^{19}&w^{7}&w^{14}\\
\hline
w^{5}&w^{10}&w^{20}&w^{9}&w^{18}&w^{26}&w^{21}&w^{11}&w^{22}&w^{13}&w^{25}&w^{19}&w^{7}&w^{14}&w^{28}&w^{6}&w^{12}&w^{24}&w^{17}&w^{3}&w^{1}&w^{2}&w^{4}&w^{8}&w^{16}&w^{30}&w^{29}&w^{27}&w^{23}&w^{15}\\
w^{10}&w^{20}&w^{9}&w^{18}&w^{5}&w^{21}&w^{11}&w^{22}&w^{13}&w^{26}&w^{19}&w^{7}&w^{14}&w^{28}&w^{25}&w^{12}&w^{24}&w^{17}&w^{3}&w^{6}&w^{2}&w^{4}&w^{8}&w^{16}&w^{1}&w^{29}&w^{27}&w^{23}&w^{15}&w^{30}\\
w^{20}&w^{9}&w^{18}&w^{5}&w^{10}&w^{11}&w^{22}&w^{13}&w^{26}&w^{21}&w^{7}&w^{14}&w^{28}&w^{25}&w^{19}&w^{24}&w^{17}&w^{3}&w^{6}&w^{12}&w^{4}&w^{8}&w^{16}&w^{1}&w^{2}&w^{27}&w^{23}&w^{15}&w^{30}&w^{29}\\
w^{9}&w^{18}&w^{5}&w^{10}&w^{20}&w^{22}&w^{13}&w^{26}&w^{21}&w^{11}&w^{14}&w^{28}&w^{25}&w^{19}&w^{7}&w^{17}&w^{3}&w^{6}&w^{12}&w^{24}&w^{8}&w^{16}&w^{1}&w^{2}&w^{4}&w^{23}&w^{15}&w^{30}&w^{29}&w^{27}\\
w^{18}&w^{5}&w^{10}&w^{20}&w^{9}&w^{13}&w^{26}&w^{21}&w^{11}&w^{22}&w^{28}&w^{25}&w^{19}&w^{7}&w^{14}&w^{3}&w^{6}&w^{12}&w^{24}&w^{17}&w^{16}&w^{1}&w^{2}&w^{4}&w^{8}&w^{15}&w^{30}&w^{29}&w^{27}&w^{23}\\
\hline
w^{26}&w^{21}&w^{11}&w^{22}&w^{13}&w^{5}&w^{10}&w^{20}&w^{9}&w^{18}&w^{6}&w^{12}&w^{24}&w^{17}&w^{3}&w^{25}&w^{19}&w^{7}&w^{14}&w^{28}&w^{30}&w^{29}&w^{27}&w^{23}&w^{15}&w^{1}&w^{2}&w^{4}&w^{8}&w^{16}\\
w^{21}&w^{11}&w^{22}&w^{13}&w^{26}&w^{10}&w^{20}&w^{9}&w^{18}&w^{5}&w^{12}&w^{24}&w^{17}&w^{3}&w^{6}&w^{19}&w^{7}&w^{14}&w^{28}&w^{25}&w^{29}&w^{27}&w^{23}&w^{15}&w^{30}&w^{2}&w^{4}&w^{8}&w^{16}&w^{1}\\
w^{11}&w^{22}&w^{13}&w^{26}&w^{21}&w^{20}&w^{9}&w^{18}&w^{5}&w^{10}&w^{24}&w^{17}&w^{3}&w^{6}&w^{12}&w^{7}&w^{14}&w^{28}&w^{25}&w^{19}&w^{27}&w^{23}&w^{15}&w^{30}&w^{29}&w^{4}&w^{8}&w^{16}&w^{1}&w^{2}\\
w^{22}&w^{13}&w^{26}&w^{21}&w^{11}&w^{9}&w^{18}&w^{5}&w^{10}&w^{20}&w^{17}&w^{3}&w^{6}&w^{12}&w^{24}&w^{14}&w^{28}&w^{25}&w^{19}&w^{7}&w^{23}&w^{15}&w^{30}&w^{29}&w^{27}&w^{8}&w^{16}&w^{1}&w^{2}&w^{4}\\
w^{13}&w^{26}&w^{21}&w^{11}&w^{22}&w^{18}&w^{5}&w^{10}&w^{20}&w^{9}&w^{3}&w^{6}&w^{12}&w^{24}&w^{17}&w^{28}&w^{25}&w^{19}&w^{7}&w^{14}&w^{15}&w^{30}&w^{29}&w^{27}&w^{23}&w^{16}&w^{1}&w^{2}&w^{4}&w^{8}\\
\hline
w^{25}&w^{19}&w^{7}&w^{14}&w^{28}&w^{6}&w^{12}&w^{24}&w^{17}&w^{3}&w^{1}&w^{2}&w^{4}&w^{8}&w^{16}&w^{30}&w^{29}&w^{27}&w^{23}&w^{15}&w^{5}&w^{10}&w^{20}&w^{9}&w^{18}&w^{26}&w^{21}&w^{11}&w^{22}&w^{13}\\
w^{19}&w^{7}&w^{14}&w^{28}&w^{25}&w^{12}&w^{24}&w^{17}&w^{3}&w^{6}&w^{2}&w^{4}&w^{8}&w^{16}&w^{1}&w^{29}&w^{27}&w^{23}&w^{15}&w^{30}&w^{10}&w^{20}&w^{9}&w^{18}&w^{5}&w^{21}&w^{11}&w^{22}&w^{13}&w^{26}\\
w^{7}&w^{14}&w^{28}&w^{25}&w^{19}&w^{24}&w^{17}&w^{3}&w^{6}&w^{12}&w^{4}&w^{8}&w^{16}&w^{1}&w^{2}&w^{27}&w^{23}&w^{15}&w^{30}&w^{29}&w^{20}&w^{9}&w^{18}&w^{5}&w^{10}&w^{11}&w^{22}&w^{13}&w^{26}&w^{21}\\
w^{14}&w^{28}&w^{25}&w^{19}&w^{7}&w^{17}&w^{3}&w^{6}&w^{12}&w^{24}&w^{8}&w^{16}&w^{1}&w^{2}&w^{4}&w^{23}&w^{15}&w^{30}&w^{29}&w^{27}&w^{9}&w^{18}&w^{5}&w^{10}&w^{20}&w^{22}&w^{13}&w^{26}&w^{21}&w^{11}\\
w^{28}&w^{25}&w^{19}&w^{7}&w^{14}&w^{3}&w^{6}&w^{12}&w^{24}&w^{17}&w^{16}&w^{1}&w^{2}&w^{4}&w^{8}&w^{15}&w^{30}&w^{29}&w^{27}&w^{23}&w^{18}&w^{5}&w^{10}&w^{20}&w^{9}&w^{13}&w^{26}&w^{21}&w^{11}&w^{22}\\
\hline
w^{6}&w^{12}&w^{24}&w^{17}&w^{3}&w^{25}&w^{19}&w^{7}&w^{14}&w^{28}&w^{30}&w^{29}&w^{27}&w^{23}&w^{15}&w^{1}&w^{2}&w^{4}&w^{8}&w^{16}&w^{26}&w^{21}&w^{11}&w^{22}&w^{13}&w^{5}&w^{10}&w^{20}&w^{9}&w^{18}\\
w^{12}&w^{24}&w^{17}&w^{3}&w^{6}&w^{19}&w^{7}&w^{14}&w^{28}&w^{25}&w^{29}&w^{27}&w^{23}&w^{15}&w^{30}&w^{2}&w^{4}&w^{8}&w^{16}&w^{1}&w^{21}&w^{11}&w^{22}&w^{13}&w^{26}&w^{10}&w^{20}&w^{9}&w^{18}&w^{5}\\
w^{24}&w^{17}&w^{3}&w^{6}&w^{12}&w^{7}&w^{14}&w^{28}&w^{25}&w^{19}&w^{27}&w^{23}&w^{15}&w^{30}&w^{29}&w^{4}&w^{8}&w^{16}&w^{1}&w^{2}&w^{11}&w^{22}&w^{13}&w^{26}&w^{21}&w^{20}&w^{9}&w^{18}&w^{5}&w^{10}\\
w^{17}&w^{3}&w^{6}&w^{12}&w^{24}&w^{14}&w^{28}&w^{25}&w^{19}&w^{7}&w^{23}&w^{15}&w^{30}&w^{29}&w^{27}&w^{8}&w^{16}&w^{1}&w^{2}&w^{4}&w^{22}&w^{13}&w^{26}&w^{21}&w^{11}&w^{9}&w^{18}&w^{5}&w^{10}&w^{20}\\
w^{3}&w^{6}&w^{12}&w^{24}&w^{17}&w^{28}&w^{25}&w^{19}&w^{7}&w^{14}&w^{15}&w^{30}&w^{29}&w^{27}&w^{23}&w^{16}&w^{1}&w^{2}&w^{4}&w^{8}&w^{13}&w^{26}&w^{21}&w^{11}&w^{22}&w^{18}&w^{5}&w^{10}&w^{20}&w^{9}
\end{array}\right]
$$

Let

$$
A=\begin{bmatrix}
w^{1}&w^{2}&w^{4}&w^{8}&w^{16}\\
w^{2}&w^{4}&w^{8}&w^{16}&w^{1}\\
w^{4}&w^{8}&w^{16}&w^{1}&w^{2}\\
w^{8}&w^{16}&w^{1}&w^{2}&w^{4}\\
w^{16}&w^{1}&w^{2}&w^{4}&w^{8}
\end{bmatrix}
\qquad
B=\begin{bmatrix}
w^{30}&w^{29}&w^{27}&w^{23}&w^{15}\\
w^{29}&w^{27}&w^{23}&w^{15}&w^{30}\\
w^{27}&w^{23}&w^{15}&w^{30}&w^{29}\\
w^{23}&w^{15}&w^{30}&w^{29}&w^{27}\\
w^{15}&w^{30}&w^{29}&w^{27}&w^{23}
\end{bmatrix}
$$

$$C=\begin{bmatrix} w^5 & w^{10} & w^{20} & w^9 & w^{18} \\ w^{10} & w^{20} & w^9 & w^{18} & w^5 \\ w^{20} & w^9 & w^{18} & w^5 & w^{10} \\ w^9 & w^{18} & w^5 & w^{10} & w^{20} \\ w^{18} & w^5 & w^{10} & w^{20} & w^9 \end{bmatrix} \quad D=\begin{bmatrix} w^{26} & w^{21} & w^{11} & w^{22} & w^{13} \\ w^{21} & w^{11} & w^{22} & w^{13} & w^{26} \\ w^{11} & w^{22} & w^{13} & w^{26} & w^{21} \\ w^{22} & w^{13} & w^{26} & w^{21} & w^{11} \\ w^{13} & w^{26} & w^{21} & w^{11} & w^{22} \end{bmatrix}$$

$$E=\begin{bmatrix} w^{25} & w^{19} & w^7 & w^{14} & w^{28} \\ w^{19} & w^7 & w^{14} & w^{28} & w^{25} \\ w^7 & w^{14} & w^{28} & w^{25} & w^{19} \\ w^{14} & w^{28} & w^{25} & w^{19} & w^7 \\ w^{28} & w^{25} & w^{19} & w^7 & w^{14} \end{bmatrix} \quad F=\begin{bmatrix} w^6 & w^{12} & w^{24} & w^{17} & w^3 \\ w^{12} & w^{24} & w^{17} & w^3 & w^6 \\ w^{24} & w^{17} & w^3 & w^6 & w^{12} \\ w^{17} & w^3 & w^6 & w^{12} & w^{24} \\ w^3 & w^6 & w^{12} & w^{24} & w^{17} \end{bmatrix}$$

Then, the matrix may be written as:

$$w=\begin{bmatrix} A & B & C & D & E & F \\ B & A & D & C & F & E \\ C & D & E & F & A & B \\ D & C & F & E & B & A \\ E & F & A & B & C & D \\ F & E & B & A & D & C \end{bmatrix}$$

This is a two-dimensional circular convolution of size (5×6). It can be converted into several subgroup circular convolutions of length 5 in terms of Agarwal and Cooley methods[B5]. The example shows that the DFT of length (2^5-1) may be computed by using a number of subgroup circular convolutions of length 5. All the subgroup convolutions have the same length 5, and each element of subgroups is the square of its predecessor. Let $c_i\ [i=0,1,\cdots,(M-1)]$ be the elements of subgroups, then:

$$c_{i+1}=c_i^2 \quad i \bmod M$$

where $i=0,1,\cdots,(M-1)$.

4.4 Fast DFT Algorithm of Length $N=p^M-1$, with N not Prime, Using Subgroup Convolutions

4.4.1 Converting the DFT of Length (p^M-1) into Subgroup Convolutions, when (p^M-1) is not a Prime

There are $\phi(N)$ elements that are prime to N on ring $Z(N)$. According to whether k is prime to N $[(k,N)=1]$ or not $[(k,N)>1]$, eqn. (4.7) may be separated in the two cases:

$$\overline{X}(k)=\sum_{n=1}^{N-1}x(n)w^{nk} \qquad (k,N)=1 \tag{4.17}$$

$$\overline{X}(k)=\sum_{n=1}^{N-1}x(n)w^{nk} \qquad (k,N)>1 \tag{4.18}$$

Based on whether n is prime to N $[(n,N)=1]$ or not $[(n,N)>1]$, eqn. (4.17) may be expressed as the sum of two parts $\overline{X}_1(k)$ and $\overline{X}_2(k)$, which are defined by:

$$\overline{X}_1(k) = \sum_{(n,N)=1} x(n)w^{nk} \qquad (k,N) = 1 \tag{4.17a}$$

$$\overline{X}_2(k) = \sum_{(n,N)>1} x(n)w^{nk} \qquad (k,N) = 1 \tag{4.17b}$$

Similarly, eqn. (4.18) can be expressed as the sum of $\overline{X}_3(k)$ and $\overline{X}_4(k)$ defined by:

$$\overline{X}_3(k) = \sum_{(n,N)=1} x(n)w^{nk} \qquad (k,N) > 1 \tag{4.18a}$$

$$\overline{X}_4(k) = \sum_{(n,N)>1} x(n)w^{nk} \qquad (k,N) > 1 \tag{4.18b}$$

$\overline{X}_1(k)$, $\overline{X}_2(k)$, $\overline{X}_3(k)$ and $\overline{X}_4(k)$ can be converted into subgroup circular convolutions as follows.

4.4.1.1 $\overline{X}_1(k)$ for both n and k are prime to $N=p^M-1$

Since $(k,N)=1$ and $(n,N)=1$ in eqn. (4.17a), from eqn. (4.4), n and k may be expressed as:

$$\begin{aligned} n&=(n_1,n_2)=g^{n_1}p^{n_2} \\ k&=(k_1,k_2)=g^{k_1}p^{k_2} \end{aligned} \tag{4.19}$$

where $n_1, k_1=0,1,2,\cdots,(m_1-1)$,

$n_2, k_2=0,1,2,\cdots,(M-1)$,

$m_1=\phi(N)/M$,

g is a unity root of order m_1, say,

$g^{m_1}=1 \bmod N$.

Using eqn. (4.19), eqn. (4.17a) may be written as follows:

$$\overline{X}_1(k_1,k_2) = \sum_{\substack{(n,N)=1 \\ (k,N)=1}} x(n)w^{nk} = \sum_{n_1=0}^{m_1-1}\sum_{n_2=0}^{M-1} x(n_1,n_2)w^{(g^{k_1+n_1}p^{k_2+n_2})} \tag{4.20}$$

$$k_1=0,1,2,\cdots,(m_1-1), \quad k_2=0,1,2,\cdots,(M-1)$$

Eqn. (4.20) may be considered as a two-dimensional circular convolution of size $(m_1 \times M)$.

4.4.1.2 $\overline{X}_2(k)$ for k is prime to N, but n is not

Since $(k,N)=1$, and $(n,N)>1$ in eqn. (4.17b), from eqn. (4.5) and (4.4), n and k may be respectively expressed as:

$$\begin{aligned} n&=(n'_1,n'_2)=g_{n'_1}p^{n'_2} \qquad n'_1=0,1,\cdots,(m_2-1) \qquad n'_2=0,1,\cdots,(d_{n'_1}-1) \\ k&=(k_1,k_2)=g^{k_1}p^{k_2} \qquad k_1=0,1,\cdots,(m_1-1) \qquad k_2=0,1,\cdots,(M-1) \end{aligned} \tag{4.21}$$

where m_1, M are the same as shown in eqn. (4.19),

$m_2=m-m_1$, m is the total number of subgroups and is shown in Table 4.1,

$d_{n'_1}$ is equal to M or divisor of M.

Using eqn. (4.21), eqn. (4.17b) may be written as follows:

$$\overline{X}_2(k_1,k_2) = \sum_{\substack{(n,N)>1 \\ (k,N)=1}} x(n)w^{nk} = \sum_{n'_1=0}^{m_2-1}\sum_{n_2=0}^{d_{n'_1}-1} x(n'_1,n'_2)[w^{(g_{n'_1}g^{k_1}}]^{p^{k_2+n'_2}} \tag{4.22}$$

where $k_1=0,1,\cdots,(m_1-1)$,

$k_2=0,1,\cdots,(M-1)$.

Eqn. (4.22) may be considered as ($m_1 \times m_2$) one-dimensional subgroup circular convolutions.

4.4.1.3 $\overline{X}_3(k)$ for k is not prime to N, n is prime to N

Since $(k,N)>1$ and $(n,N)=1$ in eqn. (4.18a) from eqns. (4.4) amd (4.5), n and k may be respectively expressed as follows:

$$\begin{aligned} n &= (n_1,n_2) = g_{n_1} p^{n_2} \quad n_1 = 0,1,\cdots,(m_1-1) \quad n_2 = 0,1,\cdots,(M-1) \\ k &= (k'_1,k'_2) = g_{k'_1} p^{k'_2} \quad k'_1 = 0,1,\cdots,(m_2-1) \quad k'_2 = 0,1,\cdots,(d_{k'_1}-1) \end{aligned} \tag{4.23}$$

m_1, M, m_2, d_i are as described by eqn. (4.21).

Using eqn. (4.23), eqn. (4.18a) may be written as follows:

$$\overline{X}_3(k'_1,k'_2) = \sum_{\substack{(n,N)=1 \\ (k,N)>1}} x(n)w^{nk} = \sum_{n_1=0}^{m_1-1}\sum_{n_2=0}^{M-1} x(n_1,n_2)[w^{(g_{n_1}g_{k'_1})}]^{p^{k'_2+n_2}} \tag{4.24}$$

It may be considered as ($m_2 \times m_1$) one-dimensional circular subgroup convolutions. It is obvious that the trigonometric function matrix $\{w\}$ in $\overline{X}_3(k)$ is the transpose of the one in $\overline{X}_2(k)$.

4.4.1.4 $\overline{X}_4(k)$ for neither n nor k are prime to N

Since $(k,N)>1$ and $(n,N)>1$ in eqn. (4.18b), from eqn. (4.5), n and k can be expressed as follows:

$$\begin{aligned} n &= (n'_1,n'_2) = g_{n'_1} p^{n'_2} \quad n'_1 = 0,1,\cdots,(m_2-1) \quad n'_2 = 0,1,\cdots,(d_{n'_1}-1) \\ k &= (k'_1,k'_2) = g_{k'_1} p^{k'_2} \quad k'_1 = 0,1,\cdots,(m_2-1) \quad k'_2 = 0,1,\cdots,(d_{k'_1}-1) \end{aligned} \tag{4.25}$$

Using eqn. (4.25), eqn. (4.18b) may be written as:

$$\overline{X}_4(k'_1,k'_2) = \sum_{\substack{(n,N)>1 \\ (k,N)>1}} x(n)w^{nk} = \sum_{n'_1=0}^{m_2-1}\sum_{n'_2=0}^{d_{n'_1}-1} x(n'_1,n'_2)[w^{(g_{n'_1}g_{k'_1})}]^{p^{k'_2+n'_2}} \tag{4.26}$$

Eqn. (4.26) may be considered as m_2^2 one-dimensional subgroup circular convolutions.

4.4.2 Example

$N=2^4-1=15=3\times 5$ is not a prime.

From eqns. (4.6) and (4.7), it follows

$$\overline{X}(k) = \sum_{n=1}^{14} x(n)w^{nk}$$
$$k=1,2,\cdots,14$$
$$w=e^{-j2\pi/15}$$

From Section 4.2.2, the non-zero elements of $Z(15)$ may be separated into 4 subgroups. There are 8 elements which are prime to 15 in $Z(15)$. 11 is a unity root of order 2. Using eqns. (4.4) and (4.5), rearrange n and k of the above example in the new ordering based on their subgroups as follows:

1,2,4,8,11,7,14,13,6,12,9,3,5,10

Then the above equation may be written in the following matrix form:

$$
\begin{bmatrix} \overline{x}(1) \\ \overline{x}(2) \\ \overline{x}(4) \\ \overline{x}(8) \\ \overline{x}(11) \\ \overline{x}(7) \\ \overline{x}(14) \\ \overline{x}(13) \\ \overline{x}(6) \\ \overline{x}(12) \\ \overline{x}(9) \\ \overline{x}(3) \\ \overline{x}(5) \\ \overline{x}(10) \end{bmatrix} = \begin{bmatrix}
w^1 & w^2 & w^4 & w^8 & w^{11} & w^7 & w^{14} & w^{13} & w^6 & w^{12} & w^9 & w^3 & w^5 & w^{10} \\
w^2 & w^4 & w^8 & w^1 & w^7 & w^{14} & w^{13} & w^{11} & w^{12} & w^9 & w^3 & w^6 & w^{10} & w^5 \\
w^4 & w^8 & w^1 & w^2 & w^{14} & w^{13} & w^{11} & w^7 & w^9 & w^3 & w^6 & w^{12} & w^5 & w^{10} \\
w^8 & w^1 & w^2 & w^4 & w^{13} & w^{11} & w^7 & w^{14} & w^3 & w^6 & w^{12} & w^9 & w^{10} & w^5 \\
w^{11} & w^7 & w^{14} & w^{13} & w^1 & w^2 & w^4 & w^8 & w^6 & w^{12} & w^9 & w^3 & w^5 & w^{10} \\
w^7 & w^{14} & w^{13} & w^{11} & w^2 & w^4 & w^8 & w^1 & w^{12} & w^9 & w^3 & w^6 & w^{10} & w^5 \\
w^{14} & w^{13} & w^{11} & w^7 & w^4 & w^8 & w^1 & w^2 & w^9 & w^3 & w^6 & w^{12} & w^5 & w^{10} \\
w^{13} & w^{11} & w^7 & w^{14} & w^8 & w^1 & w^2 & w^4 & w^3 & w^6 & w^{12} & w^9 & w^{10} & w^5 \\
w^6 & w^{12} & w^9 & w^3 & w^6 & w^{12} & w^9 & w^3 & w^6 & w^{12} & w^9 & w^3 & 1 & 1 \\
w^{12} & w^9 & w^3 & w^6 & w^{12} & w^9 & w^3 & w^6 & w^{12} & w^9 & w^3 & w^6 & 1 & 1 \\
w^9 & w^3 & w^6 & w^{12} & w^9 & w^3 & w^6 & w^{12} & w^9 & w^3 & w^6 & w^{12} & 1 & 1 \\
w^3 & w^6 & w^{12} & w^9 & w^3 & w^6 & w^{12} & w^9 & w^3 & w^6 & w^{12} & w^9 & 1 & 1 \\
w^5 & w^{10} & w^5 & w^{10} & w^5 & w^{10} & w^5 & w^{10} & 1 & 1 & 1 & 1 & w^{10} & w^5 \\
w^{10} & w^5 & w^{10} & w^5 & w^{10} & w^5 & w^{10} & w^5 & 1 & 1 & 1 & 1 & w^5 & w^{10}
\end{bmatrix} \begin{bmatrix} x(1) \\ x(2) \\ x(4) \\ x(8) \\ x(11) \\ x(7) \\ x(14) \\ x(13) \\ x(6) \\ x(12) \\ x(9) \\ x(3) \\ x(5) \\ x(10) \end{bmatrix}
$$

The above DFT may be separated into a number of subgroup convolutions as follows.

(a) Let

$$
\begin{bmatrix} \overline{X}(1) \\ \overline{X}(2) \\ \overline{X}(4) \\ \overline{X}(8) \\ \overline{X}(11) \\ \overline{X}(7) \\ \overline{X}(14) \\ \overline{X}(13) \end{bmatrix} = \begin{bmatrix} Y(1) \\ Y(2) \\ Y(4) \\ Y(8) \\ Y(11) \\ Y(7) \\ Y(14) \\ Y(13) \end{bmatrix} + \begin{bmatrix} Y'(1) \\ Y'(2) \\ Y'(4) \\ Y'(8) \\ Y'(11) \\ Y'(7) \\ Y'(14) \\ Y'(13) \end{bmatrix} + \begin{bmatrix} Y''(1) \\ Y''(2) \\ Y''(4) \\ Y''(8) \\ Y''(11) \\ Y''(7) \\ Y''(14) \\ Y''(13) \end{bmatrix}
$$

where

$$
\begin{bmatrix} Y(1) \\ Y(2) \\ Y(4) \\ Y(8) \\ Y(11) \\ Y(7) \\ Y(14) \\ Y(13) \end{bmatrix} = \begin{bmatrix}
w^1 & w^2 & w^4 & w^8 & w^{11} & w^7 & w^{14} & w^{13} \\
w^2 & w^4 & w^8 & w^1 & w^7 & w^{14} & w^{13} & w^{11} \\
w^4 & w^8 & w^1 & w^2 & w^{14} & w^{13} & w^{11} & w^7 \\
w^8 & w^1 & w^2 & w^4 & w^{13} & w^{11} & w^7 & w^{14} \\
w^{11} & w^7 & w^{14} & w^{13} & w^1 & w^2 & w^4 & w^8 \\
w^7 & w^{14} & w^{13} & w^{11} & w^2 & w^4 & w^8 & w^1 \\
w^{14} & w^{13} & w^{11} & w^7 & w^4 & w^8 & w^1 & w^2 \\
w^{13} & w^{11} & w^7 & w^{14} & w^8 & w^1 & w^2 & w^4
\end{bmatrix} \begin{bmatrix} x(1) \\ x(2) \\ x(4) \\ x(8) \\ x(11) \\ x(7) \\ x(14) \\ x(13) \end{bmatrix}
$$

This is a two-dimensional subgroup convolution of size (2×4), and it can be easily converted into 2 one-dimensional subgroup convolutions of length $4\ (=M)$. For the sake of brevity, it is not presented here.

$$\begin{bmatrix} Y'(1) \\ Y'(2) \\ Y'(4) \\ Y'(8) \\ Y'(11) \\ Y'(7) \\ Y'(14) \\ Y'(13) \end{bmatrix} = \begin{bmatrix} w^6 & w^{12} & w^9 & w^3 \\ w^{12} & w^9 & w^3 & w^6 \\ w^9 & w^3 & w^6 & w^{12} \\ w^3 & w^6 & w^{12} & w^9 \\ w^6 & w^{12} & w^9 & w^3 \\ w^{12} & w^9 & w^3 & w^6 \\ w^9 & w^3 & w^6 & w^{12} \\ w^3 & w^6 & w^{12} & w^9 \end{bmatrix} \begin{bmatrix} x(6) \\ x(12) \\ x(9) \\ x(3) \end{bmatrix}$$

It can also be written as follows:

$$\begin{bmatrix} Y'(1) \\ Y'(2) \\ Y'(4) \\ Y'(8) \end{bmatrix} = \begin{bmatrix} Y'(11) \\ Y'(7) \\ Y'(14) \\ Y'(13) \end{bmatrix} = \begin{bmatrix} w^6 & w^{12} & w^9 & w^3 \\ w^{12} & w^9 & w^3 & w^6 \\ w^9 & w^3 & w^6 & w^{12} \\ w^3 & w^6 & w^{12} & w^9 \end{bmatrix} \begin{bmatrix} x(6) \\ x(12) \\ x(9) \\ x(3) \end{bmatrix}$$

This is a circular subgroup convolution of length 4 ($=M$).

$$\begin{bmatrix} Y''(1) \\ Y''(2) \\ Y''(4) \\ Y''(8) \\ Y''(11) \\ Y''(7) \\ Y''(14) \\ Y''(13) \end{bmatrix} = \begin{bmatrix} w^5 & w^{10} \\ w^{10} & w^5 \\ w^5 & w^{10} \\ w^{10} & w^5 \\ w^{10} & w^5 \\ w^5 & w^{10} \\ w^{10} & w^5 \\ w^5 & w^{10} \end{bmatrix} \begin{bmatrix} x(5) \\ x(10) \end{bmatrix}$$

Change the ordering of $Y''(11)$ and $Y''(7)$, $Y''(14)$ and $Y''(13)$. The above can be written as follows:

$$\begin{bmatrix} Y''(1) \\ Y''(2) \end{bmatrix} = \begin{bmatrix} Y''(4) \\ Y''(8) \end{bmatrix} = \begin{bmatrix} Y''(7) \\ Y''(11) \end{bmatrix} = \begin{bmatrix} Y''(13) \\ Y''(14) \end{bmatrix} = \begin{bmatrix} w^5 & w^{10} \\ w^{10} & w^5 \end{bmatrix} \begin{bmatrix} x(5) \\ x(10) \end{bmatrix}$$

which is a circular subgroup convolution of length 2 ($=\frac{M}{2}$).

(b) It is obvious that

$$\begin{bmatrix} \overline{X}(6) \\ \overline{X}(12) \\ \overline{X}(9) \\ \overline{X}(3) \end{bmatrix} = \begin{bmatrix} w^6 & w^{12} & w^9 & w^3 \\ w^{12} & w^9 & w^3 & w^6 \\ w^9 & w^3 & w^6 & w^{12} \\ w^3 & w^6 & w^{12} & w^9 \end{bmatrix} \begin{bmatrix} x(1)+x(11)+x(6) \\ x(2)+x(7)+x(12) \\ x(4)+x(14)+x(9) \\ x(8)+x(13)+x(3) \end{bmatrix} + \begin{bmatrix} x(5)+x(10) \\ x(5)+x(10) \\ x(5)+x(10) \\ x(5)+x(10) \end{bmatrix}$$

It can be represented as a circular subgroup convolution of length 4 ($=M$).

c) $\overline{X}(5)$ and $\overline{X}(10)$ can be expressed as:

$$\begin{bmatrix} \overline{X}(5) \\ \overline{X}(10) \end{bmatrix} = \begin{bmatrix} w^5 & w^{10} \\ w^{10} & w^5 \end{bmatrix} \begin{bmatrix} x(1)+x(4)+x(7)+x(13)+x(10) \\ x(2)+x(8)+x(11)+x(14)+x(5) \end{bmatrix} + \begin{bmatrix} x(6)+x(12)+x(9)+x(3) \\ x(6)+x(12)+x(9)+x(3) \end{bmatrix}$$

It contains a circular subgroup convolution of length 2 ($=\frac{M}{2}$).

4.5 Number and Length of Cyclic Subgroups

The number and length of cyclic subgroups on ring $Z(p^M-1)$ are determined by the corresponding subsets in Galois Field GF(p^M). The finite field theory[Q1] shows that each subset GF(p^M) corresponds to one irreducible factor of the polynomial $(x^{p^M-1}-1)$. The length of each subset is equal to the degree of the corresponding irreducible factor. The total number of subsets is equal to the total number of irreducible factors of $(x^{p^M-1}-1)$. For the example in GF(2^4), their subsets and corresponding irreducible factors are as follows:

$$\begin{array}{llllcl}
\alpha & \alpha^2 & \alpha^4 & \alpha^8 & & x^4+x+1 \\
\alpha^{11} & \alpha^7 & \alpha^{14} & \alpha^{13} & & x^4+x^3+1 \\
\alpha^6 & \alpha^{12} & \alpha^9 & \alpha^3 & & x^4+x^3+x^2+x+1 \\
\alpha^5 & \alpha^{10} & & & & x^2+x+1 \\
\alpha^0 & =\alpha^{15} & =1 & & & x+1
\end{array}$$

They are in one-to-one correspondence. The elements in each subset are the roots of the corresponding irreducible polynomial, and

$$(x^{15}-1)=(x^4+x+1)(x^4+x^3+1)(x^4+x^3+x^2+x+1)(x^2+x+1)(x+1)$$

An important theory of finite fields is that the irreducible factors of the polynomial $(x^{p^M-1}-1)$ over GF(p) are the irreducible polynomials whose degrees are M or divisors of M[P1]. The correspondence between the above irreducible factors and the subgroups of ring $Z(p^M-1)$ enable the following conclusions to be stated.

(a) The length of subgroups of ring $Z(p^M-1)$ is M or divisor of M.

(b) The number of subgroups can be recursively calculated by using the equation

$$p^M-1=\sum_{d/M} d.m(d) \tag{4.27}$$

where d is the length of subgroups, d divides M, $m(d)$ is the number of subgroups with length d.

Since $m(1)=1$, $m(2)$ can be obtained by

$$2^2-1=1.m(1)+2.m(2),\ m(2)=2, \text{ etc.}$$

Actually, most subgroups have the length M, as may be seen in Table 4.1.

Table 4.1 Number and length of subgroups

M	Length of DFT	Total No. of Subgroups	No. of Subgroups with Length				
	$N=2^M-1$	m	M	$M/2$	$M/3$	$M/4$	$M/5$
2	3	1	1	—	—	—	—
3	7	2	2	—	—	—	—
4	15	4	3	1	—	—	—
5	31	6	6	—	—	—	—
6	63	12	9	2	1	—	—
7	127	18	18	—	—	—	—

(To be continued)

Table 4. 1

M	Length of DFT	Total No. of Subgroups	No. of Subgroups with Length				
	$N=2^M-1$	m	M	$M/2$	$M/3$	$M/4$	$M/5$
8	255	34	30	3	—	1	—
9	511	58	56	—	2	—	—
10	1023	106	99	6	—	—	1
11	2047	186	186	—	—	—	—

Table 4. 1 shows that most subgroups for computing the DFT of length (2^m-1) have the same length. This can give considerable program simplification.

4. 6 Conclusion

An efficient Discrete Fourier Transform Algorithm was developed in this chapter. It has proved that any DFT of length $N=(p^M-1)$, where p is prime, may be converted into several subgroup circular convolutions of length M or divisor of M. The length M circular convolution may be computed by Winograd algorithm[A8]. Most of the circular convolutions for this DFT have the same length, which leads to a more simple realisation than previous work[A4,A9].

This algorithm can be used to compute either the conventional DFT of length (p^M-1) for $w=\exp\left[-j2\pi/(2^M-1)\right]$, or the Reed—Muller polynomials when W is replaced by a primitive element α of $GF(2^M)$ in eqn. (4. 1).

Chapter 5
Minimization of Reed-Muller Logic Functions①

5. 1 Introduction

Digital signal processing systems may be represented as finite state machines in the form of state tables. The state table can be represented as a Reed-Muller polynomial. The advantage of using a Reed-Muller polynomial for realisation of a switching function is that it may be more economical than the conventional Boolean function realisation either in the number of gates or in the number of gate interconnections[D9]. Further, the Exclusive-OR realisation of switching functions can be easily tested[D10-D12]. The gains from the second of these advantages may even exceed possible disadvantages in such cases where the Exclusive-OR realisation is more costly than the equivalent vertex form. The main problem of realising modulo-2 expansions is how to minimize the number of product terms in

① The contents of this Chapter have been published in IEE Proceedings, Part E, Computers and Digital Techniques, Vol. 131, No. 5, 1984, pp. 176-186[Z3] (submitted in Dec. 1983).

the representation. This problem has long been, and still is, an open one. With the advent of LSI and VLSI techniques, the development of Reed-Muller polynomial minimization algorithms is becoming important in synthesising a function using Cellular Logic Array.

Many authors[D1-D7] have been concerned with the minimization of Reed-Muller polynomials. Saluja and Ong[D7] proposed an exhaustive algorithm to obtain all the fixed polarity modulo-2 expressions. This algorithm needs 2^n matrix multiplications and $(2^n - 1)$ permutations of output vectors. Mukhopadhyay and Schmitz[D2] gave the polarity functions of Reed-Muller polynomial coefficients. The best polarity for the minimum Reed-Muller polynomial can be obtained by finding the maximum clique of the polarity-compatibility graph. This graph possesses a very large number of vertices in most applications. Bioul, Davio and Deschamps[D3] proposed a minimization method for mixed-polarity ring-sum expansions with variable number $n \leq 5$. Papakonstantinou[D6] developed a minimization algorithm for mixed-polarity modulo-2 sum of product expressions with the number of terms $m < 6$. Recently, an efficient computer method for single-output Exclusive-OR logic design has been developed by Besslich[D15], it does not deal with multiple-output functions. Robinson and Yeh[D14] proposed a local minimization procedure for mixed-polarity R-M polynomials using the minimization polynomial coefficients with fixed polarities, but it does not guarantee global minimization. All the above work is concerned only with the single-output function, and does not deal with the minimization of the R-M polynomials over extension Galois Fields.

In Section 5.2, an efficient algorithm is presented for the minimization of single-output R-M polynomials with fixed polarities. The minimization of multi-output R-M polynomials by considering the effect of common terms is developed in Section 5.3. The minimisation of R-M polynomials over extension Galois Fields is studied in Section 5.4.

5.2 An Efficient Algorithm for Minimization of Single-Output Functions

5.2.1 Fast Reed-Muller Transformations

Any switching function of n-variables may be represented as a Read-Muller polynomial with fixed polarity as follows:

$$f(x_0, x_1, \cdots, x_{n-1}) = a_0 \oplus a_1 \overset{*}{x}_0 \oplus a_2 \overset{*}{x}_1 \oplus a_3 \overset{*}{x}_0 \overset{*}{x}_1 \oplus \cdots \oplus a_r \overset{*}{x}_0 \overset{*}{x}_1 \cdots \overset{*}{x}_{n-1} \tag{5.1}$$

where $\oplus$ denotes modulo-2 addition,

$a_i \in (0,1)$, $i = 0, 1, 2, \cdots, 2^n - 1$.

$\overset{*}{x}_i$ represents either complemented $\overline{x}_i$ for polarity 1 or uncomplemented for polarity 0 f but not both $r = 2^n - 1$.

If one knows all the possible function outputs $f_0, f_1, \cdots, f_r$, then the coefficients of the above polynomial may be obtained by[C9]

$$\underline{A} = \underline{T}_n \underline{F} \tag{5.2}$$

where $\underline{A} = (a_0, a_1, \cdots, a_r)^t$, t denotes transpose,

$\underline{F} = [f(0,0,\cdots,0), f(0,0,\cdots 1), \cdots, f(1,1,\cdots,1)]^t$ is the function output vector,

$\underline{T}_n$ is a $(2^n \times 2^n)$ matrix recursively defined by

$$\underline{T}_n=\begin{bmatrix}\underline{T}_{n-1} & 0\\ \underline{T}_{n-1} & \underline{T}_{n-1}\end{bmatrix}\qquad \underline{T}_1=\begin{bmatrix}1 & 0\\ 1 & 1\end{bmatrix}$$

It is named the Read-Muller transformation matrix.

For example, the three-variable Reed-Muller polynomial with positive polarity (x_2, x_1, x_0) may be written as:

$$f(x_2,x_1,x_0)=a_0\oplus a_1x_0\oplus a_2x_1\oplus a_3x_0x_1\oplus a_4x_2\oplus a_5x_0x_2\oplus a_6x_1x_2\oplus a_7x_0x_1x_2$$

Its coefficients can be obtained by

$$\begin{bmatrix}a_0\\a_1\\a_2\\a_3\\a_4\\a_5\\a_6\\a_7\end{bmatrix}=\begin{bmatrix}1&&&&&&&\\1&1&&&&&&\\1&0&1&&&&&\\1&1&1&1&&&&\\1&0&0&0&1&&&\\1&1&0&0&1&1&&\\1&0&1&0&1&0&1&\\1&1&1&1&1&1&1&1\end{bmatrix}\begin{bmatrix}f(0,0,0)\\f(0,0,1)\\f(0,1,0)\\f(0,1,1)\\f(1,0,0)\\f(1,0,1)\\f(1,1,0)\\f(1,1,1)\end{bmatrix}$$

The above example shows that the computation of a three-variable Reed-Muller transformation needs (3^3-2^3) modulo-2 additions. In general, a Reed-Muller transformation with n variables requires (3^n-2^n) modulo-2 additions. However, the Reed-Muller transformation matrix $\underline{T}_n$ may be factored into Kronecker products as follows:

$$\underline{T}_n=\underline{T}_1\otimes\underline{T}_1\otimes\cdots\otimes\underline{T}_1 \tag{5.3}$$

where $\otimes$ denotes Kronecker products.

This factorisation may considerably reduce the number of modulo-2 additions for the coefficient computation.

For example, the three-variable Reed-Muller transformation matrix may be factored into

$$\underline{T}_3=\begin{bmatrix}1&&&&&&&\\1&1&&&&&&\\&&1&&&&&\\&&1&1&&&&\\&&&&1&&&\\&&&&1&1&&\\&&&&&&1&\\&&&&&&1&1\end{bmatrix}\begin{bmatrix}1&&&&&&&\\&1&&&&&&\\1&&1&&&&&\\&1&&1&&&&\\&&&&1&&&\\&&&&&1&&\\&&&&1&&1&\\&&&&&1&&1\end{bmatrix}\begin{bmatrix}1&&&&&&&\\&1&&&&&&\\&&1&&&&&\\&&&1&&&&\\1&&&&1&&&\\&1&&&&1&&\\&&1&&&&1&\\&&&1&&&&1\end{bmatrix}$$

It only needs $3*2^2=12$ modulo-2 additions. The flow graph of the above example is shown in Fig. 5.1.

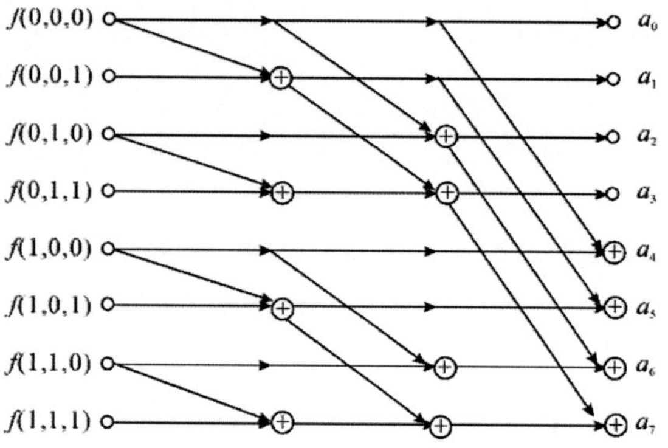

Fig. 5. 1 Flow graph of fast Reed-Muller transformation

In general, using the Kronecker factorisation, the number of modulo-2 additions for computing an n-variable Reed-Muller transformation will reduce to

$$n * 2^{n-1} \tag{5.4}$$

This algorithm is named the Fast Reed-Mutter Transformation (FRMT). It is $2(1.5^n - 1)/n$ times faster than Reed-Muller transformation. The comparison of two algorithms is shown in Fig. 5. 2.

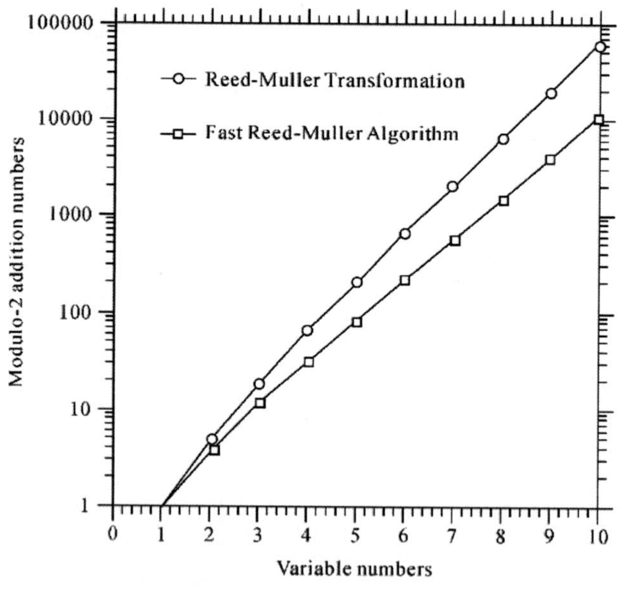

Fig. 5. 2 Comparison between R-M and fast R-M transforms

There exist 2^n different polarities for n-variables. The minimisation of Reed-Muller polynomials is the process of finding the polarity from 2^n possible polarities such that the polynomial possesses the minimum number of modulo-2 additions.

Saluja and Ong[D7] proposed that the function output vector for a new polarity is a permutation of another polarity function output vector. All the 2^n sets of polynomial coefficients may be obtained in terms of multiplying the successive modified output vectors by Reed-Muller transform matrix. One or more among them are minimum. If the FRMT algorithm is used, then the total number of modulo-2 additions for this exhaustive algorithm is

$$2^n * n * 2^{n-1} \tag{5.5}$$

5.2.2 Adjacent Polarity Mapping

It will now be shown that a set of polynomial coefficients with a new polarity may be obtained directly from another set of polynomial coefficients without permuting the output function vectors and computing eqn. (5.2). We begin with the adjacent polarity polynomials. The polarity of $(\overset{*}{x}_{n-1}, \cdots, \bar{\overset{*}{x}}_k, \overset{*}{x}_0)$ is said to be adjacent to the polarity of $(\overset{*}{x}_{n-1}, \cdots, \overset{*}{x}_k, \cdots, \overset{*}{x}_0)$, as only one variable x_k has different polarity between the two sets of variables.

Theorem 5.1

The coefficients of a Reed-Muller polynomial with n-variables may be obtained directly from the coefficients of its adjacent polarity polynomial in terms of the map with 2^{n-1} modulo-2 additions. There exist n adjacent polarity maps for an n-variable polynomial.

Proof

From eqn. (5.1), the Reed-Muller polynomial with n-variables may be rewritten as follows:

$$f(\overset{*}{x}_0, \overset{*}{x}_1, \cdots, \overset{*}{x}_{n-1}) = a_{0\cdots 00} \oplus a_{0\cdots 01}\overset{*}{x}_0 \oplus a_{0\cdots 10}\overset{*}{x}_1 \oplus a_{0\cdots 11}\overset{*}{x}_0\overset{*}{x}_1 \oplus \cdots \oplus a_{1\cdots 11}\overset{*}{x}_0\overset{*}{x}_1\cdots\overset{*}{x}_{n-1} \tag{5.6}$$

Eqn. (5.6) may be separated into two parts according to whether containing x_0 in it or not

$$f(\overset{*}{x}_0, \overset{*}{x}_1, \cdots, \overset{*}{x}_{n-1}) = f_0(\overset{*}{x}_1, \overset{*}{x}_2, \cdots, \overset{*}{x}_{n-1}) \oplus \overset{*}{x}_1 f_1(\overset{*}{x}_2, \overset{*}{x}_1, \cdots, \overset{*}{x}_{n-1}) \tag{5.7}$$

where f_0 and f_1 are not functions of the variable $\overset{*}{x}_0$

$$\begin{aligned} f_0(\overset{*}{x}_1, \overset{*}{x}_2, \cdots, \overset{*}{x}_{n-1}) &= a_{0\cdots 00} \oplus a_{0\cdots 10}\overset{*}{x}_1 \oplus \cdots \oplus a_{1\cdots 10}\overset{*}{x}_1\overset{*}{x}_2\cdots\overset{*}{x}_{n-1} \\ f_1(\overset{*}{x}_1, \overset{*}{x}_2, \cdots, \overset{*}{x}_{n-1}) &= a_{0\cdots 01} \oplus a_{0\cdots 11}\overset{*}{x}_1 \oplus \cdots \oplus a_{1\cdots 11}\overset{*}{x}_1\overset{*}{x}_1\cdots\overset{*}{x}_{n-1} \end{aligned} \tag{5.8}$$

If one changed the polarity $\overset{*}{x}_0$ into $\bar{\overset{*}{x}}_0$ and using

$$\bar{\overset{*}{x}}_0 = 1 \oplus \overset{*}{x}_0 \tag{5.9}$$

eqn. (5.7) may be changed into

$$f'(\bar{\overset{*}{x}}_0, \overset{*}{x}_1, \cdots, \overset{*}{x}_{n-1}) = f'_0(\overset{*}{x}_1, \overset{*}{x}_2, \cdots, \overset{*}{x}_{n-1}) \oplus \bar{\overset{*}{x}}_0 f_1(\overset{*}{x}_1, \overset{*}{x}_2, \cdots, \overset{*}{x}_{n-1}) \tag{5.10}$$

where

$$f'_0(\overset{*}{x}_1, \overset{*}{x}_2, \cdots, \overset{*}{x}_{n-1}) = (a_{0\cdots 00} \oplus a_{0\cdots 01}) \oplus (a_{0\cdots 10}a_{0\cdots 11})\overset{*}{x}_1 \oplus \cdots \oplus (a_{1\cdots 10} \oplus a_{1\cdots 11})\overset{*}{x}_1\overset{*}{x}_2\cdots\overset{*}{x}_{n-1} \tag{5.11}$$

$f_1(\overset{*}{x}_1, \overset{*}{x}_2, \cdots, \overset{*}{x}_{n-1})$ is not changed.

Here, $f'(\bar{\overset{*}{x}}_0, \overset{*}{x}_1, \cdots, \overset{*}{x}_{n-1})$ is the adjacent polarity polynomial of $f(\overset{*}{x}_0, \overset{*}{x}_1, \cdots, \overset{*}{x}_{n-1})$ and can be represented as follows:

$$f'(\bar{\overset{*}{x}}_0, \overset{*}{x}_1, \cdots, \overset{*}{x}_{n-1}) = a'_{0\cdots 00} \oplus a'_{0\cdots 01}\bar{\overset{*}{x}}_0 \oplus a'_{0\cdots 10}\overset{*}{x}_1 \oplus a'_{0\cdots 11}\bar{\overset{*}{x}}_0\overset{*}{x}_1 \oplus \cdots \oplus a'_{1\cdots 11}\bar{\overset{*}{x}}_0\overset{*}{x}_1\cdots\bar{\overset{*}{x}}_{n-1} \tag{5.12}$$

Compare eqns. (5.10), (5.11) and (5.12), the coefficient of $f'(\bar{\overset{*}{x}}_0, \overset{*}{x}_1, \cdots, \overset{*}{x}_{n-1})$ is as follows:

$$\begin{aligned} a'_{XX\cdots X1} &= a_{XX\cdots X1} \\ a'_{XX\cdots X0} &= a_{XX\cdots X0} \oplus a_{XX\cdots X1} \end{aligned} \tag{5.13}$$

where $XX\cdots X$ denotes $(00\cdots 0), (00\cdots 1), \cdots, (11\cdots 1)$.

This is the map of Reed-Muller polynomial coefficients from the polarity $(\overset{*}{x}_0, \overset{*}{x}_1, \cdots, \overset{*}{x}_{n-1})$ into the polarity $(\bar{\overset{*}{x}}_0, \overset{*}{x}_1, \cdots, \overset{*}{x}_{n-1})$. The map needs 2^{n-1} modulo-2 additions.

Similarly, the polarity of $\overset{*}{x}_1$ may be changed etc. So, there are n adjacent polarity maps for n

variables.

For example, the coefficients of three-variable R-M polynomials can be mapped from the polarity $(\overset{*}{x}_2\overset{*}{x}_1\overset{*}{x}_0)$ into the polarity $(\overset{*}{x}_2\overset{*}{x}_1\overline{\overset{*}{x}}_0)$ by the following adjacent polarity map:

$$\begin{bmatrix} a_0(1) \\ a_1(1) \\ a_2(1) \\ a_3(1) \\ a_4(1) \\ a_5(1) \\ a_6(1) \\ a_7(1) \end{bmatrix} = \begin{bmatrix} 1 & 1 & & & & & & \\ & 1 & & & & & & \\ & & 1 & 1 & & & & \\ & & & 1 & & & & \\ & & & & 1 & 1 & & \\ & & & & & 1 & & \\ & & & & & & 1 & 1 \\ & & & & & & & 1 \end{bmatrix} \begin{bmatrix} a_0(0) \\ a_1(0) \\ a_2(0) \\ a_3(0) \\ a_4(0) \\ a_5(0) \\ a_6(0) \\ a_7(0) \end{bmatrix}$$

where $a_i(1)$ denotes the coefficient with the polarity $(\overset{*}{x}_2\overset{*}{x}_1\overline{\overset{*}{x}}_0)$,
$a_i(0)$ denotes the coefficient with the polarity $(\overset{*}{x}_2\overset{*}{x}_1\overset{*}{x}_0)$,
$i=0,1,\cdots,7$.

Similarly, the coefficients with the polarities $(\overset{*}{x}_2\overline{\overset{*}{x}}_1\overset{*}{x}_0)$ and $(\overline{\overset{*}{x}}_2\overset{*}{x}_1\overset{*}{x}_0)$ can be mapped from the coefficient with the polarity $(\overset{*}{x}_2\overset{*}{x}_1\overset{*}{x}_0)$ by the following adjacent maps respectively:

$$\begin{bmatrix} a_0(2) \\ a_1(2) \\ a_2(2) \\ a_3(2) \\ a_4(2) \\ a_5(2) \\ a_6(2) \\ a_7(2) \end{bmatrix} = \begin{bmatrix} 1 & & 1 & & & & & \\ & 1 & & 1 & & & & \\ & & 1 & & & & & \\ & & & 1 & & & & \\ & & & & 1 & & 1 & \\ & & & & & 1 & & 1 \\ & & & & & & 1 & \\ & & & & & & & 1 \end{bmatrix} \begin{bmatrix} a_0(0) \\ a_1(0) \\ a_2(0) \\ a_3(0) \\ a_4(0) \\ a_5(0) \\ a_6(0) \\ a_7(0) \end{bmatrix}$$

$$\begin{bmatrix} a_0(4) \\ a_1(4) \\ a_2(4) \\ a_3(4) \\ a_4(4) \\ a_5(4) \\ a_6(4) \\ a_7(4) \end{bmatrix} = \begin{bmatrix} 1 & & & & 1 & & & \\ & 1 & & & & 1 & & \\ & & 1 & & & & 1 & \\ & & & 1 & & & & 1 \\ & & & & 1 & & & \\ & & & & & 1 & & \\ & & & & & & 1 & \\ & & & & & & & 1 \end{bmatrix} \begin{bmatrix} a_0(0) \\ a_1(0) \\ a_2(0) \\ a_3(0) \\ a_4(0) \\ a_5(0) \\ a_6(0) \\ a_7(0) \end{bmatrix}$$

where $a_i(2)$ and $a_i(4)$ denote respectively the coefficients of polarities

$$(\overset{*}{x}_2\overline{\overset{*}{x}}_1\overset{*}{x}_0) \text{ and } (\overline{\overset{*}{x}}_2\overset{*}{x}_1\overset{*}{x}_0) \quad i=0,1,\cdots,7$$

There exist 3 adjacent polarity maps for 3-variable Reed-Muller polynomials. Each map needs 4 modulo-2 additions.

5.2.3 Gray Code Ordering and Efficient Algorithms

Gray Code is a reflective binary code[D8], i. e. in changing from one value to the next increment only one bit is changed at a time. If all the polarities of n variables are arranged according to Gray Codes, then each polarity is adjacent to the next one. Therefore, all 2^n sets of the Reed-Muller polynomial coefficients may bo mapped by adjacent polarity maps based on Gray Code ordering. An efficient algorithm for exhaustive search of the minimum polynomial coefficients can be obtained in terms of $(2^n - 1)$ adjacent polarity maps.

For example, all the possible polarity coefficients of a 3-variable Reed-Muller polynomial may be obtained in terms of 7 adjacent polarity maps according to the following Gray Code ordering:

0 0 0	0 01	011	010	1 10	111	101	100
$x_2 x_1 x_0$	$x_2 x_1 \bar{x}_0$	$x_2 \bar{x}_1 \bar{x}_0$	$x_2 \bar{x}_1 x_0$	$\bar{x}_2 \bar{x}_1 x_0$	$\bar{x}_2 \bar{x}_1 \bar{x}_0$	$\bar{x}_2 x_1 \bar{x}_0$	$\bar{x}_2 x_1 x_0$

This efficient algorithm may be represented by the flow graph (Fig. 5. 3).

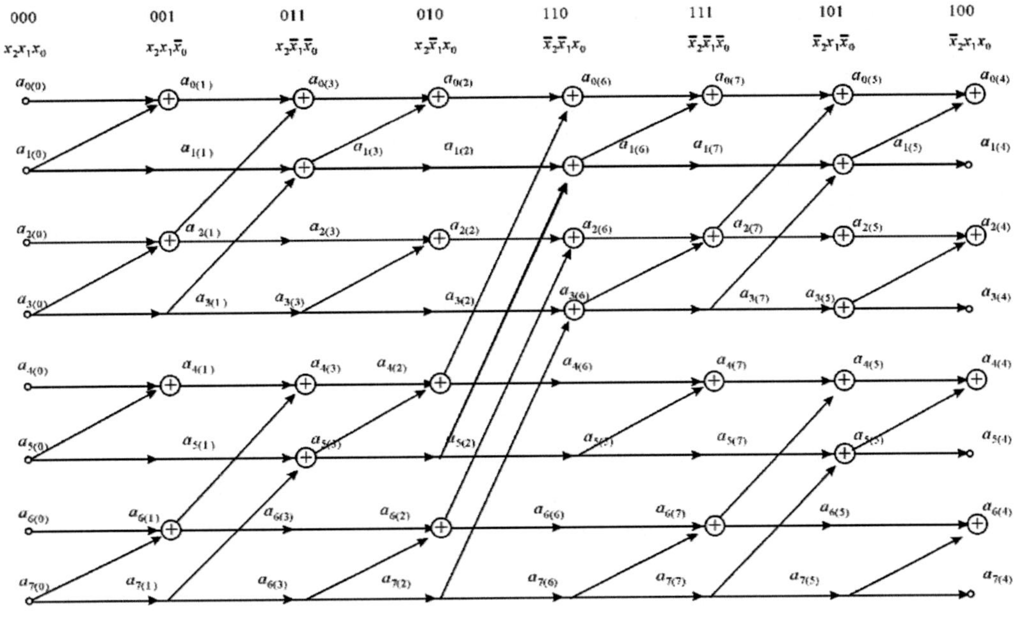

Fig. 5. 3 Flow chart of the algorithm for three variables

5. 2. 4 Complexity Comparison

From Fig. 5. 3, it is shown that $7 * 2^2 = 28$ modulo-2 additions are required to compute 7 sets of the polarity coefficients. The first set of the coefficients can be obtained from its output function vector by the FRMT algorithm, it needs 12 modulo-2 additions. The total amount of modulo-2 additions for computing all the polynomial coefficients is 40. For the case of n-variables, $(2^n - 1)2^{n-1}$ modulo-2 additions are required to compute $(2^n - 1)$ sets of the polynomial polarity coefficients. $n * 2^{n-1}$ modulo-2 additions are required to compute the first set of the coefficient by FRMT algorithm. The total amount of modulo-2 additions for computing all of the polynomial coefficients is:

$$(2^n + n - 1) * 2^{n-1} \tag{5.14}$$

One or more among the polynomials are minimum. If only the minimal set of coefficients are required, then one can use the in-place algorithm. Only 2^n memory locations are required.

From eqns. (5. 5) and (5. 14), the amount of modulo-2 additions for this efficient algorithm is

$n/(1+(n-1)2^{-n})$ times less than that of the FRMT algorithm. The comparison of two algorithms is shown in Fig. 5. 4 and Table 5. 1.

Table 5. 1 Comparison of the two algorithms

Variable Number n	No. of $\oplus$($* 2^{n-1}$)		Times
	New	FRMT	
2	5	8	1. 600
3	10	24	2. 400
4	19	64	3. 368
5	36	160	4. 444
6	69	384	5. 565
7	134	896	6. 686
8	263	2048	7. 787
9	520	4608	8. 861
10	1033	10240	9. 912
12	4107	49152	11. 967
16	65551	1048576	15. 966

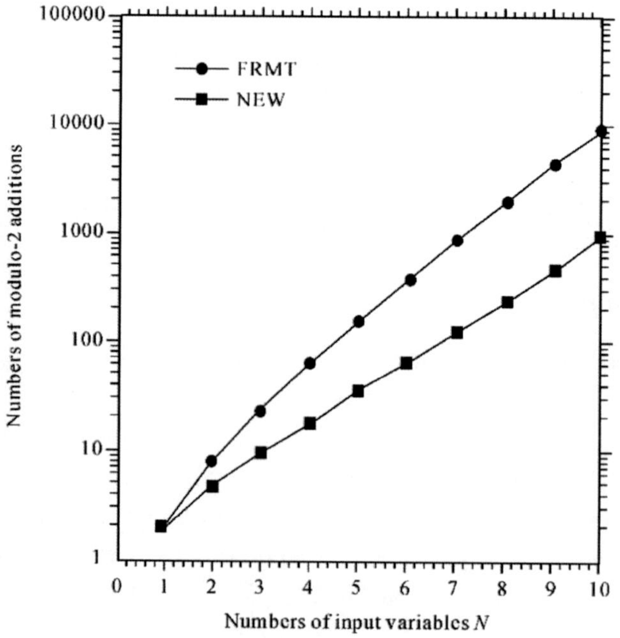

Fig. 5. 4 Comparison between FRMT and new algorithm

5.3 Minimization of Multiple-Output Functions

5.3.1 Consideration of Common Terms

The minimum polarity of a single output function can be obtained by the efficient algorithm proposed in the above section. The minimum number $w(j)$ of Exclusive-OR gates which are required to realise the function is determined by the non-zero coefficient number of the polynomial at the minimum polarity (j).

For the case of m output functions, it is very obvious that the minimum polarity for one of the output functions is not likely to be optimum for the whole m output functions, as the polarity j that makes the Exclusive-OR gate number $w_i(j)$ of the ith function minimum cannot guarantee to make the total number of Exclusive-OR gates for all the m functions minimum. Further, the polarity j which makes the sum $w = \sum_{i=0}^{m-1} w_i(j)$ minimum is also unlikely to be optimum for all the m output functions, because some product terms will be common to a number of the output functions. One common term among s output polynomials may be realised in terms of only one Exclusive-OR gate, $(s-1)$ Exclusive-OR gates may be saved. Hence the optimum polarity for m output functions should be searched with regard to all the possible common terms.

For example, the truth table of 3 output functions is shown in Table 5.2. The coefficients of the polynomials with polarity (000) may be computed by eqn. (5.2), and they are also shown in Table 5.2.

Table 5.2 The example of 3 output functions

Input			Output			Polynomial Coefficient		
x_2	x_1	x_0	f_2	f_1	f_0	c_i	b_i	a_i
0	0	0	1	0	0	1	0	0
0	0	1	0	0	0	1	0	0
0	1	0	1	1	1	0	1	1
0	1	1	0	0	0	0	1	1
1	0	0	1	0	1	0	0	1
1	0	1	1	1	1	1	1	0
1	1	0	1	1	0	0	0	0
1	1	1	0	0	0	1	1	1

Therefore, the 3 output polynomials with polarity (000) are as follows:

$$f_0 = x_1 \oplus x_0 x_1 \oplus x_2 \oplus x_0 x_1 x$$

$$f_1 = x_1 \oplus x_0 x_1 \oplus x_0 x_2 \oplus x_0 x_1 x$$

$$f_2 = 1 \oplus x_0 \oplus x_0 x_2 \oplus x_0 x_1 x_2$$

Using the algorithm in Section 5.2, the coefficients of 3 polynomials for all possible polarities may be directly mapped from the coefficients $\underline{A}(0)$, $\underline{B}(0)$ and $\underline{C}(0)$. The flow graph is shown in Fig. 5. 5.

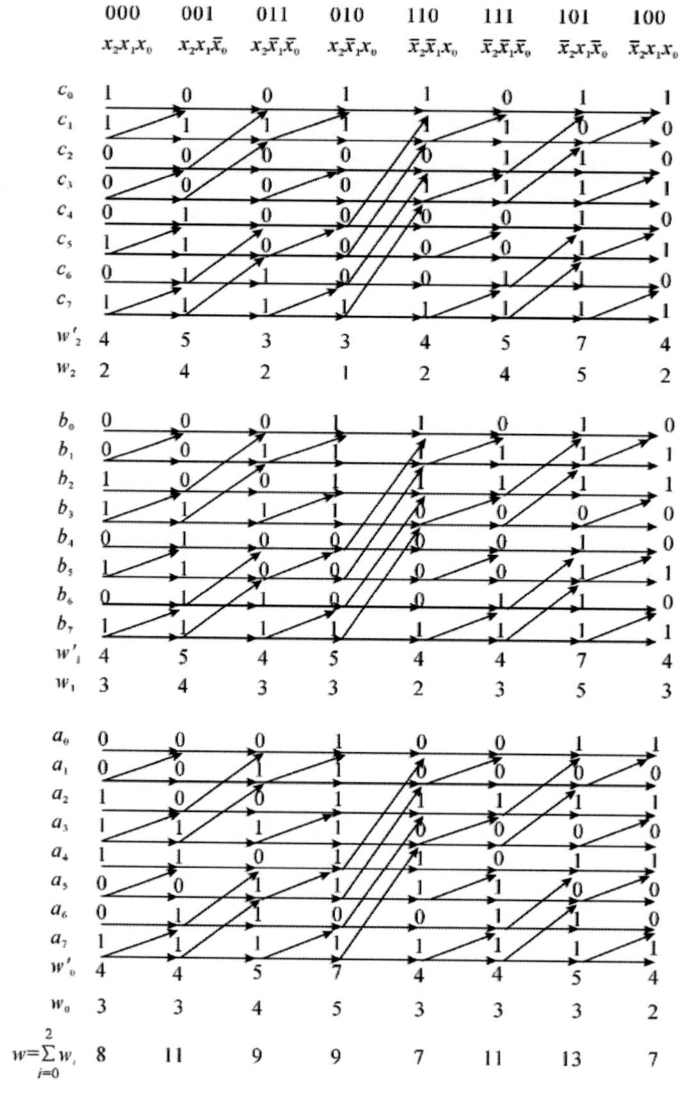

Fig. 5. 5 Flow graph for multiple-output functions

In Fig. 5. 5, w'_i denotes the number of non-zero coefficients of the ith polynomial f_i, and it is referred to as the weight of the ith polynomial, w_i denotes the number of Exclusive-OR gates for realisation of the ith polynomial. Since

$$1 \oplus f = \bar{f} \tag{5.15}$$

so, the constant term of polynomials may be realised by means of a NOT gate at the output instead of an Exclusive-OR gate in circuits. Hence, the number of Exclusive-OR gates for realisation of the ith polynomial is as follows:

$$w_i = w'_i - a_{i0} - 1 \tag{5.16}$$

where $a_{i0} \in (0,1)$ is the constant term of the ith polynomial, $w = \sum_{i=0}^{m-1} w_i$ represents the total number of the Exclusive-OR gates for realisation of m output functions.

Here, w, w_i and w'_i are all functions of polarities j ($j=0,1, \cdots, 2^n-1$). One or more polarities may make w minimum. Fig. 5. 3 shows that the sum w has the minimum value 7, when the polarity is (100) or (110). The output polynomials with input polarity (100) are as follows:

$$f_0 = x_1 \oplus \bar{x}_2 \oplus x_0 x_1 \bar{x}_2$$

$$f_1 = x_0 \oplus x_1 \oplus x_0 \bar{x}_2 \oplus x_0 x_1 \bar{x}_2$$

$$f_2 = x_0 x_1 \oplus x_0 \bar{x}_2 \oplus x_0 x_1 \bar{x}_2$$

Seven Exclusive-OR gates are required to realise the three output functions. But this is not minimum, because $(x_0 \bar{x}_2 \oplus x_0 x_1 \bar{x}_2)$ are the common terms between the functions f_1 and f_2. One Exclusive-OR gate may be commonly used by the functions f_1 and f_2, and another Exclusive-OR gate can be saved. Only 6 Exclusive-OR gates are needed to realise these 3 output functions.

Is this an optimum realisation for the example? It is unlikely to be optimum, as the other polarity polynomials may have more common terms than the above polarity polynomials. Although $w = \sum_{i=0}^{2} w_i$ is not minimum for the other polarities, the final required number of Exclusive-OR gates with regard to the common terms is possibly less than that of the polarity (100). For example, the polarity (011) has $w=9>7$, the polynomials are

$$f_0 = \bar{x}_0 \oplus \bar{x}_0 \bar{x}_1 \oplus \bar{x}_0 x_2 \oplus \bar{x}_1 x_2 \oplus \bar{x}_0 \bar{x}_1 x_2$$

$$f_1 = \bar{x}_0 \oplus \bar{x}_0 \bar{x}_1 \oplus \bar{x}_0 x_2 \oplus \bar{x}_0 \bar{x}_1 x_2$$

$$f_2 = \bar{x}_0 \oplus \bar{x}_1 x_2 \oplus \bar{x}_0 \bar{x}_1 x_2$$

where $(\bar{x}_0 \oplus \bar{x}_1 x_2 \oplus \bar{x}_0 \bar{x}_1 x_2)$ are the common terms for 3 output polynomials, 4 Exclusive-OR gates may be saved, and $(\oplus \bar{x}_0 \bar{x}_1)$ is the common term of the functions f_0 and f_1, one Exclusive-OR gate may be saved. Only $9-(4+1)=4$ Exclusive-OR gates are required to realise the 3 polynomials. Though w for the polarity (011) is larger than that for the polarity (100), the polarity (011) needs less Exclusive-OR gates as a result of the common terms.

From the above example, the optimum polarity of m output functions should be searched by considering the common terms. Theorem 5. 2 will give a method for determining the common terms among the multiple-output polynomials.

5. 3. 2 Common and Residue Function, Remainder Terms

Before giving the theorem, some definitions are introduced.

Definition 5. 1

The function consisting of the common terms among s functions is called an s-common function. For example, the 3-common function of the above example is

$$f_{012} = \bar{x}_0 \oplus \bar{x}_1 x_2 \oplus \bar{x}_0 \bar{x}_1 x_2$$

The coefficients of common functions may be directly obtained by logical AND operations among the coefficients of the s functions

$$\underline{A}_{j_1 j_2 \cdots j_s} = \underline{A}_{j_1} \wedge \underline{A}_{j_2} \wedge \cdots \wedge \underline{A}_{j_s} \tag{5.17}$$

where $\underline{A}_{j_1 j_2 \cdots j_s}$ denotes the coefficient vector of the s-common function,

$\underline{A}_{j_1}, \underline{A}_{j_2}, \cdots, \underline{A}_{j_s}$ are respectively the coefficient vectors *of* the output functions f_{j_1}, f_{j_2}, $\cdots$, f_{j_s},

$\wedge$ denotes Logical AND operations.

For the above example of polarity (011), the coefficient of the 3-common polynomial is

$$\underline{A}_{012} = \underline{A}_0 \wedge \underline{A}_1 \wedge \underline{A}_2 = (01000011)^t$$

Let $w'_{j_1 j_2 \cdots j_s}$ denote the weight of the s-common function, then the number $w_{j_1 j_2 \cdots j_s}$ of Exclusive-OR gates may be obtained from eqn. (5.16). For the above example of the polarity (011)

$$w'_{012} = 3$$

$$w_{012} = 2$$

Because $\underline{A}_{j_1}, \underline{A}_{j_2}, \cdots, \underline{A}_{j_s}$ are functions of the polarity, so the number $w_{j_1 j_2 \cdots j_s}$ of Exclusive-OR gates of the common function is also a function of the polarity. The number of s-common functions among m output functions is

$$c_m^s \qquad s = 2, 3, \cdots, m \tag{5.18}$$

The total number of common functions among m output functions is as follows:

$$c_m^2 + c_m^3 + \cdots + c_m^{m-1} + c_m^m = 2^m - (m+1) \tag{5.19}$$

For the above example, the common function numbers for 3 and 2 polynomials are respectively $c_3^3 = 1$ and $c_3^2 = 3$. The total number of common functions among 3 output functions is $2^3 - (3+1) = 4$.

In order to avoid repeated counting of the common Exclusive-OR gates in different common functions, the residue function is introduced.

Definition 5.2

The residue function of a function f_i is the sub-function which consists of the terms of f_i without $(w_{j_1 j_2, \cdots, j_s} - 1)$ common Exclusive-OR terms of the s-common function, where $i = j_1, j_2, \cdots, j_s$.

For the above example, the 3-common function is

$$f_{012} = \bar{x}_0 \oplus \bar{x}_1 x_2 \oplus \bar{x}_0 \bar{x}_1 x_2$$

The 3 residue functions may be

$$f_0^{012} = \bar{x}_0 \oplus \bar{x}_0 \bar{x}_1 \oplus \bar{x}_1 x_2$$

$$f_1^{012} = \bar{x}_0 \oplus \bar{x}_0 \bar{x}_1$$

$$f_2^{012} = \bar{x}_0$$

The coefficients of residue functions may be obtained by additions of the coefficients of the function f_i, the s-common function $f_{j_1 j_2 \cdots j_s}$ and remainder term $\underline{C}_s$ as follows:

$$\underline{A}_i^{(s)} = \underline{A}_i \oplus \underline{A}_{j_1 j_2 \cdots j_s} \oplus \underline{C}_s \tag{5.20}$$

where $\oplus$ denotes modulo-2 additions,

$\underline{A}_i$ is the coefficient vector of f_i,

$\underline{A}_{j_1 j_2 \cdots j_s}$ is the coefficient vector of s-common vector $f_{j_1 j_2 \cdots j_s}$,

$\underline{C}_s$ is a column vector, it has a single one remaining from $\underline{A}_{j_1 j_2 \cdots j_s}$ and all the other elements in $\underline{C}_s$ are zeros.

In the above example, the remainder term is $\underline{C}_3 = (01000000)^t$. The existence of $\underline{C}_s$ means that one of the common function terms is remained in the residue functions. In the above example, the common term $\bar{x}_0$ is remained in the residue functions f_0^{012}, f_1^{012} and f_2^{012}. It is named remainder term.

As the common function possesses $w_{j_1 j_2,\cdots,j_s}$ terms, so $\underline{C}_s$ has $w_{j_1 j_2,\cdots,j_s}$ forms. For the example, let $\underline{C}_3=(00000001)^t$, then the remainder term will be $x_0 x_1 x_2$, and the residue functions will be

$$f_0^{012}=\bar{x}_0 \oplus \bar{x}_0 x_2 \oplus \bar{x}_0 \bar{x}_1 x_2$$

$$f_1^{012}=\bar{x}_0 \bar{x}_1 \oplus \bar{x}_0 \bar{x}_1 x_2$$

$$f_2^{012}=\bar{x}_0 \bar{x}_1 x_2$$

It is obvious that the choice of the remainder term and the form of $\underline{C}_s$ does not change the weight and number of Exclusive-OR gates of the residue functions. Therefore, any term of the s-common function may be the remainder term. But, the remainder term of the s-common function must not be the remainder term of the $(s-1)$-common function, say

$$\underline{C}_s \neq \underline{C}_{s-1} \tag{5.21}$$

If $\underline{C}_s=\underline{C}_{s-1}$, then $\underline{C}_s \oplus \underline{C}_{s-1}=0$, the incorrect common gates may be introduced.

For the above example, if $\bar{x}_0$ is the remainder term of the 3-common function, then $\bar{x}_0$ must not be the remainder term of the 2-common functions, the latter should be $\bar{x}_0 \bar{x}_1$.

After the common function is produced, the output function should be substituted by its residue function for further operations. The coefficient $\underline{A}_i$ is substituted by $\underline{A}_i^{(s)}$ in memory after $\underline{A}_{j_1 j_2 \cdots j_s}$ is produced.

For 4-output function example, if the output functions are f_0, f_1, f_2 and f_3, then the 4-common function f_{0123} is

$$\underline{A}_{0123}=\underline{A}_0 \wedge \underline{A}_1 \wedge \underline{A}_2 \wedge \underline{A}_3$$

The 4 residue functions are

$$\underline{A}(4)_i=\underline{A}_i \oplus \underline{A}_{0123} \oplus \underline{C}_4 \qquad i=0,1,2,3$$

The 3-common function f_{012} is

$$\underline{A}_{012}=\underline{A}_i^{(4)} \wedge \underline{A}_1^{(4)} \wedge \underline{A}_2^{(4)}$$

The residue functions are

$$\underline{A}_i^{012}=\underline{A}_i^{(4)} \oplus \underline{A}_{012} \oplus \underline{C}_3 \qquad i=0,1,2$$

But, the 3-common function f_{123} is

$$\underline{A}_{123}=\underline{A}^{012} \wedge \underline{A}_2^{012} \wedge \underline{A}_3^{(4)}$$

Here $\underline{A}_1^{012}$ and $\underline{A}_2^{012}$ are substituted for $\underline{A}_1^{(4)}$ and $\underline{A}_2^{(4)}$. The residue functions are

$$\underline{A}_1^{123}=\underline{A}_1^{012} \oplus \underline{A}_{123} \oplus \underline{C}'_3$$

$$\underline{A}_2^{123}=\underline{A}_2^{012} \oplus \underline{A}_{123} \oplus \underline{C}'_3$$

$$\underline{A}_3^{123}=\underline{A}_3^{(4)} \oplus \underline{A}_{123} \oplus \underline{C}'_3$$

Here $\underline{A}_1^{(4)}$ and $\underline{A}_2^{(4)}$ were substituted by $\underline{A}_1^{012}$ and $\underline{A}_2^{012}$ in the first two equations and $\underline{C}'_3 \neq \underline{C}_3$. This substitution may be easily realised by an "in place" algorithm.

5.3.3 Minimum Number of Exclusive-OR Gates

After giving the definition and algorithm for finding common and residue functions, the number of gates which may be saved is given by the following theorem.

Theorem 5.2

The number of Exclusive-OR gates that may be saved for m output functions is as follows:

$$N_s=(m-1)w^{(m)}+(m-2)w^{(m-1)}+\cdots+2w^{(3)}+w^{(2)} \tag{5.22}$$

where N_s denotes the number of gates that may be saved,

$w^{(m)}$ is the number of common gates of m output functions,

$w^{(m-1)} = w_{01\cdots(m-2)} + w_{01\cdots(m-3)(m-1)} + \cdots + w_{12\cdots(m-1)}$ denotes the number of common gates of C_m^1 the $(m-1)$-common functions,

$w^{(3)} = \sum_{i \neq j \neq k}^{m-1} w_{ijk}$ denotes the gate number of all 3-common functions,

$w^{(2)} = \sum_{i \neq j}^{m-1} w_{ij}$ denotes the gate number of all 2-common functions.

Corollary 5.1

The total number of Exclusive-OR gates for the realisation of m output functions with regard to the common terms is as follows:

$$N_{\text{eor}} = \sum_{i=0}^{m-1} w_i - N_s \tag{5.23}$$

where N_{eor} denotes the required total number of Exclusive-OR gates,

w_i is the number of EOR gates for individual realisation of each single function f_i,

N_s is the number of EOR gates that may be saved by common terms.

Since w_i and N_s are functions of the polarities, so N_{eor} is also a function of polarities. The optimum polarity may be obtained by minimum N_{eor}.

5.3.4 Algorithm for Finding Optimum Polarity

The optimum polarity and the number of EOR gates for m output functions may be obtained by the algorithm shown in Fig. 5.6.

Step 1

Compute the common gate number $w^{(m)}$ of m functions.

(1) Compute the m-common function coefficient $\underline{A}_{01\cdots m-1}$ based on eqn. (5.17) in terms of logical AND operations.

(2) Count the number of common gates $w^{(m)}$.

(3) Choose $\underline{C}_m$. Compute m residue function coefficients $\underline{A}_i^{(m)}$ $(i=0,1,\cdots,m-1)$ based on eqn. (5.20) in terms of modulo-2 additions.

(4) Substitute $\underline{A}_i^{(m)}$ for $\underline{A}_i$ in the computer memory $(i=0,1,\cdots,m-1)$.

Step 2

Compute the common gate number $w^{(m-1)}$ for all sets of $(m-1)$ output functions.

(1) Choose the first set of $(m-1)$ output functions, as Step 1. Compute the $(m-1)$-common function coefficients $\underline{A}_{01\cdots(m-2)}$ and the number of common gates $w_{01\cdots(m-2)}$. Choose $\underline{C}_{m-1} \neq \underline{C}_m$, compute $(m-1)$ residue function coefficients $\underline{A}_i^{(m-1)}$ $(i=0,1,\cdots,(m-2))$ and substitute $\underline{A}_i^{(m)}$ by $\underline{A}_i^{(m-1)}$ in the computer memory.

(2) Choose the second set of $(m-1)$ output functions, as Step 2.1. Compute $\underline{A}_{01\cdots(m-3)(m-1)}$, $w_{01\cdots(m-3)(m-1)}$ and residue functions $\underline{A}'^{(m-1)}_i$ $(i=0,1,\cdots,(m-3),(m-1))$, and substitute $\underline{A}'^{(m-1)}_i$ for $\underline{A}_i^{(m-1)}$ and $\underline{A}_i^{(m)}$, but the second set of $(m-1)$ output function coefficients have been modified by the above step, and they will be modified by $\underline{A}'^{(m-1)}_i$ $(i=0,1,\cdots,m-3,m-1)$ again in this step.

⋮

(m) Choose the mth set of $(m-1)$ output functions and compute $\underline{A}_{12\cdots m-1}$ and $w_{12\cdots m-1}$ etc. as the above step and modify the function coefficients.

(n) Compute the total number of all the $(m-1)$-common gates by

$$w^{m-1}=w_{01\cdots(m-2)}+w_{01\cdots(m-3)(m-1)}+\cdots+w_{12\cdots(m-1)}$$

Step 3

Compute the number of common gates $w(m-2)$, $w\ (m-3)$, $\cdots$, $w^{(2)}$ as above steps.

Step 4

Compute the total number N_s of the EOR gates that may be saved based on eqn. (5. 22). Compute the total gate number N_{eor} for one polarity by eqn. (5. 23).

Step 5

Compute the total number N_{eor} of EOR gates for all possible polarities. The polarity that makes N_{eor} minimum is optimum.

For the above example, the total number of EOR gates for different polarities is shown in Table 5. 3.

Table 5. 3 The number of EOR gates for the example

Gray Code	000	001	011	010	110	111	101	100
Polarity	$x_2x_1x_0$	$x_2x_1\overline{x}_0$	$x_2\overline{x}_1\overline{x}_0$	$x_2\overline{x}_1x_0$	$\overline{x}_2\overline{x}_1x_0$	$\overline{x}_2\overline{x}_1\overline{x}_0$	$\overline{x}_2x_1\overline{x}_0$	$\overline{x}_2x_1x_0$
w_0	2	4	2	1	2	4	5	2
w_1	3	4	3	3	2	3	5	3
w_2	3	3	4	5	3	3	3	2
$w=\sum_{i=0}^{2}w_i$	8	11	9	9	7	11	13	7
$w^{(3)}$	0	2	2	1	0	2	3	0
$w^{(2)}$	2	1	1	2	1	1	1	1
N_s	2	5	5	4	1	5	7	1
N_{eor}	6	6	4	5	6	6	6	6

From Table 5. 3 it may be seen that the polarity $x_2\overline{x}_1\overline{x}_0$ is the optimum one for realisation of the 3 output functions with regard to the common terms. Only 4 EOR gates are required to realise the circuit. The logical circuit of the example is shown in Fig. 5. 7.

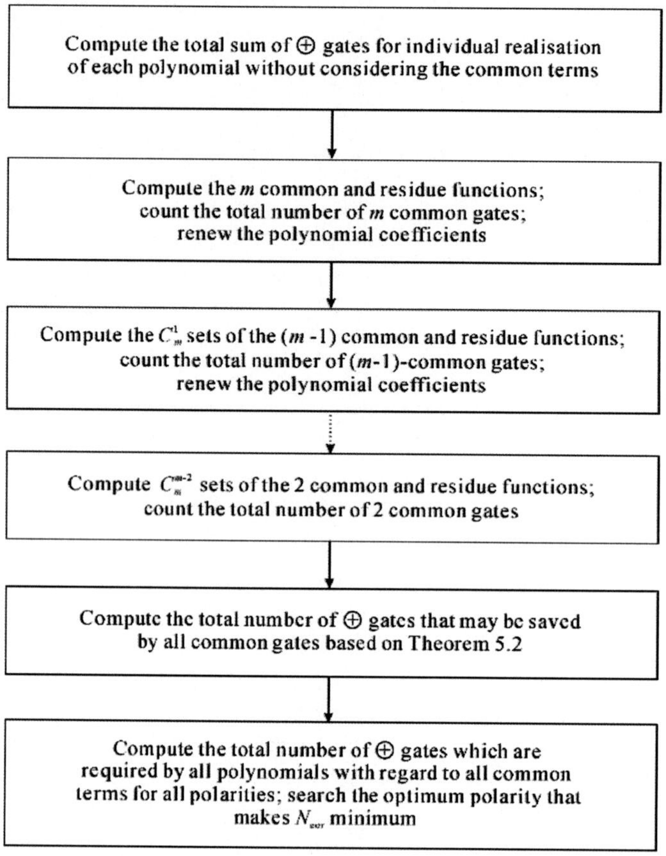

Fig. 5. 6 Flow chart of the minimization algorithm

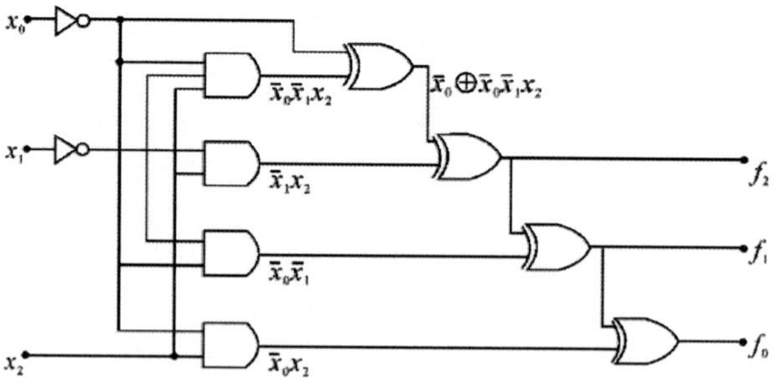

Fig. 5. 7 Optimum logical circuit for the example

5. 3. 5 Estimation of Total Number of AND and NOT Gates

The number N_{in} of input NOT gates depends on the input polarities. The number of output NOT gates and the number of AND gates are given by the following corollaries.

Corollary 5. 2

The number N_{out} output NOT gates is

$$N_{out} = \sum_{i=0}^{m-1} a_{i0} \tag{5.24}$$

where N_{out} denotes the number of NOT gates at output end,

a_{i0} is the constant term of the ith output function.

It is obvious that N_{out} is a function of the polarities.

Corollary 5.3

The number N_{and} of AND gates for realisation of m output functions is as follows:

$$N_{and} = w_d - \sum_{i=0}^{n-1} d_{2^i} - d_0 \tag{5.25}$$

where w_d is the weight of the OR polynomial $\underline{D}$ for the m output functions;

$$\underline{D} = \underline{A}_0 \vee \underline{A}_1 \vee \cdots \underline{A}_{m-1} \tag{5.26}$$

$\vee$ denotes the logical OR operation; d_{2^i} denotes the coefficients of single variable terms of the D polynomial; N_{and} denotes the number of AND gates.

For the example in Table 5.2, the number of input NOT gates, output NOT gates and EOR gates are shown in Table 5.4.

If the price of NOT, AND and EOR gates is P_1, P_2 and P_3, then the total price of the circuits is

$$P = P_1 N_{not} + P_2 N_{and} + P_3 N_{eor} \tag{5.27}$$

The optimum polarities should be obtained by minimum P.

Table 5.4 Numbers of NOT, AND and EOR gates for the example

Gray Code	000	001	011	010	110	111	101	100
Polarity	$x_2 x_1 x_0$	$x_2 x_1 \bar{x}_0$	$x_2 \bar{x}_1 \bar{x}_0$	$x_2 \bar{x}_1 x_0$	$\bar{x}_2 \bar{x}_1 x_0$	$\bar{x}_2 \bar{x}_1 \bar{x}_0$	$\bar{x}_2 x_1 \bar{x}_0$	$\bar{x}_2 x_1 x_0$
N_{in}	0	1	2	1	2	3	2	1
N_{out}	1	0	0	3	2	0	3	2
N_{not}	1	1	2	4	4	3	5	3
N_{and}	3	4	4	3	3	4	4	3
N_{eor}	6	6	4	5	6	6	6	6

5.4 Minimization of Extension Field Polynomials

5.4.1 Polarity Coefficients of Reed-Muller Polynomials over GF(2^M)

As shown in Chapter 3, any multiple-valued logic functions may be represented as a polynomial over Extension Galois Fields GF(2^M) as follows:

$$y = f(x) = \sum_{i=0}^{r} a_i x^i \tag{5.28}$$

where

$$a_i, x, y \in \mathrm{GF}(2^M) \qquad i = 0, 1, \cdots, (2^M - 1) \qquad r = 2^M - 1$$

The polynomial coefficients may be obtained from the function output vector by

$$a_0 = f(0)$$

$$a_i = \sum_{x \in GF(2^M)} f(x) x^{r-i} \tag{5.29}$$

where $i=1,2,\cdots,(2^M-1)$,

$\sum_{x \in GF(2^M)}$ denotes the sum for all the possible elements of $GF(2^M)$.

Let α be a primitive root of primitive polynomial $P_M(\alpha)$ of order M in $GF(2^M)$, then the variable x may be represented as

$$x = x_{M-1}\alpha^{M-1} + \cdots + x_1\alpha + x_0 \quad \mod P_M(\alpha) \tag{5.30}$$

where $(x_{M-1}, \cdots, x_1, x_0)$ are M input variables over ground field GF(2).

For example, Table 5.2 may be rewritten in terms of powers of α over $GF(2^3)$ as follows:

Table 5.5 The truth table over $GF(2^3)$

Input				Output			
x	x_2	x_1	x_0	y_2	y_1	y_0	y
0	0	0	0	1	0	0	α^2
1	0	0	1	0	0	0	0
α	0	1	0	1	1	1	α^5
α^2	1	0	0	1	0	1	α^6
α^3	0	1	1	0	0	0	0
α^4	1	1	0	1	1	0	α^4
α^5	1	1	1	0	0	0	0
α^6	1	0	1	1	1	1	α^5

The truth table may be represented by the polynomial over $GF(2^3)$

$$y = f(x) = \alpha^2 + \alpha^2 x + \alpha^6 x^2 + \alpha x^3 + \alpha^6 x^4 + \alpha^4 x^5 + \alpha^3 x^6 + \alpha^5 x^7$$

where

$$x, y \in GF(2^3)$$

and

$$x = x_2\alpha^2 + x_1\alpha + x_0$$

$$y = y_2\alpha^2 + y_1\alpha + y_0 \qquad \mod (\alpha^3 + \alpha + 1)$$

$$x_i, y_i \in (0,1) \qquad i = 0,1,2$$

Seven multipliers and adders of GF (2^3) are required to realise the polynomial. The multipliers over $GF(2^M)$ may be implemented by M-stage Feedback Shift Register[C8], and $GF(2^M)$ adders are implemented by M modulo-2 adders. It is desirable to reduce the number of non-zero coefficients in the polynomial (5.28), thus reducing the number of multiplications and additions.

Obviously, the number of non-zero coefficients is affected by polarities of input and output variables. For the example of Table 5.5, let the input polarity be changed to $(\bar{x}_2 x_1 \bar{x}_0)$, then the truth table is changed to Table 5.6.

Table 5.6 The truth table with new input polarity

Input				Output			
x'	$\bar{x}_2$	x_1	$\bar{x}_0$	y_2	y_1	y_0	y
α^6	1	0	1	1	0	0	α^2
α^2	1	0	0	0	0	0	0
α^5	1	1	1	1	1	1	α^5
1	0	0	1	1	0	1	α^6
α^4	1	1	0	0	0	0	0
α^3	0	1	1	1	1	0	α^4
α	0	1	0	0	0	0	0
0	0	0	0	1	1	1	α^5

Let

$$x' = \bar{x}_2\alpha^2 + x_1\alpha + \bar{x}_0 \qquad \text{mod } (\alpha^3+\alpha+1)$$

From eqn. (5.29), the polynomial is

$$y = \alpha^5 + \alpha^6 x' + \alpha^6 x'^5 + \alpha^6 x'^6 + \alpha^5 x'^7$$

This polynomial possesses 5 non-zero coefficients which are less than that of the polarity (x_2, x_1, x_0). There are 2^M different polarities for M binary variables. The optimum polarity which possesses the minimum number of non-zero coefficients may be obtained by exhaustive searching all the 2^M sets of polynomial coefficients.

Actually, the minimization of polynomials over Extension Galois Fields GF(2^M) is a special problem of the minimization of M output functions over GF(2) with the constraint of minimum number of non-zero coefficients for the Extension Galois Field polynomial. Certain properties of Galois Fields lead to an efficient algorithm for minimization of the extension field polynomials.

Equation (5.29) shows that the polynomial coefficients may be computed from its function output vectors. The direct computation of all the polarity coefficients needs $2^M(2^M-1)$ field multiplications and additions, and (2^M-1) permutations of the truth table. It will be shown that the polynomial coefficient with a new polarity may be mapped from another polarity coefficient without permuting the truth table and computing eqn. (5.29).

5.4.2 Input Polarity Mappings

Any change of input polarities corresponds to an input variable transformation as follows:

$$x' = \overset{*}{x}_{M-1}\alpha^{M-1} + \cdots + \overset{*}{x}_1\alpha + \overset{*}{x}_0 = x + b_{M-1}\alpha^{M-1} + \cdots + b_1\alpha + b_0 = x + B \qquad (5.30)$$

$$\text{where } \quad b_i = \begin{cases} 1 & \text{for } \overset{*}{x}_i = \bar{x}_i \\ 0 & \text{for } \overset{*}{x}_i = x_i \end{cases} \quad i = 0, 1, \cdots, (M-1) \qquad (5.31)$$

$$B = b_{M-1}\alpha^{M-1} + \cdots + b_1 + b_0 = \alpha^m \text{ is an element of GF}(2^M)$$

where $m = 0, 1, \cdots, (2^M - 2)$.

When $m < M$, $B = \alpha^m$ is called a lower-order element which has only one non-zero coefficient b_i in eqn. (5.31).

Substitute eqn. (5.30) into the polynomial (5.28). It follows that

$$y = f(x' + B) = \sum_{i=0}^{r} a_i (x' + B)^i = \sum_{i=0}^{r} a'_i x'^i$$

Compare the coefficients of variable x', the new polarity coefficient can be mapped from another polarity coefficient as follows:

$$\underline{A} = \underline{T}_M(B)\underline{A} \tag{5.32}$$

where

$$\underline{A}' = (a'_0 a'_1 \cdots a'_{M-1})^t \qquad \underline{A} = (a_0 a_1 \cdots a_{M-1})^t$$

$\underline{T}_M(B)$ is recursively computed by

$$\underline{T}(B) = \begin{bmatrix} \underline{T}_{M-1}(B) & B^{2^{M-1}} \underline{T}_{M-1}(B) \\ 0 & \underline{T}_{M-1}(B) \end{bmatrix} \qquad \underline{T}_1(B) = \begin{bmatrix} 1 & B \\ 0 & 1 \end{bmatrix} \tag{5.33}$$

For the polynomial over GF(2^3), eqn. (5.32) may be written in matrix form by

$$\begin{bmatrix} a'_0 \\ a'_1 \\ a'_2 \\ a'_3 \\ a'_4 \\ a'_5 \\ a'_6 \\ a'_7 \end{bmatrix} = \begin{bmatrix} 1 & B & B^2 & B^3 & B^4 & B^5 & B^6 & B^7 \\ & 1 & 0 & B^2 & & B^4 & 0 & B^6 \\ & & 1 & B & & & B^4 & B^5 \\ & & & 1 & & & & B^4 \\ & & & & 1 & B & B^2 & B^3 \\ & & & & & 1 & 0 & B^2 \\ & & & & & & 1 & B \\ & & & & & & & 1 \end{bmatrix} \begin{bmatrix} a_0 \\ a_1 \\ a_2 \\ a_3 \\ a_4 \\ a_5 \\ a_6 \\ a_7 \end{bmatrix}$$

In general, any input polarity map corresponds to a polynomial in the transformed variable $(x+B)$, B may be any element of GF(2^M). 2^M input polarities correspond to 2^M field elements. The number of field multiplications and additions of GF(2^M) for each polarity map is

$$(3^M - 2^M) \tag{5.34}$$

This number can be considerable reduced, as the above mapping matrix may be factored as Kronecker products as follows:

$$\begin{bmatrix} 1 & B & & & & & & \\ & 1 & & & & & & \\ & & 1 & B & & & & \\ & & & 1 & & & & \\ & & & & 1 & B & & \\ & & & & & 1 & & \\ & & & & & & 1 & B \\ & & & & & & & 1 \end{bmatrix} \begin{bmatrix} 1 & & B^2 & & & & & \\ & 1 & & B^2 & & & & \\ & & 1 & & & & & \\ & & & 1 & & & & \\ & & & & 1 & & B^2 & \\ & & & & & 1 & & B^2 \\ & & & & & & 1 & \\ & & & & & & & 1 \end{bmatrix} \begin{bmatrix} 1 & & & & B^4 & & & \\ & 1 & & & & B^4 & & \\ & & 1 & & & & B^4 & \\ & & & 1 & & & & B^4 \\ & & & & 1 & & & \\ & & & & & 1 & & \\ & & & & & & 1 & \\ & & & & & & & 1 \end{bmatrix}$$

Using the above factorisation, the number of field multiplications and additions for each mapping is reduced to

$$M \cdot 2^{M-1} \tag{5.35}$$

5.4.3 Efficient Algorithms

An efficient algorithm for further reducing the number of multiplications is developed by introducing adjacent polarity mappings and rearranging the polarity orderings. The polarity $(\overset{*}{x}_{M-1}, \cdots, \overline{\overset{*}{x}}_k, \cdots, \overset{*}{x}_0)$ is called the adjacent polarity of the polarity $(\overset{*}{x}_{M-1}, \cdots, \overset{*}{x}_k, \cdots, \overset{*}{x}_0)$, as only one variable $\overset{*}{x}_k$ between them is in different polarity. From eqn. (5.31), any adjacent polarity change corresponds to a low-order transformed variable of $(x+\alpha^m)(m<M)$. The most important is the adjacent polarity change of variable x_0 which corresponds to the transformed variable $(x+1)$, i. e. $B=1$. No field multiplications are required for computing this polarity mapping. If the 2^M sets of polarities can be arranged in an ordering such that transformed variable $(x+1)$ appears as often as possible, then the exhaustive search of polarity coefficients needs the minimum number of multiplications.

Let all the 2^M sets of polarities be rearranged according to the Gray Code ordering, then 2^{M-1} sets of polarity changes out of a total (2^M-1) sets of polarity changes correspond to the transformed variable of $(x+1)$. No field multiplications are needed for these 2^{M-1} mappings. The other $(2^{M-1}-1)$ sets of polarity changes correspond to low-order transformed variables. The total number of field multiplications for all (2^M-1) adjacent polarity mappings is $M * 2^{M-1}(2^{M-1}-1)$. The average number of field multiplications for one mapping is less than

$$M * 2^{M-2} \tag{5.36}$$

The average number of field additions is still

$$M * 2^{M-1} \tag{5.37}$$

For the example of GF(2^3) in Table 5.5, the flow chart of the efficient algorithm is shown in Fig. 5.8.

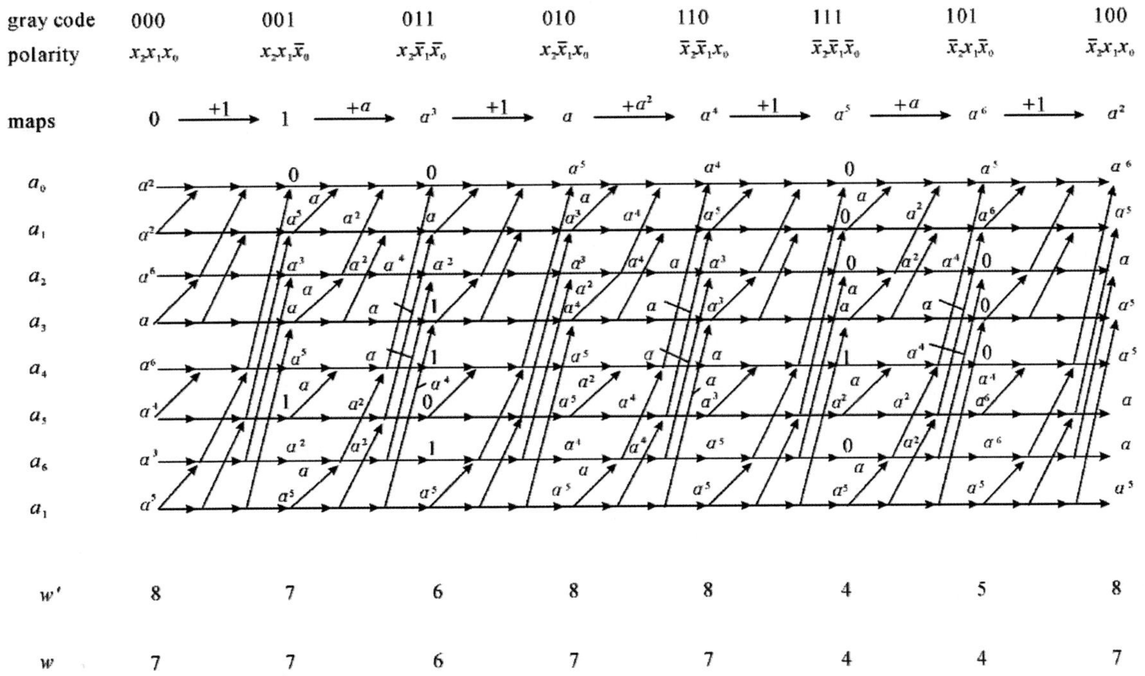

Fig. 5.8 Algorithm for the polynomial over extension field GF(2^3)

In Fig. 5. 8, w' denotes the weight of polynomial coefficients, w is the number of non-constant terms.

The figure shows that 4 sets of mappings are transformations of $(x+1)$, no multiplications are needed. The other 3 mappings require 36 multiplications. The average number of field multiplications for one mapping is 5. 2<6. The optimum polarity can be obtained by the efficient algorithm.

5. 4. 4 Output Polarity Mappings

The constant term of the output polynomial may be saved by changing the output variable polarity. For the above example, if the output polarity is changed from (y_2, y_1, y_0) into $(\bar{y}_2, \bar{y}_1, \bar{y}_0)$ then the truth table (Table 5. 6) is changed to Table 5. 7.

Table 5. 7 The truth table with new output polarity

Input				Output			
x'	$\bar{x}_2$	x_1	$\bar{x}_0$	$\bar{y}_2$	$\bar{y}_1$	$\bar{y}_0$	y'
α^6	1	0	1	0	1	1	α^3
α^2	1	0	0	1	1	1	α^5
α^5	1	1	1	0	0	0	0
1	0	0	1	0	1	0	α
α^4	1	1	0	1	1	1	α^5
α^3	0	1	1	0	0	1	1
α	0	1	0	1	1	1	α^5
0	0	0	0	0	0	0	0

Let

$$y' = \bar{y}_2\alpha^2 + \bar{y}_1\alpha + \bar{y}_0 \qquad \mod (\alpha^3 + \alpha + 1)$$

then

$$y' = y + \alpha^2 + \alpha + 1 = y + \alpha^5$$

From eqn. (5. 29), the polynomial becomes

$$y' = \alpha^6 x' + \alpha^6 x'^5 + \alpha^6 x'^6 + \alpha^5 x'^7$$

The constant term α^5 is removed by the output map (y_2, y_1, y_0) to $(\bar{y}_2, \bar{y}_1, \bar{y}_0)$. The input polarity $(\bar{x}_2, x_1, \bar{x}_0)$ and output polarity $(\bar{y}_2, \bar{y}_1, \bar{y}_0)$ are optimum. Only 4 non-zero terms are in the optimum polarity polynomial.

From the above example, it may be seen that any non-zero constant term $(a_0 \neq 0)$ in the polynomial may be removed by the output map

$$y' = y + a_0 \tag{5. 38}$$

The optimum input and output polarity may be obtained by the above algorithms.

5.5 Conclusion

An efficient algorithm for minimization of Reed-Muller polynomials with fixed polarities has been presented. The number of modulo-2 additions for this algorithm is $n/(1+(n-1)2^{-n})$ times less than that of fast Reed-Muller transformations.

With regard to the common terms, the minimization of multiple-output functions is developed. The common functions, residue functions and remainder terms are defined. An algorithm for computing the common terms of m output functions is presented. The common EOR gates may be obtained by modulo-2 additions and logical AND operations among m sets of polynomial coefficients. The number of AND gates may be computed by logical OR operations.

The minimization of Extension Galois Field polynomials is studied. An efficient algorithm that directly maps the polynomial coefficient from one polarity to an adjacent one without permuting the truth table and computing the R-M transforms has been presented. The average number of field multiplications for the exhaustive algorithm is reduced to $M * 2^{M-2}$. This algorithm is easily extended to use in the case of multiple-output polynomials over Extension Galois Fields.

This chapter is concerned only with the minimization of fixed-polarity Reed-Muller polynomials. The mixed-polarity R-M polynomials may give a smaller number of non-zero coefficients than fixed-polarity polynomials[D14]. Of course, this is at the expense of having to provide more than n inputs ($x_i, \bar{x}_i$, $i=0,1,\cdots,n-1$). So far as the author knows, no general minimization algorithms are available for the mixed-polarity polynomials when the number of variables is $n>6$[D6].

The algorithms proposed in this chapter are exhaustive, efficient and can be computed "in place". They can be efficiently used in computer-aided logic design, and only need about 2^M memory locations and are efficient in terms of the number of multiplications.

Chapter 6
Synthesis of Stable Feedback Shift Registers①

6.1 Introduction

Many finite state machines may be realised in terms of Feedback Shift Registers (FSR)[Q9, F7-F8, F2-F13]. Chapter 2 shows that the high-order infinite impulse response filter may be represented as a generalised FSR. Stable digital filters correspond to stable FSRs. Stable FSRs are also useful for coding theory and other stable sequential machines.

Golomb[Q6] systematically studied the properties of non-linear FSRs, especially the cyclic

① The contents of Section 6.3 in this Chapter have been published in International Journal of Electronics, Vol. 57, No. 1, 1984, pp. 79-84[Z2]. The contents of Sections 6.4 and 6.5 have been published in International Journal of Electronics, Vol. 57, No. 4, 1984, pp. 569-579[Z4].

properties. Gill[F9] presented the analysis and synthesis of stable linear sequential circuits, but nonlinear FSRs were not considered. The synthesis of non-linear FSRs was reported by Magleby[F2], he did not deal with the synthesis of stable FSRs which are useful for stable realisations of FSMs. Massey and Liu[F5] dealt with the stability of FSRs by using Lyapunov's methods. Mowle[F11] proposed the relation between cyclic and stable FSRs, but did not give feedback function forms. Lempel[F15] dealt with K-stable FSRs and presented a direct realisation procedure for a class of FSRs.

The definition of stable FSRs is first introduced in Section 6.2. The sufficient and necessary conditions as well as algorithms for direct synthesis of stable FSRs are proposed in Section 6.3. The synthesis methods of maximum transient FSRs from maximum period or primitive linear FSRs are presented in Section 6.4. Section 6.5 shows the indirect synthesis methods of stable FSRs.

6.2 Stable Feedback Shift Registers

The general form of a FSR is shown in Fig. 6.1.

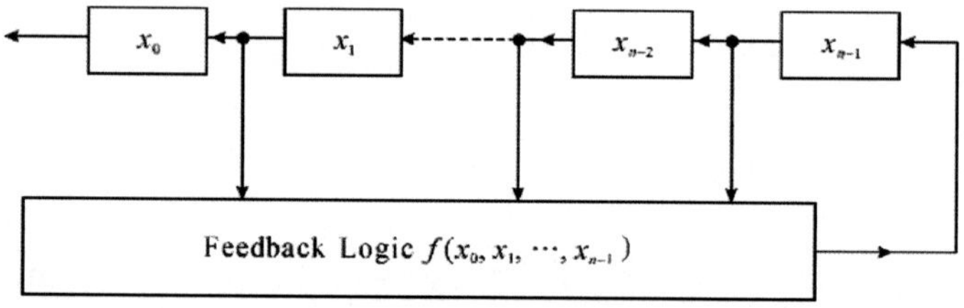

Fig. 6.1 General forms of FSRs

In Fig. 6.1, the n squares denote binary storage elements, feedback logic $f(x_0, x_1, \cdots, x_{n-1})$ is a combinatorial circuit, in general, this is non-linear.

Any FSR may be considered as a Finite State Machine (FSM). Let the present state of a finite state machine be

$$\underline{S} = (x_0, x_1, \cdots, x_{n-1}) \tag{6.1}$$

Then the next state of the machine is

$$\underline{S}' = (x'_0, x'_1, \cdots, x'_{n-1}) \tag{6.2}$$

where

$$\begin{aligned} x'_0 &= x_1 \\ x'_1 &= x_2 \\ &\vdots \\ x'_{n-2} &= x_{n-1} \\ x'_{n-1} &= f(x_0, x_1, \cdots, x_{n-1}) \end{aligned} \tag{6.3}$$

Let $\underline{N}$ be a next state operator, the next state $\underline{S}'$ may be expressed by

$$\underline{S}' = \underline{N}(\underline{S}) = [x_1, x_2, \cdots, x_{n-1}, f(x_0, x_1, \cdots, x_{n-1})] \tag{6.4}$$

The state after i shifts will be denoted as

$$\underline{N}^i(\underline{S}) \tag{6.5}$$

Definition 6.1

A FSR is <u>stable</u> if there exists an integer $q=q(\underline{S})$ for any state $\underline{S}$ of the machine, such that

$$N^q(\underline{S})=\underline{0} \tag{6.6}$$

where $\underline{0}=(0,0,\cdots,0)$ is a zero state.

The state $\underline{0}$ is the only cyclic state and there are no other loops in the stable FSM state graph.

The output sequence of a stable FSR is as follows:

$$c_0, c_1, \cdots, c_{q-1}, 0, 0, 0, \cdots \tag{6.7}$$

where

$$c_i \in (0,1) \qquad i=0,1,\cdots,q-1$$

This is called a stable sequence. After a finite sequence of length q, the stable sequence results in all zeros.

6.3 Direct Synthesis of Stable FSRs

6.3.1 Necessary and Sufficient Conditions

Any stable sequence may be generated from the initial state $(c_0, c_1, \cdots, c_{q-1})$ by an FSR of stage $n \geq q$, but in the synthesis of stable FSRs, it is desirable to design an FSR with minimum number of stages n which generates the stable sequence (6.7). If the feedback function of the required stable FSR is

$$f(x_0, x_1, \cdots, x_{n-1}) \tag{6.8}$$

then from Fig. 6.1 and eqn. (6.3) it follows

$$f(c_k, c_{k+1}, \cdots, c_{k+n-1})=c_{k+n} \qquad \text{for } k=0,1,\cdots,q-1 \qquad f(0,0,\cdots,0)=0 \tag{6.9}$$

Actually, eqns. (6.8) and (6.9) is a combinatorial function. The combination function (6.8) with minimum number of variables n may be obtained from the stable sequence (6.7) by the following theorem.

Theorem 6.1

A stable sequence $(c_0, c_1, \cdots, c_{q-1}, 0, 0, 0, \cdots)$ may be generated by a stable FSR of stage n, if and only if all $(q+1)$ sets $(c_k, c_{k+1}, \cdots, c_{k+n-1})$ of length n for $k=0,1,\cdots,q$ are distinct.

Proof

Necessity: suppose a stable FSR of n stages can generate the stable sequence (6.7), if two sets of length n are identical, say, there exists k_1 and k_2 such that

$$(c_{k1}, c_{k_1+1}, \cdots, c_{k_1+n-1})=(c_{k2}, c_{k_2+1}, \cdots, c_{k_2+n-1}) \tag{6.10}$$

where

$$k_1 \neq k_2 \qquad 0 \leq k_1, k_2 \leq q$$

Then, from eqn. (6.9) it must hold that

$$c_{k_1+n}=f(c_{k1}, c_{k_1+1}, \cdots, c_{k_1+n-1})=f(c_{k2}, c_{k_2+1}, \cdots, c_{k_2+n-1})=c_{k_2+n} \tag{6.11}$$

Therefore

$$(c_{k1}, c_{k_1+1}, \cdots, c_{k_1+n})=(c_{k2}, c_{k_2+1}, \cdots, c_{k_2+n}) \tag{6.12}$$

Similarly,

$$c_{k_1+n-1}=c_{k_2+n-1} \qquad c_{k_1+j}=c_{k_2+j} \qquad \text{for } j=n,(n+1),\cdots \tag{6.13}$$

Hence the sequence is cyclic and the FSR is unstable, this is in contradiction with the definition.

The sufficiency is obvious. If all the $(q+1)$ sets of length n from the sequence (6.7) are distinct, then a truth table of $(q+1)$ rows and $(n+1)$ columns may be set up from the sequence as follows.

Table 6.1 Truth table of distinct sets of sequences

x_0	x_1	$\cdots$	x_{n-1}	f
c_0	c_1	$\cdots$	c_{n-1}	c_n
c_1	c_2	$\cdots$	c_n	c_{n+1}
$\vdots$	$\vdots$	$\cdots$	$\vdots$	$\vdots$
c_{q-n-1}	c_{q-n}	$\cdots$	c_{q-2}	c_{q-1}
c_{q-n}	c_{q-n+1}	$\cdots$	c_{q-1}	0
$\vdots$	$\vdots$	$\cdots$	$\vdots$	$\vdots$
c_{q-1}	0	$\cdots$	0	0
0	0	$\cdots$	0	0

Since the $(q+1)$ sets of $(x_0, x_1, \cdots, x_{n-1})$ in Table 6.1 are distinct, so the combinatorial function f can be uniquely determined by the above truth table. From Fig. 6.1, this function is the feedback function of stable FSRs which generate the output sequence (6.7).

6.3.2 Algorithm of Direct Synthesis

Theorem 6.1 presented not only the proof of necessary and sufficient conditions, but also a procedure for synthesis of stable FSRs. A direct synthesis algorithm of stable FSRs from the stable sequence (6.7) is shown below. Step 1 is for finding the minimum stage n of the FSR. Step 2 is used in finding the feedback functions.

Step 1

Let $n=2,3,\cdots$, determine the minimum n such that any sets of length n, $(c_k, c_{k+1}, \cdots, c_{k+n-1})$ for $k=0,1,2,\cdots,q$, appear once and only once in the stable sequence $(c_0, c_1, \cdots, c_{q-1}, 0, 0, 0, \cdots)$. The flow chart is shown in Fig. 6.2.

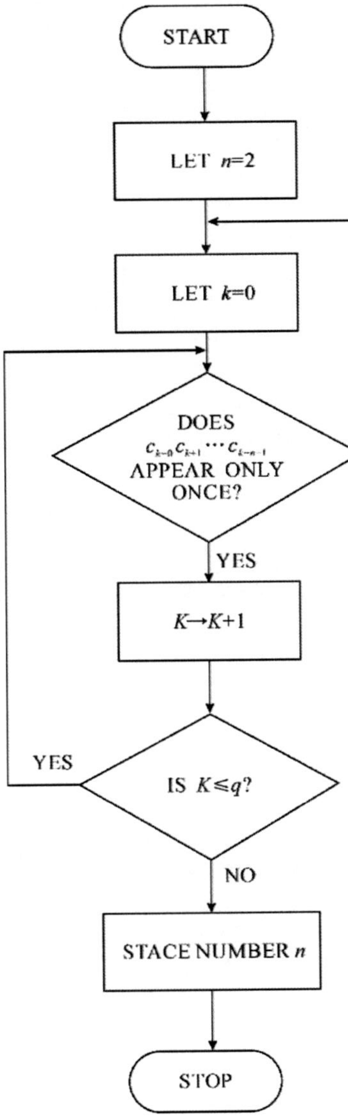

Fig. 6. 2 Flow chart of determining minimum stage number *n*

Step 2

After finding the minimum number of stages n, the truth table of the feedback function may be obtained as shown in Table 6. 1. Since

$$c_q = c_{q+1} = \cdots = 0 \tag{6.14}$$

the feedback function may be computed from the first $(q-n)$ rows of the truth table as follows:

$$f(x_0, x_1, \cdots, x_{n-1}) = \bigvee_{k=0}^{q-n-1} c_{k+n} x_0^{c_k} x_1^{c_{k+1}} \cdots x_{n-1}^{c_{k+n-1}}$$

where

$$x_i^{c_{k+i}} = \begin{cases} x_i & \text{for } c_{k+i}=1 \\ \overline{x}_i & \text{for } c_{k+i}=0 \end{cases} \qquad i=0,1,\cdots,(n-1)$$

$$c_{k+n}=0 \text{ or } 1 \qquad k=0,1,\cdots,(q-n-1)$$

$\vee$ denotes Logical OR operation.

6.3.3 Example

A stable sequence of length 12 is shown below:

1	1	1	0	1	1	0	0	0	1	0	1	0	0	0	…
c_0	c_1	c_2	c_3	c_4	c_5	c_6	c_7	c_8	c_9	c_{10}	c_{11}	c_{12}	c_{13}	c_{14}	…

Step 1

Find the minimum stage number n.

$n=2$ (1 1)$=(c_0\ c_1)=(c_1\ c_2)=(c_4\ c_5)$ appears 3 times

$n=3$ (1 1 1) appears once

(1 1 0)$=(c_1\ c_2\ c_3)=(c_4\ c_5\ c_6)$ appears twice

$n=4$ (1 1 1 0), (1 1 0 1), (1 0 1 1), (0 1 1 0), (1 1 0 0) appears once

(1 0 0 0)$=(c_5\ c_6\ c_7\ c_8)=(c_{11}\ c_{12}\ c_{13}\ c_{14})$ appears twice

$n=5$ (1 1 1 0 1), (1 1 0 1 1), (1 0 1 1 0), (0 1 1 0 0), (1 1 0 0 0), (1 0 0 0 1), (0 0 0 1 0), (0 0 1 0 1), (0 1 0 1 0), (1 0 1 0 0), (0 1 0 0 0), (1 0 0 0 0), (0 0 0 0 0)

Each of 13 sets of length 5 appears only once in the sequence, so the minimum stage number of the FSR which can generate the above sequence is 5.

Step 2

Determine the feedback function.

From Table 6.1 the first (12−5)=7 rows of the truth table are as follows.

Table 6.2 Truth table of the example

x_0	x_1	x_2	x_3	x_4	f
1	1	1	0	1	1
1	1	0	1	1	0
1	0	1	1	0	0
0	1	1	0	0	0
1	1	0	0	0	1
1	0	0	0	1	0
0	0	0	1	0	1

The feedback function is as follows:

$$f(x_0, x_1, x_2, x_3, x_4) = x_0 x_1 x_2 \overline{x}_3 x_4 \vee x_0 x_1 \overline{x}_2 \overline{x}_3 \overline{x}_4 \vee \overline{x}_0 \overline{x}_1 \overline{x}_2 x_3 \overline{x}_4$$

6.3.4 Optimum Synthesis

In the above example, 3 AND gates of 5 inputs, 2 OR gates, 5 NOT gates and 5 binary storage elements are required to implement the stable FSR. It is easy to prove that if all the $(q+1)$ sets of *length* n are distinct for the stable sequence (6. 7), then all the $(q+1)$ sets of length larger than n are also distinct. The stable sequence (6. 7) can also be implemented by a stable FSR of stage greater than n. For the above example, all the 13 sets of length 6 are distinct, the truth table consisting of the first $(12-6)=6$ rows from the sequence is as follows.

Table 6. 3 Truth table of 6 variables

x_0	x_1	x_2	x_3	x_4	x_5	f
1	1	1	0	1	1	0
1	1	0	1	1	0	0
1	0	1	1	0	0	0
0	1	1	0	0	0	1
1	1	0	0	0	1	0
1	0	0	0	1	0	1

The feedback function with 6 variables for the example is

$$f(x_0, x_1, x_2, x_3, x_4, x_5)=\overline{x}_0 x_1 x_2 \overline{x}_3 \overline{x}_4 \overline{x}_5 \vee x_0 \overline{x}_1 \overline{x}_2 \overline{x}_3 x_4 \overline{x}_5$$

2 AND gates of 6 inputs, 1 OR gate, 6 NOT gates and 6 binary storage elements are required to realise the stable FSR. The optimum stage number may be obtained by comparing the number and price of AND, Exclusive-OR, NOT gates and binary storage elements.

It is easy to prove from Theorem 6. 1 that for any stable binary sequences of length q, if m is the minimum integer which satisfies that

$$2^{m-1}<q\leq 2^m \tag{6.16}$$

then the possible number of stages n of the required stable FSRs is bounded by;

$$m\leq n\leq q \tag{6.17}$$

6. 4 Synthesis Methods of Maximum Transient FSRs

6. 4. 1 Maximum Transient FSRs

The Maximum Transient FSR is a special class of stable FSRs. It is defined by:

Definition 6. 2

If a stable FSR of n stages generates a stable sequence of length $q=2^n$, then it is called a Maximum Transient Feedback Shift Register, the output being the longest stable sequence among all the stable FSRs of n stages. The state graph of maximum transient FSRs is a maximum single tree of length 2^n which contains all the 2^n distinct states of FSRs of n stages.

The maximum transient FSR may be considered as a stable generator of field elements over GF(2^n), and can be used to synthesise the other stable FSRs (Section 6. 5). Before giving the synthesis methods of maximum transient FSRs, the following lemma is introduced.

Lemma 6. 1

The possible successors of state $(x_0, x_1, \cdots, x_{n-1})$ for any FSRs are the states $(x_1, x_2, \cdots, x_{n-1}, 0)$ or

$(x_1, x_2, \cdots, x_{n-1}, 1)$.

It is easily proved from eqn. (6.4) and Fig. 6.1 by using $f=0$ or 1. For example,

$$\underline{N}(0,0,\cdots,0,0)=(0,0,\cdots,0,0) \text{ or } (0,0,\cdots,0,1) \quad (6.18)$$

$$\underline{N}(1,0,\cdots,0,0)=(0,0,\cdots,0,0) \text{ or } (0,0,\cdots,0,1) \quad (6.19)$$

where $\underline{N}$ is the next state operator.

6.4.2 Synthesis from Maximum Period FSRs

The maximum period FSR is a cyclic FSR whose state graph has a single cycle of length 2^n, and no branches and other cycles. If the state graph of a maximum period FSR can be broken at point A, and made a cycle of length 1 at the state $\underline{0}$ as shown in Fig. 6.3, then a maximum transient FSR may be obtained from a maximum period FSR.

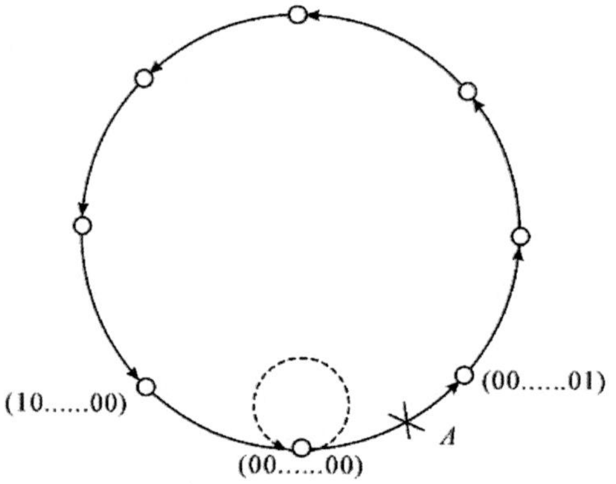

Fig. 6.3 State graph of maximum period feedback shift registers

Theorem 6.2

If $f_0(x_0, x_1, \cdots, x_{n-1})$ is a feedback function of maximum period FSRs of order n, then the feedback function of the corresponding maximum transient FSR is as follows:

$$f(x_0, x_1, \cdots, x_{n-1}) = f_0(x_0, x_1, \cdots, x_{n-1}) \oplus x_0 \overline{x}_1 \cdots \overline{x}_{n-1} \quad (6.20)$$

where $\oplus$ denotes modulo-2 addition.

Proof

Since the state graph of a maximum period FSR possesses only a single cycle of period 2^n, so no state is a successor of itself, from eqn. (6.18), the successor of the state $(0, 0, \cdots 0, 0)$ for a maximum period FSR must be $(0,0, \cdots,0,1)$, namely, $x'_{n-1}=1$, from eqn. (6.3) it follows

$$f_0(0,0,\cdots,0,0)=1 \quad (6.21)$$

Since

$$\overline{x}_0 \overline{x}_1 \cdots \overline{x}_{n-1} = \begin{cases} 1 & \text{only for } x_0=1, \text{ and } x_1=\cdots=x_{n-1}=0 \\ 0 & \text{otherwise} \end{cases} \quad (6.22)$$

From eqns. (6.21) and (6.22), eqn. (6.20) becomes

$$f(x_0, x_1, \cdots, x_{n-1}) = \begin{cases} 0 & \text{for } x_0=1, \text{ and } x_1=\cdots=x_{n-1}=0 \\ f_0(x_0, x_1, \cdots x_{n-1}) & \text{otherwise} \end{cases} \quad (6.23)$$

Say, the state graph of the maximum period FSR with feedback function f_0 is broken at the point A

(Fig. 6. 3), and forms a cycle of length 1 at the state $\underline{0}$, all the states form a maximum single tree.

De Bruijn[Q6] shows that there exists $2^{2^{n-1}-n}$ maximum period FSRs of order n. So, there is the same number of maximum transient FSRs. But, it is still an open question to find all the feedback functions of maximum period FSRs. The linear FSRs are well understood[Q5]. Some maximum transient FSRs can be obtained from linear FSRs.

6. 4. 3 Synthesis from Primitive Linear FSRs

First, the primitive LFSR may be obtained from the following lemma.

Lemma 6. 2

If $p(x)=1+a_1+a_2x^2+\cdots+a_nx^n, a_n=1$, is a primitive polynomial of degree n over GF(2), then the linear FSR with the feedback function

$$p(x_0,x_1,\cdots,x_{n-1})=a_nx_0\oplus a_{n-1}x_1\oplus\cdots\oplus a_1x_{n-1} \tag{6.24}$$

is a primitive linear FSR of order n, $\oplus$ denotes modulo-2 addition.

Its proof can be found in Stone's book[Q1]. A primitive LFSR has two cycles in its state graph: all the non-zero states on a pure cycle of length (2^n-1), the state $\underline{0}$ forms a pure cycle of length 1. No branch is in the state graph (Fig. 6. 4). A large class of feedback functions for maximum transient FSRs may be derived from primitive linear FSRs by the following theorem.

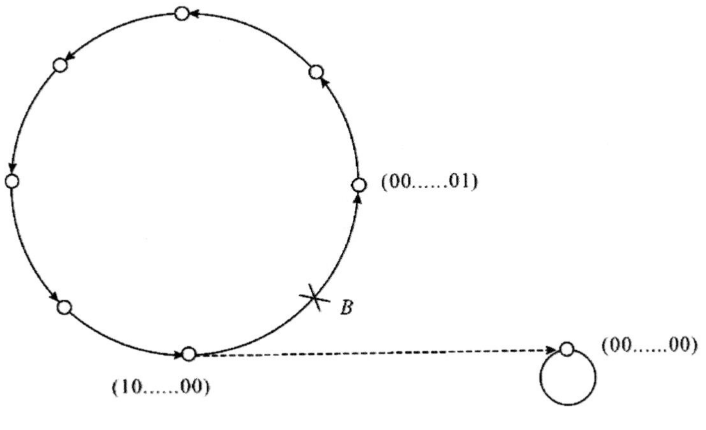

Fig. 6. 4 State graph of primitive linear FSRs

Theorem 6. 3

If $p(x_0,x_1,\cdots,x_{n-1})=a_nx_0\oplus a_{n-1}x_1\oplus\cdots\oplus a_1x_{n-1}$ is a primitive polynomial over GF(2), where

$$a_n=1, a_i\in(0,1) \qquad i=1,2,\cdots,n$$

then a large class of maximum transient FSRs may be obtained by the following feedback functions:

$$f(x_0,x_1,x_{n-1})=a_nx_0\oplus a_{n-1}x_1\oplus\cdots\oplus a_1x_{n-1}\oplus x_0\overline{x}_1\cdots\overline{x}_{n-1} \tag{6.25}$$

Proof

Since $p(x)$ is a primitive polynomial of degree n over GF(2), from Lemma 6. 2, the linear FSR with feedback function $p(x_0,x_1,\cdots,x_{n-1})$ is a primitive linear FSR. The state $\underline{0}$ forms a pure cycle of length 1. Its predecessor is only itself. From eqn. (6. 19), the successor of the state $(1,0,\cdots,0)$ must be $(0,0,\cdots,1)$, say, $x'_{n-1}=1$. From eqn. (6. 3), it holds

$$p(1,0,\cdots,0,0)=1 \tag{6.26}$$

Since

$$x_0\bar{x}_1\cdots\bar{x}_{n-1}=\begin{cases}1 & \text{only for } x_0=1\text{, and } x_1=x_2=\cdots x_{n-1}=0\\ 0 & \text{otherwise}\end{cases} \tag{6.27}$$

Using eqns. (6.26) and (6.27), eqn. (6.25) becomes

$$f(x_0,x_1,\cdots,x_{n-1})=\begin{cases}0 & \text{for } x_0=1\text{, and } x_1=x_2=\cdots=x_{n-1}=0\\ p(x_0,x_1,\cdots x_{n-1}) & \text{otherwise}\end{cases} \tag{6.28}$$

The large cycle in the state graph of the primitive linear FSR is broken at point B (Fig. 6.4), and the state $(1,0,\cdots,0,0)$ is connected to the state $(0,0,\cdots,0,0)$; all the states form a maximum single tree.

For example, the primitive polynomial of degree 4 over GF(2) is given by:

$$p(x)=1+x+x^4$$

Then the following feedback function

$$f(x_0,x_1,x_2,x_3)=x_0\oplus x_3\oplus x_0\bar{x}_1\bar{x}_2\bar{x}_3$$

generates a maximum transient FSR of stage 4, its state table and state graph are shown in Table 6.4 and Fig. 6.5 respectively.

Table 6.4 State table of the example

S	x_0	x_1	x_2	x_3	$f(x_0,x_1,x_2,x_3)$	x'_0	x'_1	x'_2	x'_3	S′
S_0	0	0	0	0	0	0	0	0	0	S_0
S_1	0	0	0	1	1	0	0	1	0	S_3
S_2	0	0	1	0	0	0	1	0	1	S_4
S_3	0	0	1	1	1	0	1	1	0	S_7
S_4	0	1	0	0	0	1	0	0	1	S_8
S_5	0	1	0	1	1	1	0	1	0	S_{11}
S_6	0	1	1	0	0	1	1	0	1	S_{12}
S_7	0	1	1	1	1	1	1	1	0	S_{15}
S_8	1	0	0	0	0	0	0	0	1	S_0
S_9	1	0	0	1	0	0	0	1	0	S_2
S_{10}	1	0	1	0	1	0	1	0	1	S_5
S_{11}	1	0	1	1	0	0	1	1	0	S_6
S_{12}	1	1	0	0	1	1	0	0	1	S_9
S_{13}	1	1	0	1	0	1	0	1	0	S_{10}
S_{14}	1	1	1	0	1	1	1	0	1	S_{13}
S_{15}	1	1	1	1	0	1	1	1	0	S_{14}

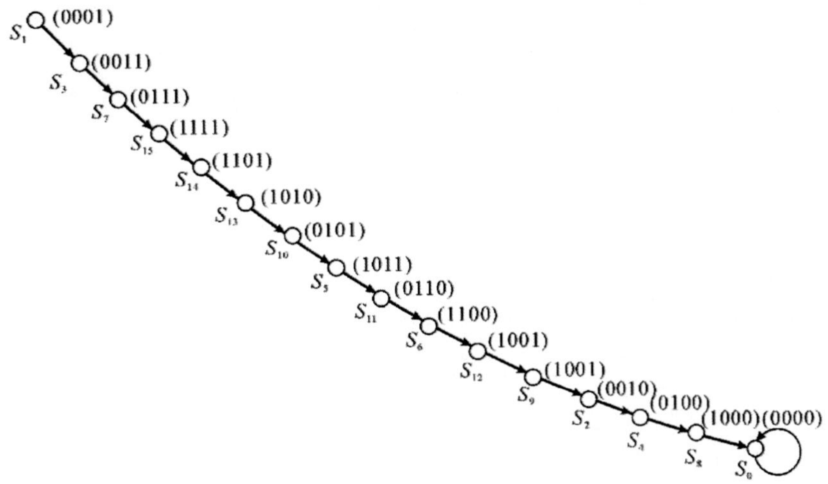

Fig. 6.5 Example of maximum transient FSRs

This is a maximum single tree beginning with the state (0 0 0 1) and ending at the 0 state. The number of maximum transient FSRs of stage n generated by Theorem 6.3 is equal to the number of primitive polynomials of degree n as follows:

$$\phi(2^n-1)/n \tag{6.29}$$

where ϕ is Euler's function.

6.4.4 A Large Class of Field Element Generators

As each state of an n-stage FSR represents one element of GF(2^n), the maximum transient FSR of stage n may be used as a stable generator of field elements over GF(2^n). It outputs each of the (2^n-1) non-zero elements once and only once, and ends in the all zero state. From Theorem 6.3, a large class of stable generators of field elements can be obtained from primitive LFSRs, and their feedback function is shown in Table 6.5.

Table 6.5 Feedback functions of stable generator of field elements

n	Feedback Functions $f(x_0, x_1, \cdots, x_{n-1})$	Length
2	$x_0 \oplus x_1 \oplus x_0 \bar{x}_1$	4
3	$x_0 \oplus x_2 \oplus x_0 \bar{x}_1 \bar{x}_2$	8
4	$x_0 \oplus x_3 \oplus x_0 \bar{x}_1 \bar{x}_2 \bar{x}_3$	16
5	$x_0 \oplus x_3 \oplus x_0 \bar{x}_1 \bar{x}_2 \bar{x}_3 \bar{x}_4$	32
6	$x_0 \oplus x_5 \oplus x_0 \bar{x}_1 \bar{x}_2 \bar{x}_3 \bar{x}_4 \bar{x}_5$	64
7	$x_0 \oplus x_6 \oplus x_0 \bar{x}_1 \bar{x}_2 \bar{x}_3 \bar{x}_4 \bar{x}_5 \bar{x}_6$	128
8	$x_0 \oplus x_4 \oplus x_5 \oplus x_6 \oplus x_0 \bar{x}_1 \bar{x}_2 \bar{x}_3 \bar{x}_4 \bar{x}_5 \bar{x}_6 \bar{x}_7$	256

(To be continued)

Table 6.5

n	Feedback Functions $f(x_0, x_1, \cdots, x_{n-1})$	Length
9	$x_0 \oplus x_5 \oplus x_0 \bar{x}_1 \bar{x}_2 \bar{x}_3 \bar{x}_4 \bar{x}_5 \bar{x}_6 \bar{x}_7 \bar{x}_8$	512
10	$x_0 \oplus x_7 \oplus x_0 \bar{x}_1 \bar{x}_2 \bar{x}_3 \bar{x}_4 \bar{x}_5 \bar{x}_6 \bar{x}_7 \bar{x}_8 \bar{x}_9$	1024
11	$x_0 \oplus x_9 \oplus x_0 \bar{x}_1 \bar{x}_2 \bar{x}_3 \bar{x}_4 \bar{x}_5 \bar{x}_6 \bar{x}_7 \bar{x}_8 \bar{x}_9 \bar{x}_{10}$	2048
12	$x_0 \oplus x_6 \oplus x_8 \oplus x_{11} \oplus x_0 \bar{x}_1 \bar{x}_2 \bar{x}_3 \bar{x}_4 \bar{x}_5 \bar{x}_6 \bar{x}_7 \bar{x}_8 \bar{x}_9 \bar{x}_{10} \bar{x}_{11}$	4096
13	$x_0 \oplus x_9 \oplus x_{10} \oplus x_{11} \oplus x_0 \bar{x}_1 \bar{x}_2 \bar{x}_3 \bar{x}_4 \bar{x}_5 \bar{x}_6 \bar{x}_7 \bar{x}_8 \bar{x}_9 \bar{x}_{10} \bar{x}_{11} \bar{x}_{12}$	8192
14	$x_0 \oplus x_9 \oplus x_{11} \oplus x_{13} \oplus x_0 \bar{x}_1 \bar{x}_2 \bar{x}_3 \bar{x}_4 \bar{x}_5 \bar{x}_6 \bar{x}_7 \bar{x}_8 \bar{x}_9 \bar{x}_{10} \bar{x}_{11} \bar{x}_{12} \bar{x}_{13}$	16384
15	$x_0 \oplus x_{14} \oplus x_0 \bar{x}_1 \bar{x}_2 \bar{x}_3 \bar{x}_4 \bar{x}_5 \bar{x}_6 \bar{x}_7 \bar{x}_8 \bar{x}_9 \bar{x}_{10} \bar{x}_{11} \bar{x}_{12} \bar{x}_{13} \bar{x}_{14}$	32768
16	$x_0 \oplus x_{11} \oplus x_{13} \oplus x_{14} \oplus x_0 \bar{x}_1 \bar{x}_2 \bar{x}_3 \bar{x}_4 \bar{x}_5 \bar{x}_6 \bar{x}_7 \bar{x}_8 \bar{x}_9 \bar{x}_{10} \bar{x}_{11} \bar{x}_{12} \bar{x}_{13} \bar{x}_{14} \bar{x}_{15}$	65536

6.5 Indirect Synthesis of Stable FSRs

The direct synthesis algorithm of stable FSRs was presented in Section 6.3. The minimum number of stages n and feedback function of stable FSRs which can generate the required output sequences may be obtained in terms of the proposed algorithm. However, for some stable binary sequences, the obtained stage n from the direct algorithm is rather large. For example, consider a stable sequence of length 15:

$$1000,0000,0000,0010,0000 \cdots \tag{6.30}$$

from the direct synthesis method, the stage n of the required FSR should be 15. A smaller stage can be obtained by the indirect synthesis.

6.5.1 FSRs with Output Mappings

The maximum transient FSR possesses the global minimum number of stages n for generating the stable sequences of length 2^n. It may be used as a stable generator of field elements. Any stable sequences of length less than 2^n may be generated from a maximum transient FSR of stage n by the output mapping as shown in Fig. 6.6.

Theorem 6.4

If a stable binary sequence of length q is as follows:

$$d_0, d_1, \cdots, d_{q-1}, 0, 0, \cdots \tag{6.31}$$

where

$$d_i \in (0,1) \qquad i=0,1,\cdots,(q-1)$$

and a maximum transient sequence of length 2^n $(2^{n-1} < q < 2^n)$ as follows:

$$c_0, c_1, \cdots, c_{2^n-1}, 0, 0, 0, \cdots \tag{6.32}$$

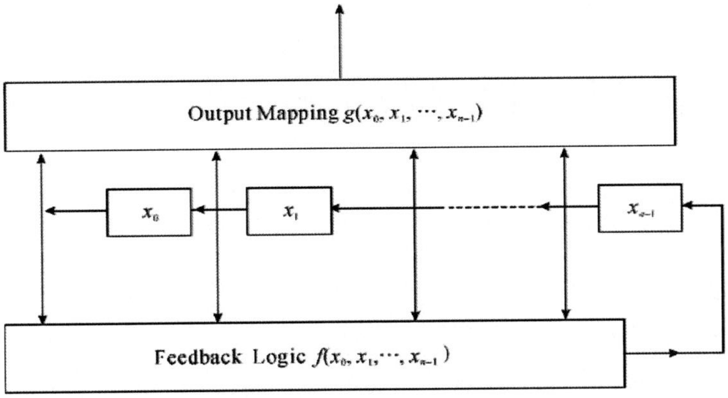

Fig. 6.6 FSRs with output mapping

where

$$c_i \in (0,1) \qquad i=0,1,\cdots,(2^n-1)$$

then the stable sequence (6.31) may be generated by using the following output mapping logic:

$$g(x_0,x_1,\cdots,x_{n-1}) = \sum_{k=0}^{q-1} d_k x_0^{c_k} x_1^{c_{k+1}} \cdots x_{n-1}^{c_{k+n-1}} \qquad (6.33)$$

where

$$x_j^{c_{k+j}} = \begin{cases} x_j & \text{for } c_{k+j}=1 \\ \overline{x}_j & \text{for } c_{k+j}=0 \end{cases} \qquad (6.34)$$

$$j=0,1,\cdots,(n-1)$$

Proof

From eqn. (6.1), the state $\underline{s}_k$ of maximum transient FSRs may be expressed by

$$\underline{s}_k=(c_k,c_{k+1},\cdots,c_{k+n-1}) \qquad (6.35)$$

where

$$k=0,1,\cdots,(2^n-1)$$

Because a maximum transient FSR is stable, its final state ($k=2^n-1$) is the $\underline{0}$ state and forms a cycle of length 1

$$\underline{s}_{2^n-1}=(0,0,\cdots,0) \text{ and } c_{2^n-1}=0 \qquad (6.36)$$

and

$$\underline{N}(0,0, \cdots, 0)=(0,0,\cdots,0) \qquad (6.37)$$

Since the 2^n states of maximum transient FSRs are distinct, a truth table of $(n+i)$ columns and 2^n rows can be set up from the sequences (6.31) and (6.32).

Table 6.6 Truth table of output mappings

x_0	x_1	$\cdots$	x_{n-1}	g
c_0	c_1	$\cdots$	c_{n-1}	d_0
c_1	c_2	$\cdots$	c_n	d_1
$\vdots$	$\vdots$	$\cdots$	$\vdots$	$\vdots$

Table 6.6

x_0	x_1	$\cdots$	x_{n-1}	g
c_{q-1}	c_q	$\cdots$	c_{q+n-2}	d_{q-1}
c_q	c_{q+1}	$\cdots$	c_{q+n-1}	0
$\vdots$	$\vdots$	$\cdots$	$\vdots$	$\vdots$
c_{2^n-2}	0	$\cdots$	0	0
0	0	$\cdots$	0	0

As 2^n sets of variables $(x_0, x_1, \cdots, x_{n-1})$ are distinct in Table 6.6, so the combination function $g(x_0, x_1, \cdots x_{n-1})$ can be uniquely determined from the above truth table in the form of eqn. (6.33). From the last row of Table 6.6, it follows that $g(0,0,\cdots,0)=0$. So, from eqn. (6.37), the output of g is stable.

6.5.2 Algorithm of Indirect Synthesis

The proof of Theorem 6.4 also presented a procedure for generating a stable binary sequence from a maximum transient FSR. If a stable binary sequence of length q is given by eqn. (6.31), then the algorithm for indirect synthesis of the required sequence from a maximum transient FSR is as follows.

Step 1

Choose stage n of maximum transient FSRs such that

$$2^{n-1} < q < 2^n \tag{6.38}$$

Step 2

Choose a primitive polynomial $p(x)$ of degree n over GF(2)

$$p(x) = 1 + a_1 x + a_2 x^2 + \cdots + a_n x^n \tag{6.39}$$

Step 3

Design a maximum transient FSR with feedback function f as follows:

$$f(x_0, x_1, \cdots, x_{n-1}) = a_n x_0 \oplus a_{n-1} x_1 \oplus \cdots \oplus a_1 x_{n-1} \oplus x_0 \bar{x}_1 \cdots \bar{x}_{n-1} \tag{6.40}$$

Step 4

Generate the output sequence of the maximum transient FSR

$$c_0, c_1, \cdots, c_{2^n-2}, 0, 0, 0, \cdots \tag{6.41}$$

Step 5

Construct a truth table for the output mapping function from the sequences (6.31) and (6.41), design the output mapping function g as follows:

$$g(x_0, x_1, \cdots, x_{n-1}) = \sum_{k=0}^{q-1} d_k x_0^{c_k} x_1^{c_{k+1}} \cdots x_{n-1}^{c_{k+n-1}} \tag{6.42}$$

The flow chart of this indirect algorithm is shown in Fig. 6.7.

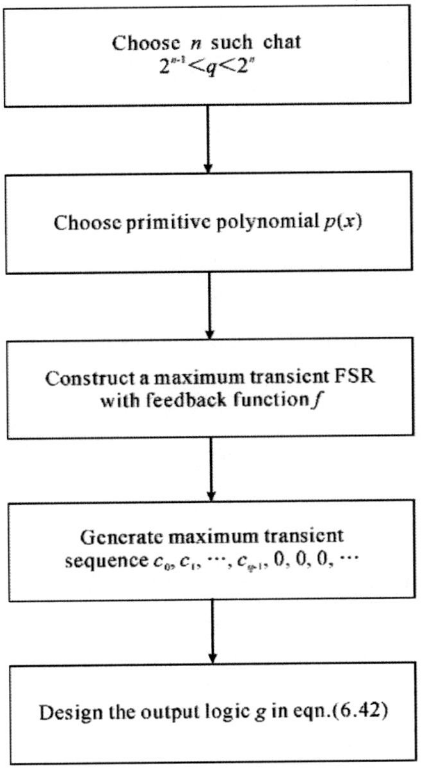

Fig. 6. 7 Flow chart of indirect algorithm

6.5.3 Example

Consider the stable binary sequence of length 15 as follows:

$$1\,0\,0\,0, 0\,0\,0\,0, 0\,0\,0\,0, 0\,0\,1\,0, 0\,0\,0\cdots$$

where

$$d_0 = d_{14} = 1, d_1 = d_2 = \cdots = d_{13} = 0$$

Step 1

Choose $n=4$, as

$$2^3 < 15 < 2^4$$

Step 2

Choose a primitive polynomial of degree 4 over GF(2) as follows:

$$p(x) = 1 + x + x^4$$

Step 3

From Theorem 6.3, design a maximum transient FSR with the feedback function

$$f(x_0, x_1, x_2 x) = x_0 \oplus x_3 \oplus x_0 \overline{x}_1 \overline{x}_2 \overline{x}_3$$

Step 4

The output sequence of the above maximum transient FSR is

$$0\,0\,0\,1, 1\,1\,1\,0, 1\,0\,1\,1, 0\,0\,1\,0, 0\,0\,0\cdots$$

Step 5

As $d_0 = d_{14} = 1$, the corresponding sets $s_0 = (0\ 0\ 0\ 1)$ and $S_{14} = (1\ 0\ 0\ 0)$. From Theorem 6.4, the required sequence may be obtained from the following output mapping:

$$g(x_0, x_1, x_2, x_3) = x_0 \overline{x_1} \overline{x_2} \oplus \overline{x_1} \overline{x_2} \overline{x_3}$$

6.5.4 Discussion

The indirect synthesis of stable binary sequences uses a maximum transient FSR which possesses minimum stage n, so it requires minimum binary memory elements. But, it needs extra external logic to realise the output mapping. This increases the complexity of circuits. The direct synthesis of stable sequences does not need an output mapping, but it usually requires more stages than the indirect method. The optimum synthesis may be obtained by comparing the complexities of output logic and binary memory circuits.

6.6 Conclusion

The direct and indirect synthesis algorithms of stable FSRs were proposed in this chapter. The sufficient and necessary conditions for directly generating a stable binary sequence by means of a stable feedback shift register have been proved. A direct algorithm is presented for synthesis of stable FSRs with minimum stage n. The optimum synthesis may be obtained by certain constrained conditions.

The maximum transient FSR having n stages generates the longest stable sequences of length 2^n and forms a maximum single tree in its state graph. It can be synthesised by breaking the single loop in the state graph of maximum period FSRs, or by connecting and breaking the two pure cycles in the state graph of primitive LFSRs. The maximum transient FSR can be used as a stable generator of field elements. A large class of field element generators have been presented. Any stable sequences of length q ($2^{n-1} < q < 2^n$) may be indirectly generated from a maximum transient FSR of stage n in terms of an output mapping. The algorithm for synthesis of the output mapping has been developed. In general, the stable FSR designed by indirect procedure possesses a smaller number of stages than that designed by a direct algorithm for synthesis of the same stable sequences. But the former needs an external circuit to realise the output mapping. The optimum synthesis depends on the price and complexity of the two methods.

Chapter 7
Linearization and Stabilization of Non-Linear Machines①

7.1 Introduction

Many finite state machines are unstable. When the inputs of machines are zero, they do not result in all-zero outputs. The FSM which represents a digital signal processing system is required to be stable. The stability of systems should be studied. Linear FSMs have been studied extensively[Q5]. If a non-linear FSM can be linearized, then it may be dealt with by using the theory of linear machines. The concept of linearization was proposed by Gill[Q5]. Fukunaga[F4] proposed that a non-linear autonomous machine may be linearized by increasing the number of variables and the z-transform may be applied to the linearized machine. However he did not give the representation of the linearization matrix. Gossel and Poschel[G20] gave necessary and sufficient conditions of linearizability of automata. Herman[G7, G8] and Reush[G15] studied the possibility of linear realization. Recently, Rayner[H10] developed a method for linearizing non-linear machines over GF(q) and gave a similarity transformation of the linearization matrix. But, the discrete Fourier transform (DFT) property of the similarity transformation matrix is not presented. The linearization of the non-linear machines over extension Galois Field GF(2^M) and the properties of this linearization representation are discussed in Section 7.2.

If a FSM is known as unstable, then efficient methods which can improve the machine and make it stable should be proposed. An analytical method for breaking the cycles in state graph and stabilizing an unstable machine is presented in Section 7.3.

7.2 Linearization of Non-Linear Machines

7.2.1 Properties of Linear FSMs

If the output and the next state of a finite state machine are linear functions of present states and inputs, then the machine is called the linear finite state machine. The general forms of linear finite state machines are as follows:

$$\begin{aligned} \underline{s}' &= \underline{As} + \underline{Bu} \\ \underline{g} &= \underline{Cs} + \underline{Du} \end{aligned} \tag{7.1}$$

where $\underline{s}$, $\underline{u}$ and $\underline{g}$ are the states, input and output column vectors with M, N and K dimension respectively. $\underline{A}$, $\underline{B}$, $\underline{C}$ and $\underline{D}$ are $(M \times M)$, $(M \times N)$, $(K \times M)$ and $(K \times N)$ matrices respectively.

The linear FSMs are well understood. The stability of a linear machine may be investigated by considering its autonomous machine

① The contents of Section 7.3 in this chapter have been submitted to IEEE Transactions on Computers, 1984[Z6].

$$\underline{s}' = \underline{As} \tag{7.2}$$

Using a similarity transformation, the matrix $\underline{A}$ can be transformed into its companion matrix form[Q5] which is a block-diagonal matrix, and each block factor is similar to a linear feedback shift register matrix

$$\begin{bmatrix} 0 & 1 & 0 & \cdots & 0 & 0 \\ 0 & 0 & 1 & \cdots & 0 & 0 \\ \vdots & \vdots & \vdots & \ddots & \vdots & \vdots \\ 0 & 0 & 0 & \cdots & 0 & 1 \\ a_0 & a_1 & a_2 & \cdots & a_{n-2} & a_{n-1} \end{bmatrix} \tag{7.3}$$

The linear feedback shift register has a cyclic structure isomorphic to the original machine (7.2). Its stability can be specified by its characteristic polynomial

$$C_A(\lambda) = \det|\lambda \underline{IA}| = \lambda^n - a_{n-1}\lambda^{n-1} - a_{n-2}\lambda^{n-2} - \cdots a_1\lambda - a_0 \tag{7.4}$$

If $C_A(\lambda)$ is primitive over $GF(2^M)$, then the machine is unstable and has a maximum period of length $(2^M - 1)$. A linear finite state machine is stable nilpotent, if and only if its characteristic polynomial $C_A(\lambda)$ is of the form λ^n[F9].

If a non-linear machine may be represented in linear forms, then the properties of linear machines can be used to study the stability of non-linear machines.

7.2.2 Linearization Method

We deal with only the one-dimensional autonomous machine which is a single-variable machine, as shown in Section 3.6, any multiple-variable machines can be converted into a single-variable machine.

Consider the polynomial representation of a single-variable machine over $GF(2^M)$

$$s' = f(s) = \sum_{i=0}^{r} a_i s^i \tag{7.5}$$

where

$$s', s, a_i \in GF(2^M), r = 2^M - 1$$

Since the field $GF(2^M)$ is finite, powers of the next state s' may also be expressed as a polynomial of present states s and in general

$$s'^k = f^k(s) = \sum_{i=0}^{r} a_{ki} s^i \tag{7.6}$$

It should be noted that the coefficients a_{ki} in eqn. (7.6) are functions of the coefficients a_i in eqn. (7.5) and should not be considered as independent variables. Let k and i in eqn. (7.6) be rearranged in the following order:

$$0, r, 1, 2, \cdots, (r-1) \tag{7.7}$$

Then eqn. (7.6) may be written in the matrix form:

$$\begin{bmatrix} 1 \\ s'^r \\ s'^1 \\ s'^2 \\ \vdots \\ s'^{r-1} \end{bmatrix} = \begin{bmatrix} 1 & 0 & 0 & \cdots & 0 \\ a_{r0} & a_{rr} & a_{r1} & \cdots & a_{r(r-1)} \\ a_{10} & a_{1r} & a_{11} & \cdots & a_{1(r-1)} \\ a_{20} & a_{2r} & a_{21} & \cdots & a_{2(r-1)} \\ \vdots & \vdots & \vdots & \ddots & \vdots \\ a_{(r-1)0} & a_{(r-1)r} & a_{(r-1)1} & \cdots & a_{(r-1)(r-1)} \end{bmatrix} \begin{bmatrix} 1 \\ s^r \\ s^1 \\ s^2 \\ \vdots \\ s^{r-1} \end{bmatrix} \tag{7.8}$$

or

$$\underline{s}' = \underline{L}\underline{s} \tag{7.9}$$

Equation (7.9) has the same form as eqn. (7.2) for the linear autonomous machine. It can be viewed as a multi-dimensional linear machine over GF(2^M). However, there are some important differences. The vectors $\underline{s}$ in eqn. (7.2) have elements which are the individual state variables in multiple-dimensional linear machines, whereas the elements of the vectors in the formulation of the non-linear machine are not independent variables. In spite of this, the form of eqn. (7.9) allows the non-linear machine to be studied as a linear machine in an extended state space. $\underline{L}$ is named a linearization matrix. A major disadvantage of this method is that the elements of the linearization matrix $\underline{L}$ have to be calculated for each power of s'. However, this computation may be included in the state transition matrix formulation by the following similarity transformation.

7.2.3 Similarity Transformation of the Linearization Matrix $\underline{L}$

Let the elements of GF(2^M) be denoted by

$$B = 0, 1, \alpha, \alpha^2, \cdots, \alpha^{r-1} \qquad r = 2^M - 1 \tag{7.10}$$

where α is a primitive element of GF(2^M).

If the present state s in eqn. (7.8) takes successively the value of each field element, then

$$[\underline{s}'(0)\underline{s}'(1)\underline{s}'(\alpha)\cdots\underline{s}'(\alpha^{r-1})] = \underline{L}\begin{bmatrix} 1 & 1 & 1 & \cdots & 1 \\ 0 & 1 & 1 & \cdots & 1 \\ 0 & 1 & \alpha & \cdots & \alpha^{r-1} \\ 0 & 1 & \alpha^2 & \cdots & \alpha^{2(r-1)} \\ \vdots & \vdots & \vdots & \ddots & \vdots \\ 0 & 1 & \alpha^{r-1} & \cdots & \alpha^{(r-1)^2} \end{bmatrix} \tag{7.11}$$

where $\underline{s}'(B) = [1\ f^r(B)\ f^1(B) \cdots f^{r-1}(B)]^t$ is the next state column vector corresponding to the present state column vector $\underline{s}$ when $s = B$.

Since $\underline{s}(B) \in$ GF(2^M) which is finite, so that $\underline{s}'$(B) will be identical to one of the columns of the post-multiplying matrix on the right-hand side of eqn. (7.11). Therefore, eqn. (7.11) may be written as

$$\begin{bmatrix} 1 & 1 & 1 & \cdots & 1 \\ 0 & 1 & 1 & \cdots & 1 \\ 0 & 1 & \alpha & \cdots & \alpha^{r-1} \\ 0 & 1 & \alpha^2 & \cdots & \alpha^{2(r-1)} \\ \vdots & \vdots & \vdots & \ddots & \vdots \\ 0 & 1 & \alpha^{r-1} & \cdots & \alpha^{(r-1)^2} \end{bmatrix} [\underline{c}_0 \underline{c}_r \underline{c}_1 \cdots \underline{c}_{r-1}] = \underline{L}\begin{bmatrix} 1 & 1 & 1 & \cdots & 1 \\ 0 & 1 & 1 & \cdots & 1 \\ 0 & 1 & \alpha & \cdots & \alpha^{r-1} \\ 0 & 1 & \alpha^2 & \cdots & \alpha^{2(r-1)} \\ \vdots & \vdots & \vdots & \ddots & \vdots \\ 0 & 1 & \alpha^{r-1} & \cdots & \alpha^{(r-1)^2} \end{bmatrix} \tag{7.12}$$

where $\underline{c}_i$ is a column vector having only one element equal to unity and all other elements equal to zero. The position of the unity element is such that the vector $\underline{c}_i$ selects that vector from the pre-multiplying matrix in eqn. (7.11), which is identical to the corresponding $\underline{s}(B)$. $\underline{c}_L = [\underline{c}_0 \underline{c}_r \underline{c}_1 \cdots \underline{c}_{r-1}]$ is named a connection matrix. Let

$$\underline{p}=\begin{bmatrix} 1 & 1 & 1 & 1 & \cdots & 1 \\ 0 & 1 & 1 & 1 & \cdots & 1 \\ 0 & 1 & \alpha & \alpha^2 & \cdots & \alpha^{r-1} \\ 0 & 1 & \alpha^2 & \alpha^4 & \cdots & \alpha^{2(r-1)} \\ \vdots & \vdots & \vdots & \vdots & \ddots & \vdots \\ 0 & 1 & \alpha^{r-1} & \alpha^{2(r-1)} & \cdots & \alpha^{(r-1)^2} \end{bmatrix} \quad (7.13)$$

From eqn. (7.12), the linearization matrix $\underline{L}$ may be represented as the following similarity transformation form:

$$\underline{L}=\underline{p}\underline{c}_L\underline{p}^{-1} \quad (7.14)$$

Using the Galois field properties in Lemma 3.2 (Chapter 3), the inverse matrix p^{-1} may be obtained as follows:

$$\underline{p}^{-1}=\begin{bmatrix} 1 & 1 & 1 & 1 & \cdots & 1 \\ 0 & 1 & 1 & 1 & \cdots & 1 \\ 0 & 1 & \alpha^{-1} & \alpha^{-2} & \cdots & \alpha^{-(r-1)} \\ 0 & 1 & \alpha^{-2} & \alpha^{-4} & \cdots & \alpha^{-2(r-1)} \\ \vdots & \vdots & \vdots & \vdots & \ddots & \vdots \\ 0 & 1 & \alpha^{-(r-1)} & \alpha^{-2(r-1)} & \cdots & \alpha^{-(r-1)^2} \end{bmatrix} \quad (7.15)$$

From eqns. (7.14), (7.13) and (7.15), a theorem is obtained as follows.

Theorem 7.1

The linearization matrix $\underline{L}$ of a non-linear machine may be expressed as the similarity transformation of its connect matrix $\underline{c}_L$ as follows:

$$\underline{L}=\begin{bmatrix} 1 & 1 & 1 & 1 & \cdots & 1 \\ 0 & 1 & 1 & 1 & \cdots & 1 \\ 0 & 1 & \alpha & \alpha^2 & \cdots & \alpha^{r-1} \\ 0 & 1 & \alpha^2 & \alpha^4 & \cdots & \alpha^{2(r-1)} \\ \vdots & \vdots & \vdots & \vdots & \ddots & \vdots \\ 0 & 1 & \alpha^{r-1} & \alpha^{2(r-1)} & \cdots & \alpha^{(r-1)^2} \end{bmatrix} \underline{c}_L \begin{bmatrix} 1 & 1 & 1 & 1 & \cdots & 1 \\ 0 & 1 & 1 & 1 & \cdots & 1 \\ 0 & 1 & \alpha^{-1} & \alpha^{-2} & \cdots & \alpha^{-(r-1)} \\ 0 & 1 & \alpha^{-2} & \alpha^{-4} & \cdots & \alpha^{-2(r-1)} \\ \vdots & \vdots & \vdots & \vdots & \ddots & \vdots \\ 0 & 1 & \alpha^{-(r-1)} & \alpha^{-2(r-1)} & \cdots & \alpha^{-(r-1)^2} \end{bmatrix} \quad (7.16)$$

Since α is a primitive element of $GF(2^M)$, the similarity transformation matrix $\underline{p}$ contains a DFT of length (2^M-1), and $\underline{p}^{-1}$ contains an inverse DFT.

7.2.4 Important Property of the Connection Matrix $\underline{c}_L$

From the definition of the connection matrix $\underline{c}_L$, it is shown that the state transition direction of a machine is from the position of each diagonal element c_{ii} to the position of the element "one" in the same column $\underline{c}_i$. Using this property, the connection matrix $\underline{c}_L$ can be obtained directly from the state transition table of machines without any computation.

Example 7.1

The state table of an autonomous machine and its field element assignment (as shown in Section 3.2) over $GF(2^3)$ are given in Table 7.1.

Table 7.1 State Table of Example 7.1

s	$s'=f(s)$	s	$s'=f(s)$
0	0	0	0
1	-1	$\alpha^7=1$	α^5
2	-1	α	α^5
3	-2	α^3	α^4
-4	2	α^2	α
-3	2	α^6	α
-2	1	α^4	1
-1	1	α^5	1

This is a non-linear machine. Its next state function is

$$f(s)=\alpha^4 s^7+\alpha s+\alpha s^2+\alpha^2 s^3+\alpha^2 s^4+\alpha^6 s^5+\alpha^2 s^6$$

The connection matrix $\underline{c}_L$ can be obtained from the above property as follows:

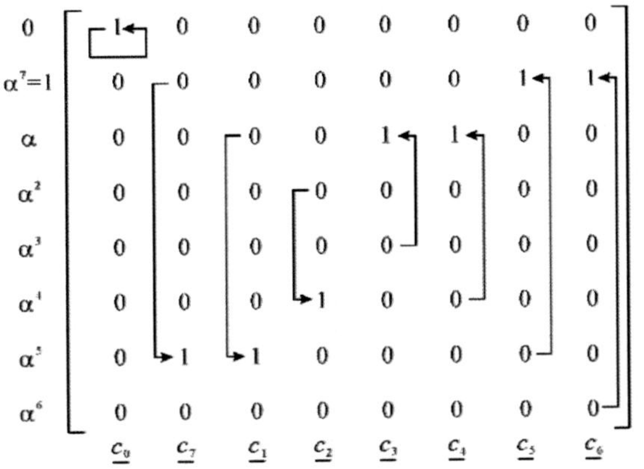

where the array denotes the state transition direction. This property of the connection matrix will be used to detect and locate limit cycles of IIR filters in the next chapter.

The linearization matrix $\underline{L}$ can be obtained from the connection matrix $\underline{c}_L$ in terms of the similarity transformation in eqn. (7.16) and the non-linear may be represented as follows:

$$\begin{bmatrix} 1 \\ s'^7 \\ s' \\ s'^2 \\ s'^3 \\ s'^4 \\ s'^5 \\ s'^6 \end{bmatrix} = \begin{bmatrix} 1 & 0 & 0 & 0 & 0 & 0 & 0 & 0 \\ 0 & 1 & 0 & 0 & 0 & 0 & 0 & 0 \\ 0 & \alpha^4 & \alpha & \alpha & \alpha^2 & \alpha^2 & \alpha^6 & \alpha^2 \\ 0 & \alpha & \alpha^4 & \alpha^4 & \alpha^5 & \alpha^2 & \alpha^4 & \alpha^4 \\ 0 & \alpha^5 & \alpha & \alpha^6 & \alpha^3 & \alpha^3 & \alpha^3 & 1 \\ 0 & \alpha^2 & \alpha^4 & \alpha & \alpha & \alpha^4 & \alpha & \alpha^3 \\ 0 & \alpha^6 & \alpha^3 & \alpha^5 & 1 & \alpha^4 & \alpha^5 & \alpha^5 \\ 0 & \alpha^3 & \alpha^6 & \alpha^2 & \alpha^6 & \alpha^5 & 1 & \alpha^6 \end{bmatrix} \begin{bmatrix} 1 \\ s^7 \\ s \\ s^2 \\ s^3 \\ s^4 \\ s^5 \\ s^6 \end{bmatrix}$$

This may be viewed as a multiple-dimensional linear machine.

Since eqn. (7.16) is a similarity transformation, the characteristic polynomial of the linearization matrix $\underline{L}$ is equal to that of connection matrix $\underline{c}_L$. The stability of a non-linear machine may be specified by its connection matrix.

7.3 Stabilization of FSMs

7.3.1 State Graph of Stable Machines

An autonomous machine is stable if it results in all zeros after a finite state sequence. Let $\underline{N}$ be a next state operator, $\underline{N}^i(\underline{s})$ denote the state after i intervals of time. For a stable autonomous machine at any initial state, an integer q can be found such that

$$\underline{N}^q(\underline{s})=\underline{0} \tag{7.17}$$

$\underline{0}$ denotes zero state.

The state graph of a stable machine is shown in Fig. 7.1.

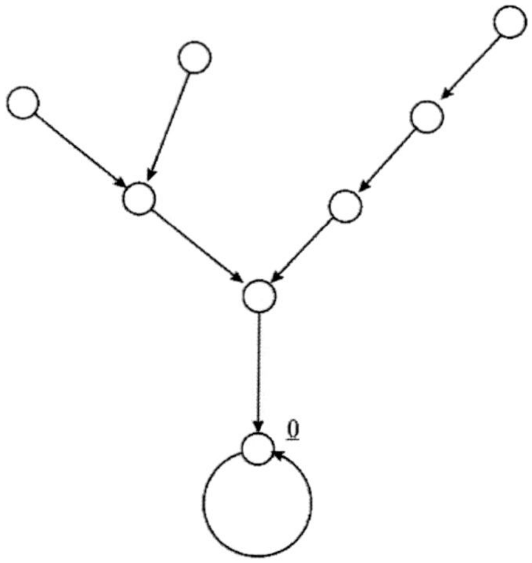

Fig. 7.1 State graph of a stable machine

From Fig. 7.1, it may be shown that the state graph of a stable autonomous machine is a tree. The state $\underline{0}$ forms a cycle, it is named the root of the tree.

For constant input, the state graph of a stable machine is also a tree, but the root is the expected non-zero constant state.

For an unstable machine, its state graph is not a tree and possesses some cycles which are not at the root state.

The stabilization of an unstable machine is a process of breaking the unexpected cycles in the state graph and connecting them as a tree.

7.3.2 Analytical Method for Breaking a Cycle

Fig. 7.2 shows the state graph of some unstable machines. If the state a is on a cycle and its successor can be changed from a' to b', then the cycle can be broken at point A. This can be realised in terms of an analytical method as follows.

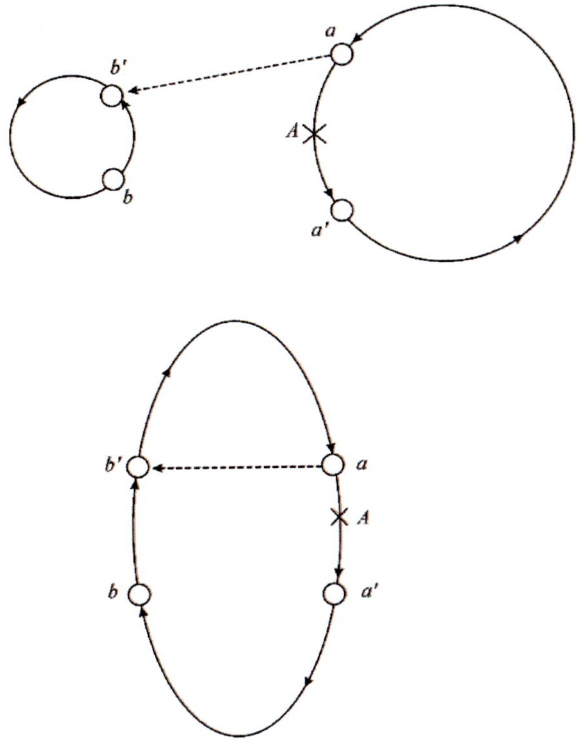

Fig. 7. 2 State graph of some unstable machines

Theorem 7. 2

If $s' = f(s) = \sum_{i=0}^{r} a_i s^i$ is the polynomial representation of an unstable machine over GF(2^M), then the cycle in the state graph can be broken at point A and the state a on the cycle can be connected to another state b' by modifying its next state polynomial as follows:

$$f'(s)=f(s)+[b'-f(a)][1-(s-a)^r] \tag{7.18}$$

where

$$a, b' \text{ and } s\in \mathrm{GF}(2^M),\ r=2^M-1$$

Proof

Using the property of Galois Field in eqn. (3. 2), it follows

$$(s-a)^r=\begin{cases}0 & \text{for } s=a\\ 1 & \text{otherwise}\end{cases} \tag{7.19}$$

It is easy to prove that

$$f'(s)=\begin{cases}b' & \text{for } s=a\\ f(s) & \text{otherwise}\end{cases} \tag{7.20}$$

It means that the function $f'(s)$ has the same state transition graph as that of the function $f(s)$ except that the cycle is broken at the point A, and the state a is connected to the state b'.

Any unstable machine may be stabilised by breaking all the unexpected cycles and connecting them as a tree, using the above analytical method.

For Example 7. 1, its state graph is shown in Fig. 7. 3. This is an unstable machine. The state

α^5 and 1 form a cycle of length 2. If the cycle can be broken at point A and the state 1 can be connected to the state 0 (as shown in Fig. 7.3), then the machine becomes stable. Using Theorem 7.2, this may be realised by modifying its polynomial as follows:

$$f'(s)=f(s)-f(1)[1-(s-1)^7]=\alpha^6 s+\alpha^6 s^2+\alpha^3 s^3+\alpha^3 s^4+\alpha s^5+\alpha^3 s^6+s^7$$

Another important example is the primitive Linear Finite State Machine (LFSM).

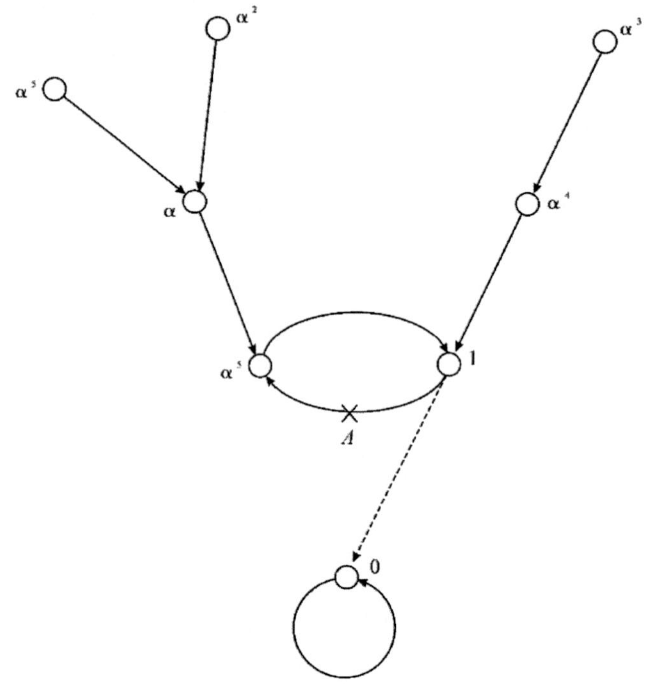

Fig. 7.3 State graph of Example 7.1

7.3.3 Stabilization of Primitive LFSMs

A primitive linear FSM over GF(2^M) is unstable. As shown in Fig. 7.4, its state graph consists

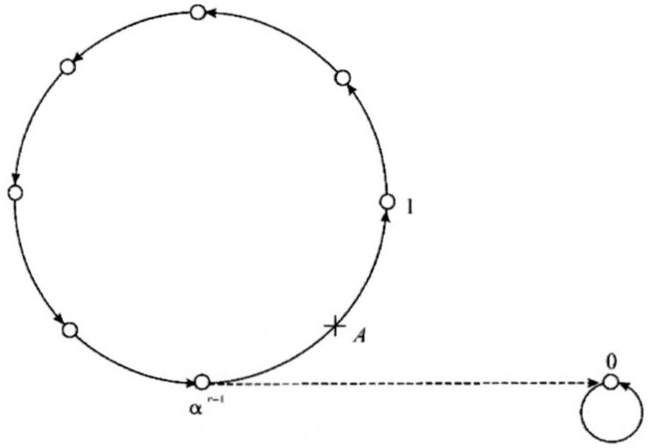

Fig. 7.4 Primitive LFSMs

of two cycles: all the non-zero states form a large cycle of length (2^M-1) and the zero state forms a small cycle of length 1 which is the expected root state. From any non-zero states, the outputs never reach the zero state.

If the large cycle can be broken at any point, and connected to the root state, then the machine is stabilized.

From eqn. (3.9), the primitive LFSM over GF(2^M) may be represented by

$$s'=\alpha^n s, \qquad (n,r)=1, r=2^M-1 \tag{7.21}$$

where α is a primitive element of GF(2^M).

Let $s'=\alpha s$, then the large cycle may be broken at point A (Fig. 7.4) and the state α^{r-1} may be connected to the root state 0 by using eqn. (7.18) as follows:

$$s'=\alpha s+[1-(\alpha^{r-1}-s)^r] \tag{7.22}$$

This is a stable non-linear machine. Its state graph forms a maximum single tree. The maximum single tree may be used as a stable generator of field elements. For example, the primitive LFSM $s'=\alpha s$ over GF(2^3) may be stabilized by

$$s'=\alpha s+[1-(\alpha^6-s)^7]$$

where α denotes the primitive element of GF(2^3). Its state graph is a maximum tree of length 8.

7.3.4 Formation of a Root Cycle

The state graph of a stable machine is a tree. Its root state should form a cycle of length 1. For some unstable machines, the expected root state does not form a cycle of length 1. See the machine in Fig. 7.2, if the state b is the expected root state, obviously, it does not form a cycle of length 1. The cycle of length 1 at the state b can be formed by the following theorem.

Theorem 7.3

If $s' = f(s) = \sum_{i=0}^{r} a_i s^i$ is the polynomial representation of the machine over GF(2^M), then the cycle of length 1 at the state b can be formed by modifying the polynomial as follows:

$$f_0(s)=(s-b)^r[f(s)-b]+b \tag{7.23}$$

where

$$s,b\in \mathrm{GF}(2^M), \qquad r=2^M-1$$

Proof

Using the property of Galois Field in eqn. (3.2), it follows

$$(s-b)^r=\begin{cases}0 & \text{for } s=b\\ 1 & \text{otherwise}\end{cases} \tag{7.24}$$

Equation (7.23) becomes

$$f_0(s)=\begin{cases}b & \text{for } s=b\\ f(s) & \text{otherwise}\end{cases} \tag{7.25}$$

Since $f_0(b)=b$, the state b is connected to itself and forms a cycle of length 1. It may be used as a root state.

An important example is the maximum period machine whose state graph is a single cycle with the maximum length 2^M. If the polynomial representation of a maximum period machine is

$$s' = f(s) = \sum_{i=0}^{r} a_i s^i \tag{7.26}$$

where

$$s', s, a_i \in GF(2^M), r = 2^M - 1$$

and the state $\underline{0}$ is the expected root state. From eqn. (7.23) a cycle of length 1 at the state $\underline{0}$ can be formed by

$$f_0(s) = s^r f(s) = \sum_{i=1}^{r-1} a_i s^i + (a_0 + a_r) s^r \qquad (7.27)$$

The maximum cycle is broken at point B (Fig. 7.5) and a root is formed at the state $\underline{0}$. The maximum period machine is changed into a maximum single tree.

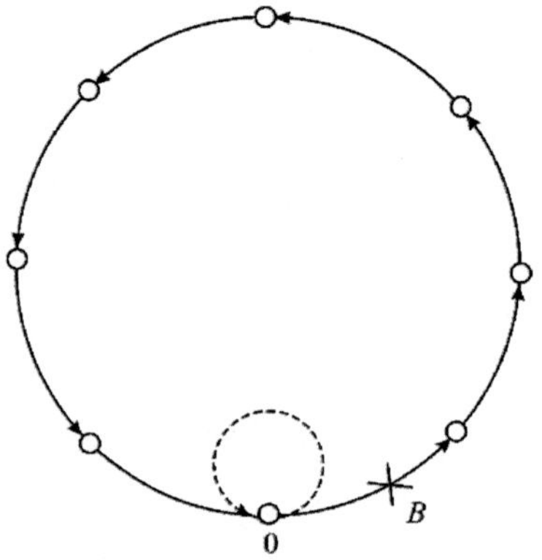

Fig. 7.5 Maximum period machines

7.3.5 Generation of Maximum Period Machines

Some maximum period machines may be obtained from the other machines which have two or more pure cycles.

As shown in Fig 7.2, using the analytical method of Section 7.3.2, the state b can be connected to the state a'. Two cycles can be connected as a large cycle and vice versa in terms of the following equation:

$$f'(s) = f(s) + (a' + b')[(s-a)^r - (s-b)^r] \qquad (7.28)$$

where

$$a' = f(a) \qquad b' = f(b)$$

For example, the primitive linear FSM $s' = \alpha s$ in GF(2^M) may be merged as a maximum period machine in terms of the following equation:

$$g(s) = \alpha s + [(s - \alpha^{r-1})^r - s^r] \qquad (7.29)$$

where

$$s \in GF(2^M), \qquad f = 2^M - 1$$

α is a primitive root of GF(2^M).

Its state graph is shown in Fig. 7.6.

In Fig. 7.6, the state α^{r-1} is connected to the state 0, and the state 0 is connected to the state 1. Its state graph forms a single cycle which contains all the possible states and has a period of 2^M. The maximum period machine may be used as a cyclic generator of field elements.

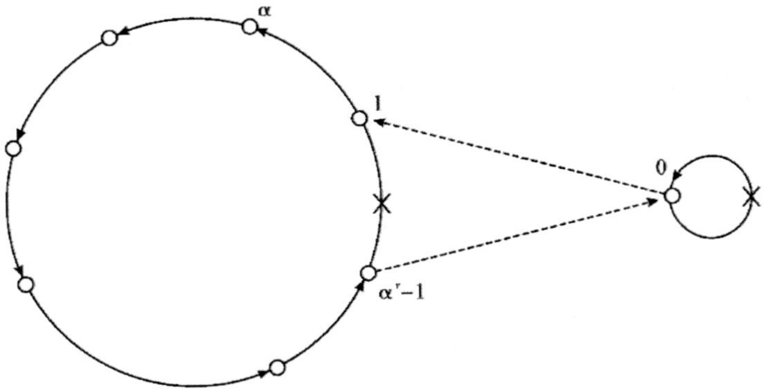

Fig. 7.6 Formation of maximum period machines

For example, if α is a primitive element of GF(2^3), then

$$f(s)=\alpha s+[(s-\alpha^6)^7-s^7]$$

forms a maximum period machine of length 8 and can be used as a cyclic generator of the elements over GF(2^3).

7.4 Conclusion

The methods for linearization and stabilization of non-linear FSMs have been presented in this chapter. Any non-linear machine over GF(2^M) may be represented as a linear machine by increasing the dimensionality. The linearization matrix $\underline{L}$ may be obtained by using a similarity transformation of the connection matrix $\underline{c}_L$. This similarity transformation possesses DFT properties. The connection matrix $\underline{c}_L$ represents the state transition direction and can be obtained directly from the state table. The stability of a non-linear machine may be studied by means of its linearization representation.

An analytical method has been developed for breaking the cycles in unstable machines. This method can be used to remove the cycles and stabilize an unstable machine. A maximum single tree may be obtained from a primitive linear machine by using this method. The method will be used to remove the limit cycles of IIR digital filters in the next chapter.

Another analytical method for forming a cycle of length 1 in a machine has also been presented. This method can be used to form a root cycle in the machine. The above two methods can be used to break the cycles and form a root in the state graph of a machine, and the state graph can be "modified" as a tree. The maximum single tree can be obtained from a maximum cycle, and the maximum cycle may be formed from two or more pure cycles in terms of the above analytical methods.

Chapter 8
Stable Realization of IIR Digital Filters

8.1 Introduction

An infinite impulse response digital filter may be represented as a finite state machine in the form of state tables or polynomials. Sometimes, the system is unstable. This dynamic behaviours of a system can be clearly shown in terms of its state graph.

Example 8.1

A first order IIR digital filter is given by

$$w(n) = -0.8w(n-1) + x(n)$$

Its state table for input $x(n)=1$ is shown in Table 8.1.

Table 8.1 State table of Example 8.1

$\overline{w}(n-1)$	$w(n)$	$\overline{w}(n)$	s	$f(s,1)$
0	1.0	1	0	1
1	0.2	0	1	0
2	−0.6	−1	α	α^5
3	−1.4	−1	α^3	α^5
−4	4.2	3	α^2	α^3
−3	3.4	3	α^6	α^3
−2	2.6	3	α^4	α^3
−1	1.8	2	α^5	1

where α is a primitive element of $GF(2^3)$. The polynomial representation of state table 8.1 is

$$s' = f(s,1) = 1+\alpha^5 s+\alpha^4 s^2+\alpha^4 s^3+\alpha^4 s^3+\alpha^4 s^4+s^5+\alpha s^6+\alpha^3 s^7$$

where

$$s', s \in GF(2^3)$$

The state graph of the machine for input 1 is shown in Fig. 8.1.

Figure 8.1 shows that the system is unstable. The system cannot reach a stable output from any initial state. This behaviour is a constant input limit cycle oscillation (LCO). Example 8.1 exhibits two limit cycles: a limit cycle of period 2 and amplitude −1 to 2, another one of period 2 and amplitude 0 to 1. For a high order filter, limit cycles of period greater than 2 may be exhibited.

FSMs may be efficiently used to analyse the existence and properties of limit cycles. The state graph can clearly show the amplitude and period of limit cycles. However, the state graph of a complex system is too large to show. Other methods should be developed for detection and location of limit cycles in the system. These methods are presented in Section 8.2. Section 8.3 presents the method for determining the root state. For practical purposes, the FSM realization of IIR filters

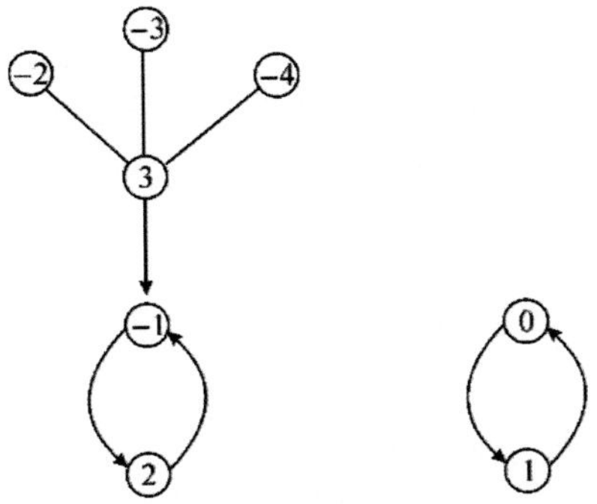

Fig. 8. 1 State graph of Example 8. 1

should be stable For any constant inputs, the system should result in the required stable output. All the limit cycles should be removed from the system with minimum increase in noise.

Rayner[H10] has described a method, developed by Proudler, for the optimal breaking of limit cycles in terms of minimum spanning tree algorithms[Q21]. The optimum method is discussed in Section 8. 4 and the analytical method is applied to obtain the polynomial for the stabilization machine.

8. 2 Detection and Location of Limit Cycles

8. 2. 1 Cutting of Branches

Limit cycles of IIR digital filters may be shown as cycles in the state graph of machines. The states which are not on the cycles form a number of branches. Each state on a cycle has at least one predecessor. However, one or more states on a branch have no predecessors. These states are called the top of branches.

If all the states which are in branches can be removed from the state graph or state table, then the remaining states show the behaviour of limit cycles. This may be realised by cutting off all the states at the tops of branches. These states have no predecessors and are not on any cycles. After cutting these tops of branches, a new graph is obtained which may have some new tops of branches. Continue cutting the new tops of branches, until all the states in branches are cut off. Finally, only the states on cycles are left. This cutting can be realised in terms of the connection matrix.

8. 2. 2 Connection Matrix Methods

The connection matrix is an $N \times N$ matrix, where $N = 2^M$ is total number of states of FSM. Its elements consist of "0" and "1". Each column of the connection matrix has only a single "1", all other elements are zeros. In each column, the edge from the position of the diagonal element to the position of element "1" represents the state transition direction.

For Example 8. 1, its connection matrix is

$$\begin{array}{r} 0 \\ 1 \\ 2 \\ 3 \\ -4 \\ -3 \\ -2 \\ -1 \end{array}\begin{bmatrix} 0 & 1 & 0 & 0 & 0 & 0 & 0 & 0 \\ 1 & 0 & 0 & 0 & 0 & 0 & 0 & 0 \\ 0 & 0 & 0 & 0 & 0 & 0 & 0 & 1 \\ 0 & 0 & 0 & 0 & 1 & 1 & 1 & 0 \\ 0 & 0 & 0 & 0 & 0 & 0 & 0 & 0 \\ 0 & 0 & 0 & 0 & 0 & 0 & 0 & 0 \\ 0 & 0 & 0 & 0 & 0 & 0 & 0 & 0 \\ 0 & 0 & 1 & 1 & 0 & 0 & 0 & 0 \end{bmatrix}$$

It is easy to find that each column of the matrix has only a single "one", but some rows have no "one". For the above example, there is no "one" in rows -4, -3 and -2. This means that the states -4, -3 and -2 have no predecessors in the state graph. They are on the tops of some branches and not on any cycles. For the purpose of detecting limit cycles, the states -4, -3 and -2 may be cut off from the state graph without any effect on determining the amplitude and period of limit cycles. This procedure may be realised by deleting the rows -4, -3 and -2 and the columns -4, -3 and -2 on computers. Continue the above procedure and delete the rows which have no "1" element and the corresponding columns in the remaining matrix until every row of the remaining matrix has a single "1". Finally, only the states which form pure cycles are left. This method can be described by the following steps.

Step 1

If there is no "1" in the jth row, delete the jth row and the jth column from the connection matrix. For Example 8.1, there is no "1" in the rows -4, -3 and -2. Delete the columns and rows -4, -3 and -2 from the connection matrix. A 5×5 matrix is left as follows.

$$\begin{array}{r} 0 \\ 1 \\ 2 \\ 3 \\ -1 \end{array}\begin{bmatrix} 0 & 1 & 0 & 0 & 0 \\ 1 & 0 & 0 & 0 & 0 \\ 0 & 0 & 0 & 0 & 1 \\ 0 & 0 & 0 & 0 & 0 \\ 0 & 0 & 1 & 1 & 0 \end{bmatrix}$$

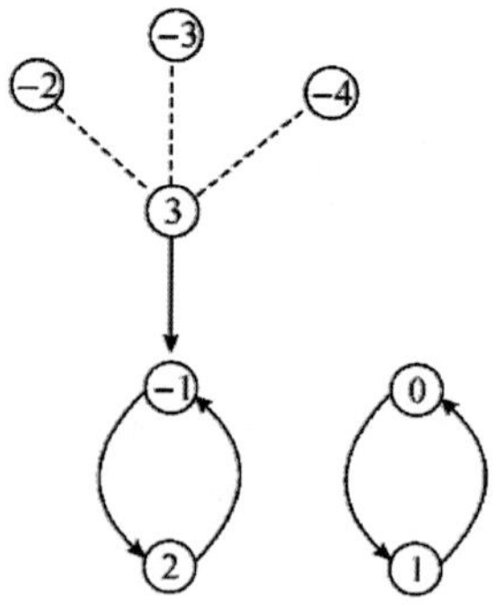

Fig. 8. 2 State graph of Step 1

Step 2

Check the new matrix, delete the rows having no "1", and the corresponding columns. For the above example, in the new matrix there is no "1" in row 3. Delete the row 3 and the column 3. A 4×4 matrix remains:

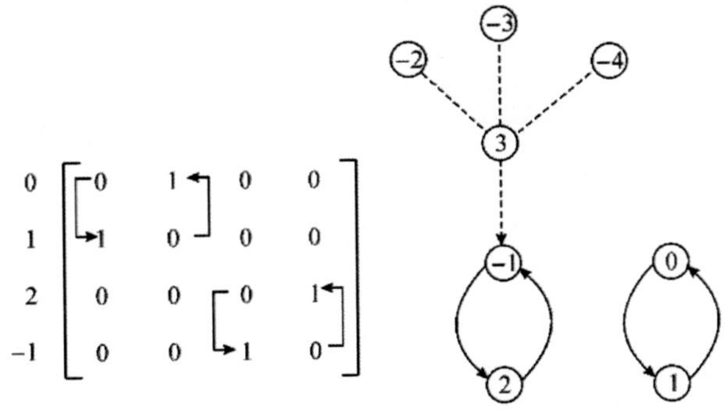

Fig. 8. 3 State graph of Step 2

Step 3

Continue the above steps until every row has a "1" in the matrix. This matrix represents pure cycle state in the state graph.

Step 4

The final matrix may be expressed as a block-diagonal matrix form by rearranging the order of states. Each submatrix block denotes a limit cycle. The number of blocks is equal to the number of limit cycles. The dimension number of each submatrix denotes the period of each limit cycle. The amplitude of limit cycles is determined by the state assignments on a cycle. For the above example, the final matrix may be represented as a block-diagonal matrix with two 2×2 submatrices; the system has two limit cycles of period 2. From the state values on the cycles, it is shown that the amplitudes of the two limit cycles are respectively 0 to 1 and -1 to 2.

The above method is called the connection matrix method and is easily programmed on computers. The connection matrix is a $(2^M \times 2^M)$ matrix. However, it possesses only 2^M ones, the other elements of the matrix are zeros. Therefore, only 2^M memory locations are needed for this method.

8.3 Determination of the Root State

The state graph of a stable machine is a tree. The root of the tree represents the stable outputs of an IIR digital filter for constant inputs. The stable outputs of IIR filters for zero inputs should be zeros. Therefore, the root state of stable autonomous machines must be the zero state which forms a single cycle of length 1 in the tree graph.

The stable outputs of IIR filters for constant inputs are constant. The stable output value can be obtained from the filter's difference equation. If the stable input of an all-pole filter is x_c, then its stable output w_c may be calculated from eqn. (2. 2) as follows:

$$w_c = \sum_{i=1}^{m} k_i w_c + x_c \tag{8.1}$$

or

$$w_c = \frac{1}{1 - \sum_{i=1}^{m} k_i} x_c \tag{8.2}$$

For Example 8.1, the stable output of the first order filter for the constant input 1 is

$$w_c=\frac{1}{1+0.8}=0.5556\doteq 1$$

The root state of the stable machine for inputs {1} should be the state "1".

As shown in Fig. 8.1, the state "1" does not form a cycle of length 1 in the state graph. A cycle of length 1 at the state "1" can be formed by the polynomial method shown in Section 7.3.4. The polynomial representation of Example 8.1 is

$$f(s,1)=1+\alpha^5 s+\alpha^4 s^2+\alpha^4 s^3+\alpha^4 s^4+s^5+\alpha s^6+\alpha^3 s^7$$

From Theorem 7.3, a cycle of length 1 at the state "1" can be formed and the cycle (0,1) can be broken at point B (see Fig. 8.4) by the following polynomial

$$f_0(s,1)=(s-1)^7[f(s,1)-1]+1=1+\alpha^4 s+\alpha^5 s^2+\alpha^5 s^3+\alpha^5 s^4+\alpha^3 s^6+\alpha s^7$$

Its state graph is shown in Fig. 8.4.

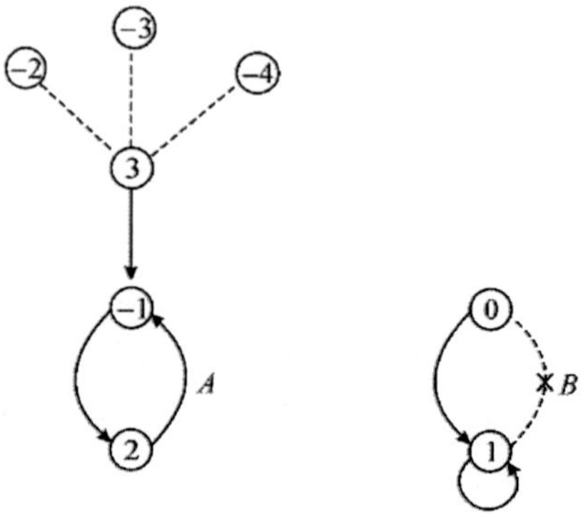

Fig. 8.4 State graph of $f_0(s,1)$

8.4 Removal of Limit Cycles

If the limit cycles of an IIR filter have been located and the root cycle has been formed, then the limit cycles can be broken and a stable system with a tree graph can be obtained in terms of the analytical methods in Section 7.3. For the above example, if the limit cycle (−1,2) can be broken at point A and the state "2" is connected to the state "0", then a tree graph may be formed.

However, as shown in Fig. 8.4, the state "2" can also be connected to the state "1". Another tree graph may be formed as well. Which is the best successor of the state "2"? This will be discussed in Section 8.4.1. On the other hand, there are two states "−1" and "2" on the cycle. The other state "−1" can also be used as a predecessor such that the cycle can be broken at another edge and connected to the root. Which state on a cycle is the best predecessor for breaking the cycle? This will be discussed in Section 8.4.2.

8.4.1 Sub-optimum Successor

A limit cycle of IIR digital filters can be broken at one edge of the cycle and the predecessor state of the edge may be connected to another successor state which is not on the cycle. This method causes an increase in the errors of system representations.

The error from the predecessor to the successor in the FSM representation of IIR filters may be defined by

$$e=[w(n)-s']^2 \tag{8.3}$$

where $w(n)$ is the infinite accuracy output of the system calculated from the filter difference equation (2.2), s' denotes the assignment of the given successor state.

The optimum successor of a state is the state which has the minimum error among all the possible states of machines. The next state obtaining from the quantization equation (2.4) is an optimum successor. The optimum successor of each state of a cycle is also on the cycle, and the error can be calculated by

$$e_0=[w(n)-\overline{w}(n)]^2 \tag{8.4}$$

where $\overline{w}(n)$ denotes the quantization value of $w(n)$ in eqn. (2.4).

Breaking of a limit cycle requires reconnection of a state of the cycle to another state which is not on the cycle. A number of states in a machine are not on the cycle. One of them has the minimum error from the state of the cycle. This one is named a sub-optimum successor of that state of the cycle. "Sub-optimum" means that this successor of the state of the cycle possesses the minimum error among all the successors which are not on the cycle, but its error is still larger than that of the optimum successor which is on the cycle.

For the above example, if the state "2" on the cycle is considered as a predecessor, then all its possible successors and errors are shown in Fig. 8.5.

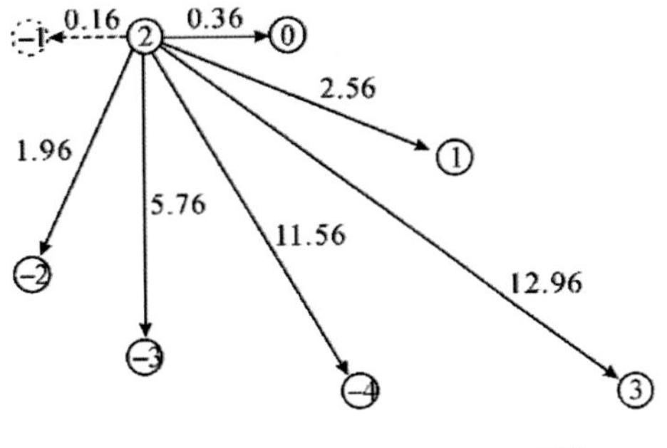

Fig. 8.5 Successors of the state "2"

In Fig. 8.5, the error is allocated to each edge. The error values may be calculated in terms of eqn. (8.3). For example, the system output $w(n)$ from $w(n-1)=2$ is calculated by the difference equation (for $x_c=1.0$)

$$w(n)=-0.8w(n-1)+1.0=-0.6$$

The error e in making the state transition "2" to "1" is:

$$e=(-0.6-1)^2=2.56$$

Fig. 8.5 shows that the state "−1" is the optimum successor which has an error 0.16 and is one of the states on the cycle (−1, 2). The state "0" is the sub-optimum successor which has the minimum error 0.36 among the successors not on the cycle.

8.4.2 Optimum Breaking of a Limit Cycle

As shown in the above example, there are two states: the states "−1" and "2" on the cycle. Each state of the cycle has its sub-optimum successor. Any state of the cycle may be considered as a predecessor for breaking the cycle and can be connected to its own sub-optimum successor. In general, a limit cycle possesses n states. Therefore, there are n distinct points for breaking a limit cycle.

However, the breaking of a cycle at a distinct point may cause distinct error. If the state a is on a cycle (Fig. 8.6), the state a' is its optimum successor (on the same cycle) with error e_0 and the state b is its sub-optimum successor (not on the same cycle) with error e_s, then breaking the cycle at point A and connecting the state a to its sub-optimum successor b may cause an increase in errors as follows:

$$\Delta e = e_s - e_0 \tag{8.5}$$

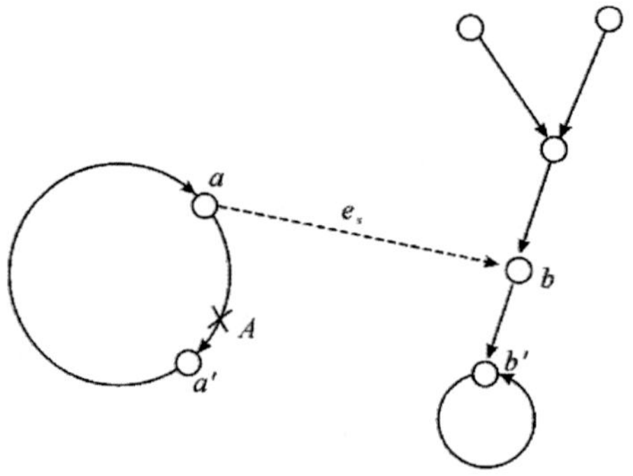

Fig. 8.6 Error increase for breaking a cycle

The distinct predecessor on the cycle for breaking the cycle causes distinct increase in errors. The optimum breaking of a limit cycle can be obtained by minimum error increase in eqn. (8.5).

For Example 8.1, there are two states "−1" and "2" on the cycle. For the state "2", the state "0" is its sub-optimum successor with an error of 0.36. The state "−1" is its optimum successor with an error of 0.16. If the state "2" is connected to its sub-optimum successor "0" and the cycle is broken at point A as shown in Fig. 8.7, then the increase in errors may be obtained from eqn. (8.5) as follows:

$$\Delta e = 0.36 - 0.16 = 0.20$$

The state "−1" is another state on the cycle. Similarly, the state "1" is its sub-optimum successor with an error of 0.64, and the state "2" is its optimum successor with an error of 0.04. If the state "−1" is connected to its suboptimum successor "1" and the cycle is broken, then an additional error is introduced as follows:

$$\Delta e = 0.64 - 0.04 = 0.60$$

Compare the increases in errors between these two breaks. It is obvious that breaking at point A is the optimum breaking of the limit cycle, only an increase in errors of 0.20 is added.

From Theorem 7.2, the state "2" can be connected to the state "0", and the cycle can be broken at point A by the following polynomial

$$f'(s)=f_0(s)+[0-\alpha^5][1-(s-\alpha)^7]=1+\alpha^2s^2+\alpha^3s^3+\alpha^6s^4+s^5+\alpha^4s^6+\alpha^6s^7$$

where $s\in GF(2^M)$, α is a primitive element of $GF(2^3)$.

The above polynomial represents a stable machine which realises the IIR filter $w(n)=-0.8w(n-1)+x(n)$ for input 1.0.

8.4.3 Discussion

The optimum breaking of a limit cycle is shown in the above subsection. A state of the cycle can be connected to its sub-successor with the minimum increase in errors and the cycle is broken. However, this connection may generate another new cycle in the state graph. Therefore, the detection and breaking of limit cycles in an IIR digital filter should be repeatedly carried out until all the cycles, old and new, are removed.

In eqn. (8.3), the square error is used as an error function. Of course, other error functions can also be used.

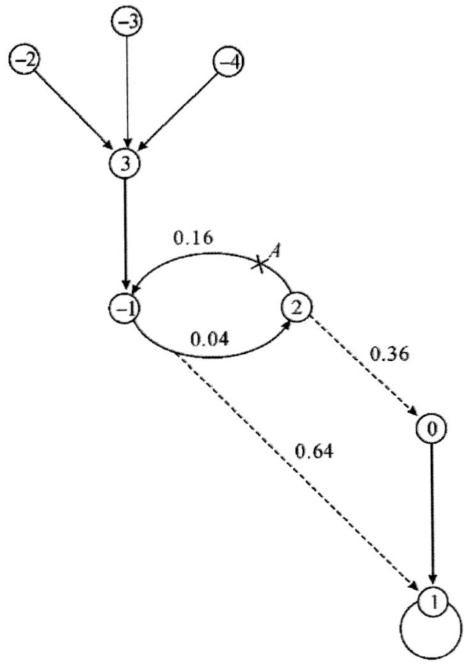

Fig. 8.7 Optimum breaking of limit cycles

8.5 Conclusion

A connection matrix method is presented for detecting and locating the limit cycles in FSM representation of IIR digital filters. The period and amplitude of limit cycles can be determined by using this method on computers.

A limit cycle can be broken by connecting a state of the cycle to the sub-optimum successor of the state. This causes an increase in errors. The optimum breaking of a limit cycle may be obtained by minimizing this error.

The optimum method can remove all the limit cycles with the minimum increase in errors. A stable machine whose state graph is a tree can be obtained. The removal of limit cycles can be carried out either on the state table by re-locating the states or in the polynomial by modifying the coefficients. This is unlikely to increase the complexity of system implementations.

Chapter 9
Conclusions and Suggestions for Further Research

9.1 Conclusions

A new and more general realization of digital signal processing systems—finite state machine realization — has been studied in this dissertation. A FSM can more accurately approximate an infinite precision DSP system, and the problems caused by finite word length in DSP systems can be efficiently solved by this realization (e. g. limit cycle oscillations can be eliminated with minimum increase in noise of systems). The FSM has been analytically represented and efficiently minimized. It can be implemented in terms of feedback shift registers or other hardware, e. g. VLSI, ROM, PLA and microprocessors, etc. They are very fast and programmable.

The work containing in this dissertation may be separated into three parts: system representation, system realization and system stability.

(a) System representation

It includes the FSM representation of DSP systems and the finite field representation of FSMs.

A modified canonic form of IIR filters has been presented, so that an n-order IIR filter can be represented as an n-dimensional Mealy Machine by quantizing its infinite accuracy outputs. This representation has only $\pm 1/2$ bit error from the ideal system. The dynamic range and word length choice has been discussed. The DFT of prime length can be represented as a two-dimensional Moore Machine with an input permutation. The coefficients with optimum S/N ratio have been presented.

An analytical method has been presented for transforming a state table of FSMs into a polynomial over $GF(2^M)$. A DFT of length (2^M-1) can be used to compute the coefficients of polynomials by rearranging the ordering of inputs and outputs. A least common extension field has been introduced, so that the state table whose inputs and outputs have different word lengths can be represented as an extension field polynomial. The cyclic subgroup property of these polynomials has been presented and can considerably reduce the size of memory and the amount of computation. A one-dimensional FSM can be represented as a two-variable polynomial by a two-dimensional DFT and 2 one-dimensional DFTs. An efficient method has been developed for converting a multiple-variable polynomial into a single-variable one and vice versa. All the above Galois Field polynomials can also be used to represent other logical circuits.

(b) System realization

A new fast algorithm for computing the DFT of length (2^M-1) has been developed by using subgroup cyclic convolutions. The properties of subgroup cyclic convolutions have been presented. Most of the subgroups have the same length which can considerably simplify the program structures. This algorithm can be employed in determining the polynomial coefficients over GF(2^M), or computing the conventional DFT over the complex field.

The main problem of system realizations is minimization. An efficient algorithm has been developed for minimization of Exclusive-OR functions. This algorithm is $n/[1+(n-1)\ 2^{-n}]$ times faster than the fast Reed-Muller transformation. With regard to the common gates, the minimization of multiple-output functions has been presented. A method for determining the common gates has been developed in terms of certain logical operations among the multiple-output function coefficients. An efficient algorithm has been developed for minimization of the extension field polynomials over GF(2^M). The average number of multiplications for this algorithm is reduced to $M * 2^{M-2}$. All these algorithms can be easily programmed on computers and can also be used in computer-aided logic circuit design.

Another method for realization of FSMs is feedback shift register realization. The direct and indirect methods for synthesis of stable FSRs have been proposed. The sufficient and necessary conditions for direct synthesis of FSRs have been proved. A direct synthesis algorithm was presented. The synthesis of maximum transient FSRs has been developed. A large class of maximum transient FSRs was presented. They can be used as stable generators of field elements in indirect synthesis. The indirect synthesis algorithm which generates minimum stage FSRs has been presented. The above algorithms and results can also be used in code theory.

(c) System stability

A linearization method has been introduced, so that the stability of non-linear machines can be investigated by means of linear system theory. The linearization matrix can be represented as a similarity transformation of the connection matrix. The DFT property of the similarity transformation matrix over GF(2^M) has been presented by arranging the ordering of field elements. The stabilization methods of an unstable machine have been developed. The analytical methods were presented for breaking cycles, forming a root, and connecting all the branches as a tree.

Theory and methods of the stability of FSMs can be used in IIR digital filters. A connection matrix method has been proposed to employ in detecting and locating limit cycles. The polynomial methods have been used to remove limit cycles with minimum increase in noise of systems.

9.2 Suggestions for Further Research

There are a lot of subjects in this field for further research. The FSM realization of DSP systems is a complex problem. The main problem is complexity. The minimization of fixed-polarity Reed-Muller polynomials have been successfully studied in Chapter 5. The mixed-polarity polynomial may give fewer terms than the fixed-polarity one[D1, D14]. However, the minimization of mixed-polarity Reed-Muller polynomials has long been, and still is, and open problem.

Feedback shift registers are the simplest realization of FSMs and have been systematically studied in Chapter 6. Another approach to simplication realization of FSMs is to consider factorisation of FSMs. A complex machine may be factorised as a number of small and simple machines in the forms of cascade, parallel or their combinations. This factorisation may lead to a simplification realization of machines.

An IIR filter has been represented as an FSM in the forms of state tables from the filter's difference equation (Chapter 2). However, the coefficients of the difference equations are determined by the filter's specifications in frequency or time domain. A method for obtaining the FSM representation directly from the filters specifications should be studied. The Finite Field has been successfully used to represent a FSM (Chapter 3) and to study the stability of machines (Chapter 7, 8). Another finite mathematical structure is the finite Ring. It has been used to study a linear FSM and its stability[G9, G14]. The finite ring representation of non-linear machines should be an object of further research

BIBLIOGRAPHY

SECTION (I): PAPERS

(A) FAST FOURIER TRANSFORM ALGORITHM

[A1] Cooley, J. W. and Tukey, J. W., "An algorithm for machine calculation of complex Fourier series", Math. Comput., Vol. 19, No. 90, April 1965, pp. 297-301

[A2] Rader, C. M., "Discrete Fourier transform when the number of data samples is a prime", Proc. IEEE, Vol. 56, No. 6, June 1968, pp. 1107-1108

[A3] Singleton, R. C., "An algorithm for computing the mixed-radix fast Fourier transform", IEEE Trans. Audio Electroacoustics, Vol. AU-17, June 1969, pp. 93-103

[A4] Good, I. J., "The relationship between the two fast Fourier transforms", IEEE Trans. on Computers, Vol. C-20, No,3, March 1971, pp. 310-317

[A5] Pollard, J. M., "The fast Fourier transform in a finite field", Math. Comput., Vol. 15, No. 114, April 1971, pp. 365-374

[A6] Winograd, S., "On computing the discrete Fourier transform", Proc. Acad. Sci., U. S. A., Vol. 73 (Mathematics), No. 4, April 1976, pp. 1005-1006

[A7] Rader, C. M. and Brenner, N. M., "A new principle for the fast Fourier transformation", IEEE Trans., Vol. ASSP-24, No. 3, June 1976, pp. 264-266

[A8] Winograd, S., "On computing the discrete Fourier transform", Math. Comput., Vol. 32, No. 114, January 1978, pp. 175-179

[A9] Silveman, H. S., "An introduction to programming the Winograd Fourier transform algorithm", IEEE Trans., Vol. ASSP-25, No. 2, April 1977, pp. 152-165

[A10] Burrus, C. S., "Index mappings for multidimensional formulation of DFT and convolutions" IEEE Trans., Vol. ASSP-25, No. 3,June 1977, pp. 239-242

[A11] Kolba, D. P. and Parks, T. W., "A prime factor FFT algorithm using high-speed convolution", IEEE Trans., Vol. ASSP-25, No. 4, August 1977, pp. 281-294

[A12] Winograd, S., "Some bilinear forms whose multiplicative complexity depends on the fields of constants", Mathematical Systems Theory, Vol. 10, 1977, pp. 169-180

[A13] Dubois, E. and Venetsanopoulos, A. N., "A new algorithm for radix 3 FFT", IEEE Trans., Vol. ASSP-26, No. 3, June 1978, pp. 222-225

[A14] Nussbaumert H. J. and Quandalle, P., "Fast computation of discrete Fourier transforms using polynomial transforms", IEEE Trans., Vol. ASSP-27, No. 2, April 1979, pp. 69-181

[A15] Arambepola, B. and Raynerf P. J. W., "Multidimensional fast Fourier transform algorithm", Electronics Letters, Vol. 15, No. 13, June 1979, pp. 382-383

[A16] Arambepola, B., "Fast computation of multidimensional discrete Fourier transforms", Proc. IEE, Vol. 127, Pt. F., No. 1, February 1980, pp. 49-53

[A17] Burrus, C. S. and Eschenbacher, P. W., "An in-place, in-order prime factor FFT algorithm", IEEE Trans., Vol. ASSP-29, No. 4, August 1981, pp. 806-817

[A18] Curts, T. E. and Wickenden, J. T., "Hardware-based Fourier transform: algorithm and architectures", IEE Proceedings, Vol. 130, Pt. F, No. 4, August 1983, pp. 423-432

(B) CONVOLUTION ALGORITHM AND POLYNOMIAL TRANSFORM

[B1] Rayner, P. J. W., "A fast cyclic convolution algorithm", presented at the Symp. Digital Filtering, Imperial College, London, August 1977

[B2] Terrel, P. M. and Rayner, P. J. W., "Improved algorithm for low-order digital circular convolution", Electronics Lett., Vol. 10, No. 11, May 1974, pp. 224-225

[B3] Agarwalf R. C. and Burrus, C. S., "Fast one-dimensional digital convolution by multidimensional techniques", IEEE Trans., Vol. ASSP-22, No. 1, February 1974, pp. 1-10

[B4] Reed, I. S. and Truong, T. K., "The use of finite fields to compute convolutions", IEEE Trans., Vol. IT-21, No. 2, March 1975, pp. 208-213

[B5] Agarwal, R. C. and Cooley, J. W., "New algorithm for digital convolution", IEEE Trans., Vol. ASSP-25, No. 5, October 1977, pp, 392-410

[B6] Nussbaumer, H. J. and Quandalle, P., "Computation of convolutions and discrete Fourier transforms by polynomial transforms", IBM, J. Res. Develop., Vol. 22, No. 2, March 1978, pp. 134-140

[B7] Arambepola, B. and Rayner, P. H. W., "Efficient transforms for multidimensional convolutions", Electronics Lett., Vol. 15, No. 6, March 1979, pp. 189-190

[B8] Nussbaumer, H. J., "Fast polynomial transform algorithms for digital convolutions", IEEE Trans., Vol. ASSP-28, No. 2, April 1980, pp. 205-215

[B9] Arambepola, B. and Raynerf P. J. W., "Discrete transforms over polynomial rings with applications in computing multidimensional convolutions", IEEE Trans., Vol. ASSP-28, No. 4, August 1980, pp. 407-414

[B10] Arambepola, B., "Discrete transforms and computational algorithm for digital signal processing", Ph. D. thesis, Cambridge University, Engineering Department, 1980

[B11] Nussbaumer, H. J., "New polynomial transform algorithm for multidimensional DFTs and convolutions", IEEE Trans., Vol. ASSP-29, No. 1, February 1981, pp. 74-83

[B12] Reed, I. S., Shao, H. M. and Truong, T. K., "Fast polynomial transform and its implementation by computer", Proc. IEE, Vol. 128, Pt. E, No. 1, March 1981, pp. 50-60

[B13] Truong, T. K., Reed, I. S., Lipes, R. G. and Wu, C., "On the application of a fast polynomial transform and Chinese Remainder Theorem to compute a two-dimensional colvolution", IEEE Trans., Vol. ASSP-29, No. 1, February 1981, pp. 91-97

[B14] Cooley, J. W., "Rectangular transforms for digital convolution on the research signal processor", IBM,J. Res. Develop., Vol. 26,No. 4,July 1982, pp. 424-430

[B15] Yen, C. S., Reed, I. S. and Truong, T. K., "Parallel architectures for computing cyclic convolutions", IEE Proceedings, Vol. 130,Pt. F, No. 5, August 1983, pp. 409-416

(C) GALOIS FIELDS, FINITE RING AND LOGIC FUNCTIONS

[C1] Reed, I. S. and Stewart, R. M., "Note on the existence of perfect maps", IRE Transactions on Information Theory, Vol. IT-8, No. 1, January 1962, pp. 10-12

[C2] Booth, T. L., "An analytical representation of signals in sequential networks". Proceedings of Symposium on the Mathematical Theory of Automata, Polytechnic Institute of Brookley, April 1962, pp. 301-340

[C3] Bartee, T. C. and Schneider, D. I., "Computation with finite fields", Information and Control, Vol. 6, 1963, pp. 79-98

[C4] Bollman, D. A., "Some periodicity properties of transformations on vector spaces over residue class rings", J. Soc. Indust. Appl. Math., Vol. 13, No. 3, September 1965, pp. 902-912

[C5] Richalet, Je, "Operational calculus for finite rings", IEEE Trans. on Circuit Theory/Vol. CT-12, No. 4, December 1965, pp. 558-570

[C6] Gill, A. and Jacob, J. P., "On a mapping polynomial for Galois fields", Quart. Appl. Math., Vol. 24, No. 1, 1966, pp. 57-62

[C7] Berlekampr E. R., "Factoring polynomials over finite fields" The Bell System Technical Journal, Vol. 46, No. 8, October 1967, pp. 1853-1859

[C8] Tannak, H., Kasahara, M., Tezuka, Y. and Kasahara, Y., "Computation over Galois fields using shift registers", Information and Control, Vol. 13, 1968, pp. 75-84

[C9] Green, D. H. and Taylor, I. S., "Modular representation of multiple-valued logic systems", IEE Proceedings, Vol. 121, No. 6, June 1974, pp. 409-418

[C10] Lempel, A., "Matrix factorization over GF(2) and trace-orthogonal bases over GF(2^n)", SIAM J. Comput., Vol. 4, No. 2,June 1975, pp. 175-186

[C11] Sontag, E. D., "On linear systems and noncommutative rings", Mathematical Systems Theory, Vol. 9, No. 4, 1975, pp. 327-344

[C12] Pradhan, D. K. and Patel, A. M., "Reed-Muller like canonic forms for multivalued functions", IEEE Trans. on Computers, Vol. C-24, No. 2, February 1975, pp. 206-210

[C13] Benjauthrit, B. and Reed, I. S., "Galois switching functions and their applications", IEEE Trans. on Computers, Vol. C-25,No. 1, January 1976, pp. 78-86

[C14] Pradhan, D. K., "A theory of Galois switching functions"., IEEE Trans. on Computers, Vol. C-27, No. 3, March 1978, pp. 239-248

[C15] Benjanthrit, B. and Reed, I. S., "On the fundamental structure of Galois switching functions", IEEE Trans. on Computers, Vol. C-27, No. 8, August 1978, pp. 757-762

[C16] Imamura, K., "A method for computing addition tables in GF(p^n)", IEEE Trans. on Information Theory, Vol. IT-26, No. 3, May 1980, pp. 367-369

[C17] English, W. R., "Synthesis of finite state algorithms in a Galois field GF(p^n)", IEEE Trans. on Computers, Vol. C-30, No. 3, March 1981, pp. 225-229

[C18] Fleisher, H., Tavel, M. and Yeager, J., "Exclusive-OR representation of Boolean functions", IBM, J. Res. Develop., Vol. 32, No. 4, July 1983, pp. 412-416

[C19] Yeh, C. S., Reed, I. S. and Truong, T. K., "Systolic multipliers for finite fields GF (2^m)", IEEE Trans. on Computers, Vol. C-33, No. 4, April 1984, pp. 357-360

(D) MINIMIZATION OF REED-MULLER POLYNOMIALS

[D1] Even, S., Kohavi, I. and Paz, A., "On minimal modulo-2 sums of products for switching functions", IEEE Trans. Electronic Computers, Vol. EC-16, October 1967, pp. 671-674

[D2] Mukhopadkyay, A. and Schmitz, G., "Minimization of Exclusive-OR and logical equivalence switching functions", IEEE Trans. on Computers, Vol. C-19, No. 2, February 1970, pp. 132-140

[D3] Bioulf G., Davio, M. and Deschamps, J. P., "Minimization of ring-sum expansions of Boolean functions", Philips Res. Repts. 28, 1973, pp. 17-36

[D4] Marincovic, S. B. and Tosic, Z., "Algorithms for minimal polarized form determination", IEEE Trans. on Computers, Vol. C-23, December 1974, pp. 1313-1315

[D5] Kodandapani, K. L. and Sethur, R. V., "A note on minimal Reed-Muller canonical forms of switching functions", IEEE Trans. on Computers, Vol. C-26, March 1977, pp. 310-313

[D6] Papakonstantinou, G., "Minimization of modulo-2 sum of products", IEEE Trans. on Computers, Vol. C-28, No. 2, February 1979, pp. 163-167

[D7] Saluja, K. K. and Ong, E. H., "Minimization of Reed-Muller canonic expansion", IEEE Trans. on Computers, Vol. C-28, No. 7, July 1979, pp. 535-537

[D8] Wang, M. C., "An algorithm for Gray-to-binary conversion", IEEE Trans. on Electronic Computers, Vol. EC-18, August 1966, pp. 659-660

[D9] Hellerman, L., "A measure of computational work", IEEE Trans. on Computers, Vol. C-21, No. 5, May 1972, pp. 439-446

[D10] Reddy, S. M., "Easily testable realization for logic functions", IEEE Trans. on Computers, Vol. C-21, November 1972, pp. 1183-1188

[D11] Kopandapani, K. L., "A note on easily testable realizations for logic functions", IEEE Trans. on Computers, Vol. C-23, 1974, pp. 332-333

[D12] Saluja, K. K. and Reddy, S. M., "Fault detecting test sets for Reed-Muller canonic networks", IEEE Trans. on Computers, Vol. C-24, 1975, pp. 995-998

[D13] Wu, X., Chen, X. and Hurst, S. L., "Mapping of Reed-Muller coefficients and minimisation of Exclusive-OR switching functions", Proc. IEE, Vol. 129, Pt. E, No. 1, 1982, pp. 15-20

[D14] Robinson, J. P. and Yeh, C.-L., "A method for modulo-2 minimization", IEEE Trans. on Computers, Vol. C-31, No. 8, August 1982, pp. 800-801

[D15] Besslich, Ph. W., "Efficient computer method for ExOR logic design", Proc. IEE., Vol. 130, Pt. E, No. 6, November 1983, pp. 203-206

(E) MINIMIZATION OF BOOLEAN FUNCTIONS

[E1] McCluskey, Jr. E. J., "Minimization of Boolean functions", Bell System Technical Journal, Vol. 35, No. 6, November 1956, pp. 1417-1444

[E2] Su, Y. H. and Dietmeyer, D. L., "Computer reduction of two-level multiple-output switching circuits", IEEE Trans. on Electronic Computers, Vol. EC-18, No. 1, 1969, pp. 58-63

[E3] Hung, S. J., Cain, R. G. and Ostapko, D. L., "Mini: A heuristic approach for logic minimization", IBM, J. Res. Develop., Vol. 18, No. 5, 1974, pp. 443-458

[E4] Roth, J. P., "Programmed logic array optimization", IEEE Trans. on Computers, Vol. C-27, No. 2, 1978, pp. 174-176

[E5] Gupta, S. C., "A fast procedure for finding prime implicants of a Boolean function", Int. J. Electronics, Vol. 51, No. 5, 1981, pp. 701-709

(F) NON-LINEAR FEEDBACK SHIFT REGISTERS

[F1] Bryant, P. R. and Killick, R. D., "Non-linear feedback shift registers", IRE Trans. on Electronic Computers, June 1962, pp. 410-412

[F2] Magleby, K. M., "The synthesis of non-linear feedback shift registers", Technical Report No. 6207-1, Stanford University, October 1963

[F3] Booth, T. L., "Nonlinear sequential networks", IEEE Trans., Vol. CT-10, 1963, pp. 279-281

[F4] Fukunaga, K., "A non-linear autonomous sequential net using z-transforms", IEEE Trans. on Electronic Computers, Vol. EC-13, 1964, pp. 310-312

[F5] Massey, J. L. and Liu, R. W., "Application of Lyapunov's direct methods to the error-propagation effect in convolutional codes", IEEE Trans. on Information Theory, Vol. IT-10, July 1964, pp. 248-250

[F6] Massey, J. L. and Liu, R. W., "Equivalence of nonlinear shift registers", IEEE Trans. on Information Theory, Vol. IT-10, 1964, pp. 378-379

[F7] Liu, C. L., "Sequential-machine realization using feedback shift registers", 5th Int. Symp. on Switching Circuit Theory & Logical Design, Princeton, N. J., 1964, pp. 209-227

[F8] Davis, W. A., "On shift register realizations for sequential machines", IEEE Conf. on Switching Circuits Theory and Logical Design, Ann Arbor, 1965, pp. 71-83

[F9] Gill, A., "Analysis and synthesis of stable linear sequential circuits", Journal of the Association for Computing Machinery, Vol. 12, No. 1, January 1965, pp. 141-149

[F10] Friedman, A. D., "Feedback in synchronous sequential switching circuits", IEEE Trans. on Electronic Computers, Vol. EC-15, No. 3, June 1966, pp. 354-367

[F11] Mowle, F. J., "Relations between P_n cycles and state feedback shift registers", IEEE Trans. on Electronic Computers, Vol. EC-15, July 1966, pp. 375-377

[F12] Haring, D. R., "On shift-register realization of sequential machine and finite-state universal sequential machines", IEEE Trans. on Computers, Vol. C-17, No. 4, April 1968, pp. 309-312

[F13] Davisi-W. A., "Single shift-register realization for sequential machines", IEEE Trans. on Computers, Vol. C-17, No. 5, May 1968, pp. 421-431.

[F14] Cohn, M. and Even, S., "The design of shift register generators for finite sequences", IEEE Trans. on Computers, Vol. C-18, No. 7, July 1969, pp. 660-662

[F15] Lempel, A., "On *k*-stable feedback shift registers", IEEE Trans. on Computers, Vol. C-18, No. 7, July 1969, pp. 652-660

[F16] Green, D. H. and Dimond, K. R., "Polynomial representation of non-linear feedback registers", Proceedings IEE, Vol. 117, No. 1, January 1970, pp. 56-60

[F17] Green, D. H. and Dimond, K. R., "Nonlinear product-feedback shift registers", Proceedings IEE, Vol. 117, No. 4, April 1970, pp. 681-686

[F18] Green, D. H. and Kelsch, R. G., "Some polynomial compositions of nonlinear feedback shift registers and their sequence-domain consequences", Proceedings IEE, Vol. 117, No. 9, September 1970, pp. 1750-1756

[F19] Mowle, F. J., "Readily programmable procedures for the analysis of non-linear feedback shift registers", IEEE Trans. on Computers, Vol. C-18, September 1969, pp. 824-829

[F20] Kain, R. Y., "Nonlinear sequential circuits", IEEE Trans. on Computers, Vol. C-19, March 1970, pp. 249-254

[F21] Lempel, A., "On a homomorphism of the de Bruijin Graph and its applications to the design of feedback shift registers", IEEE Trans. on Computers, Vol. C-19, No. 12, December 1970, pp. 1204-1209

[F22] Lempel, A., "Analysis and synthesis of polynomials and sequences over GF(2)", IEEE Trans. on Information Theory, Vol. IT-17, No. 3, May 1971, pp. 297-303

[F23] Hemmati, F., "A large class of nonlinear shift register sequences", IEEE Trans. on Information Theory, Vol. IT-28, No. 2, March 1982, pp. 355-359

[F24] Kalouptsides, N. and Manolarakis, M., "Sequences of linear shift registers with nonlinear-feedforward logic", IEE Proceedings, Vol. 130, Pt. E, No. 5, September 1983, pp. 174-176

(G) FINITE STATE MACHINE AND LINEARIZATION

[G1] Mealy, G. H., "A method for synthesizing sequential circuits", Bell System Technical Journal, Vol. 33, September 1955, pp. 1054-1079

[G2] Cadden, W. J., "Equivalent sequential circuits", IRE Trans. on Circuit Theory, Vol. CT-6, March 1959

[G3] Gill, A., "The minimization of linear sequential circuits", IEEE Trans. on Circuit Theory, Vol. CT-12, June 1965, pp. 292-294

[G4] Messey, J., "Note on finite-memory sequential machines", IEEE Trans. on Electronic Computers, Vol. EC-15, August 1966, pp. 658-659

[G5] King, IV, W. F., "Analysis of iterative NOR autonomous sequential machines", IEEE Trans. on Electronic Computers, Vol. EC-15, No. 4, August 1966, pp. 569-577

[G6] Curtis, H. A., "Polylinear sequential circuits realizations of finite automata", IEEE Trans. on Computers, Vol. C-17, No. 3, March 1968, pp. 251-259

[G7] Herman, G. T., "When is a sequential machine the realization of another?", Math. System Theory, Vol. 5, 1971, pp. 115-127

[G8] Herman, G. T., "Every finite sequential machine is linearly realizable", Journal of Computer

Science, Vol. 5, 1971, pp. 489-510
[G9] Matluk, M. M. and Gill, A., "Decomposition of linear sequential circuits over residue class rings". Journal of Franklin Institute, Vol. 294, 1972, pp. 167-180
[G10] Skornyakov, L. A., "On algebraic automata", UDC 519-1, translated from Kibernetika, No. 2, March-April 1974, pp. 31-34
[G11] Mikhailov, G. I., "Sets and automata having a finite memory", UDC 007-52, translated from Avtomatika i Telemekhanika, No. 5, May 1975, pp. 121-128
[G12] Kapitonova, Y. V., "Discrete systems and problems of their realization I", UDC 62-5: 681. 3: 007, translated from Kibernetika, No. 4, July-August 1975, pp. 7-10
[G13] Kapitonovar Y. V., "Discrete systems and problems of their realization II", UDC 62-5: 681. 3:007, translated from Kibernetika, No. 5, September-October 1975, pp. 21-26
[G14] Magidin, Mr and Gill, A., "Singular shift register over residue class rings", Mathematical Systems Theory, Vol. 9, No. 4, 1976, pp. 345-358
[G15] Reush, B., "Linear realization of finite automata", Journal of Computer and Systems Theory, 15, 1977, pp. 146-168
[G16] Porter, W. A. and Zalecka-Melamed, A. S., "Large-scale finite state machines", Int. J. Control, Vol. 25, No. 2, 1977, pp. 185-200
[G17] Porter, W. A., "Continuous state models for finite state machines", Int. J. Control, Vol. 25, No. 2, 1977, pp. 165-183
[G18] Porter, W. A., "Inverses and adjoints of finite state machines", Int. J. Control, Vol. 25, No. 2, 1977, pp. 201-211
[G19] Ejima, T. and Kimura, M., "Structural stability of linear space automata as models of generalized linear systems", The Transactions of IECE of Japan, Vol. E-60, No. 9, September 1977, pp. 487-488
[G20] Gosselt M. and Poschel, R., "On the characterization of linear and linearizable automata by a superposition principle", Mathematical System Theory, 11, 1977, pp. 61-76
[G21] Roitberg, M. A., "Automata circuits and the mappings realized by them", translated from Avtomatika i Telemekhanika, No. 4, April 1978, pp. 151-160
[G22] Baranov, S. I. and Keevallik, A. E., "Synthesis of control automata using graph-schemes of algorithms", Digital Process 6, 1980, pp. 149-165
[G23] Russo, G. V. and Paland, G., "Minimization of incompletely specified sequential machine", Digital Process 6, 1980, pp. 199-206
[G24] Beckhoff, G. F., "A novel approach to the state reduction problem", Digital Process 6, 1980, pp. 305-314
[G25] Emre, E. and Khargonekar, "Regulation of split linear systems over rings: coefficient-assignment and observers", IEEE Trans. on Automatic Control, Vol. AC-27, No. 1, February 1982, pp. 104-113
[G26] Papachristou, C. A. and Sarma, D., "An approach to sequential circuit construction in LSI programmable arrays", IEE Proceedings, Vol. 130, Pt. E, No. 5, September 1983, pp. 159-164

(H) APPLICATION OF FINITE STATE MACHINES

[H1] Wang, P. K. C., "A method for approximating dynamical processes by finite-state systems", Int. J. Control, Vol. 8, No. 3, 1968, pp. 285-286

[H2] Depeyrot, M., Marmorat, J. P. amd Mondelli, J., "An automation-theoretic approach to the Fast Fourier Transform", Proceedings of the Symposium on Computers and Automata, Polytechnic Institute of Brooklyn, April 1971, pp. 359-377

[H3] Kornoushenko, E. K., "Finite-automaton approximation to the behavior of continuous plants", UDC 007. 52, translated from Avtomatika i Telemekhanika, No. 12, December 1975, pp. 150-157

[H4] Kornoushenko, E. K., "Finite automaton approximation of the behavior of linear-stationary continuous plants of the first order", UDC 62-501, 42, translated from Avtomatika i Telemekhanika, No. 7, July 1976, pp. 123-131

[H5] Kornoushenko, E. K., "Finite-automaton approximation of behavior of linear stationary continuous plants", UDC 62-501. 4, translated from Avtomatika i Telemekhanika, No. 7, July 1977, pp. 180-191

[H6] Kornoushenko, E. K., "Finite-automaton approximation of behavior of linear stationary continuous systems with arbitrary inputs", UDC 62-507: 62-501. 42, translated from Avtomatika i Telemekhanika, No. 3, March 1978, pp. 137-145

[H7] Rayner, P. J. W., "The application of finite arithmetic structures to the design of digital processing systems", Proceedings of International Conference on Digital Signal Processing, Florences Italy, September 1978

[H8] Albison, L. J. and Rayner, P. J. W., "Finite automata representation of digital filters". Proceedings of International Conference on Digital Signal Processing, Florence, Italy, September 1981, pp. 432-439

[H9] Gaarder, N. T. and Slepian, D., "On optimal finite-state digital transmission systems", IEEE Trans. on Information Theory, Vol. IT-28, No. 2, 1982, pp. 167-186

[H10] Rayner, P. J. W., "Number theoretic and finite field processing", lecture of NATO Symposium on Impact of Digital Signal Processing to Communications, France, July 1983

(I) HARDWARE IMPLEMENTATION

[I1] Fleisher, H. and Maissel, L. I., "An introduction to array logic", IBM J. Res. Develop Vol. 19, March 1975, pp. 98-109

[I2] Jones, J. W., "Array logic macros", IBM J. Res. Develop., Vol. 19, March 1975, pp. 120-126

[I3] Logne, J. C., Brickman, N. F., Howley, F., Jones, J. W. and Wu, W. W., "Hardware implementation of a small system in programmable logic arrays", IBM J. Res. Develop., March 1975, pp. 110-119

[I4] Almaini, A. E. A., "Sequential machine implementations using universal logic modules", IEEE Trans. on Computers, Vol. C-27, No. 10, October 1978, pp. 951-960

[I5] Mintzer, F. and Peled, A., "A microprocessor for signal processing, the RSP" IBM J. Res. Develop., Vol. 26, No. 4, July 1982, pp. 413-423

(J) COEFFICIENT QUANTIZATION

[J1] Kaiserr J. F., "Some practical considerations in the realization of linear digital filters", Proceedings of 3rd Allerton Ann. Conf. on Circuits and System Theory, Nonticello, 1965, pp. 621-631

[J2] Rader, C. M. and Gold, B., "Effects of parameter quantization on the poles of a digital filter", Proc. IEEE, Vol. 55, May 1967, pp. 688-689

[J3] Knowles, J. B. and Olcayto, E. M., "Coefficient accuracy and digital filter response", IEEE Trans. on Circuit Theory, Vol. CT-15, March 1968, pp. 31-41

[J4] Jackson, L. B., "On the interaction of roundoff noise and dynamic range in digital filters", Bell System Technical Journal, Vol. 49, No. 2, February 1970, pp. 159-184

[J5] Jackson, L. B., "Roundoff-noise analysis of digital filters realized in cascade or parallel form", IEEE Trans. on Audio and Electroacoustics, Vol. AU-18, No. 2, June 1970, pp. 107-122

[J6] Steiglitz, K., "Designing short-word recursive digital filters", Proceedings of the 9th Allerton Conf. on Circuit and System Theory, October 1971, pp. 778-788

[J7] Liu, B., "Effect of finite word length on the accuracy of digital filters—a review", IEEE Trans. on Circuit Theory, Vol. CT-18, No. 6, November 1971, pp. 670-677

[J8] Avenhaus, E., "On the design of digital filters with coefficients of limited word length", IEEE Trans. on Audio and Electroacoustics, Vol. AU-20, August 1970, pp. 206-212

[J9] Oppenheim, A. V., "Effect of finite register length in digital filtering and fast Fourier transform", Proceedings of IEEE, Vol. 60, August 1972, pp. 957-976

[J10] Chan, D. S. K. and Rabiner, L. R., "Analysis of quantization errors in the direct form for finite impulse response digital filters", IEEE Trans. on Audio and Electroacoustics, Vol. AU-21, August 1973, pp. 354-366

[J11] Charalambousr C. and Best, M. J., "Optimization of recursive digital filters with finite word lengths", IEEE Trans. on Acoustics Speech and Signal Processing, Vol. ASSP-22, December 1974, pp. 424-431

[J12] Crochiere, R. E., "A new statistical approach to the coefficient word length problem for digital filters", IEEE Trans. on Circuits and Systems, Vol. CAS-22, March 1915, pp. 190-196

[J13] Clansen, T. A. C. M., Meckelenbrauker, W. F. G. and Peek, J. B. H., "Quantization noise analysis for fixed-point digital filters using magnitude truncation for quantization", IEEE Trans. on Circuits and Systems, Vol. CAS-22. No. 11, November 1975, pp. 887-895

[J14] Butterweck, H. J., "On the quantization noise contributions in digital filters which are uncorrelated with the output signal", IEEE Trans. on Circuits and Systems, Vol. CAS-26, No. 11, November 1979, pp. 901-911

[J15] Mcleod, M., "Digital signal processing systems with discrete parameters", Ph. D. dissertation, University of Cambridge, 1979

[J16] Butterweck, H. J., "Quantization effects in various second-order digital filters: a comparative study", Proceedings of the 1981 European Conference on Circuits Theory and Design August 1981, pp. 863-864

[J17] Liu, B. and Ansari, R., "Quantization effects in computationally efficient realization of recursive filters", 1982 International Symposium on Circuits and Systems, Rome, Italy, Vol. 2, May 1982, pp. 716-720

[J18] Rader, C. M., "The application of dynamic programming to the optimal ordering of digital filter sections". Proceedings of International Conference on ASSP, Paris, France, May 1982, pp. 73-76

(K) OVERFLOW AND LIMIT CYCLE OSCILLATIONS

[K1] Ebert, P. M., Mazo, J. E. and Taylor, M. C., "Overflow oscillations in digital filters", Bell System Technical Journal, Vol. 48, November 1969, pp. 2999-3020

[K2] Sandberg, I. W., "A theorem concerning limit cycles in digital filters", Proceedings of the 7th Allerton Conference on Circuits and System Theory, 1969, pp. 63-68

[K3] Jackson, L. B., "An analysis of limit cycles due to multiplication rounding in recursive digital (sub) filters", Proceedings of 7th Allerton Conference on Circuits and System Theory, 1969, pp. 69-78

[K4] Parker, A. T. and Hess, S. F., "Limit cycle oscillations in digital filters", IEEE Trans. on Circuit Theory, Vol. CT-18, No. 6, November 1971, pp. 687-697

[K5] Sandberg, I. W. and Kaiser, J. F., "A bound on limit cycles in fixed point implementations of digital filters", IEEE Trans. on Audio and Electroacoustics; Vol. AU-20, June 1972, pp. 110-112

[K6] Brubaker, T. A. and Cowdy, J. N., "Limit cycles in digital filters", IEEE Trans. on Automatic Control, Vol. AC-17, October 1972, pp. 675-677

[K7] Long, J. L. and Tick, T. N., "An absolute bound on limit cycles due to roundoff errors in digital filters", IEEE Trans. on Audio and Electroacoustics, Vol. AU-21, February 1973, pp. 27-30

[K8] Kaneko, T., "Limit-cycle oscillation in float-point digital filters", IEEE Trans. on Audio and Electroacoustics, Vol. AU-21, April 1973, pp. 100-106

[K9] Kieburtz, R. B., "An experimental study of roundoff effects in a tenth-order recursive digital filter", IEEE Trans. on Communications, Vol. COM-21, No. 6, June 1973, pp. 757-763

[K10] Claasen, T. A. C. M., Mecklenbrauker, W. F. G. and Peek, J. H. B., "Some remarks on the classification of limit cycles in digital filters", Philips Research Report, Vol. 28, April 1973, pp. 297-305

[K11] Claasen, T., Mecklenbraukerf W. F. G. and Peek, J. B. H., "Frequency domain criteria for the absence of zero-input limit cycles in nonlinear discrete-time system with application to digital filters", IEEE Trans. on Circuits and Systems, Vol. CAS-22, March 1975, pp. 232-239

[K12] Claasen, T. A. C. M. and Kristiansson, L. O. G., "Necessary and sufficient conditions for the absence of overflow phenomena in a second order recursive digital filter", IEEE Trans. on Acoustics Speech and Signal Processing, Vol. ASSP-23, December 1975, pp. 509-515

[K13] Chang, T. L., "A note on upper bounds on limit cycles in digital filters", IEEE Trans. on Acoustics Speech and Signal Processing, Vol. ASSP-24, February 1976, pp. 99-100

[K14] Rahman, M. H., Maria, G. A. and Fahmy, M. M., "Bounds on zero input cycles in all-pole digital filters", IEEE Trans. on Acoustics Speech and Signal Processing, Vol. ASSP-24, April 1976, pp. 189-192

[K15] Claasen, T. A. C. M., Mecklenbrauker, W. F. G. and Peek, J. B. H., "Effects of quantization and overflow in recursive digital filters", IEEE Trans. on Acoustics Speech and Signal Processing, Vol. ASSP-24, December 1976, pp. 517-529

[K16] Jackson, L. B., "Limit cycles in state-space structures for digital filters", IEEE Trans. on Circuits and Systems, Vol. CAS-26, January 1979, pp. 67-68

[K17] Abu-El-Haiyap A. I., "A tight bound on $\sum_{N}^{\infty} = 0 \mid h(n) \mid$ for general second-order $H(z)$", IEEE Trans. on Circuits and Systems, Vol. CAS-29, No. 7, July 1982, pp. 492-496

[K18] Storzbach, W. H., "Forced oscillations in recursive digital filters", Proc. Int. Symp. on Circuit Theory, North Hollywood, California, April 1972, pp. 233-236

[K19] Samueli, H. and Willson, Jr., A. N., "Almost period P sequences and analysis of forced overflow oscillations in digital filters", IEEE Trans. on Circuits and Systems, Vol. CAS-29, No. 8, August 1982, pp. 510-515

[K20] Butterweck, H. J., Lucassen, F. H. R. and Verkroost, G., "Subharmonics and related quantization effect in periodically excited recursive digital filters", Proc. Second European Digital Signal Processing Conf., Erlangen-Nurnberg, West Germany, September 1983, pp. 57-59

(L) SUPPRESSION OF LIMIT CYCLES

[L1] Claasen, T. A. C. M., Mecklenbrauker, W. F. G. and Peek, J. B. H., "Second-order digital filter with only one magnitude-truncation quantiser and having practically no limit cycles", Electronic Lett., Vol. 9, November 1973, pp. 531-532

[L2] Butterweck, H. J., "Suppression of parasitic oscillations in second-order digital filters by means of a controlled-rounding arithmetic", A. E. U., Electronics and Communication, Vol. 29, 1975, pp. 371-374

[L3] Verkroof, G., "A general second-order digital filter with controlled rounding to exclude limit cycles for constant input signals", IEEE Trans. on Circuits and Systems, Vol. CAS-24, No. 8, August 1977, pp. 428-431

[L4] Lawrence, V. B. and Mina, K. V., "Control of limit cycle oscillations in second-order recursive digital filters using constrained random quantization", IEEE Trans. on Acoustics Speech and Signal Processing, Vol. ASSP-26, April 1978, pp. 127-134

[L5] Mitra, D. and Lawrence, V. B., "Controlled rounding arithmetics for second-order direct-form digital filters that eliminate all self-sustained oscillations", IEEE Trans. on Circuits and Systems, Vol. CAS-28, No. 9, September 1981, pp. 894-905

[L6] Buttner, M., "Elimination of limit cycles in digital filters with very low increase in quantization noise", IEEE Trans. on Circuits and Systems, Vol. CAS-24, June 1977, pp. 300-304

[L7] Fettweis, A. and Meerkotter, K., "Suppression of parasitic oscillations in wave digital filters", IEEE Trans. on Circuits and Systems, Vol. CAS-22, No. 3, March 1975, pp. 239-246

[L8] Barnes, C. W. and Fam, A. T., "Minimum norm recursive digital filters that are free of overflow limit cycles", IEEE Trans. on Circuits and Systems, Vol. CAS. 24, No. 10, October 1977, pp. 569-574

[L9] Change, T. L., "Suppression of limit cycles in digital filters designed with one magnitude-truncation quantizer", IEEE Trans. on Circuits and Systems, Vol. CAS-28, No. 2, February 1981, pp. 107-111

[L10] Abu-El-Haiya, A. I. , "Error-feedback digital filters with minimum limit cycle oscillations", Proc. Second European Signal Processing Conf. , Erlangen-Nurnberg, West Germany, September 1983, pp. 111-114

[L11] Renfors, M. , Sikstrom, B. and Wanhammar, L. , "LSI implementation of limit-cycle-free digital filters using error-feedback techniques", Proc. Second European Digital Signal Processing Conf. , Erlangen-Nurnberg, West Germany, September 1983, pp. 107-110

[L12] Butterweck, H. J. , Van Meer, A. C. P. and Verkroost, G. , "New second-order digital sections without limit cycles", Proc. Second European Digital Signal Processing Conf. , Erlangen-Nurnberg, West Germany, September 1983, pp. 97-98

[L13] Szczupak, J. and Mitra, S. K. , "Detection, location and removal of delay-free loops in digital filter configurations", IEEE Trans. on Acoustics Speech and Signal Processing, Vol. ASSP-23, No. 6, December 1975, pp. 558-562

[L14] Liu, H. M. and Ackroyd, M. H. , "Suppression of limit cycles in digital filters by random dither", Radio and Electronic Engineer, Vol. 53, No. 6, June 1983, pp. 235-240

[L15] Turner, L. E. , "Elimination of constant-input limit cycles in recursive digital filters using a generalised minimum norm", IEE Proceedings, Pt. G, Vol. 130, No. 3, June 1983, pp. 69-77

[L16] Sridharan, S. and Williamson, D. , "Comments on suppression of limit cycles in digital filters designed with one magnitude-truncation quantizer", IEEE Trans. on Circuits and Systems, Vol. CAS-31, No. 2, February 1984, pp. 235-236

(M) OPTIMUM QUANTIZATION

[M1] Max, J. , "Quantizining for minimum distortion", IRE Trans. on Information Theory, Vol. IT-6, March 1960, pp. 7-12

[M2] Bluestein, L. I. , "Asymptotically optimal quantizers and optimum analogue to digital conversion for continuous signals", IEEE Trans. on Information Theory, Vol. IT-10, July 1964, pp. 242-246

[M3] Roe, G. M. , "Quantizing for minimum distortion", IEEE Trans. on Information Theory, Vol. IT-10, October 1964, pp. 384-385

[M4] Wood, R. C. , "On optimum quantization", IEEE Trans. on Information Theory, Vol. IT-15, March 1969, pp. 248-252

[M5] Elias, P. , "Bounds on performance of optimum quantizers", IEEE Trans. on Information Theory, Vol. IT-16, March 1970, pp. 172-184

[M6] Sharman, D. K. , "Design of absolutely optimal quantizers for a wide class of distortion measures", IEEE Trans. on Information Theory, Vol. IT-24, November 1978, pp. 693-702

[M7] Manersberger, W. , "Experimental results on the performance of mismatched quantizer", IEEE Trans. on Information Theory, Vol. IT-25, July 1979, pp. 381-386

[M8] Lloyd, S. P. , "Least squares quantization in PCM", IEEE Trans. on Information Theory, Vol. IT-28, March 1982, pp. 129-137

SECTION (Ⅱ): BOOKS

(P) DIGITAL SIGNAL PROCESSING

[P1] Rabiner, L. R. and Gold, Be., "Theory and application of digital signal processing", Prentice-Hall, Englewood Cliffs, N. J., 1975

[P2] Oppenheim, A. V. and Schafer, R. W., "Digital signal processing", Prentice-Hall, Englewood Cliffs, N. J., 1975

[P3] Goldr B. and Rader, C., "Digital processing of signals", McGraw-Hill, New York, 1969

[P4] McClellan, J. H. and Rader, C. M., "Number theory in digital signal processing", Prentice-Hall, Englewood Cliffs, N. J., 1979

[P5] Nussbaumer, H. J., "Fast Fourier Transform and convolution algorithms", Springer-Verlag, Berlin, 1981

[P6] Jury, E. I., "Theory and application of the z-transform method", John Wiley & Sons, New York, 1964

[P7] Brigham, E. O., "The Fast Fourier Transform", Prentice-Hall, Englewood Cliffs, N. J., 1974

[P8] Ahmed, N. and Rao, K. R., "Orthogonal transforms for digital signal processing", Springer-Verlag, Berlin, 1975

[P9] Peled, A. and Liu, B., "Digital signal processing theory, design, and implementation", John Wiley & Sons, New York, 1976

[P10] Oppenheim, A. V. (editor), "Applications of digital signal processing", Prentice-Hall, Englewood Cliffs, N. J., 1978

[P11] Willsky, A. S., "Digital signal processing and control and estimation theory", MIT Press, Massachusetts, 1979

[P12] Papoulis, A., "Signal analysis", McGraw-Hill, New York, 1977

(Q) DIGITAL SYSTEMS AND DISCRETE MATHEMATICS

[Q1] Stone, H. S., "Discrete mathematical structures and their applications", Science Research Associates, Chicago, 1973

[Q2] Davio, M., Deschamps, J. and Thayse, A., "Discrete and switching functions", Geogi Publishing, Switzerland, McGraw-Hill, New York, 1978

[Q3] Friedman, A. D. and Menon, P. R., "Theory and design of switching circuits", Bell Telephone Laboratories Inc. and Computer Science Press Inc., Woodland Hills, California, 1975

[Q4] Gill, A., "Applied algebra for computer sciences", Prentice-Hall, Englewood Cliffs, N. J., 1976

[Q5] Gill, A., "Linear sequential circuits", McGraw-Hill, New York, 1966

[Q6] Golomb, S. W., "Shift register sequences", Holden-Day, San Francisco, 1967

[Q7] Hennie, I. C., "Finite-state models for logical machines", John Wiley & Sons, New York, 1968

[Q8] Berlekamp, E. R., "Algebraic coding theory", McGraw-Hill, New York, 1968

[Q9] Martin, R. L., "Studies in feedback-shift-register synthesis of sequential machines", MIT

Press, Massachusetts, 1969

[Q10] Gill, A., "Introduction to the theory of finite state machines", McGraw-Hill, New York, 1962

[Q11] Moore, E. F. (editor), "Sequential machines selected papers", Addison-Wesley, Massachusetts, 1964

[Q12] Harrison, M. A., "Introduction to switching and automata theory", McGraw-Hill, New York, 1965

[Q13] Hartmanis, J. and Stearns, R. E., "Algebraic structure theory of sequential machines", Prentice-Hall, Englewood Cliffs, N. J., 1966

[Q14] Ginzburg, A., "Algebraic theory of automata", Academic Press, New York, 1968

[Q15] Tou, J. T., "Applied automata theory", Academic Press, New York, 1968

[Q16] Knuth, D. E., "The art of computer programming, Vol. 2: Semi-numerical algorithms", Addison-Wesley, Massachusetts, 1969

[Q17] Karpovsky, M. G., "Finite orthogonal series in the design of digital devices", John Wiley, New York, 1976

[Q18] Lind, L. F. and Nelson, J. C. C., "Analysis and design of sequential digital systems", Macmillan Press, London, 1977

[Q19] Konavi. Z., "Switching and finite automata theory", (second edition), McGraw-Hill, New York, 1978

[Q20] Booth, T. L., "Digital networks and computer systems", (second edition), John Wiley & Sons, New York, 1978

[Q21] Even, S., "Graph algorithm", Computer Press, Photomac M. D., U. S. A., 1979

[Q22] Hamming, R. H., "Coding and information theory", Prentice-Hall. Englewood Cliffs, N. J., 1980

SECTION(Ⅲ): PUBLISHED WORKS USED IN THE DISSERTATION

[Z1] Zhang, Y. Z, and Rayner, P. J. W., "Fast DFT algorithm using subgroup convolutions", Signal Processing Ⅱ: Theories and Applications, edited by H. M. Schussler, Elsevier Science Publishers B. V. (North-Holland), 1983, pp. 719-724

[Z2] Zhang, Y. Z., "A direct algorithm for synthesis of stable feedback registers", International Journal of Electronics, Vol. 57, No. 1, 1984, pp. 79-84

[Z3] Zhang, Y. Z., and Rayner, P. J. W., "On minimization of Reed-Muller polynomials with fixed polarity", IEE Proceedings, Part E, Computers and Digital Techniques, Vol. 131, No. 5, 1984, pp. 176-186.

[Z4] Zhang, Y. Z., "Indirect synthesis of stable binary sequences", International Journal of Electronics, Vol. 57, No. 4, 1984, pp. 569-579.

[Z5] Zhang, Y. Z., "Galois Field representation of finite state machines", Technical Research Reports of Cambridge University Engineering Department, CUED/B-Elect/TR70, February 1984, pp. 1-44

[Z6] Zhang, Y. Z., "Stabilisation of finite state machines over GF (2^M)", submitted for publication in IEEE Transactions on Computers

[Z7] Zhang, Y. Z., " Optimum quantization for maximum entropy", submitted for publication in IEEE Transactions on Information Theory
[Z8] Zhang, Y. Z., "Optimum coefficients for recursively computing DFTs of prime length", accepted for presentation in the seventh European Conference on Circuit Theory and Design, Prague, Czechoslovakia

Editor's Note

The book contains part of academician Zhang Yanzhong's research papers published in English. It represents the development of aerospace systems engineering and infor-mation technology over a comparatively longer span of time. Regarded as the crucial reference materials of related disciplines, it falls into three categories, namely, information technique, aeronautical engineering, as well as strategy issue of development, with each paper arranged in the light of its importance.

The style of writing and layout of the papers included in the book may differ to some extent in virtue of up to 50 years of writing duration, varied publications and conferences. In this connection, we merely modify a touch of words and expressions, punctuation, unit of measurement, equations and variables according to modern publishing standards in order to maintain its original style while achieving a general consistence of layout in the book.

Due to time constraints, omissions and mistakes would inevitably exist in the editing process. As a result, your corrections and suggestions are mostly welcome and appreciated.

Y. Zhang, *Footprints in Cambridge and Aviation Industries of China*,
https://doi.org/10.1007/978-981-16-3176-4

Prof. ZHANG Yanzhong

Ph. D. Cam, Academician of Chinese Academy of Engineering (CAE)

Prof. ZHANG Yanzhong is an expert of aeronautical system engineering and signal processing.

Mr. Zhang was born in Shaanxi Province, China. He graduated from Northwestern University in 1962. In 1981, he went to Trinity College, University of Cambridge, U. K. to study signal processing, and received a Ph. D. degree in 1984. Prof. Zhang was elected as an academician of CAE in 2001, afterwards, the head of the Division of Mechanical & Vehicle Engineering, and a board director of CAE.

He was appointed as a professor of Beijing University of Aeronautic & Astronautics in 1986, a visiting professor at the University of Sydney in 1989, an honorary professor for Technical University of Macau in 2009, the honorary president of the US Sound & Vibration Institute in 2013, and a *Qiushi* Chair Professor of Zhejiang University in 2016.

Prof. Zhang has been involved in the fields of mechanical vibration, acoustic, signal processing and aeronautical engineering for about 60 years. He received 2 international awards and more than 11 national and ministerial awards. Plus, he has published 12 books, an English-Chinese dictionary, and more than 200 papers.

Dr. Zhang used to be a chief engineer of the Ministry of Aviation Industry and the Ministry of Aero-Space Industry, president of Chinese Aeronautical Establishment, president of China Aviation Industry Cooperation II (AVIC II), and CEO & Chairman of board of directors for AviChina. At present, he is the head of expert group for COMAC (Commercial Aircraft Corporation of China Ltd.).

Prof. Zhang has been a council member of the International Council of the Aeronautical Sciences (ICAS), and was the chairman of the organizing committee for the 18th congress of ICAS. He used to be president of Chinese Society of Signal Processing, vice president for Chinese Society of Aeronautics & Astronautics, and vice president for Vibration Engineering Society, etc.

Y. Zhang, *Footprints in Cambridge and Aviation Industries of China*,
https://doi.org/10.1007/978-981-16-3176-4